George Rolleston, William Turner

Scientific Papers and Addresses

Vol. 2

George Rolleston, William Turner

Scientific Papers and Addresses
Vol. 2

ISBN/EAN: 9783337416065

Printed in Europe, USA, Canada, Australia, Japan

Cover: Foto ©berggeist007 / pixelio.de

More available books at **www.hansebooks.com**

SCIENTIFIC PAPERS

AND

ADDRESSES

BY

GEORGE ROLLESTON, M.D., F.R.S.

LINACRE PROFESSOR OF ANATOMY AND PHYSIOLOGY
AND FELLOW OF MERTON COLLEGE, OXFORD

ARRANGED AND EDITED BY

WILLIAM TURNER, M.B., Hon. LL.D., F.R.S.

PROFESSOR OF MEDICINE AND ANATOMY IN THE UNIVERSITY OF EDINBURGH

WITH A BIOGRAPHICAL SKETCH BY

EDWARD B. TYLOR, Hon. D.C.L., F.R.S.

KEEPER OF THE MUSEUM, OXFORD

WITH PORTRAIT, PLATES, AND WOODCUTS

VOL. II

Oxford

AT THE CLARENDON PRESS

M.DCCC.LXXXIV

ZOOLOGY.

XXVIII.

ON THE DOMESTIC CATS, *FELIS DOMESTICUS* AND *MUSTELA FOINA*, OF ANCIENT AND MODERN TIMES.

Having recently had occasion to study the habits and anatomy of the English Marten Cats, *Mustela martes* and *Mustela foina*, and having looked into the history of these animals and of certain other carnivora which have borne the same name as they in ancient and modern times, I have come to think that the latter of the two creatures specified, namely, the white-breasted marten, *Mustela foina*, was the animal which the ancient Greeks and Romans employed for the same domestic purposes for which we employ the *Felis domesticus;* whilst this latter animal has been employed as at present in Western Europe for probably a considerably longer period than the last thousand years. Other writers to whom I have referred, and amongst them notably Dureau de la Malle, have adopted the former of these conclusions; but upon premises more or less inadequate or incorrect, or both. I am well assured that this question has for the philosophic naturalist much more than a merely archaeological interest, for in the resolving of it we may have light cast upon the working of the principle which Mr. Darwin has shown to be more potent than any other in regulating the distribution of species, and which Van Beneden, with the well-known physiological[1] aphorisms of Wolff and Treviranus before his eyes, has formulated in the words[2], 'les êtres qui composent une Faune sont solidaires entre eux comme les organes d'un être vivant.' It is true that no such close interdependence as exists between man

[1] Paget's 'Surgical Pathology,' p. 17, ed. Prof. Turner, 1863; Lewes, 'Physiology of Common Life,' i. p. 286; Wolff's 'Theoria Generationis,' p. 108, § 236.

[2] 'Recherches sur la Faune Littorale de Belgique—Cétacés,' p. 4.

and certain others of the domesticated animals, even to the extent of the interchange of disease and death by the intermediation of cysticerci and echinococci, has been shown to obtain between him and any of the animals of which I shall here have to write; but readers of the 'Origin of Species' will recollect that more intricate, if less mournful net-works, may bind up the presence in a particular country of a domestic animal with something so apparently distant from the sphere of operation as the general colouration of the landscape. Col. Newman [1] showed how the number of humble-bees in a district depended on the presence of man, whose domestic cats kept down the numbers of the honey-loving and devouring field-mice; and Mr. Darwin demonstrated in the way of experiment, that the presence of the humble-bee was a pre-requisite for the fertilisation of the heart's-ease and red clover. The dependence of animal life upon the presence of particular forms of vegetable life is familiar enough to us; our own comfort depends too directly in these over-peopled days and countries upon the adequate abundance of our flocks and herds, and that adequate abundance again depends too directly upon that of the turnip crops, unknown to our forefathers, to suffer us to forget it; but it is none the less true that the peculiar character of the vegetation, and consequently of the landscape of a country, depends very frequently upon the peculiar characters of its animal inhabitants. The antelopes [2], by carrying and dispersing the seeds of grapes, change the characters of the African desert; and the red glow of the clover field which gratifies the eyes of the artisan depends ultimately on pollen-carrying insects, and at second-hand from them upon the cats which spend their daylight hours in the same murky atmosphere that he does. It is possible that some such secret bond may exist between the widely spread family of the *Mustelidae* [3], and that of the *Abietineae*, the geographical range of which so nearly coincides with theirs. In this country the geographical distribution of the fir is now all but exclusively dependent upon man's artificial aid; where it does take place independently of him, it takes place in great measure by the aid of the squirrel which carries off the cones or seeds, and, having buried them, forgets, or is unable to dig them

[1] 'Origin of Species,' p. 84, 4th edit.

[2] Livingstone's 'Travels and Researches in South Africa,' 1857, p. 99.

[3] Wagner, 'Abhand. Akad. Wiss. München,' Bd. iv. pp. 26 and 107, 1846.

up again. It is not likely that any direct alliance subsists between any musteline and any rodent, but the more arboreal and egg-loving mustelines, such as the martens, may by robbing the nests of the rapacious birds, such as the carrion crows, the great enemies of the squirrels, indirectly but most efficiently favour the spread of these latter creatures. But the subject of the interdependence of the two kingdoms of animal and vegetable life is a much larger one than the one I propose to deal with here; and I have but introduced the mention of it to show what broad and distant views may be gained by attentive gazing through what may seem to be but narrow casements. I shall arrange what I have to say under two heads. I shall first attempt to prove that, though the ancient Greeks and Romans had not domesticated the cat, *Felis domesticus*, in classical times, this animal was nevertheless domesticated in Western Europe at an earlier period than is commonly assigned[1]. And secondly, I shall address myself to showing that the white-breasted marten, *Mustela foina*, which is known also as the 'Beech Marten' or 'Stone Marten,' was functionally the 'Cat' of the ancients. A list of the different works and memoirs which I have consulted or read upon the topics I have written upon will be found appended to this article. Either my conclusions or my premises, or both, will be found by any one who will verify the references I have given to differ more or less from those of most or all the writers I have quoted, in all or most of the points which we treat of in common; but I have abstained from the invidious task of specifying in detail the various errors, small and great, into which, I think, my predecessors, great and small, have, Sir G. C. Lewis not excepted, fallen.

A few words may be employed at the outset in elucidating one of the few universally conceded points in this history, and showing that the classical writers knew nothing of any domesticated Felis in their own countries, and that Mr. George Scharf[2] has consequently fallen into an anachronism in introducing a figure of our cat into his vignette in illustration of the telling of Lord Macaulay's tale of the Battle of the Lake Regillus. In Egypt the cat was domesticated

[1] Link, 'Die Urwelt und das Alterthum,' pp. 199–201, Berlin, 1821; Klemm, 'Culturgeschichte,' i. 1, quoting Link; Isidore Geoffroy St. Hilaire, 'Hist. Nat. Gen.' tom. iii. 96.

[2] 'Lays of Ancient Rome,' p. 78, 1860.

from an early period, as we know from Herodotus and also from the Book of Baruch[1]; and Cuvier[2] could not discover any specific difference between the mummied remains of the domesticated Felis of those days and the similar structures in our own pet cats. Varro (fl. B.C. 28) and Columella (fl. A.D. 20) both speak of the *Felis* as Appian (fl. A.D. 190) does of the *αἴλουροι κακοεργοί* as animals to be kept out of poultry-pens, but none of these writers speak of the animal as being domesticated; and Cicero when he speaks, or hints, *more suo*, that he would speak but for a reluctance to be tedious, *de felium utilitate*, speaks in the same connexion of the ichneumon and the crocodile, and shows us thereby that he had Egypt and not Italy in his eye. The two lines in the Batrachomyomachia, 51–52,

πλεῖστον δὴ γαλέην περιδείδια ἥτις ἀρίστη,
ἣ καὶ τρωγλοδύοντα κατὰ τρώγλην ἐρεείνει,

do not seem to compel us to translate *γαλέην* by *felem*, as *κατὰ τρώγλην* may very well mean '*along and throughout my hole*,' and indeed the line

ἐχθίστη γαλέη τρώγλης ἔκτοσθεν ἑλοῦσα

may seem to show that the catching *outside* was a rare occurrence, and the inside or musteline method the commoner[3].

Neither do I think that certain others of my predecessors have been more correct in translating the words in the fifteenth Idyll of Theocritus, the Adoniazusae, line 28,

αἱ γαλέαι μαλακῶς χρῄσδοντι καθεύδεν,

as though the domestic cat was alluded to by Praxinoe. For though our proverb speaks of the catching of a weasel asleep as a matter of some difficulty, anybody who has watched the way in which one of

[1] 'Epistle of Jeremy, Book of Baruch' (Apocrypha), chap. vi. verse 22.

[2] 'Annales du Muséum,' An. xi. (1802), p. 234; 'Ossemens Fossiles,' Discours Préliminaire, p. lxii, ed. 1821.

[3] It must be recollected, however, that though a carnivore's body will always pass tolerably easily through a foramen which will allow of the passage of its head, the *μῦες* of the Batrachomyomachia were not *rats*, Parliamentary institutions not having been introduced into the kingdom of Artemisia, where it is supposed to have been composed; and that as the thirty-first of Babrius' Fables, or indeed measurement with a pair of compasses of even our smallest weasel's interzygomatic diameter, will show, many a mouse-hole will admit a *mouse* which will not let his enemy in after him. But the γαλῆ was essentially troglodytic, though some holes were too small for it, else my argument would fall to the ground, and Sannyrion would not have written—

τί ἂν γενόμενος εἰς ὀπὴν ἐνδύσομαι;
φέρ' εἰ γενοίμην γαλῆ;

the larger mustelines when tame or at least in captivity composes itself to sleep, and the very evident reluctance with which it unrolls itself, when awaked, out of the dog-like convolution into which it has curled itself up, will feel the force of the more correct rendering. Some of the Alexandrians, again, who were contemporary with Theocritus, used the same word for the tame cats (with which they were more familiar than he, a mere occasional visitor in Egypt, could have been) that Herodotus uses. Callimachus, in his 'Hymn to Ceres,' has, line 111,

καὶ τὰν αἴλουρον τὰν ἔτρεμε θηρία μικκά,

where the Scholiast *in loco* gives κάττος as the synonym.

On the other hand, the compiler of the collection of foolish stories (which is ascribed falsely to Aristotle, and called the 'De Mirabilibus Auscultationibus'), who is supposed to have lived about the same time as Callimachus and Theocritus, speaks, as does Ælian after him, N. A. 15. 26, section 28, of a kind of Cyrenian mouse as being πλατυπρόσωπος ὥσπερ αἱ γαλαῖ; and it is difficult to think that γαλαῖ is not intended to stand in this passage for *Feles*. Section 28 however may have been introduced into this treatise in later times, or possibly the words ὥσπερ αἱ γαλαῖ may by themselves have found their way there as a gloss from the pen of some wise Byzantine to whom in a later age, when the γαλῆ of the classic times had resigned both office and name to the αἴλουρος from Egypt, *curae fuit ejusmodi quisquilias conscribere*. The Batrachomyomachia, it should be observed, is quite free from any taint of Alexandrian or Byzantine impurity, and the use of the word γαλῆ by writers from these localities does not bear therefore upon its employment in the verses above quoted from it. But though there is no reason for supposing that the *Felis domesticus* was domesticated in any other country than Egypt before the Christian era, there are many reasons for demurring to the statement ordinarily[1] made to the effect that this animal was first spread throughout Europe at the end of the period of the Crusades. On looking into Ducange's Glossarium, under the words 'Catta' and 'Cattinae Pelles,' I find that Caesarius, who was the physician-brother of Gregory, the theologian of Nazianzus, and who died A.D. 369, having been the friend of the second Constantine and Constantius, speaks of ἔνδρυμοι

[1] For example, 'Conversations-Lexicon,' Bd. viii. p. 735; Bahr, Hdt. ii. 66.

κάτται; and I presume that the use of this expression shows with some degree of probability that tame cats were in use by this time in Constantinople. The date of Palladius is somewhat uncertain, though supposed with a good deal of probability to have been about the same as that of Caesarius, but as his local habitation appears to have been Italy, his words, iv. 9. 4, 'Contra talpas prodest catos frequenter habere in mediis conductis, mustelas habent plerique mansuetas,' are of importance. They show that the two kinds of cat were both in use as domesticated animals side by side and at the same time, in Italy, nine hundred years before the first of the Crusaders reached Constantinople, and in the days of Gratian and Theodosius, not in those of Godfrey and Tancred.

From the same authority, Ducange, I find that Evagrius (fl. 536) many years later, indeed almost a couple of centuries after the date ordinarily assigned to Palladius, still recognises *αἴλουρος* as the more correct denomination for the *Felis domesticus*, saying, as though the word *κάττα* were a somewhat trivial and over-familiar designation, *αἴλουρον ἣν κάτταν ἡ συνθεία λέγει*, lib. vi. 24. The word 'bird' stood once, I believe, in much the same relation to the word 'fowl:' and the householders of this country show often a greater precision and correctness, more *prisca fides* in short in this matter of the use of these two words, than the upper Ten Thousand do.

I have had a reference given me to a work of the period of Eustathius, i.e. about 1100 A.D., viz. the *Γαλεωμυομαχία* of Theodorus Prodromus, in which the word *γαλῆ* may be found[1] and proved to be used for the cat as we understand the word. I have not however been able to discover or borrow any reference to the employment of the word *γαλῆ* in the sense of Felis or in any other

[1] The following are a few lines from this work, taken from an edition printed at Basle, 1518, by Frobenius, without numbering to the lines. They are, I think, conclusive:—

A. *τίς ἐστιν αὕτη; μὴ φθονήσῃς τοῦ λέγειν.*
B. *ἣν γαλέην ὠνόμαζεν ἀνθρώπων γένος,*
αὕτη γὰρ αἰεὶ χηραμὸν περιβλέπει,
καὶ μῦς ἐρευνᾷ . . .
ἡμᾶς δυχνεῖ λυγκικὸν βλέπουσά τι.
. . . .
πολλὴν καταγνοὺς τῶν πάλαι πεφυκότων
τὴν ἀσθένειαν καὶ κακίστην δουλείαν
ἀνθ' ὧνπερ εἴσω φωλεῶν μυχαιτάτων
μένοντες οὐκ ἄτολμον εἶχον καρδίαν.
καὶ δὴ κατεῖχε τοῖς ὄνυξιν ἀγρίως,
καὶ σὺν τάχει βέβρωκε τὸν νεανίαν.

in the writings of any author who flourished, if authors did flourish, in the years which rolled so drearily away between the days of Evagrius and those of the author of this tragi-comedy. The word *mustela* seems, I may say, in leaving this part of my subject, nearly always, if not quite, to stand for a weasel of one kind or another; though Phædrus does once, namely iv. 1. 9, use the word in a passage homologous with one of those in which we find *αἴλουρος* used in the fables ascribed to Æsop. This latter word, on the other hand, seems always to stand for a *Felis domesticus* or *catus;* whilst *Felis* in the Latin writers does seem, according to Facciolati, to be used indifferently, or nearly so, for either Feline or Musteline. The argument to show that our white-breasted Marten, *Mustela foina*, was used for the same domestic purposes by the ancients as the *Felis domesticus* is by ourselves, may be briefly stated thus. An animal called *γαλῆ* by Aristotle ('Hist. Anim.' ii. 3. 5, vi. 30. 2, viii. 27. 2, ix. 2. 9, ib. 7. 4), and repeatedly referred to by Aristophanes[1] and other Greek writers of the best ages, as well as by the Scholiasts[2], under this title and in more or less completely proverbial expressions, is spoken of as destroying mice, snakes, lizards, birds and birds' eggs, as being the reverse of odoriferous, as

[1] Aristophanes, 'Acharn.' 255 :—

κἀκποιήσεται γαλᾶς
σοῦ μηδὲν ἧττον βδεῖν.

'Plutus' 693 :—

ὑπὸ τοῦ δέους βδέουσα δριμύτερον γαλῆς.

'Vesp.' 363, 4 :—

ὥσπερ με γαλῆν κρέα κλέψασαν
τηροῦσιν ἔχοντ' ὀβελίσκους.

Ibid. 1182 :—

οὕτω ποτ' ἦν μῦς καὶ γαλῆ.

Ibid. 1186 :—

μῦς καὶ γαλᾶς μέλλεις λέγειν ἐν ἀνδράσιν;

'Thesmoph.' 558 :—

ὥς τ' αὖ τὰ κρέ' ἐξ 'Απατουρίων ταῖς μαστροποῖς διδοῦσαι,
ἔπειτα τὴν γαλῆν φάμεν.

'Eccles.' 792 :—

ἢ διᾴξειεν γαλῆ.

[2] Simonides of Amorgos, fl. B.C. 660; Stobæus, 'Florilegia,' vol. iii. ed. Gaisford, p. 63, T. 73. 61 :—

κλέπτουσα δ' ἔρδει πολλὰ γείτονας κακά,
ἄθυτα δ' ἵερα πολλάκις κατεσθίει.

Cf. also *γαλῆ χιτώνιον, γαλῆ κροκωτόν*; Babrius 27, *γαλῆ θηρῶσα μῦς τε καὶ σαύρας.* Mr. Max Müller tells me that in the 'Hitopadesa' an animal, called *Nakula*, kills the serpent under the same provocation and with the same reward as the hound kills the wolf in our story of Llewellyn. It very probably may have been a marten.

being addicted to stealing, and also being so common an animal as to be, like our cat, a convenient scapegoat for the blame due to the thefts of other and not quadrupedal animals, and finally, as being like in its colour and its pelage (except that this latter is a little less thick) and its general appearance and its moral characteristics to the animal called in antithesis to it γαλῆ ἀγρία, but ordinarily ἴκτις, which is a little larger, which loves honey, which kills birds, and is very susceptible of being tamed. It is impossible to think that any great mistake can attach to the interpretation of statements so consentient, so numerous, and relating so eminently to matters of every-day life and constantly observable occurrence. We have two sets of resemblances and differences detailed to us as existing between two animals, the γαλῆ and the γαλῆ ἀγρία or ἴκτις; these two sets of resemblances and differences are just those which exist between our white-breasted marten and our yellow-breasted marten, and as I believe it is impossible to find a second pair of animals to which this comparison will apply, I apprehend that the point is proved. Both the British martens are, as I know from my own observations, and information gathered from persons in the habit of hunting them, great destroyers of mice, birds, and snakes; they are both stated in the ordinary works of natural history to be fond of honey, which the ferret[1] and the weasel will not touch; the fur of both is valuable, but that of the larger species is the more valuable.

The colouration of the polecat, *Mustela putorius*, puts it out of the field into which it has so often been wrongfully introduced, as does also the not altogether[2] unimportant fact that it is not certain that it is found in the extreme south of Europe. If any one who has not had, or perhaps does not care to have, proof that the common north-country name for the marten, viz. 'Sweet Mart' as opposed to 'Foumart' or 'Foul Mart,' an *alias* which the polecat has earned, may not after all be so distinctive as to make us think that we cannot have in the white-breasted marten the same creature as that alluded to in the first two passages I have quoted from Aristophanes, wishes to have this scruple removed at easy cost, he may consult Gesner ('Hist. Anim.' p. 866), who quotes something, *in loco*, to the purpose from Alexander Aphrodisiensis. The stoat,

[1] Buffon, vii. 213; Wagner, 'Säugethiere,' i. 500.

[2] Blasius, 'Säugethiere Deutschlands,' p. 224.

Mustela herminea, could not have failed to have had the well-marked club-shaped black tip of its much larger tail mentioned in contradistinction to that of the common weasel, *Mustela vulgaris*, if these two animals had been the pair contrasted as γαλῆ and ἴκτις; and it may further be remarked that the comparatively small bulk of these animals, as also of the Sardinian weasel, *Mustela boccamela*, which has likewise had its claims advocated for the title of γαλῆ, would have very sufficiently prevented them from being the 'fine thieves' which we know from Simonides, Aristophanes and Babrius, the γαλαῖ were. In looking over my notes of the anatomy of the last marten, a male which came into my hands, I find that there was upon its *linea alba* a space of two inches in length almost bare of hair to direct attention to the fact mentioned by Aristotle, 'Hist. Anim.' ii. 3. 5, καὶ γὰρ ἡ γαλῆ ὀστοῦν ἔχει τὸ αἰδοῖον, and to explain how he came to class this comparatively small animal as regards this carnivorous peculiarity with animals so much larger as the fox and the wolf.

I will conclude the question of the marten cat with the lines from Nicander, the contemporary of Theocritus, which Schneider gives, l. c., iv. p. 49, and I will add to them, what Schneider does not *in extenso*, viz. the commentary of the Scholiast as given in the Editio Princeps of Aldus Manutius, printed 1499, before the art of printing was sixty years old. I give the lines and the scholium, as both are much to the points just discussed and dismissed, and neither are hackneyed in the literature of the subject. Nicander, 'Theriaca,' l. 196:—

> Μορφὴ δ' ἰχνεύταο κινωπέτου οἷον ἀμυδρῆς.
> ἴκτιδος ἥ τ' ὄρνισι κατοικιδίῃσιν ὄλεθρον
> μαίεται, ἐξ ὕπνοιο καθαρπάξασα πετεύρων.

To these lines are laterally apposed in the cramped contractions of the edition specified, the following words of the Scholiast[1]: ἀμυδρῆς

[1] I hope I shall not be considered presumptuous for saying that having seen the Pine Marten or Yellow-breasted Marten escape 'like a shadow,' as Longfellow describes the hare as doing, from the midst of an assemblage of men and dogs of all dimensions when smoked out from under some rocks, I believe I have a better idea of what Nicander, who may have been similarly privileged, meant by ἀμυδρῆς than the Scholiast had. Great as was the creature's need, its agility was more than commensurate with it; and whilst the words *abiït, excessit, evasit*, and alas, for the interests of anatomical investigation, *effugit* also, are but weak symbols to express its speed, it takes a whole hexameter-ful of imagery to give the picturesque effect which its lithe abstraction of itself from jaws and paws produced upon me. It fled

'Par levibus ventis volucrique simillima somno.'

δὲ ἤτοι μικρᾶς ἢ τῆς δεινῆς καὶ ὀργίλου. ἴκτις δὲ ἡ λεγομένη ἀγρία γαλῆ. καὶ Ὅμηρος κρατὶ δ' ἐπικτιδέην ὡς ἀπὸ τῆς ἴκτιδος τοῦ ζώου.

Thirdly: Strabo's words, iii. 386, quoted by Schneider, l. c., iii. p. 524, and relating to the taking of rabbits, *καὶ δὴ καὶ γαλᾶς ἀγρίας ἃς ἡ Λιβύη τρέφει φέρουσιν ἐπιτηδές, ἃς φιμώσαντες παρίασιν εἰς τὰς ὀπάς. αἱ δ' ἐξέλκουσιν ἔξω τοῖς ὄνυξιν*, prove, I suppose, beyond a question, that *γαλῆ* stood for ferret, *Mustela furo*, as well as for the martens, as early as the Christian era.

Fourthly: as one kind of *γαλῆ* was known as the *Ταρτησσία γαλῆ*, and as the Scholiasts tell us[1] *Ταρτησσίαν* was used *ἀντὶ τοῦ μεγάλην*, and as Herodotus informs us, iv. 192, that *γαλαῖ* very like these Tartessian (or larger) *γαλαῖ* were to be seen in Libya, where we know any animal like and larger than a marten would be a viverra, I think we may with some little probability suppose that the *Ταρτησσία γαλῆ* was nothing less or else than the Viverra genetta, which is found all over Africa and also in the South of Spain and France, where it is domesticated even now, here and there, and acts as a tolerable cat. I have not been able to find that this animal is known in Greece, it is not included by Blasius in his 'Säugethiere Mitteleuropas;' whence we can the better understand why it was called *Ταρτησσία γαλῆ*; though it is found in Asia Minor, whither it may have found[2] its way from Egypt. If the Herpestes widringtoni, recently found[3] in the Sierra Morena, had been as well known in the region of Tartessus as the Viverra genetta, its claims to be considered as the *Ταρτησσία γαλῆ* would perhaps have been as much greater as its size is. But the Pharaoh's Rat, which would then have been the *γαλῆ* of Herodotus as being the representative of the Spanish Herpestes, would have been contrasted with it in the point of being considerably smaller, which however is not the case. I append measurements of these and of certain other of the animals of which I have spoken in this paper, the point of size being the point in which the Tartessian or Spanish Marten is contrasted with the commoner one by the Scholiast, and being, as it seems to me, sufficiently great to mask the difference which the lateral and the caudal striping of the Genet also constitutes between it and the *γαλαῖ* of Greece.

[1] See Aristoph. 'Ranae,' ed. Bothe, adnot. ad l. 440, sub vocc. *Ταρτησσία μύραινα*, ibique citata.

[2] Ainsworth, cit. A. Wagner, 'Abhandl. Akad. Wiss. München,' iv. 107.

[3] Gray, 'Ann. and Mag. Nat. Hist.' 1842, ix. 50.

	Length of Body.		Length of Tail.
Common Genet, *Viverra genetta*	20″		16″
Pharaoh's Rat, *Herpestes ichneumon*	18″		18″
Spanish Ichneumon, *Herpestes widdringtoni*	22″		20″
White-breasted Marten, *Mustela foina*	16″		8″
Yellow-breasted Marten, *Mustela martes*	18″		12″
Polecat and Ferret, *Mustelæ putorius et furo*	18″		6″
Sardinian Weasel, *Mustela boccamela*	8½″		4″
Stoat, *Mustela erminea*	10″		4″
Common Weasel, *Mustela vulgaris*	7½″		2″

The upshot of this paper then is to show that in classical times the word γαλῆ was used by the Greeks to denote the musteline marten and ferret, but not the polecat probably, though probably the genet, and that in later times, but not till later times, it was used also for the *Felis domesticus*. The word *mustela* does not seem to have been transferred together with the office when the latter was handed over from the marten to the felis, in Italy. In the East the felis took both the name and the work of the rival it supplanted. It did succeed in supplanting the marten as the domestic mouse-killer, probably partly by virtue of its greater attachment to man and to place, partly by virtue of its less pronounced tendency to burglary and petty larceny, partly by virtue of its more even temper, and partly by its greater cleanliness and less offensiveness. The very points also in which as a wild animal it is inferior make it superior as a domestic one to a musteline. Its constitution being less plastic it cannot fit itself as easily as the latter to varying climates, in many of which, as Reugger has shown of Paraguay, it cannot run wild. Its range of foods is more limited, and its faculty for, and its courage in adopting new methods of purveying for itself, less conspicuous than theirs. Hence 'the poor cat of the adage' being more dependent on man, has been obliged to render itself more useful to him than the marten, and it has very successfully turned its inferiority to 'commodity.'

The question as to how, in the trivial language of two different nations, English and Greek, in modern as in ancient times, Viverrines[1], Mustelines, and Felines have each had a representative called by the same name as a couple of animals, one in each of the two other families, is a little harder to understand for the anatomist, or for the anatomical artist, than it is for anybody who, being devoid of either of these accomplishments, will stand inside

[1] We speak of a civet *cat* as well as of a marten *cat* and common *cat*.

the half-shed half-house for the 'Small Carnivora' at the Zoological Gardens, and listen there to the remarks of people who overlook the little differences upon which scientific zoology is founded. 'They are all cats,' I heard one of these authorities[1] say there one day, albeit there were then plenty of the eminently annuloid viverrines as well as a very typical felis, the *Felis chaus*, to be compared and contrasted at a single glance and within a few feet of each other. It is not hard to see how the mustelines and viverrines come to be classed together, seeing that so many members of both families are so markedly elongate, vermiform, tapering and low on their limbs. But the relative proportions in the sides, in the trapezium which four lines, corresponding one to the fore-legs, one to the hind, one to the line of the back, and the fourth to the ground on which the creature stands, make up respectively in a feline and in a viverrine or musteline viewed from the side, are so very different, to say nothing of the all but equally striking differences in the proportions of the skull and jaw diameters, longitudinal and transverse, *inter se*, firstly, and in relation to the cervical region, secondly, that we must look to points of habit rather than of structural arrangement to account for the imposition of this common name upon creatures to our eyes so different. And I suppose the springy yet silent lightness of their step when placid, and the lightning-like pounce of their attacking step, correlated as they are with a more or less similar armature in tooth and claw, are the points which 'imaginationem ferientia aut intellectum vulgarium notionum nodis astringentia' have caused the imposition of the common name these animals have had given them. The arboreal and nocturnal habits again, correlated with certain modifications in the organs of special sense, are common points to the feles and the mustelae, and especially though not exclusively to the martens. Both alike take to trees when pressed by hounds, but since the invention of fire-arms this single device of the 'cats' is no longer worth as in the old fable more than all the tricks of the fox. The phrase 'up a tree' was not, perhaps could not have been, anterior to that of 'as sure as a gun.' The pine marten indeed will, Blasius informs us, sit still on the same place on a bough

[1] Strabo, however, uses much the same language in speaking of what must. I think, in all probability have been the common genet, *Viverra genetta*. Writing of Mauritania he says, xvii. c. 3. p. 827 A, Casaubon :—*φέρει δὲ καὶ γαλᾶς αἰλούροις ἴσας καὶ ὁμοίας, πλὴν ὅτι τὰ ῥύγχη προπέπτωκε μᾶλλον.*

after having been shot at once and missed. The cat has, Reugger informs us ('Säugethiere Paraguay,' p. 214), learnt to kill the rattlesnake in Paraguay, and I have read that the felis acquired this self-same snake-killing dexterity in the island of Naxos, but I have not the reference at hand at this moment. Herein it has by practice under the stimulus of constant provocation come to resemble the mustelines in what is instinctive to them; but though it will steal cream, as Falstaff told us, it will never, like the martens, steal eggs nor honey nor take to burrows in the way of refuge.

I am aware that there are both scholars and men of science to whom disquisitions such as these will seem but the *strenua inertia hominis male feriati.* Critics such as Pope, and, I regret to have to add, such as Hallam (see 'Literature of Modern Europe,' i. 277), speak of such attempts to preserve the unities of time and place in Faunae as in dramas, the one with the cynical sneering giggle, the other with the elevated and refrigerating yet half-compassionate contempt congenial to their respective schools of literature and of politics. But to the scholar I would say that, though in these matters as in many others by increasing knowledge we increase also sorrow, or at least our susceptibility for annoyance, it is rare indeed to find a writer of the classical periods making blunders in the way of putting animals into places which they never were found in, except in connexion with the circus of olden or the menagerie of modern times, which are so rife in all but our very best modern writers. Modern catalogues of African mammals show that Virgil did not deserve the criticism as to the presence of the stag there which Pope in the 'Martinus Scriblerus' puts into the mouth of Bentley as unworthy of any one else; and that Lipsius need not have explained away, as he does ('Elect.' ii. 4), the phrase *Libystidos ursae.* The placing of the lion by Theocritus, i. 72, *τῆνον χὡ 'κ δρυμοῖο λεών ἀνέκλαυσε θανόντα*, is in fact the only anachronism or anatopism of the kind which my memory furnishes me with from the writers of the best periods of Greek and Roman literature.

To the man of science I may say in the words of Goethe :—

'Müsset im Naturbebrachten
Immer eins wie alles achten
Nichts ist drinnen, nichts ist draussen :
Denn was innen, das ist aussen
So ergreifet ohne Säumniss
Heilig öffentlich Geheimniss.'

BIBLIOGRAPHY.

Aristotle, 'Die Thierarten,' von C. J. Sundevall. Stockholm, 1863. 'De Animalibus Historiæ,' Schneider, Lipsiæ, 1811.
Sir George Cornewall Lewis, 'Notes and Queries,' 2nd Series, viii. Oct. 1. 1859.
Ducange, 'Glossarium.'
Link, 'Die Urwelt und das Alterthum.' Berlin, 1821.
J. Geoffroy St. Hilaire, 'Hist. Nat. Gen.' Paris, 1862.
De Blainville, 'Ostéographie Felis,' p. 68.
Dureau de la Malle, 'Ann. Sci. Nat.' tom. xvii. 1829.
Bazin, 'Actes Soc. Linn.' Bordeaux, 1843.
Lenz, 'Zöologie der alten Griechen und Römer.' Gotha, 1856.

XXIX.

ON THE CAT OF THE ANCIENT GREEKS.

In a book, with the existence of which I became acquainted a few days ago, by a reference in Aubert and Wimmer's new edition of Aristotle's 'Historia Animalium,' and which, through the kindness of Mr. Sclater, I have had put into my hands this day, I have come upon certain statements which confirm not only certain of the conclusions, but certain also of the conjectures put forward by me in the 'Journ. of Anat. and Phys.,' Nov. 1867 (Article XXVIII). This book is Dr. Erhard's 'Fauna der Cycladen,' Leipzig, 1858. From it I learn that the *Mustela foina*, the white-breasted marten, the animal which in my paper I strove to show was the domestic mouse-killer of the ancient Greeks, is common now in all the Cyclades, and in some of them actually has the old Greek name ἴκτις at the present day. The polecat, *Mustela putorius*, and the ferret are not members of this fauna; neither could Dr. Erhard find the genet there. I need not point out the bearing of these statements upon those advanced by me in that article. But I will take this opportunity of saying that Dr. Erhard's little volume deserves to be better known than it is at present in England. Besides giving us an excellent example and a 'simple case' for the study of the rationale of the Distribution of Species, it teaches us the very important, and not a little needed lesson of caution, in receiving catalogues of indigenous animals of any area, however small and accessible, as being necessarily exhaustive. Though the vegetation of the Cyclades is (p. 7) of such a character that a hare can hardly hide itself from the eye of the eagle, and though at first Dr. Erhard was (p. 8) inclined to think their mammalian fauna was as exclusively Adullamite as that of a coral island, he has, *after an investigation of several years*, given us a list of no less than sixteen land mammals, amongst which there is one new and previously undescribed species, the *Ægoceros pictus*, the Ibex of the Cyclades. The very general distribution of the rabbit, which in its wild state here is as large as the hare or larger, goes some way towards showing that it was indigenous in the area of the Cyclades, as it is supposed to have been in the Balearic Islands before they were broken up into an archipelago. But at the same time it is the harder to understand how Aristotle and how the Greek *gourmets* who ζῶντες ἐν πᾶσι λαγῴοις must have known how different a rabbit's flesh was from a hare's, if they had ever eaten it, could have failed to distinguish the one animal from the other, the rabbit being now most abundant there, and having made the Myconos, so often mentioned by these ancients, into a honeycomb with their burrows. But in their days these islands were richer in population, an occasional massacre of Melians having been as nothing to the constant operation of Turkish barbarism; and if, as I have striven to show, this larger population had in domestication, house by house, such an enemy of the rabbit as the marten is, we have some sort of an explanation of the absence of the mention by them of an animal as existing in the Cyclades in classic, which must all but certainly have existed there in geological, as it does also in our own times. The islands, I may add, were probably or certainly better wooded than now; and trees favour the multiplication of the rabbit less than they do that of its many and various enemies.

2. In my previous paper I said, Article XXVIII. p. 511, 'In the East the felis took both the name and the work of the rival it supplanted.' It is possible that I should have been right in making this statement more extensive, and in saying that the cat of the Egyptians took not only the name γαλῆ, but also the name ἴκτις of its predecessor in the Greek house. For I find from a passage of Tzetzes, 'Chil.' v. 8, quoted by Bochart, 'Hierozoicon,' i. 986, 57, that this authority, if so we may call him, called the αἴλουρος by the name of ἴκτις. The passage from Bochart runs thus: 'In Hesychio voce κτιδέα, κτίς ἐστι ζῷον ὅμοιον γαλῇ, viverra est animal feli simile. *Proïnde putavit* Tzetzes esse felis speciem, quod his verbis diserte asserit Chiliadis quinti capite octavo,

ἴκτις δὲ ζῷον καὶ αὐτὸ τελεῖ (l. πέλει) ὀρνιθοφάγον
χερσαῖον καὶ τετράπουν μὲν, ὃ αἴλουρον καλοῦμεν.'

But I am inclined, as I think Bochart was also, to consider these lines to be so worthy of their author as to be unworthy of any attention from us. And it is interesting to note that, in a Basle edition, of the year 1546, of Tzetzes' works, the first, and to the credit of human nature it should be added, also, up to 1826, the last independent edition of this portion of his works, some of which happily still remain unprinted, there stands opposite these lines the following Latin note: '*Alii mustelam rusticam* seu viverram.'

3. If I understated my case in this instance, I overstated it in another, when I said (p. 513) that the cat will never take to burrows in the way of refuge. I have already said, in this paper, that in drawing up catalogues and making faunae, it is wise and well to avoid universal negatives. It would have been well if I had had this precept before my eyes in a more generalised form when I wrote my last paper; for I have since been informed by two good observers that they have each of them known a cat take to troglodytism. In neither of these cases, however, did the animal profit much by doing what it could not have been expected to do.

XXX.

ON THE DOMESTIC PIG OF PREHISTORIC TIMES IN BRITAIN,

AND ON THE MUTUAL RELATIONS OF THIS VARIETY OF PIG AND 'SUS SCROFA FERUS,' 'SUS CRISTATUS,' 'SUS ANDAMANENSIS,' AND 'SUS BARBATUS.'

Portions of two skeletons of domestic pigs having been put into my hands by the Rev. William Greenwell, F.S.A., from an interment of the so-called late Celtic period, i.e. of the ultimate or penultimate century before the Roman conquest of this country, I determined to compare them with such other specimens of *Suidae* as might by any possibility be genetically connected with them. Among these other specimens I may mention, first, several specimens of the wild boar, *Sus scrofa*, var. *ferus*, from the alluvial deposits of this neighbourhood, and now in the Geological Series of the Oxford Museum, under the charge of Professor Prestwich, F.R.S.; secondly, five specimens of the Indian wild hog, *Sus cristatus*, kindly lent me by Sir Walter Elliot, K.C.S.I., F.L.S.; thirdly, two skulls of *Sus andamanensis*, presented to the Oxford University Museum by my friend Prof. J. Wood-Mason, of the Indian Museum, Calcutta; and fourthly, four skulls of *Sus barbatus* from Borneo. The extensive series of skulls of *Suidae* contained in the British Museum, those in the Royal College of Surgeons of London, and the specimens of wild- and domestic-swine skulls contained in our own collection were also used for this comparison.

It may be well at the outset to specify the several points of wide and general interest upon which such an enquiry as the ensuing may be brought to bear. First among these I would mention its bearings upon the now so commonly discussed questions relating to the early migrations of our own species. The pig was one of the earliest, possibly the very earliest, of animals which man domesticated; and the question of the source or sources whence it was derived has consequently an 'ethnographisch-archäologische Bedeutung' (to use the words of Fischer [1], in his analogous investigation as to the sources whence the jade and nephrite of early European times were procured) of the first importance. Gibbon [2] has remarked that 'man is the only animal which can live and multiply in every country from the equator to the poles;' and he has proceeded to aver that 'the hog seems to approach nearest to our species in that privilege.' As a matter of fact, Gibbon here, as so often elsewhere, was very nearly though not quite exact: the northward limit of the range of the wild boar may perhaps be taken as somewhere between 55° [3] and 60° N., and that of the tame pig as 64° N., whilst that of the common fowl [4] extends probably a little, and that of the dog certainly much, further northward. The mention of the Australian dingo suggests another amendment of the somewhat cynical remark of our great historian.

Secondly, a very useful light is thrown, or may be thrown, upon the question of the extent to which the influences of civilisation act upon our own species, by the analogous enquiry into the effects which domestication has been able to produce upon an animal linked so closely with ourselves in a many-sided commensalism. The least pleasant aspect of that commensalism, that, namely, which is presented to us by the facts of our solidarity with swine in the maintenance of the alternation of the forms of life of *Taenia* and *Trichina*, must, it may be remarked, upon any view of the origination of the four species concerned, force upon our attention

[1] H. Fischer, 'Nephrit und Jadeit nach ihren mineralogischen Eigenschaften.' Stuttgart, 1875. For similar investigations as to the sources of the cultivated plants and the weeds of prehistoric times, see Keller's 'Lake-Dwellings,' translated by Lee, pp. 303 and 343.

[2] 'Decline and Fall of the Roman Empire,' chap. ix. note 9. p. 352. Smith's edition.

[3] For the northward range of the wild boar and the tame pig, see Brandt and Ratzeburg, 'Med. Zool.' p. 89; Fitzinger, 'Sitzungsberichte d. Akad. d. Wiss. Wien,' 1864, p. 387; Radde, 'Reise im Süden von Ostsibirien,' i. 236; Zimmermann, 'Geographische Geschichte,' i. 189, 1778. Middendorf, in his 'Sibirische Reise,' p. 1062, 1867, gives 56° N. lat. as the extreme actual northward limit of the wild boar. (Since I wrote as above, Mr. H. N. Moseley has procured for me a skull of a tame pig from Stene i Bö, Lofoten Islands, 2° above the arctic circle.)

[4] For the northward range of the domestic fowl, *Gallus domesticus*, see Brandt and Ratzeburg, *l. c.* p. 150.

considerations of the very greatest gravity. To take somewhat lower ground, from such a study as that of the variations of the pig under domestication, we may obtain safe criteria for estimating the relative effects of food, whether scanty or abundant, of early or late exercise of the sexual functions, and of intercrossing, upon the formation of facial characteristics. The importance of not overlooking the influences of sex and of age is nowhere more forcibly pressed upon our attention than in an examination of a series of skulls of *Suidae*. It has often been overlooked in disquisitions on the skulls of *Hominidae*.

Thirdly, whether the question at issue as regards man between the Polygenists and the Monogenists will, as has been predicted (Darwin, 'Descent of Man,' ed. 2nd, p. 180, 1874), 'die a silent and unobserved death' or not, there can be no doubt that illustrations of the argumentations whereby that question has been, or ought to have been, dealt with, can be furnished nowhere more fitly and fully than in an enquiry into the distinctness or non-distinctness of the various races of swine.

Three distinct views have been advocated as to the relationship of the domestic to the wild swine, *Sus scrofa*, var. *ferus*. We may take as the first of these that advocated by Professor Steenstrup[1], and stated by him in the following plain words :—

'Il n'y a pas de transition à observer entre les sangliers et les plus anciens cochons domestiques.'

In this view Professor Steenstrup will find Mr. Samuel Sidney coincide in the opening sentences of his work 'On the Pig;' and a view very closely similar to it was put tentatively forward in the year 1821 by a *savant* who combined the functions of a Professor of Materia Medica with those of Director of the Botanical Garden at Berlin, Professor Link, in the following words, to be found in his work 'Die Urwelt und das Alterthum,' i. p. 192 :—

'Das zahme Schwein stammt nach allen Naturforschern von den wilden Schweinen ab, und auch die Alten waren schon dieser Meinung[2]; doch scheint mir die Sache keineswegs entschieden. Die Stärke, Grösse und Farbe des wilden Schweines würden keinen Unterschied machen, da die wilden Thiere stärker, grösser, und dünkler gefärbt sind, als die zahmen, aber die grossen Hauer des wilden Ebers scheinen doch nicht bloss Vergrösserung zu sein. Die Feltdecke des zahmen Schweines findet sie

---

[1] 'Bulletin du Congrès International d'Archéologie préhistorique à Copenhague en 1869,' p. 163.

[2] Varro, 'De Re Rustica,' l. ii. c. l.

niemals auch bei dem fettesten wilden Schweine, die gestreifte Farbe des Frischlings ist sogar merkwürdig, die Stirn der wilden Schweine ist mehr gewölbt, die Ohren sind kürzer, mehr zugerundet, der Rüssel länger, andere Verhältnisse an den innern Theilen zu geschweigen. Es fehlt ganz an Beispielen, dass die Zähmung solche Veränderungen hervorgebracht. Vielleicht stammt das zahme Schwein von einer orientalischen Art ab, welche gross aber unschädlich sein soll, und hin und wieder in einigen Reisebeschreibungen erwähnt wird[1]. Doch erfordert die Sache noch eine genauere Untersuchung. Das siamische Schwein aus dem östlichen Asien abstammend, ist ohne Zweifel eine besondere Art.'

Precisely the opposite view was held by Mr. Youatt, who at p. 35 of his work on 'The Pig,' London, 1847, says, when speaking of the wild boar,

'No one can for a moment doubt that it is the parent stock from which the domesticated breeds of swine originally sprang.'

Blasius ('Säugethiere,' 1857, p. 509), Fr. Cuvier ('Hist. des Mamm.,' 1824), and Giebel ('Säugethiere,' p. 225, 1859), would, I apprehend, agree with the extreme view enunciated by Mr. Youatt. An intermediate view is put forward by Rütimeyer in his inestimable work, 'Die Fauna der Pfahlbauten,' pp. 186–190: according to him, what, owing to the slow spread in Germany of improved breeds, may still be called *das gemeine Hausschwein* has originated from *Sus scrofa*, var. *ferus;* whilst the Berkshire breeds, he thinks, may owe their origin to *Sus celebensis;* and his *Sus scrofa,* var. *palustris*, 'das Torfschwein,' a domestic pig known as 'das Bündtner Schwein,' and *Sus indicus* may represent a distinct *stock*, if not species. Nathusius, in his 'Schweineschädel,' p. 175, agrees with Rütimeyer as to the origin of the large-eared race common in Central Europe, but suggests *Sus vittatus*, of the islands of Java, Borneo, Amboyna, and Batchian, as the parent stock of the widely-spread domestic breed known as *Sus indicus*. In this latter point he agrees with S. Müller. Fitzinger (l. c.) differs from Nathusius in supposing, without perhaps adequate reasons, *Sus leucomystax* to be the parent stock of the Chinese, Cape, Portuguese, and Cleveland breeds, whilst *Sus cristatus*, the wild pig of Hindostan, he suggests as the parent stock of the Siamese and Sardinian races. Temminck ('Fauna Japonica,' p. 57, pl. xx. cit. Nathusius, l. c. p. 167) suggests that *Sus leucomystax* may be the parent of the domestic *Sus indicus;* and though Nathusius, p. 167, demurs to

[1] Otter, 'Voyage en Persie,' t. i. p. 207; D. Maillet, 'Description de l'Egypte,' t. ii. p. 176.

the enunciation of this statement as being definitely proved, his objection amounts to little more than saying that *Sus leucomystax* is probably not specifically distinct from *Sus vittatus*, which Nathusius himself holds to be the parent stock of *Sus indicus.*

My own views, as based upon the data available to me, are to the following effect :—The prehistoric domestic swine which have come into my hands appear to me to be more nearly affined to *Sus scrofa* than to any of the Asiatic wild swine with which I am acquainted; secondly, without wishing to affirm absolutely that too much weight has been laid by Nathusius upon the shortness of the lacrymal for differentiating *Sus indicus*, as we now see it, from *Sus scrofa* and its progeny, I am inclined to think that sub-equality, if not actual equality, between the malar and the orbital borders of the lacrymal bone may be found in prehistoric skulls of the *Sus scrofa*, var. *ferus*, and especially in the female skulls of that variety of *Sus* which, in other points, such as the slenderness of the snout, differ from the 'Torfschwein,' the representative in those times of *Sus indicus*, according to Rütimeyer; and, thirdly, I think it is possible to show that, whilst *Sus cristatus*, *Sus leucomystax*, *Sus vittatus*, and *Sus timorensis* form a closely connected group of *Suidae non verrucosi*, with which, again, *Sus andamanensis* and *Sus papuensis* are to be allied, all these sub-species differ in points of considerable if not of specific value from *Sus verrucosus* of Java, from *Sus celebensis*, and, finally, from *Sus scrofa* of the Palaearctic region as well as from the non-verrucose *Sus barbatus* of Borneo.

It may be well to begin with this latter point first, and to show that the group of Eastern pigs, of which the wild pig of India, the Malay peninsula, and the Lancay Islands, *Sus cristatus*, may be taken as a type, is always distinguishable from *Sus scrofa*, var. *ferus*, of Europe, and Asia north of the Himalayas. This view is not equivalent to one which should lay it down as certain that they are specifically distinct, a question which it is not proposed to raise here. I should agree, however, with Mr. Jerdon, whose book on the Mammals of India, 1874, came into my hands subsequently to the formation of my opinion, in holding that the Indian wild hog was 'as worthy of specific distinction as many other recognised species' (l. c. p. 241), though this is not to say much. I should not, however, entirely accept his statement to the effect that the head of the Indian wild hog was longer and more pointed than that of the European wild boar, though this has been laid down by Colonel Sykes ('Proc. Zool. Soc.' 1831, p. 30) as being the state of the case; for in measuring the relative lengths of the nasal and fronto-parietal regions of the Indian wild hog, I have come to think that precisely the reverse of this statement is usually, though not invariably, the case, the Indian hog having the nasal bones shorter relatively to the rest of the roof-bones of the skull than *Sus scrofa*, var. *ferus*. Neither do I agree with Colonel Sykes in holding that the straightness of the plane of the forehead will differentiate the Indian at least from modern European wild boars: some concavity is produced in the mesial contour-line of large prehistoric wild boars, not by any angulation at the junction of the facial with the cranial bones, as in modern tame swine of highly cultivated breeds, but by the upgrowth of the back portion of the skull roof-bones and the occipital transverse crest; modern European wild boars, however, which are much inferior in size to their prehistoric and, indeed, to their mediaeval predecessors, have the fronto-parietal and nasal lines forming one unbroken straight line.

Colonel Sykes's words, 'Tail never curled or spirally twisted,' appear to me (l.c. p. 11) to be said of the tame variety of the Indian hog; but though Fitzinger ('Sitzungsberichte Akad. Wiss. Wien,' 1864) specifies the form in which the caudal vertebrae are carried as one of the specific marks in each of his descriptions of the *Suidae*, and though Linnaeus uses the words (Ed. xiii. 'Syst. Nat.' p. 65) 'Cauda sinistrorsum recurvata' for differentiating *Canis familiaris* from *Canis lupus*, s. *Canis cauda incurvata*, I am inclined to think too much weight may be laid upon this point.

Thoroughly trustworthy figures of the European wild boar, such as that given by Schreber ('Säugethiere,' taf. cccxx), or that in Buffon's 'Hist. Nat.' v. pl. xiv, and Fréd. Cuvier, 'Hist. Mamm.,' represent it as having the root and tip of its tail lying evenly between two points; though the Vienna zoologist just referred to says of this appendage in this animal,

> 'Schwanz geringelt, kurz, nicht ganz bis zum Fersengelenke reichend.'

And as an indication of the trifling value of such a point as this, it may be remarked that of two female specimens of the very well-marked species, *Sus barbatus*, one young, the other old, figured by

S. Müller, 'Verhand.' i. taf. 30, one has the tail curled, and the other, the elder one, has it straight.

Mr. Blyth (*cit.* Jerdon, *l. c.*) holds *S. cristatus* to be only a variety of the wild boar of Europe, but still to be a well-marked race. De Blainville ('Ostéog. Sus,' p. 129) sees no differences of morphological importance between any of the Asiatic swine and the European wild boar, and says :—

'La première espèce que le squelette nous permet de distinguer par des caractères susceptibles d'être lus et exposés est celle qui se trouve dans toute l'Afrique au delà de l'Atlas et jusqu'à son extrémité la plus méridionale et même au delà dans la grande île de Madagascar, et qui est connue sous le nom de *Sus larvatus* (*Potamochoerus africanus*, *Porcus madagascariensis*, nobis').

Giebel follows De Blainville in this. Dr. Gray, in 'Proc. Zool. Soc.' 1852, p. 130, said that, after examining ten skulls of the European wild boar and its offspring from this country, from the Gambia, and from the Cape, as also twelve skulls of the wild hog from Continental India, he could not discover any constant easily-described character by which the European and the Indian kinds could be distinguished. And, he adds, 'this is the case in the many other genera allied to the pigs.' It is true, no doubt, that many animals, such as—

'the lion and the tiger, the fox and the jackal, the ass and the zebra, are far more strikingly differentiated by their pelage than by their skulls,'

as Professor Huxley ('Prehistoric Remains of Caithness,' p. 132) has taught us; still it yet remains to be proved that differences which, though only skin-deep, are constant and permanent, will not ultimately be found to be correlated with more or fewer differences of the deeper-lying parts, either of a purely qualitative or of a quantitative kind.

In a disquisition the ultimate object of which is the attainment of clearer views as to the origin of our tame pigs, the question meets us at the outset whether there exists any marked difference between the wild stocks under comparison as regards their susceptibility of domestication. Upon this point I have to say that I find, in opposition to what Mr. Samuel Sidney has laid down in the first chapter of his work on the pig, that *Sus scrofa*, var. *ferus*, is credited by most trustworthy authors with as great a capacity for domestication as any wild animal, including its wild Asiatic congeners, upon which observations are recorded as to this particular susceptibility. Pallas's words in his 'Zoographia,' i. p. 269, as to the wild boar of the Palaearctic region, called by him, not inconveniently, *Sus europaeus,* in contradistinction to the China or Siam pig, called by him *Sus indicus,* says plainly and emphatically 'porcelli cicurari assuescunt facile et cum domesticis generant.' Radde's utterances ('Reisen im Süden von Ost-Sibirien,' Bd. i. p. 236) are even more to the point, as they affirm the like of wild swine of greater age. They run thus:—

'So muss ich gestehen dass sie sehr friedlicher Natur sind, und es mir mehrmals passirte mittelalte Wildeschweine sich mir bis auf vier Faden Weite nahen zu sehen.'

The Asiatic pigs, secondly, of the group represented by *Sus cristatus,* though not, within my knowledge, those known as *Sus verrucosus,* nor those known as *Sus barbatus,* have very similar and very numerous testimonies borne to their educability and capacity for attachment to man. Fitzinger, indeed, says of the domesticated Chinese pig, his '*Sus leucomystax sinensis,*' that it resembles the domesticated European pig generally in its habits and character, but that it shows much more attachment than the European farm-pig to the persons who take care of it, and will even follow them about, although it is otherwise troublesome and obstinate. ('Sitzungsberichte d. Akad. Wiss. Wien,' 1858, Bd. xxx. p. 235.) In Formosa, when the Dutch first became acquainted with it, in the beginning of the seventeenth century, every native woman, we are informed, on the authority of Ogilby ('Atlas Chinensis,' ii. p. 8, *cit.* by Swinhoe, 'Proc. Zool.' 1870, p. 643), had 'a great pig running after her, as we use to have a dog.' A closer intimacy than this has been observed to exist between *Homo sapiens asiaticus*, Linn., and *Sus sinensis*, Linn., by Professor Huxley (*cit.* Galton, 'Trans. Ethnol. Soc.' vol. iii. 1865, p. 127), who has seen sucking-pigs nursed at the breasts of women, apparently as pets, in islands of the New-Guinea group. As regards the wild races, Sir Walter Elliot tells me, in a letter of date May 15, 1876:—

'I have seen the young of *Sus cristatus*, which had been captured by some of the Indian nomad communities, and reared by them, running about among the domestic stock; so that it would be hard to say where the line should be drawn [1].'

---

[1] It would appear that this difficulty has been felt by others to be a very real one. Colonel Walter Campbell tells us, at p. 325 of his 'Indian Journal,' 1864, that he fears 'the young gentlemen of the present day have taken to spearing village pigs instead of wild boars,' and that he has 'seen the thing done before now.'

The wild *Sus papuensis* has been found by Europeans, as we are informed by M. R. P. Lesson ('Voyage de la Coquille,' 1826, vol. i. p. 176), to be more thoroughly domesticable than the half-wild state in which some of the natives are content to leave it would have led us to expect, and to set up relations of mutual amity, not only with the human, but also with canine companions.

As regards the Aethiopian region, the wild hog is reported to us by Dr. Barth[1] as consorting on terms of perfect amity, and, indeed, intimacy, with other domesticated animals than the dog, and also with the natives. These are his words relating to a district in Central Africa :—

'Naked young lads were splashing and playing about in the water together with wild hogs in the greatest harmony ; never in any part of Negroland have I seen this animal in such numbers as here about the Shárí. Calves and goats were pasturing in the fields with wild hogs in the midst of them.'

It is impossible to be perfectly certain what wild hog this may have been ; still it can scarcely have been any other than *Sus sennaariensis*, which has been supposed (see Darwin, 'Domestication,' i. p. 71, 2nd ed.) by J. W. Schütz to have been the parent stock of *Sus scrofa*, var. *palustris*, of Rütimeyer, but underneath the entry of which, in the British Museum Catalogue of 1869, p. 338, I find the following note :—

'Dr. Murie says he has often seen and eaten the true wild boar of the genus *Sus* in Africa, as well as the *Potamochoerus*, on the west coast. I have never seen any, or the skull of one.'

Like Dr. Gray, I have had no quite satisfactory opportunities for forming an opinion, such as inspection of authentic skulls gives, as to the relations of *Sus sennaariensis* to *Sus scrofa ;* but whilst going over the series of skulls of Suidae in the British Museum, I came upon one which, though entered under the head of *Potamochoerus africanus*, ought, I make no doubt, to be entered under the head of *Sus sennaariensis*, or, as I should prefer to call it, *Sus scrofa*, var. *africanus*. This skull is numbered 715 *a* ; and of it we have the following history from Dr. Gray ('Proc. Zool. Soc.' 1868, p. 35 ; Brit. Mus. Catalogue, 1869, p. 342) :—

'A skull without its lower jaw (715 *a*) was brought home by Captain Alexander from his expedition to Damara, and presented to the British Museum. It is recorded in Mr. Gerrard's "Catalogue of the Bones in the British Museum" as *Sus capensis*

[1] Referred to by Nathusius, p. 147, *l. c.* ; or see his 'Travels in Central Africa,' vol. iii. p. 311, 1857.

(p. 277). It is the skull of an adult animal, with the crown of the grinders much worn. It is probably the skull of a female, as it agrees with all the characters of *Potamochoerus;* but it has only a well-marked ridge across the upper part of the base of the sheath of the upper canine, and the upper margin of the nose is not dilated nor swollen.'

When I took this skull into my hands I was uncharitable enough to suggest to Mr. Gerrard that it was a skull of *Sus cristatus* wrongfully assigned to Damara as its habitat. Leaving this low ground, I came to think that it might have come from some descendant of pigs of the *Sus indicus* breed which had run wild at the Cape and reverted to the *Sus cristatus* form. But I have now no doubt that this is a skull of such a wild boar as those of which Dr. Murie speaks; and if this be so, the *Sus* of Africa is not readily to be distinguished from *Sus cristatus*, at least by cranial characters. Dr. Gray, under the heading *Sus sennaariensis* ('Proc. Zool. Soc.' *l. c.* p. 32, and British Mus. 'Catalogue,' 1869, p. 338), suggests that the skull figured by De Blainville ('Ostéographie,' tab. v) may have belonged to a *Sus sennaariensis*. It is described as '*Sus scrofa aegypti*' by De Blainville; but, as far as I can judge from the drawing, it has all the characters, to be hereinafter detailed, which distinguish *Sus scrofa*, var. *ferus*, from *Sus cristatus*, and consequently from such a skull as that labelled 715 *a*, and brought home by Captain Alexander. Of course there is no *à priori* difficulty in the way of our supposing that the wild boar either of Palestine or of Algiers, both well-known animals, may have extended into Egypt, a country which has so much both of the Palaearctic and of the Aethiopian fauna in occupation of its territory. Anyhow Dr. Murie and Dr. Barth are sufficient witnesses to the fact that a true *Sus* is found in Africa south of the Atlas and Sahara. I cannot, therefore, accept Mr. Wallace's statement ('Geograph. Distrib. of Animals,' vol. i. pp. 253, 286–322) to the effect that a true *Sus* is not to be found in the Aethiopian region. The mistake made, and handsomely acknowledged, by Fitzinger ('Sitz. Acad. Wiss. Wien,' 1864, Bd. 49, i. p. 389), in naming certain young specimens of *Sus sennaariensis* by the name which Fr. Cuvier had given to the masked boar, supposing them to be the young of that species, would not have occurred with adult specimens. What Fr. Cuvier called *Sus larvatus*, we know as *Potamochoerus africanus;* to save further confusion of names and errors of fact, it would be well to drop the name *Sus larvatus* altogether. *Potamochoerus,* which I observed to resemble *Sus cristatus* in having a large lacrymo-frontal ridge, is, of course, specifically distinct from it. *Sus sennaariensis*, if, as I think is most likely, closely allied to *Sus cristatus*, is another instance of the wide distribution of 'Pachyderms,' a point on which Rütimeyer insists, in a different tone ('Herkunft unserer Thierwelt,' p. 34) from that adopted by Gibbon, but not less categorically, nor, indeed, less strikingly, remarking, as he does, that at the present day the *Hyrax* and the *Hippopotamus* are the only genera of 'Pachyderms' confined to one quarter of the globe.

The Aethiopian region therefore must be held to possess a true *Sus*; and as to domesticability, the Palaearctic, the Oriental, and the Aethiopian Suidae have possibly equal claims [1].

[1] Pigs fulfil excellently well the six conditions enumerated by Mr. Francis Galton, *l. c.*, as necessary for domestication: viz. 1. That the animal should be hardy; 2. That it should have an inborn liking for man; 3. That it should be comfort-loving; 4. That it should be found useful; 5. That it should breed freely; 6. That it should be gregarious. Aesop, Aelian, and Lactantius (*cit.* by Bochart, 'Hierozoicon,' ii. 698) have, in various ways, remarked on the peculiarity of the pig as contrasted with other domestic animals, in that it is useful only when dead, giving neither milk, as does the cow, nor wool as does the sheep. With this peculiarity is connected the fact, useful for the often difficult task of deciding whether a particular skull came from a wild or a domestic breed, that domestic pigs are usually made useful while young. Rütimeyer, indeed (*l. c.* p. 52), gives it as one of his reasons for supposing *Sus scrofa*, var. *palustris*, to be represented by a wild as well as by a tame stock, that its remains are usually those either of very old or quite young individuals. A pig will father while quite young; and whilst gaining nothing in its capacity of manufacturer of food in its own body, it loses in its capacity of a breeding animal with increase of age. This is not the case with the cow; and the discovery, therefore, of remains of very old individuals of this species only justifies us in inferring that the cow was a scarce and valuable animal in the period and place to which it belonged. See Rütimeyer, 'Fauna der Pfahlbauten,' p. 10.

The special value of the pig as a domesticated animal is commonly expressed in an estimate that 'twice the weight of food may be obtained from hogs than can be obtained from the same cost of food by means of any other animals' (Richardson, *l. c.* p. 42). In a little more detail, it is to be remarked that the pig, as a meat producer, stands at an advantage (to the consumer), first, in the smaller relative weight of its 'offal' as compared with the entire weight of its body, but secondly, and chiefly, in the large proportion of fat, the kind of food which is eminently the hardest for a savage or for the poor to procure, which it will store up upon almost any dietary. For this, see Lawes and Gilbert's invaluable Paper in the Royal Society's Transactions for 1859, 'On the Composition of some of the Animals fed and slaughtered as Human Food,' page 565 for relative proportion of offal, pages 513 and 543 for storing up of fat. It may be here remarked that De Blainville ('Ostéographie, G. *Sus*,' Introduction, p. 107) may very likely be right in suggesting that the pig may have furnished animal food to the earliest races of man before either cow or sheep, but that he could not have had our knowledge of the very various kinds of animals which, even in these days, furnish lower races of men with animal food, when he supposes

There would be little weight to be laid upon mere differences in size, even if much greater differences did exist between *Sus scrofa*, var. *ferus*, and *Sus cristatus* than a comparison of the measurements given in the British Museum 'Catalogue of Carnivorous Pachydermatous and Edentate Mammalia,' 1869, pp. 334, 337, 338, or a comparison of the skulls themselves shows to be the case; for Pallas's[1] words and Nathusius's[2] are both clear to the effect that differences of as much as 50 per cent. do exist between individuals of the *Sus scrofa*, var. *ferus;* and a comparison of what I believe to be the skull of an old wild sow, from the alluvium of the Thames, near Oxford, with that of a modern wild boar, will put the matter beyond all question.

According to Nathusius[2], the differences in size between the largest and the smallest wild swine are so considerable as to have caused them to be distinguished by such names as 'Hauptschweine' and 'Kümmerer' respectively; the latter of which terms has an equivalent in the word 'Wreckling,' applied in some parts of England to the supernumerary pig in a litter, i. e. to the one which makes the litter exceed the number of available teats, and fares accordingly. Two or three of the commonly reported facts[3] as to the pairing and period of reproduction of the wild swine account very sufficiently for these great differences in their size. Though the males are monogamous, severe battles nevertheless take place between them for the possession, it is said, of the largest females; the smallest females consequently are left for the vanquished, which will usually be the smallest males. Hence a great difference in the two sets of offspring would be reasonably expected. But, further, it is known that the wild, like the tame[4] swine, will breed long

that animal food must have been supplied by one or other of the three animals named. And his reason for such truth as his conclusions contain is, it is to be feared, but a poor one; it runs thus, 'le Cochon étant l'animal qui sympathise le moins par ses qualités affectives avec l'espèce humaine a dû être celui qu'elle aura le moins répugné à tuer de sangfroid pour s'en nourrir; ce qui aura eu lieu plus tard et avec bien plus de répugnance pour le Mouton et pour le Bœuf.'

[1] 'Zoographia,' 1831, p. 267.

[2] 'Schweineschädel,' 1864, p. 65.

[3] See Richardson, 'Domestic Pigs,' pp. 18, 19 (Warne, London); Samuel Sidney, 'The Pig,' p. 4 (Routledge, London); Blasius, 'Säugethiere,' p. 509, 1857; Wagner, 'Säugethiere,' p. 426, 1835; Brandt and Ratzeburg, 'Medizinische Zoologie,' p. 88, 1829.

[4] For the period at which tame sows will breed, see Sidney, *l. c.* p. 61; Low, 'On the Domesticated Animals of the British Islands,' p. 415.

before the period of maturity; and that the offspring of such unions, whether both of the parents or only one be immature, are likely to be smaller in size as well as fewer in number, needs no argument.

Whilst no *à priori* probability can be gathered from any greater domesticability in favour of the claims of either European, Asiatic, or African *Sus* to be the exclusive source of our domestic pigs, and whilst mere size equally fails to differentiate these races, the point of the relation between the length and the height respectively of the lacrymal bone on which Nathusius has laid such weight ('Schweineschädel,' passim et pp. 9, 10, 83, 91, 92, 175), and to which Mr. Darwin has assigned so much importance ('Animals and Plants under Domestication,' vol. i. p. 70, ed. 1875), though in the immense majority of cases enabling us at once to differentiate the skulls of *Sus cristatus*, as indeed of the other Asiatic pigs without facial warts, from those of *Sus scrofa*, var. *ferus*, also sometimes fails us. Having measured a very considerable number of skulls of *Sus cristatus* from very various parts of India, and having invariably found them to have the orbital border of the lacrymal shorter than, or at most only just equal in length to the malar, and bearing in mind the constant reference made by Nathusius to the morphological and classificatory value of this proportional difference in the tame variety, the so-called *Sus indicus*, I was entirely unable to understand how that author could say (*l. c.* p. 185) that two skulls of *Sus cristatus* furnished to him by Mr. E. Gerrard differed from skulls of *Sus scrofa*, var. *ferus*, only in being smaller altogether. But after measuring the skulls from Sir Walter Elliot's collection and those in the British Museum, with the result of feeling certain that, from the contour, proportions, and, in adult males, the texture and sculpturing, together with the lacrymal of *Sus cristatus*, it was always possible to distinguish such skulls from those of our wild boar, I came, in the Royal College of Surgeons, upon a skull which, whilst possessing certain other peculiarities (to be hereafter detailed) as distinctive, more or less, of *Sus cristatus*, did combine with them the long lacrymal of *Sus scrofa*, var. *ferus*. This skull is numbered 3251 *a*, and was pointed out to me by Professor Flower, with his often experienced kindness, as being a skull of *Sus cristatus*, which, together with two others, with the same appearance as to textural condition, had been brought out of the stores of the Museum into its series. Both these other skulls possessed the short lacrymal so characteristic, as the subjoined tables of measurements will show, of *Sus cristatus* and its allies, together with the other points usually observed in that animal's skull. The principal points, besides those of general facies and proportions, which appeared to me to justify this assignment of skull 3251 *a* (Royal College of Surgeons) were, in the absence of the shortness of the lacrymal,—first, the great prominence of what may be called the lacrymo-frontal ridge, that part of the frontal bone, to wit, which lies between the channel for the supraorbital nerve mesially and the upper border of the lacrymal bone outwardly; secondly, the great relative development of the part of the third molar which is posterior to the two anterior bicuspidate lobes of that tooth; thirdly, the absence of convexity backwards in the naso-frontal suture. These three points are usually present in *Sus cristatus*, and they are usually found to be accompanied by the fourth peculiarity of a short lacrymal. One or other of these characters may be absent; but in an undoubted specimen of an adult male *Sus cristatus* I have never seen more than one of these missing; whilst it is rare for the second, and very rare for either the first or the third, to be found in undoubted specimens of the Palaearctic wild boar. In a skull figured by Mr. Richardson (*l. c.* p. 50), from

'an excavation in an island on Loch Gur, a lake in the neighbourhood of Limerick,'

and

'found in company with skulls of oxen, goats, sheep, red deer, reindeer, and our extinct gigantic deer, sometimes erroneously styled the Irish elk,'

but considered by Nathusius (p. 150) to have belonged to a domestic animal, it is true that the lacrymo-frontal ridge is represented as of great size; but we must set against the assigning of much importance to this fact the considerations that the drawing is taken from a reconstructed skull, that it is obviously inaccurate in some points, as, for example, like a drawing in S. Müller's 'Verhandl.' (Taf. 28 bis, fig. 3), in having an extra tooth posteriorly to its canines, and that it may consequently be supposed to be likely to be inaccurate in other particulars also. The fronto-lacrymal ridge is, of course, in the adult underlain by a prolongation of the frontal sinuses; it is, however, visible enough in very young specimens of domestic pigs, which show other points of affinity to the *Sus indicus* long before the frontal sinuses are fully developed; and I am inclined to think it may sometimes, though certainly not always, be detected in very young specimens of *Sus scrofa*, var. *ferus*, such as the one figured by Nathusius, *l. c.* (Taf. i. fig. 1, Taf. iii. fig. 13). Though the fulness of this region is due in the adult partly to its being underlain by frontal sinuses, which are relatively small in the early stages of the animal's life, there is still some justification for regarding this structure as an instance of the retention by the adult of an early structural arrangement; for it is easy to understand that the contour described by the external tables in early youth may be carried out conformably by the blood-vessels of the scalp as the animal grows older. A parallel to such a process is furnished to us very frequently, though by no means so nearly universally, by the retention in the adult *Sus cristatus* of that fulness and convexity of the vertical aspect of the fronto-parietal region which is characteristic of *Sus scrofa* both wild and domestic, as well as of *Sus cristatus* and *Sus indicus*, at birth[1]. Lieutenants W. E. Baker and H. V. Durand, in their paper on 'Subhimalayan Fossil Remains of the Dádupúr Collection,' in the 'Journal of the Asiatic Society of Bengal,' vol. v. 1836, p. 664, observe of the two fossil skulls which they describe, that in both

'there is in the frontal plane a total absence of convexity. As this plane ascends there is a tendency to concavity, in consequence of the parietal crests being more strongly marked than in the existing species, and thus producing the appearance of a gentle hollow where, in the common wild hog, there would be a gentle swell.'

Further on, in the same paper, the authors remark :—

'From the form of the cranium, the shape of the canines and incisors, and the other points in which the fossil differs from the existing species of the country, a specific difference may be inferred; for the dissimilarity, although less than that which occurs between the *Babyrussa*, the *S. larvatus*, and the *Sus scrofa*, or common hog, is too remarkable, particularly in the shape of the canines of the lower jaw, to admit of the fossil being considered as a mere variety of the *Sus scrofa*.'

Sir Walter Elliot, however, to whom I owe this reference as well as other things, writes me to the effect that the skulls sent by him to me

'do not seem to differ much from the Subhimalayan fossil specimens figured and described,'

as above specified. And it is worthy of being put upon record that

[1] See Nathusius, pl. i. figs 1 and 3, pl. iii. fig. 13, and pp. 3 and 13. Compare the mesial fulness in the frontals of *S. papuensis*.

of five skulls of modern wild Indian hogs thus sent by Sir Walter Elliot, three show the upgrowth of parietal crests, which Lieutenants Baker and Durand had supposed to be characteristic of the fossil animal, and to contribute towards justifying its claim to be considered specifically distinct. These three skulls have the following labels and histories appended to them by Sir Walter Elliot, which, when coupled with the localities assigned below to the British Museum specimens (specimens not, so far as I can see, different in any essential point from Sir W. Elliot's), bear importantly on the question of the unity of *Sus cristatus*:—

No. 71. Large boar, killed near Rajkote, in Kattywar, June 4, 1832. He was with a large sounder, and ripped two horses severely. Rajkote is in the extensive open plains of the Kattywar peninsula.

No. 330. Nilgherry Hills, 1840.

No. 428. Jaggiapettah, 1851. On the east side of the Madras Presidency, in the Masulipatam district, on the high road from Masulipatam to Hyderabad.'

These three skulls agree in having their third molars considerably worn and their canines large, their muscular insertion-surfaces marked with polygonal reticulations[1] in some places, and with arborescent markings in others, and, thirdly, in the spar-like hardness and density of the bones generally; and they must be supposed consequently to have belonged to old and powerful male animals. In all of these points they differ more or less from the other two skulls, also of male but of younger and less powerful animals. But such differences as these are far from being of specific value, either in comparison of modern races with fossil ones, or in comparison of modern races *inter se*. All the five skulls, however, lent to me by Sir Walter Elliot possess the lacrymo-frontal ridge developed into a very considerable prominence; and though every now and then I have had occasion, in going over the extensive series of skulls of Indian wild hogs which have been available to me, to note an almost, I have never noted an entirely complete disappearance of it. The fronto-lacrymal ridge is not, even in the aged hog, a mere expansion or dilatation of the frontal bone: it is underlain, it is true, by an arm of the frontal sinuses; but it has thick and independent walls of its own. Though not quite exactly homologous with the supraciliary ridges of human anatomy, it is nevertheless very closely comparable with them. In

[1] For similar reticulation in *Bos primigenius*, see Rütimeyer, 'Fauna der Pfahlbauten,' taf. iii. fig. 3.

some varieties of the human species the supraciliary ridges, whether underlain by sinuses or not, are very constantly developed into great prominences. In a classification, indeed, of human crania proposed by Dr. Williamson[1], 'skulls with the supraciliary ridge so prominent as to overshadow the face' formed one out of four classes into which all crania were divided. It is not necessary to go so far as this for acknowledging that this character may become, in the Suidae at least, a race character[2].

The second in order of the three points which, taken together, as they usually exist together, with the fourth, that relating to the lacrymal, help us to identify *Sus cristatus*, is complexity of its third molar, at least in male specimens, and specially the complexity of that part of the tooth which lies posteriorly to the two primary bicuspidate lobes.

The division of the third molar which lies posteriorly to the two anterior lobes is, in the male *Sus cristatus*, equal sometimes in size to and sometimes even greater than both those other lobes taken together. This is not often, though it is occasionally, the case in *Sus scrofa*, var. *ferus*, of modern times. Nor is it the case in the females of *Sus cristatus*; so that the greater development observed in the males may perhaps be due to the working of the law of sexual battle. The large size of the canines postulates a large determination of blood to the jaw; and the large size of the third molar, a tooth evolved at the period of sexual maturity, when the animal 'venerem et praelia tentat,' may be, to use a word suggested long ago by myself[3], 'tautogeneous' with it. Still the fact that the third molars are small both in *Sus barbatus* and *Sus barbirussa* shows that this smallness cannot always be explained on physiological grounds.

The third point, that of the straightness of the naso-frontal suture, appears to have some classificatory value, whether we look at it with the light furnished by its importance and validity in the

[1] 'Observations on Human Crania in Museum, Fort Pitt, Chatham,' 1857.

[2] It is observable in the pigmy Nepal pig, *Porcula salviania*, as also in the African *Potamochoerus*. It does not appear to me that male Suidae possess it more markedly than females; and herein it differs from the supraciliary-ridge development as seen in our own species, as also from the race-mark to be next mentioned. An analogous eminence, rudimentary in *S. papuensis*, occupies the middle frontal line over an area homologous with the human glabella in each of four pigs' skulls brought from the Admiralty Islands by H.M.S. Challenger.

[3] 'Nat. Hist. Rev.' Oct. 1861, p. 486, Article IV, p. 57.

classification or identification of other animals, or in that furnished by the facts of its own developmental history.

The nasal and the frontal bones together form a roof over the ethmoid and the turbinal bones; and as there is no apparent physiological reason why they should contribute in different proportions towards the securing of this end, the fact that they do so is of so much the greater morphological value. In *Sus cristatus* the naso-frontal suture very ordinarily runs (also in *Sus andamanensis*) straight across the roof of the ethmoid, at right angles to the long axis of the skull; or the frontals may intrude themselves mesially between the nasals, making thus the contour of the suture to be convex forwards. Precisely the reverse is the case in the adult European wild boar. Some weight has been laid[1] upon a similar conformation in the skull of the tiger, as being of service in differentiating it from the skull of the lion; and though it is not pretended that an equally great distinctness can be supposed to exist between the two animals now under comparison, still the structural differences in the two sets of cases are analogous.

But when we come to look at the skulls of developing pigs, we see that real value attaches, from this point of view also, to the relatively greater or less extension backwards of the nasal bones, and the contour described consequently by the naso-frontal suture. A tape crossed at right angles to the long axis of the skull from one infra-orbital foramen to the other, passes very closely in front of, and often parallel with the naso-frontal suture in the very young pig; the suture gets removed further and further away from it as the pig increases in age. Nathusius (Taf. iii. figs. 11 and 13) has figured the skulls of a young wild boar and of an adult wild boar upon the same plate; and the straight line of the suture of the former contrasts most instructively with the backwardly arching contour of the latter. The straightness, therefore, of the naso-frontal suture may be supposed to illustrate the principle that climatic or other conditions may cause structural arrangements to be permanently retained in certain races whilst they are obliterated in others. The retention of the prolongation of the sagittal suture over the frontal region is believed, with much reason, to be hereditarily transmitted in our own species; and I incline to think that the persistence of the frontal tubera with

[1] 'Ost. Catalog. Royal College of Surgeons of England,' 4506, p. 706.

very much of their infantile eminence, which we see not rarely in adult men, may be taken as furnishing another parallel to the retention in some degree, by the adult *Sus cristatus*, of the characters of the young animal's naso-frontal relations.

In the fourth place, as regards the value of the relative shortness of the lacrymal bone as a means for differentiating the skulls of *Sus cristatus* and its allies from those of *Sus scrofa*, var. *ferus*, I was for a long time of opinion that the same might be said of this all but invariably observed peculiarity of *Sus cristatus*, which I have already said, however shortly, of the straightness of its naso-frontal suture; and that all this may, *mutatis mutandis*, be repeated as to the lacrymal's peculiarities in the two sub-species there can be no doubt. Anybody who will examine the figures given in Spix's 'Cephalogenesis,' or the reproduction of them in Erdl's 'Tafeln der vergleichenden Anatomie des Schädels,' 1841, can convince himself of the fact that the malar border of the lacrymal is very short as compared with its orbital border in the young pig [1]; and if he extend his observation to the various adult Suidae, he will find that this side of the bone goes on growing from a condition of permanent inferiority in *Sus andamanensis* till it reaches subequality or entire equality with the orbital length in some Suidae—such as *Sus cristatus*, *Sus vittatus*, *Sus leucomystax*, *Sus taivanus*, *Sus timoriensis*, *Sus papuensis*, and *Sus barbatus*, all Suidae without facial warts—and finally exceeds the orbital border in length considerably in *Sus scrofa*, var. *ferus*, and ordinarily in *Sus verrucosus* (and *Sus celebensis* ?), and disproportionately, it may be added, in the African wart-hog (*Phacochoerus*).

[1] Though I do not suppose that it would be possible to say in 1876 what Dr. Gray said ('Proc. Zool. Soc.' 1868, p. 19) in 1868, to the effect that Nathusius's works were not to be had either in the library of the British Museum or in the library of either the Zoological, Royal, or Linnean Society, it may nevertheless be convenient to give here the results of his measurements of the borders of the lacrymal bones at different periods of the life of *Sus scrofa*. They stand thus at p. 10 of his work of 1864:—

Lacrymal bone.	Newly born. *millim.*	2 months old. *millim.*	6 months old. *millim.*	Adult. *millim.*
Height	7	13	18	21
Frontal border	11	20	42	60
Malar border............	3	6	25	35
Ratio of height.........	1	1·9	2·6	3
Ratio of frontal border	1	1·8	3·8	5·5
Ratio of malar border	1	2	8·3	11·6

There is no doubt that the younger the pig the greater is the distance separating its lacrymal's length from equality with or superiority to its lacrymal's height; but the subjoined tables show that there is no constant relation to be observed between the growth of the entire facial skeleton out of the short proportion of early days into the elongated muzzle of the adult, and the longitudinal evolution of the lacrymal factor of that snout. The longest snout, such as that of the adult *Sus barbatus*, may, as Nathusius has remarked with surprise (p. 167), show a lacrymal with the proportions of the domestic *Sus indicus*, or the immature *Sus scrofa*, var. *ferus*; and I venture to suggest that the elongation of the lacrymal, which Professor Owen taught us to call a 'mucodermal bone,' may be correlated with the evolution of the facial warts which are found in all the Suidae with such lacrymals. If these two structures are thus correlated, we come to be able to explain how it is that the length of the lacrymals is, though not very variable, still as variable as we have found it to be; for the facial warts themselves are a variable structure, as we should from several analogies expect them to be. If they were not so, it would be difficult to explain how it is that in many zoological descriptions[1] of our European wild boar, no mention is made of the presence of warts (small ones, it is true) immediately below the eyes; and inasmuch as they are so variable, and probably more liable[2] to disappear in the female sex, we can, on the hypothesis of the evolution of the lacrymal in length being correlated with their presence, understand how the lacrymal is sometimes found to be shorter than we should have expected it to be in skulls such as those of the wild sows from prehistoric deposits in this country, measurements of which are given p. 546.

[1] De Fatio, in his 'Faune des Vertébrés de la Suisse,' 1869, p. 354, goes, if I understand him rightly, further than this, by using the words 'Pas de saillies sur la face en dessous des yeux,' in his definition of 'Le Sanglier ordinaire, das wilde Schwein.'

[2] Dr. Gray, in his description of *Potamochoerus africanus* ('P. Z. S.' 1868, p. 34), says of the male animal's face, that it 'is swollen and often warty on the sides in front,' and of the females, that 'the side of the nose is simple.' Fitzinger, however, whose descriptions of the external characters appear to be carefully done, does not say that any such sexual difference exists in this species or in any other of the Suidae with warts. There are other reasons, however, for the suggestions in the text, of which the sexual limitations of the facial callosities in the orangs may be taken as an example.

As the interpretation I suggest of the proportions observed to exist between the length and height of the lacrymal bone depends upon the method of gradations, as applied not only to several varieties, but indeed to several species of Suidae, as well as upon a history of the development of the bone, it will be well now to give a short list of those species—pointing out the particulars which justify their claim to that rank, and contrasting them with the aggregate of characters which do enable us, as a matter of fact, to differentiate such forms as *Sus cristatus* from *Sus scrofa*, possibly without justifying us in considering them distinct species.

And first of *Sus barbatus*. I am in no way inclined to give too much weight to differences in colouring or in character of hair or bristles; still a glance at the drawing of this animal in S. Müller's 'Verhandl.' taf. 30, showing its half-black, half-tawny, wavily, not crisply, curling beard, its ochraceous dorsal stripe, and its tail ending in a considerable brush, impresses one with the idea that it is impossible that the bony substructures should not make some approach at least to a similar diversity from other forms of Suidae. This anticipation was fully borne out by an examination of four skulls, two of which are in the British, and one in the Oxford Museum—and of Schlegel's figure of the skull described by him, and S. Müller, 'Verhandl.' pp. 173, 179–181, and taf. 31. figs. 4 and 5.

The skull of *Sus barbatus* is absolutely longer than the skulls either of *Sus scrofa*, var. *ferus*, or *Sus cristatus*; and relatively to the body its length is considerably greater, being, as given by Fitzinger, no less than $\frac{3}{8}$ of the length of the trunk. The contour described by its sagittal suture is strikingly different from that described by the corresponding suture in most other Suidae; its highest point is some way in front of the plane of the occipital squama, and occupies a level far above the plane occupied by the anterior half of the frontal or by the nasal bones; and these latter bones make up more than one half of the entire length of the skull, resembling herein the typical *Sus scrofa*, whilst the naso-frontal suture resembles that of *Sus cristatus*. Its maximum inter-zygomatic width is in the middle of that arch, not at its posterior border as in most other Suidae.

In addition to these larger points, the following may be mentioned as having a great morphological importance, though relating to smaller structural peculiarities. The third molar consists of three lobes; but, large as is the jaw and the canine armature of this *Sus*, the most posteriorly placed of the three lobes is more simple than even the very simple posterior lobe of the pigmy *Sus andamanensis*, having only one cusp prominently marked in the upper, and four in the lower jaw, whilst the entire posterior lobe is little, if at all, greater in antero-posterior extent, and much smaller in transverse, than either of the anterior, differing thus altogether from *Sus cristatus*.

A second point, relating to a small structure as measured by the callipers, which is a very large one, however, to the morphologist, is the permanent retention by *Sus barbatus* of the mesopterygoid of Parker ('Phil. Trans.' 1874, p. 324, pl. xxxvi, fig. 4. *ms.pg*) as a distinct bone. This peculiarity was observed in all the four skulls of *Sus barbatus* examined by me; and its obvious general significance is increased by the fact that it is in the area of the mesopterygoid that the great basicranial cavities are excavated in the *Babirussa* and *Phacochoerus*, and are represented rudimentarily in *Sus vittatus*.

Knowing what we do[1] of the affinity of the fauna of the subregion of Ceylon and South India to that of Malaya, there is no *a priori* improbability in a view which should accept the *Sus ceylonensis* of Blyth[2] ('Journal of the Asiatic Society of Bengal,' xx. p. 173) as identical with the *Sus barbatus* of Borneo. Mr. Blyth's words, *l. c.*, are as follows:—

> '*Sus zeylanensis*. Skull longer than that of the Indian boar, nearly straight in profile, very much contracted at the vertex. Palate contracting posteriorly to less than 1″ from the magnitude of the last molar, which is considerably larger in both jaws than in the wild boar of India, the upper measuring $1\frac{3}{4}''$ long by $\frac{15}{16}''$ broad anteriorly. Vertex narrowing to 1″ only in breadth. Total length of skull from vertex to tips of nasals $16\frac{1}{4}''$. Altogether the skull approximates closely in contour to the figures of the skulls of *Sus barbatus* by Dr. S. Müller and M. Temminck.'

Dr. Gray appears to have had access, which I have not had, to a photograph of this *Sus zeylanensis*, and says that, judging from it,—

> 'The skull is much shorter and thicker than the skull of *S. barbatus*. The photograph is much more like that of *Sus verrucosus*.'

For my own part, I cannot think Mr. Blyth would ever have

[1] Tennant, 'Nat. Hist. of Ceylon,' pp. 61–68, 186; Wallace, 'Geographical Distribution,' i. p. 328.

[2] See Gray, 'Proc. Zool. Soc.' 1868, p. 24; 'Brit. Mus. Cat.' 1869, p. 331.

spoken of a real male *Sus barbatus* as having a skull with a straight profile; Müller and Schlegel figured a female skull, whereas Mr. Blyth's, I am sure, was a male's. As regards the contraction of the vertex, which shows Mr. Blyth's skull to have been a male's, this is sometimes exceeded by the old and strongly muscular Indian wild hog, such as Sir Walter Elliot's No. 71 (for history of which see p. 533), where it is only 8″, whilst in the skull No. 72 it is 1·8″, showing an oscillation, owing to the varying action of the temporal muscles, which entirely deprives it of any morphological value; and the measurements of the molars, finally, assure me that there is no need to add the words 'und wahrscheinlich auch Ceylon' to the word 'Borneo' as the 'Vaterland' of *Sus barbatus*, as Fitzinger has done, *l. c.* p. 393. Müller and Schlegel, p. 179, give Borneo, if I rightly read their words, as the 'habitat' of this well-marked species.

Secondly, of *Sus verrucosus*. The soft and perishable parts of *Sus verrucosus* are even more interesting and of greater importance than the bones; for they show us that wild pigs do have appendages —warts, to wit, covered with long bristles, and attached to the corners of the lower jaw, like those of the Irish greyhound pig, once so plentiful in Galway; and they thus do away with one of the objections to Mr. Darwin's views, stated fairly (by himself) in the work on 'Animals and Plants under Domestication,' i. p. 79, ed. 1875. If it is a profitable thing to lay Müller and Schlegel's figure of *Sus verrucosus* (tab. 28. *l. c.*) alongside of Richardson's figure of the 'old Irish greyhound pig' (*l. c.* p. 49; Darwin, *l. c.* p. 79), it is profitable also to read the Dutch letterpress[1] of the two former authors, from which we learn, when helped by the translation given by Fitzinger, that the young of *Sus verrucosus* are never striped, and hereby have it suggested to us how it may be that so many of our domestic breeds never have their young striped—none, in fact, except the Westphalian and the Turkish. Rütimeyer (*l. c.* p. 187) has suggested that the so-called 'Berkshire breed' of domestic pigs may have its parent stock in *Sus celebensis*, as figured by Schlegel

[1] The words of the two Dutch naturalists are, 'Verhandl.' p. 177: 'De deer jonge voorwerpen dezer sort zijn niet gestreept, en onderschieden zich daardoor van de jongen van *Sus vittatus* en van de meeste, ja, misschien van alle overige soorten.' It is much to be regretted that Dr. Gray did not enter a note of this most important fact in his papers above referred to, in which the 'Verhandlungen' are often cited.

and S. Müller, which resembles it in the great transverse development of its skull, the verticality of its forehead, of its occiput, and its temporal fossae, in the height of its zygomatic arch, and specially in the remarkable height and massiveness of its lower jaw. There is no doubt that *Sus celebensis* is very closely allied to *Sus verrucosus*, having not only the same general facies but the same peculiarities, though on a smaller scale, as those whence the Java pig has taken its name. The chief point of difference which Rütimeyer points out between the figured *Sus celebensis* and the Berkshire breed depends upon the artist, I apprehend—being, as it is, a greater length of the molar series, which, however, consists of eight teeth in the figure of *Sus celebensis* (Verhandlungen, pl. 28 *bis*, fig. 3); and it is possible enough that the young of *Sus celebensis* (the contrary not having been definitely recorded) may be like the young of *Sus verrucosus* in not being striped, and that they may thus have resembled the young of the Irish greyhound pig, as when old they actually do resemble it in the development of the mandibular warts.

As regards the skulls of *Sus verrucosus*, figured tab. 32 by Müller and Schlegel, the lacrymal in fig. 4 appears to have the favourable relation of length to height which is characteristic usually of the other Suidae with facial warts; whilst in fig. 1, which was taken from a very old individual, the height and length would appear to be subequal, as in the skull (fig. 6) of *Sus vittatus*, on the same plate. I should have set less store by this variation in the two drawings than I do, if Nathusius had not distinctly recorded (p. 179, *l. c.*) that this latter is the relation in the *Sus verrucosus* skull in his possession. On the other hand, all the undoubted *Sus verrucosus* skulls which I have been able to measure have presented the elongated form of this bone; and the closely allied species *Sus celebensis*, as figured in Müller and Schlegel's fig. 3, tab. 28 *bis*, may or may not have the long lacrymal usual in that pig. But, on the whole, I should certainly accept the position that the elongation of the lacrymal was not constant in *Sus verrucosus*, though it has been so in the skulls I have seen; and with this position I should hesitate to assign to this peculiarity all the value which Nathusius's insistence might lead one to assign to it.

Taking, in the second place, this variability in the proportions of the lacrymal bone, and coupling with it, first, the absence of striping in the young, and secondly the presence of mandibular warts, such as one of our domestic breeds, viz. the Galway greyhound pig, still retains, or did till quite lately retain, we may be tempted to think that some form allied to *Sus verrucosus* or *Sus celebensis* may have been the single parent stock of all our domestic breeds, except, of course, such as the Westphalian and Turkish, which are striped in their youth, and would be referred to the indigenous *Sus scrofa*, var. *ferus*, as their parent stock. In favour of such a view, we have, of course, the general principle 'Entia non sunt multiplicanda praeter necessitatem.'

But the wide range[1] of *Sus vittatus*, over Java side by side with *Sus verrucosus*, over Borneo with *Sus barbatus*, and, as stated, also over Amboyna, Macassar, Banka, and Sumatra, gives it special claims to attention. This pig appears to me to be very closely allied indeed to *Sus cristatus*, and to be similarly and readily distinguishable from our European wild boar. In this latter point I differ from Rütimeyer (*l. c.* p. 187), whilst I should agree with him in considering it all but identical with *Sus leucomystax*. The claims of *Sus vittatus* and *Sus leucomystax*, and *Sus taivanus* of Formosa (occurring, as they do, in an area comprising Japan as well as Java), to have given origin to *Sus indicus*, the domestic Chinese pig, in days sufficiently far off to have allowed the tendency to striping of the young to become eliminated, are very strong, and can scarcely be considered antagonistic.

Closely allied as *Sus cristatus* is to these races, its severe struggle for existence entailed by its *habitat* on a continental area tenanted

[1] It is interesting to compare with these statements, as to the geographical distribution of the Asiatic Suidae, the following words from M. Gabriel de Mortillet's memoir in the 'Revue d'Anthropologie,' iv. 4. 1875, p. 653, as to the origin of bronze: 'Reste le groupe de l'extrême Orient Asiatique. C'est là évidemment où il faut chercher l'origine du bronze. Les principaux gisements sont dans la presqu'île de Malacca et surtout dans l'île de Banca, mais ils s'étendent dans d'autres îles de la Sonde et remontent jusque dans l'empire Birman où l'étain est encore exploité actuellement dans le district de Merguy. Ce minéral, dans tous ces gisements, se recueille de la manière la plus simple et la plus facile dans les alluvions. Ce sont bien certainement les alluvions les plus riches du monde en étain et celles qui occupent la plus grande étendue. Il est donc tout naturel que ce soit celles qui les premières aient attiré l'attention de l'homme. Le cuivre se rencontre dans les mêmes régions. Tout le monde connaît les gisements de cuivre des îles de la Sonde, Timor, Macassar, Borneo. La Birmanie anglaise présente des mines de cuivre à côté de ses exploitations d'étain. Le pays se trouve donc dans les meilleures conditions pour avoir vu naître l'industrie du bronze.' *Gallus bankiva*, the parent stock of our common fowl, is found over the same area.

by the tiger (*Felis tigris*), as well as other Carnivora unfriendly to Suidae, appears to me to have specialised it, especially as regards its dental armature and the bones which carry it, into divergence from the probable line of parentage of the inconveniently so-called *Sus indicus*. *Sus andamanensis*, on the other hand, and, I am inclined to think, one or two other of the Asiatic Suidae, show, from a precisely opposite cause, that of restriction to a very confined area, the same divergence from what I have imagined the present stock of the Chinese breed may have been. But against any such speculations about what we do not see in the darkness of past ages we have to set what we can see by travel in the broad daylight of the present, viz. that almost all Suidae are readily domesticable by savages in almost every quarter of the globe; and what savages do now they may very well have done formerly.

I have not been able to institute any satisfactory examination into the relations of the Aethiopic to the Asiatic Suidae; and I should welcome an opportunity of examining skulls and skins of the true *Sus* seen by Dr. Murie and Dr. Barth in Central Africa. It would be additionally satisfactory if investigations could be set on foot as to the existence or non-existence in this *Sus* of the Cystic form of *Taenia solium*, which certainly exists in *Sus cristatus*. Dr. Cobbold informs me that *Taenia mediocannellata* is the *common* tapeworm of Indian as of other patients; but I apprehend that, as it has been so very definitely laid down by others[1] that the pig is at least one source whence inhabitants and sojourners in India become infested with tapeworm, it would be premature to conclude the reverse even from his authoritative statements. It must be very difficult to prove a negative here.

The facts of most direct importance, however, in investigations as to the relationships of prehistoric to modern races are those of the great variations observed—first, in the entire size of the individual animals, and, secondly, in the proportions of particular bones, and notably of the lacrymal bones, in specimens from the same species or sub-species of Suidae. In view on the one side of this twofold variability which is explicable upon acknowledged physiological principles, and affects the wild races both of Europe and

[1] Notably by Dr. Charles A. Gordon, 'Medical Times and Gazette,' May 2, 1857, p. 429; and by Dr. T. R. Lewis, Appendix B. to 'Eighth Annual Report of the Sanitary Commissioner with the Government of India,' 1871.

Asia, and, on the other, of the aptness for domestication possessed by *Sus scrofa*, var. *ferus*, it appears to me to be unsafe to postulate for prehistoric British swine any other parent stock than the one just named. On the other hand, such is the diffusibility and transportability of *Sus* that it is not impossible, nor inconceivable, that the domestic European pig, even of the Stone Age, may have had an Asiatic or African origin. As regards the Bronze Age, indeed, if its tin and copper did really come from the East, such a view cannot be said to be even improbable. But the acceptance of it does not seem to me to be necessitated by the facts.

Measurements of the Orbital and Malar Borders of the Lacrymal Bone[1].

I. *Sus cristatus* and its allies.

Skull of Indian wild boar, *Sus cristatus*, No. 72 in Sir Walter Elliot's collection. Killed about 1830, at Haugul, S. W. corner of Dharwar province, in South Mahratta country.

Height of lacrymal 1 in.; malar border 1 in.

Skull of Indian wild boar, *Sus cristatus*, No. 46, Sir Walter Elliot's collection.

---

[1] To this paper I append a number of measurements, taken from various Suidae, of the length and height of the lacrymal bone, this point having had great weight laid upon it by Nathusius, and holding good within very considerable limits. I have not given any measurements relating to the widening of the palate (a second point insisted upon by Nathusius), because, as regards the wild races, I have found that this widening of the interpremolar, as opposed to the intermolar transverse diameter of the palate, is sometimes found in specimens which undoubtedly belong to wild European boars, whilst, on the other hand, the two measurements are usually subequal in the other wild races, such as *Sus cristatus*, which I have measured. In taking the length of the lacrymal along its malar border, some little ambiguity is caused by the fact that in many specimens the lip of the orbit is a little everted, so as to resemble to some extent the spout of a mortar rather than the rim of a cup at the line of junction of the malar with the orbital border of the lacrymal. Where the differences between the heights and lengths are measured by tenths or twentieths of an inch or by millimètres, this structural arrangement may make the measurements vary importantly. There is, however, always a line separating the part of the lacrymal which is to be considered as belonging to the inner aspect of the orbit from that which is strictly facial; and from this line the measurements have been taken. The frontal border of the lacrymal, again, often bends downwards just before reaching its orbital edge, just posteriorly to the plane of the lacrymal canals, thereby curtailing the height of the bone for the distance corresponding with this deflection. My measurements have been taken in the plane occupied by the lacrymal canals, so as to avoid this source of fallacy.

Shot at Dendelly April 12, 1831, in the great forest tract between Dharwar and Goa.
Height of lacrymal 1 in.; malar border 1 in.

Skull of large Indian boar, *Sus cristatus*, No. 71, Sir Walter Elliot's collection. Killed near Rajkote, in Kattywar, June 4, 1832.
Height of lacrymal 1·0 in.; malar border 0·8 in.

Skull of Indian wild boar, *Sus cristatus*, No. 428, Sir Walter Elliot's collection. From Jaggiapettah, on the east side of the Madras Presidency, on the high road from Masulipatam to Hyderabad. An old male with some obliteration of sutures.
Height of lacrymal 0·95 in.; malar border 0·95 in.

Skull of young female, *Sus cristatus*. British Museum, 716 *m*. 'Terai Nepal; Dr. Oldham.'
Height of lacrymal 1·1 in.; malar border 0·75 in.

Skull of adult female, *Sus cristatus*. British Museum, 716 *n*. 'India; Professor Oldham.'
Height of lacrymal 0·1 in.; malar border 0·7 in.

Skull of young female, *Sus cristatus*, about nine months old. British Museum, 716 *q*. 'Nepal; B. H. Hodgson.'
Height of lacrymal 0·9 in.; malar border 0·75 in.

Indian pig, *Sus cristatus*. British Museum, 716 *g*. Under five months of age, very convex frontals, slightly tumid lacrymo-frontal ridges.
Height of lacrymals 0·5 in.; malar border 0·25 in.

Indian pig, *Sus cristatus*. British Museum, 716 *v*. 'Nearly adult female, Nilgherris.'
Height of lacrymal 0·95 in.; malar border 0·6 in.

Young male of *Sus indicus*; in collection of British Museum, not catalogued; with third molar just coming into use—that is to say, about eighteen months old.
Height of lacrymal 0·9 in.; malar border 0·9 in.

Skull of Indian wild hog, *Sus cristatus*, No. 3251 *a*, Royal College of Surgeons of England. This is the skull of an adult male probably, the complexity of the posterior lobe of the third molars being very great. The naso-frontal suture resembles that of *Sus scrofa* rather than that of *Sus cristatus*. The lacrymo-frontal ridge is very largely developed.
Height of lacrymal 0·1 in.; malar border 1·25 in.

Skull of Indian wild hog, *Sus cristatus*, No. 3251 *b*, R. C. S. England.
Skull of Indian wild hog, *Sus cristatus*, No. 3251 *c*, R. C. S. England.
Height of lacrymal 0·1 in.; malar border 0·95 in., in both.
These two skulls were probably from the same district as No. 3251 *a*; but they differ from it, and agree with all the other specimens of *Sus cristatus* in having the height and the malar length of the lacrymal subequal.

Skull of wild sow from Andaman Islands, *Sus andamanensis*, No. 1514 *b*, Oxford University Museum.
Height of lacrymal 17 mm.; malar border 8 mm.: adult.

Skull of *Sus vittatus* from Amboyna. Alfred Russel Wallace, Esq. British Museum, 1362 *c*. An adult male with a nearly straight naso-frontal suture, and a convex fronto-parietal region.
Height of lacrymal 0·55 in.; malar border 0·8 in.

Sus vittatus. British Museum, 1362 *f*. 'Skull of animal developing the hinder molar.' The lacrymo-frontal ridges are convex, the frontal region is flat, the naso-frontal suture, however, is straight, as in *S. cristatus*.
Height of lacrymal 1·05 in.; malar border 0·6 in.

II. *Sus scrofa*, var. *ferus*, and its allies.

'*Sus libycus*.' British Museum, 713 *a*. '*Hab.* Asia Minor, Xanthus (Sir Charles Fellowes). The skull is very distinct from all the skulls of the wild boars from Germany in the British Museum.' The animal was just about eighteen months old: its interectorbital region is still convex; but it has no very marked lacrymo-frontal ridge; its naso-frontal suture is convex backwards, as in *Sus scrofa*, var. *ferus*; its nasals are 8 in. long, as against 14½ of total skull length; and the height of the lacrymal is less than its inferior length. Here it resembles *Sus scrofa* rather than *Sus cristatus*.
Height of lacrymal 1 in.; malar border 0·15 in.

Wild boar from Germany, *Sus scrofa*, var. *ferus*, 3253 *a*, R. C. S. England. Adult.
Height of lacrymal 0·9 in.; malar border 1·5 in.

Wild boar from Germany, *Sus scrofa*, var. *ferus*. O. U. Museum, 1513 *a*. All teeth in place.
Height of lacrymal 1 in.; malar border 1·5 in.

Wild boar from Germany, *Sus scrofa*, var. *ferus*. O. U. Museum, 1513. Presented by Professor Max Müller. With all the teeth, except third molar, in use; sixteen months old.
Height of lacrymal 0·75 in.; malar border 1·05 in.

Young wild boar, *Sus scrofa*, var. *ferus*. R. C. S. England. With first molar only through, therefore between five and six months old, with frontals still very convex in vertical aspect, but with flat lacrymo-frontal ridge.
Height of lacrymal 0·7 in.; malar border 0·7 in.; frontal border 1·5 in.

Skull of old wild sow, *Sus scrofa*, var. *ferus*, from alluvium near Oxford. Figured plate xli. fig. 2, from the Geological Collection, O. U. Museum.
Height of lacrymal 17 millims.; malar border 18 millims.

Skull, from peat, of wild sow, *Sus scrofa*, var. *ferus*. British Museum. With third molar through.
Height of lacrymal 0·8 in.; malar border 1·05 in.

Skull from silt of old bed of Lea river, shown to me in British Museum Geological Collection by Mr. Davis. Under eighteen months.
Height of lacrymal 0·6 in.; malar border 0·5 in.

Second skull from same locality, and in British Museum, with all the teeth in place, and the third molar with considerable complexity in its posterior lobe.
Height of lacrymal 0·6 in.; malar border 0·8 in.

Skull of *Sus scrofa*, var. *ferus*? British Museum, 713 *i*. The animal was about sixteen months old, all the teeth being in place except the third molars above and below. There is still some convexity across the frontals, though flattening is beginning on either side towards the ectorbital processes. The lacrymo-frontal ridge is not markedly convex; the nasals are convex backwards in the middle line.
Height of lacrymal 0·9 in.; malar border 0·75 in.

Skull of *Sus scrofa*, var. *ferus*? British Museum, 713 *k*. The frontals are convex from the level of the supraorbital foramina backwards; the fronto-nasal suture is straight, nearly as it is represented in the young German wild boar (Nathusius, *l. c.* Taf. iii. 12). Lacrymo-frontal ridge is slightly convex.
Height of lacrymal 0·75 in.; malar border 0·5 in.

III. *Sus verrucosus*, var. *ceramica*.

Brit. Mus. 712 *d*. 'Skull, adult. A wild boar from Ceram. Collected by Mr. Wallace.' Of this skull I have noted that its third molar is comparatively simple, whereas the other specimens of *Sus verrucosus* in the British Museum have it complex. All the other specimens of *Sus verrucosus* catalogued p. 330, Catalogue 1869, as *Sus verrucosus*, with the exception of No. 1362 *d*, which was received from Mr. Wallace as *Sus vittatus*, have been noted by me as having the lacrymal long in relation to its height, and as having complex molars. As, however, Nathusius, *l. c.* p. 179, has noted the lacrymal to be short in a specimen obtained by him from Java, and as, of two skulls figured by Solomon Müller ('Verhandl.' tab. 32. figs. 2 and 4), one has the lacrymal short and the other long, it is clear that this species varies in this matter. The closely allied species *Sus celebensis* cannot be said quite certainly, from Müller's pl. 28 *bis*, fig. 3, to have, or not to have, the long lacrymal usually found in the pigs with facial warts.

Height of lacrymal 1·0 in.; malar border 0·75 in.

IV. *Sus barbatus*, Borneo.

Skull of *Sus barbatus*. 1519 *d*, O. U. Museum. Figured plate V. fig. 7.
Height of lacrymal 1·2 in.; malar border 1·2 in.

Skull of *Sus barbatus*. Procured in Borneo by the Rev. C. Spencer Bubb.
Height of lacrymal 1·2 in.; malar border 1·2 in.

Skull of *Sus barbatus*. British Museum, 712 *a*; called *Euhys barbatus*, Hand-list, p. 58, 1873.
Height of lacrymal 1·15 in.; malar border 1 in.

In *Sus barbatus* the lacrymo-frontal ridge is not very prominent, though recognisable: in the shortness of the lacrymal, which is a constant character in this species, so far as recorded in S. Müller's figures and elsewhere, and in the characters of the naso-frontal suture, it resembles *Sus cristatus* and its allies rather than *Sus scrofa*.

BIBLIOGRAPHY.

Baker and Durand, 'Subhimalayan Fossil Remains,' Journal As. Soc. Bengal, vol. v. 1836, p. 664.
Bartlett, 'Proc. Zool. Soc.' 1861, p. 264.
Blainville, 'Ostéographie,' *Sus*.
Blasius, 'Naturgeschichte der Säugethiere Deutschlands,' 1857.
Bochart, 'Hierozoicon,' 1663, i. pp. 696, 978.
Brandt und Ratzeburg, 'Medizin. Zoologie,' 1829, pp. 80–150.
Buffon, 'Hist. Nat.' tom. v. pp. 131–137, pl. xxiv. fig. 2, pl. xv.
Campbell, Colonel Walter, 'My Indian Journal,' 1864, p. 325.
Crawford, J., 'Trans. Ethn. Soc.,' ii. p. 439, 1863.
Cuvier, F. et Geoffroy, 'Hist. des Mammifères.'
Cuvier, G., 'Oss. Foss.' ii. p. 119.
Daubenton, Buffon, *l. c.* p. 189.
Erdl, 'Tafeln der vergleich. Anatomie des Schädels,' 1841.
Eyton, 'Proc. Zool. Soc.' 1837.

Fitzinger, 'Sitzungsberichte d. Akad. Wiss. Wien,' Bd. 29, 1858, p. 361; 50, 1864, p. 383.
Forrest, 'Voyage to New Guinea,' 2nd ed. 1780, p. 97.
Galton, Francis, 'Ethnological Society's Trans.' iii. p. 127, 1865.
Giebel, C. G., 'Säugethiere,' p. 225, 1859.
Gordon, C. A., Esq., M.D., 'Medical Times and Gazette,' May 2, 1857, p. 429.
Gray, 'Proc. Zool. Soc.' 1852, p. 130, 1868, Jan. 9; Brit. Mus. Cat. 'Mammals,' 1869, p. 325; 'Hand-list,' 1873, p. 57.
Hodgson, B., 'Journal Asiatic Society Bengal,' x. p. 911.
Jerdon, 'Mammals of India,' 1874, p. 241.
Link, 'Urwelt,' i. p. 192, 1821.
Linnæus, 'Systema Naturæ,' ed. 13, p. 217.
Low, 'Domesticated Animals of the British Islands.'
Müller, Solomon, 'Verhandlungen,' taf. 28–32.
Nathusius, 'Die Racen des Schweines,' 1860; 'Die Schweineschädel,' 1864.
Pallas, 'Zoographia Rosso-Asiatica,' p. 267, 1831.
Parker, W. K., 'Phil. Trans.' 1874, p. 324.
Pictet, 'Origines Indo-Européennes,' 1859, p. 369.
Radde, 'Reisen im Süden von Ost-Sibirien,' i. p. 236.
Richardson, 'Domestic Pigs,' London, Warne.
Rütimeyer, 'Fauna der Pfahlbauten,' 1861.
Schreber, J. C. D. von, 'Die Säugethiere,' 1835, taf. 320, &c.
Sclater, 'Proc. Zool. Soc.' 1861, p. 63; 1863, p. 122.
Sidney, S., 'The Pig,' London, Routledge.
Spix, 'Cephalogenesis.'
Swinhoe, 'Proc. Zool. Soc.' 1862, p. 361; 1864, p. 382; 1870, p. 369.
Sykes, 'Proc. Zool. Soc.' 1831, p. 104.
Temminck, 'Fauna Japonica,' taf. 20.
Wagner, A., 'Götting. Gel. Anzeigen,' lx. p. 535, 1839.
Youatt, 'The Pig,' 1847.
Zimmermann, 'Geogr. Geschichte,' i. p. 189, 1778.

Postscript, *March* 22, 1877.

Since the publication of an abstract of the foregoing paper in 'Nature' for July 20, 1876, I have been enabled to compare the undermentioned skulls and figures of Suidae, and to consult the undermentioned volumes and memoirs which had not previously been accessible to me. I have also been favoured with valuable information by the letters of a number of scientific correspondents named below. The information thus gained, in addition to that upon which my paper was based, has been in the main confirmatory of the views I had come to independently of it; the very important fact, however, that the young of *Sus celebensis* are striped, as communicated to me by Dr. A. B. Meyer, of the Royal Zoological Museum at Dresden, is directly contradictory of the suggestion which I threw out at p. 541, as is also the fact published by him as to the striping of the young of the nearly allied *S. verrucosus* directly contradictory of the statement of Müller and Schlegel given above, p. 540.

LIST OF SOURCES OF INFORMATION ON 'SUIDAE' MADE AVAILABLE TO ME SINCE JULY, 1876.

I. A skull of *S. celebensis* ♀, kindly lent me by Dr. A. B. Meyer, Director of the Royal Zoological Museum at Dresden.

II. A skull of a *Sus*, sp. ?, ♀, from Ternate; also lent me by Dr. A. B. Meyer.

III. Two skulls of the domestic pig of Bengal (*S. cristatus*, var. *domesticus*), presented to the Oxford University Museum by E. Lockwood, Esq., Monghyr, Bengal.

IV. Two skulls of *S. scrofa*, var. *ferus*, presented by Dr. Fiedler, of Dresden.

V. A skull of *S. scrofa*, var. *ferus*, presented by M. le Marquis de l'Aigle, Château de Compiègne, Pise, France.

VI. A skull of *S. scrofa*, var. *domesticus*, from Stene i Bö, Lofoten Islands, 2° above Arctic Circle; presented by H. N. Moseley, Esq., Fellow of Exeter College, Oxford.

VII. Two bronze statuettes of *S. scrofa*, var. *ferus*, from the Gallo-Roman period in France; presented by John Evans, Esq., F.R.S.

VIII. A Memoir by Professor L. Rütimeyer, 'Neue Beiträge zur Kenntniss des Torfschweins,' in the 'Verhandlungen der naturforschenden Gesellschaft in Basel,' iv. 1, 1864, pp. 132–186.

IX. The Inaugural Dissertation 'Zur Kenntniss des Torfschweins,' by Johann Wilhelm Schutz, Berlin, May 4, 1868, referred to by Mr. Darwin in 'Animals and Plants under Domestication,' vol. i. p. 71, 2nd ed., 1875.

X. A Memoir by Professor R. Hartmann, 'Verbreitung der in nordöstlichen Afrika wild lebenden Säugethiere,' in the Berlin 'Zeitschrift für Erdkunde,' Bd. iii. 1868, in which (at pp. 349–352) there is an account of *S. sennaariensis*, which the Professor and Dr. Schutz consider to be the parent stock of *S. scrofa*, var. *palustris*.

XI. Letter from Dr. A. B. Meyer, of the Royal Zoological Museum, Dresden.

XII. Letter from Professor R. Hartmann, of Berlin.

XIII. Letter from Professor Fitzinger, of Vienna.

XIV. Letter and drawing from Professor Busk, of London.

XV. Letter from the Rev. C. Spencer Bubb, late Missionary in Borneo.

XVI. Letter from G. Dobson, Esq., B.A., M.B., F.L.S.

A comparison of the skull of the old male of *S. celebensis* lent me by Dr. A. B. Meyer, and the measurements of it given below, with the skulls of *S. andamanensis* and with the descriptions of the skull of *S. verrucosus* given by Nathusius, Rütimeyer, and Müller and Schlegel, will show that while agreeing with these skulls in general contour and proportions it is intermediate between them in actual size. The highest point in the sagittal contour is, as in the Suidae above named, a little way anterior to the upper edge of the occipital squama; the transverse arc from one orbital process of the frontal to the other is even more convex than in any other Asiatic pig; the zygomatic arch is short, deep, and roughly sculptured; the lower canine's alveolus reaches back to a prominent external protuberance in the plane of the anterior lobe of the third molar, a point further backwards than it usually attains to. The posterior lobes of the third molars, though consisting of 4–5 cusps, have not the great antero-posterior length relatively to the rest of the tooth which we find in specimens of the same sex of *S. cristatus*.

As in *S. verrucosus*, the bony palate is nearer to the basicranial bones than in most other Suidae.

Its measurements (being those adopted by Professor Rütimeyer, 'Verhandlungen Gesellschaft Basel,' 1864, p. 163, from Nathusius, 'Abbildungen von Schweineschädeln,' 1864, p. 6) are as follows:—

MEASUREMENT OF SKULL OF 'SUS CELEBENSIS.'

	millim.		inches.
Length from foramen magnum to tip of intermaxillaries ...	281	=	10·3
Length of occipital crest to tip of intermaxillaries	325	=	12·8
Length from occipital crest to tip of nasals	320	=	12·55
Length of nasals	154	=	6·1
Anterior part of frontal and sagittal sutures	56	=	2·2
Posterior part of frontal and sagittal suture	97	=	3·8
From occipital foramen to vomer	47	=	1·7
From occipital foramen to palate	72	=	2·85
Dentigerous part of maxilla	120	=	4·7
Incisive part of palate	61	=	2·4
Greatest interzygomatic width	140	=	5·5
Greatest frontal width	88	=	3·5
Smallest facial width	31	=	1·2
Intermaxillary width	37	=	1·45
Least nasal width	24	=	1·95
Intermolar width	43	=	1·7
Interpremolar width	41	=	1·6
Occipital height	112	=	4·7

	millim.		"		"	
Lacrymal orbital border ...	22	=	·9	compare	1	in *Sus verrucosus*, Rütim.
Lacrymal malar border ...	17	=	·7	"	1·13	
Lacrymal upper border ...	31	=	1·4	"	1·84	

	millim.		"
Length of *S. celebensis*, sec. Dr. A. B. Meyer ...	1160	=	45·67
" " Müller and Schlegel	950	=	37·40
Length of tail of *S. celebensis*, sec. Dr. A. B. Meyer	250	=	9·84
" " Müll. and Schleg.	150	=	59
Height of *S. celebensis*, sec. Dr. A. B. Meyer	550–570	=	21–22
" " Müller and Schlegel	560	=	22

Dr. A. B. Meyer's specimen was more brown than Müller's and Schlegel's figure, pl. 28 *bis;* the tuft at the angle of the mouth red-brown, and the hair on the nape longer. As to his measurements of *S. celebensis*, Dr. A. B. Meyer observes that they are taken from a full-grown animal, and, being larger than those of Müller and Schlegel, bring *S. celebensis* more nearly on to a level with *S. verrucosus.* Its smaller size suggests the dwarfing action of a confined insular *habitat* undergone in some of the upheavals or subsidences of Celebes.

Dr. A. B. Meyer informs me that the young of *S. celebensis* are striped, in contradiction to the suggestion which I (see p. 541, *supra*), being impressed with the closeness of the affinity between this *Sus* and *S. verrucosus* (which Müller and Schlegel had declared not to be striped when young), had thrown out. He informs me also that he has the authority of General von Schierbrant, who has lived thirty years in Java and is a first-rate sportsman, for saying that Müller and Schlegel (*l. c.* p. 177) are not correct in what they say is the case with *S. verrucosus.* If this be so, the claim which *S. verrucosus* would have had to be considered the parent stock of our improved breed of pigs falls to the ground. On the other hand, I cannot, knowing the great modificatory power which domesticating influences of one kind or other have been proved to possess over the highly plastic porcine organism, and bearing in mind the similarity between the Irish greyhound pig and *S. verrucosus*, agree with Professor Rütimeyer (p. 184, *l.c.*, 1864) in excluding this *Sus* from consideration when we are speculating as to the parentage of our domestic pigs.

The skull of the *Sus* from Ternate lent me by Dr. A. B. Meyer belonged to a young male, the third molars of which were just coming into place, and which may be supposed therefore, according to the analogy of *S. scrofa*, var. *domesticus*, if that may be taken as any guide, to have been about eighteen months old (see Nathusius, *l. c.* p. 21). It differs from skulls of *S. cristatus* in the flatness of its lacrymo-frontal ridge, in the convexity backwards of its naso-frontal suture, and in being somewhat smaller than male skulls of that variety of *Sus* usually are. The posterior lobes of its third molars are not quite so large as they usually are in *S. cristatus;* but to this point I am not inclined to assign as much importance as some other writers do (see Rütimeyer, 'Fauna der Pfahlbauten,' pp. 33 and 188, and, *per contra*, Nathusius, *l. c.* pp. 49 and 103). In the straightness of its vertical contour it contrasts with *S. celebensis; S. andamanensis, S. vittatus*, as figured by Müller and Schlegel, *l. c.* plate 32. fig. 6, and with *S. barbatus;* and resembles *S. cristatus*, as it does also in its general facies. Dr. A. B. Meyer thinks that this *Sus* may possibly deserve to be considered a new species, *S. ternatensis;* but having had abundant experience of the facility with which the wild pigs of that region cross considerable arms of the sea, he would, I gather, express himself with much caution as to its relations to *S. papuensis* and the Suidae of the neighbouring islands, volcanic and other, tenanted by swine. The colouring, however, of the head of this *Sus* differs from that of any other *Sus* seen by Dr. A. B. Meyer, or figured by Schlegel—the head being covered all over with long black hair, except in the region occupied by a broad yellowish brown streak beginning between the eyes and descending to the snout, where it broadens.

MEASUREMENTS OF SKULL OF 'SUS TERNATENSIS.'

	millims.		inches.
Length from foramen magnum to tip of intermaxillaries ...	290	=	11·4
„ occipital squama „ „	355	=	14·1
„ „ to tip of nasals	345	=	13·5
Length of nasals	17	=	16·7
Anterior part of line of frontal and sagittal sutures	86	=	3·2
Posterior part of same line	95	=	3·7
Molar length	143	=	5·65
Incisive bone length	57	=	2·2
Greatest interzygomatic width	137	=	5·65
Greatest facial width	95	=	3·8
Least facial width	36	=	1·45
Intermaxillary width	40	=	1·6
Least nasal width	28	=	1·1
Intermolar width	49	=	1·9
Interpremolar width	44	=	1·75
Occipital height	111	=	4·4
Lacrymal orbital border	25	=	1
„ malar border	20	=	·8
„ upper border	34	=	1·3

A comparison of the two skulls of the domestic pig of Monghyr, Bengal [1]—which animals, as Mr. Lockwood informs me, are kept in large numbers by the pariah castes of his district, feed on human ordure, and are sold out of the district to the Chinamen at

[1] For which I am indebted to the kindness of E. Lockwood, Esq.

Calcutta—with the skulls of the wild pig of Hindostan (*S. cristatus*) shows that they differ from them in little else than their smaller size. They may be spoken of, therefore, as *S. cristatus*, var. *domesticus*; and they are as distinguishable from any variety of *S. scrofa*, var. *domesticus*, or indeed of *S. indicus*, as is the wild *S. cristatus* from other wild Suidae.

It is noteworthy that one of the skulls belonged to a very old sow. In these, as in the days when Juvenal spoke (Sat. vi. 160) of another Eastern country as a place where

'Vetus indulget senibus clementia porcis,'

it is a rare thing for a domestic pig to be allowed among Western nations to live long enough to wear down its third molars. Mr. H. N. Moseley, however, informs me that domestic pigs are kept in the Chinese Buddhist monasteries till they die of old age and infirmities; and of India, Meiners, in his 'Allgemeine kritische Geschichte der Religionem,' 1806, Bd. i. p. 193, says, 'In Asien war von jeher Hindostan, wie in Afrika, Egypten, der Thron des Thierdienstes.'

It is noteworthy, secondly, that with the worn-down molars of this aged domesticated pig were correlated abscesses on both sides of the lower jaw, much as might have been the case (see J. R. Mummery, Esq., 'Trans. Odont. Soc.' ii. 2, 1869, p. 72) in ill-fed human beings with similarly worn-down teeth.

Mr. Lockwood informs me that the young of the domestic pigs of his district are striped like the young of wild pigs.

The skull of a wild boar, from France, presented by the Marquis de l'Aigle, shows that the posterior lobe of the third molar in the lower jaw may attain the same proportions in *S. scrofa*, var. *ferus*, that it does in *S. cristatus*.

The skull of *S. scrofa*, var. *domesticus*, procured for me by H. N. Moseley, Esq., from the Lofoten Islands, two degrees within the Arctic Circle, has the long, low, lacrymal characteristic of *S. scrofa*, var. *ferus*, a fact of particular interest when coupled with the information, also procured for me by him, to the effect that the young domestic pigs of that region are occasionally born with stripes, and with his observation that the old pigs have a very wild-boar-like appearance. These facts should be borne in mind as telling against the views propounded by M. André Sanson, in his memoir 'Sur la prétendue transformation du Sanglier en Cochon domestique,' in the 'Journal de l'Anatomie et de la Physologie,' tom. iv. 1867, p. 38.

In Professor Hartmann's memoir above referred to, it is stated (p. 350) that certain Negro tribes, who disobey Mahommedan precepts by eating as well as domesticating *S. sennaariensis,* excuse themselves by saying it was formerly the custom to do so, a fact which goes some little way to disprove the view that this true *Sus* can be merely a feral variety of *S. scrofa*, imported by Europeans. Professor Hartmann, in a letter of date Sept. 28, 1876, says,—

'*Sus sennaariensis* ist ein kleines dem europäischen Torfschwein (der Pfahlbauten) ähnliches Schwein, echtes *Sus,* welches wild durch einen grossen Theil von Mittel-Afrika vorzukommen scheint . . . *Sus scrofa ferus* in der Sahara und in Aegypten nicht selten, findet sich *angeblich* ebenfalls in Sennaar, indess weiss ich hierüber *nichts völlig Sicheres.*'

Professor Busk, in a letter to me of Dec. 17, 1876, informs me that in the Etruscan Museum at Florence, amongst numerous little bronze articles extracted from the ancient Etruscan tombs, there were many figures of animals, one in particular being a very well made statuette of a pig, which to his eye very closely resembles the Berkshire breed, the only point in which it differed being the comparatively large eye, whilst the rest of the contour was quite what we might expect to see at an ordinary cattle-show. The animal was represented apparently as having a close curled tail. With it were a good many statuettes of stags, the horns of which were of the type of the pliocene *Cervus ctenoceros.* In the very perfectly restored Etruscan tomb erected in the Museum, with all its original contents and frescoes, Professor Busk noticed, amongst other figures of animals, one very well drawn of a monkey climbing up a tree or pole.

I should suggest that the monkey and the pig, both alike, are representations of animals from the same quarter of the globe as that whence the kinsmen of the Etruscans in the time of Solomon brought, every three years, into Mediterranean regions, the ivory, apes, and peacocks, the Sanskrit names of which still remain to speak to their *habitat.* See Max Müller, 'Lectures on the Science of Language,' 1861, ser. i. p. 190, and ser. ii. p. 234, for the source whence these animals and copper came to Europe.

Professor Busk sent me by the same post odontograms of the teeth of *S. cristatus, S. scrofa*, var. *ferus*, and *S. scrofa*, var. *domes-*

ticus, which show that in *S. cristatus* the antero-posterior length of the third molars is much greater in relation to that of the rest of the molar series than it is in either of the other Suidae named. I have above, p. 534, noted that the sexual differences in this matter are very considerable in *S. cristatus;* and those produced by domestication are also not insignificant—points which somewhat impair the value of these statements of relative proportions, though no less an authority than Rütimeyer avers, 'Fauna der Pfahlbauten,' p. 188, that

'*mit viel grösserer Zähigkeit das Gebiss den Species-Typus äusseren Einflussen gegenüber aufrecht halt als die Schädelbildung.*'

(Cf. however Nathusius, *l. c.* pp. 49, 103, and Studer, 'Mittheil. Ant. Gesell. Zürich,' xix. 3, 1876, p. 67.) Professor Busk also observes that in *S. cristatus* nearly all the teeth, except molar 3, are wider in both jaws in proportion to their length than they are in the other two, and that from this we may suppose that the Indian pig is more exclusively herbivorous than the tame or wild animals with which he has compared it. Jerdon speaks of *S. cristatus, l.c.* p. 243, as being in general almost entirely 'vegetable feeders.' Captain Baldwin's views ('The Game of India,' pp. 154–5, 1876) are to the same effect.

Professor Busk's drawing and description of this Etruscan pig, coupled with a similar figure of a pig found in the ruins of Herculaneum (see Nathusius, *l. c.* p. 142, fig. 33, and Darwin, *l. c.* p. 71), furnishes a good illustration of Rütimeyer's saying ('Fauna der Pfahlbauten,' p. 190), that though the modern breed of domestic pigs is recent enough to have been introduced by steamboats, it nevertheless had been represented continuously from former ages by the Bündtner-Schwein in the valleys of the Grisons. A figure of a sow suckling three young ones may be seen on an Umbrian medal of probably the third century B.C. at latest, figured by M. Sambon, in his 'Monnaies Antiques de l'Italie,' 1870, pl. v. 5, with the short snout bent on the chanfrin, the pendent ears, and the mane limited to the nuchal region, which justify us in considering it to have been intended to represent a tame variety. A pig, figured *ibidem*, pl. iv. 4, from Etruria has a snout of such slenderness as to correspond very closely with the description given of *S. scrofa*, var. *palustris*, whilst it contrasts in other points very strikingly with the wild boars represented from Apulia and Lucania, on the other side of Italy (*ibidem*, plates xiv. 1, xv. 17 and 37, xix. 4), which have as close a zoological as legendary connexion with the wild boar of Calydon, on the other side of the Adriatic (see 'Thesaurus,' Brandenburg, i. pp. 318 and 464, 1696, and 'Thesaurus Numismaticus,' i. p. 400, and tab. xl).

Of the two bronze statuettes given me by Mr. John Evans, one has the long slender snout, and the mane reaching the whole length of the convex back, from the prominent ears to the curled tail, which may justify us in considering it as intended for a wild boar; the other combines the sturdy straddle, and the long and large erect mane, beginning on the forehead and in front of the erect ears, characteristic of a wild boar, with a snout as disproportionately short, and tusks as reduced as we ever see them in the highest-bred modern Chinese pig. The characters of the wild and tame varieties, however, being thus inaccurately and inartistically combined in these statuettes, cause them to contrast disadvantageously with the Italian works of art just mentioned; but they furnish us with a conclusive answer to the weak reasoning of De Blainville ('Ostéographie,' Sus, p. 170), expressed in the following words:—

> 'Du temps de César, il paraît cependant qu'elle (la culture du cochon) n'était pas encore parvenue dans les Gaules, car il n'est nullement question de cet animal dans ses Commentaires; elle s'y est donc propagée depuis la conquête, d'où elle a passé en Angleterre, qui ne possédait pas même de sanglier dans ses forêts!'

If further answer were required to this astounding statement, a reference to Mr. Evans's work on British Coinage would furnish it—figures of the boar, some of which are exceedingly characteristic, being given there on pls. vi, viii, xi, xii, and xiii, from those ancient coins. But all well-informed antiquaries are aware that the wild boar is one of the earliest animals figured in Celtic works of art (see 'Horae Ferales,' p. 185, pl. xiv; Montellier, 'Mémoires sur les Bronzes Antiques,' Paris, 1865; and Stephens, 'Literature of Kymry,' p. 230).

Professor Rütimeyer's paper in the 'Verhandlungen der naturforschenden Gesellschaft in Basel,' iv. 1, 1864, is, I apprehend, referred to by Mr. Darwin when he says ('Animals and Plants under Domestication,' i. p. 71, 2nd ed. 1875) that 'Rütimeyer himself seems now to feel some doubt' as to whether the 'Torfschwein' existed as a wild animal during the first part of the stone period. No reference is given *l.c.* to any memoir of Professor Rütimeyer's; and I became, after writing the foregoing paper, acquainted with the one in question from a mention of it made by Herr Edmund Naumann in 'Archiv für Anthropologie,' Bd. viii. 1, 1875, p. 19, in a discussion on 'Die Fauna der Pfahlbauten im Starnberger See.' From a perusal of this paper of Professor Rütimeyer, I am inclined to think that he would regard the skull (fig. 2) from the alluvium of the Thames valley in the Oxford Univ. Museum as a skull of *S. scrofa*, var. *palustris*. I have spoken of it as the skull of a wild sow, considering, as said above, p. 530, that early breeding may, in a species admitting of such a wide range of structural oscillation, and notably in the matter of mere size, account for a very great distance between its male and female representatives (see for a similar view as regards our own species 'Journal of Anthropological Institute,' vol. ii. 1875, p. 122). I am not inclined to withdraw from this view even after reading Prof. Rütimeyer's and Herr J. W. Schutz's memoirs. For in the former of these I find (p. 151) that the *five* skulls used for description are acknowledged to be skulls of sows, and four of them to have been skulls of *old* sows, and the measurements given at p. 163 have been taken exclusively from skulls confessedly female (see p. 161). And I learn from Herr Schutz's essay (p. 44) that Steenstrup (cit. Wiegmann's 'Archiv,' xxvii. n. 112) had distinctly stated that the 'Torfschwein' was (as I had hinted without any knowledge of his views, see p. 254, *supra*) simply the female representative of *S. scrofa*, var. *ferus*. Professor Rütimeyer had not at the time of writing his 'Fauna der Pfahlbauten,' 1861 (see p. 33), a single perfect skull of the 'Torfschwein' available to his comparison; the almost perfect skulls treated of in his memoir of 1864 (see p. 150) have caused him to modify the view put forward at p. 190 of the earlier work, and in the later one he allows (pp. 158–160) that, both in the matter of the length of the lacrymal and in the absence of widening of the palate anteriorly, the *S. scrofa*, var. *palustris*, was more nearly affined to *S. scrofa*, var. *ferus*, than to *S. indicus*. To show this was the reason for giving a great number of my measurements, *supra*, pp. 544–547. I take this opportunity of saying that the very small breed of the Scottish highlands and islands, with suberect ears, usually of a dusky brown colour, with an arched back and coarse bristles along the neck and spine, spoken of by Professor Low in his 'Domesticated Animals of the British

Islands,' p. 429, may perhaps be such a breed as Professor Rütimeyer suggests, in his later paper (p. 168, see also p. 148), should be looked for in Eastern Europe or Western Asia or in fossil forms [1].

Mr. G. Dobson, F.L.S., in a letter to me of date Jan. 15, 1877, writes that the young of the pig of the Andaman and of the Nicobar Islands are striped.

Tho Rev. C. Spencer Bubb informs me that the young Borneo domestic pig is sometimes striped and sometimes not, whilst the young Chinese pig is never striped; and he adds that there are certainly two domestic breeds in Borneo.

DESCRIPTION OF PLATES IV AND V.

Fig. 1. Orbito-lacrymal region of partially reconstructed skull of *Sus scrofa*, var. *domesticus*, wanting nasals and intermaxillaries, from late Celtic interment at Arras, East Riding of Yorkshire. Oxford Museum.

It is, as Nathusius has well pointed out, l. c. p. 147, by no means always easy to be absolutely certain as to the question whether a particular pig's skull belonged to a wild or to a domesticated individual. The difficulty is increased when the animal is young, as in this case, the last molar having only just come into use, and the animal consequently being only about 18 months old, and when as the small size generally, and especially the small size of the third molar and the canine, may be taken, I believe, to indicate that it is of the female sex. It is true that the paucity of cusps in the molars has been taken as indicating the wild state; a comparison, however, of the male and female molar series in *Sus cristatus* has suggested to me that the greater size of the molars really depends upon a greater supply of blood, such as the male molars would get, by virtue of sharing the greater supply lavished on the canines, and such as a well-fed domesticated animal's molars would get in common with all its other structures and organs. The comparatively vertical occipital squama is one main anatomical point in favour of this skull having belonged to a domesticated specimen; the pterygoid, on the other hand, has much of the obliquity characteristic of the wild *Sus scrofa*, var. *ferus*. When we consider, however, that this skull was found in an interment containing a human body, together with portions of another skull of a pig of the same age, the probability that it belonged to a tame individual appears to be very great. The subjoined measurements show that the lacrymal bone, though not

---

[1] I suspect that latitude has more to do with the production of such varieties than longitude, and still operating causes more than geological. One of these causes is suggested by the words of Varro, ii. 4. 13, cited by Dureau de la Malle, 'Economie politique des Romains,' ii. p. 149: 'Porcique nati hieme, fiunt exiles propter frigora.'

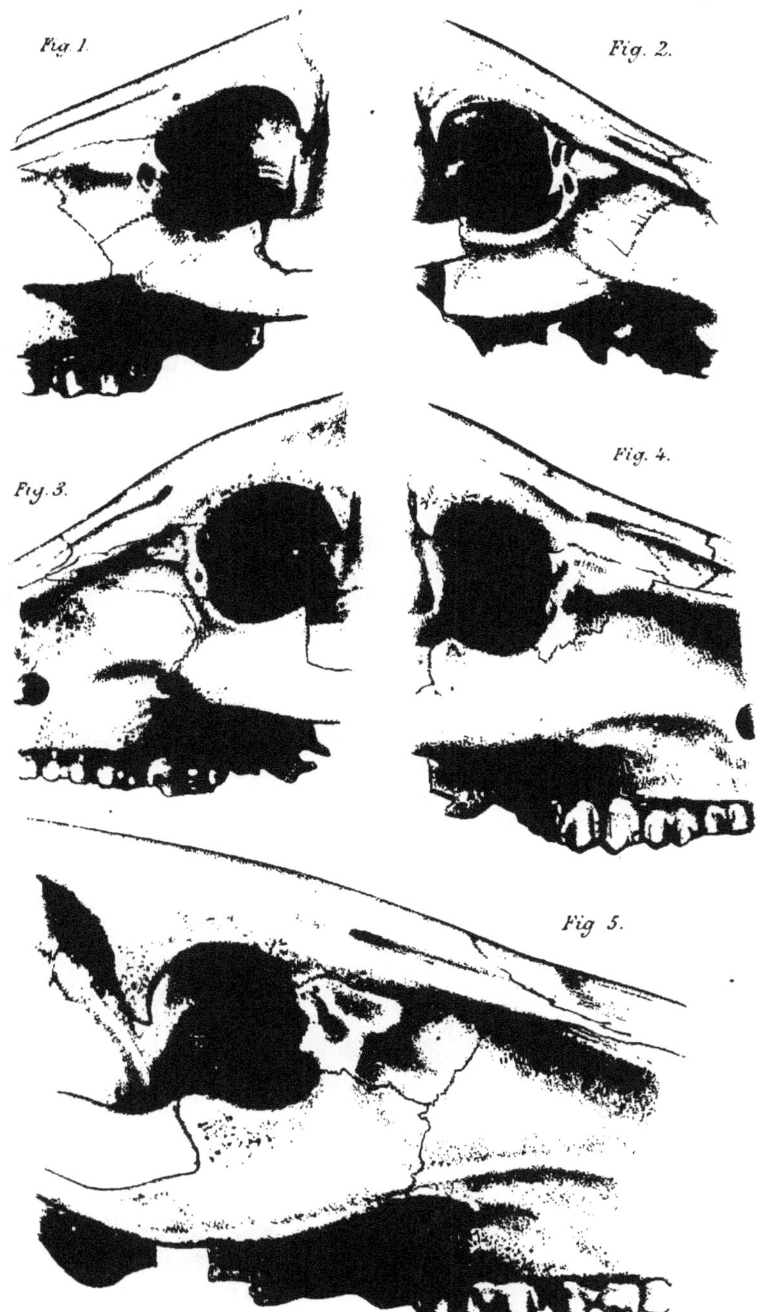

F Huth, Lith' Edinr

DOMESTIC PIGS.

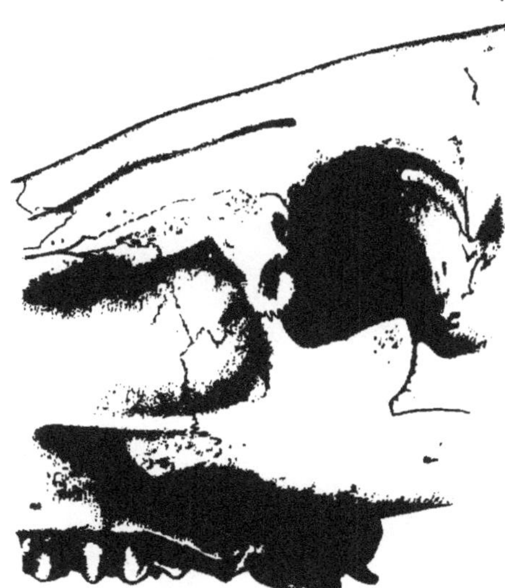

Fig. 6.

Fig. 8.

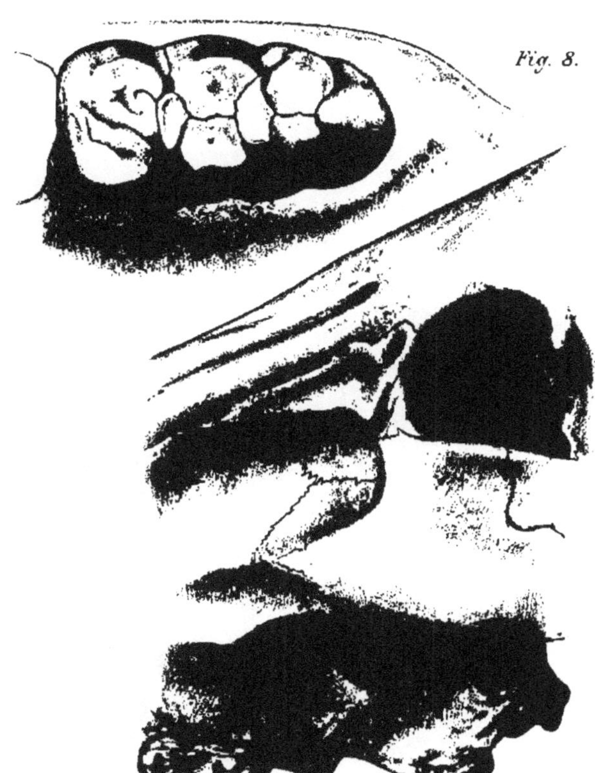

Fig. 7.

F Huth, Lith^r Edin^r

DOMESTIC PIGS.

so long relatively to its height as is often the case in the non-domesticated *Sus scrofa*, is yet longer than it is in *Sus indicus*, or in any of the Eastern pigs from which *Sus indicus* can with any probability be supposed to have descended.

	inches.		inches.
Length from anterior external angle (apex) of frontal to middle of occipital ridge .	5·9	Length from anterior border of orbit to posterior	1·4
Length from apex of frontal to apex of maxilla	4·1	Greatest width of occiput . .	2·4
Length from anterior border of orbit to temporal ridge . .	3·8	Length of molar series . .	2·3
		Height of occiput	3·8
		Height of lacrymal	0·7
		Length of lacrymo-malar suture .	0·8

The difference here noted between the height of the lacrymal and the length of its lower border amounts but to a tenth of an inch; but in the female of *Sus indicus*, with the same stage of dentition, fig. 4, the height of the lacrymal, instead of being one tenth of an inch less, is 0·35 inch greater in length than its lacrymomalar border.

Fig. 2. Orbito-lacrymal region of *Sus scrofa*, var. *ferus*, old ♀, from alluvium of Thames valley, obtained for the Oxford University Museum from the cutting for the drainage works near Iffley, 1876, by Professor Prestwich, F.R.S.

This skull combines the general contour and the slender snout of *Sus scrofa*, var. *ferus*, with a lacrymal bone differing little in its proportions from the lacrymals so characteristic of the Asiatic pigs, less *Sus verrucosus* (and *celebensis*). The fronto-parietal region does not form one continuous slope in the same plane as is the case in the Wild Boar of Germany (Pl. IV. fig. 5); but this difference may be observed in skulls certainly of Wild Boars from the Thames-valley deposits; whilst the great wear of the teeth and the slenderness and length of the naso-facial region are much in favour of considering this specimen to have belonged to the wild race. I have placed side by side with the measurements of this skull the measurements given by Rütimeyer, l. c. pp. 45 and 183, of his 'Torfschwein,' *Sus scrofa*, var. *palustris*. Unhappily, Rütimeyer has never been able to procure (see pp. 43 and 45, note 1) a skull of this variety of *Sus* with the facial bones in connexion with the brain-case, nor has Nathusius (see p. 149, l. c.) ever been able to see an uninjured lacrymal bone from the same animal. These facts, whilst making the value of this skull (the opportunity of figuring which I owe to the kindness of Professor Prestwich) greater, make the value of the comparison of its measurements less.

The instructive observations of Nathusius (l. c. pp. 99–101), to the effect that ill-nourished pigs have the entire length of their skulls greater as measured from the occipital crest to the apex of the snout, whilst the portion of that length made up by the frontal and parietal is somewhat shorter, and the nasal portion proportionally longer, when coupled with the fact of the great wear of the teeth in this specimen, enable us to explain the one great point of inferiority, that of the length of the fronto-parietal region, which this skull's measurements show us, compared with those of the 'Torfschwein.' It may be added, that the true explanation of Dr. Gray's statement 'Brit. Mus. Catal.' 1869, p. 329), that the nasal bones of the skull elongate as Suidae increase in age, 'and especially as they reach adult and old age,' is probably that he had in his mind's eye skulls of old and ill-fed wild pigs, such as this specimen. On the other hand, when we are comparing such skulls as this with the 'Torfschwein' of Rütimeyer, we must recollect that he represents this latter variety of pig, and what he supposes to be its still surviving representative, the 'Bündtner-Schwein,' as having a short snout (see pp. 42, 45, 181–185). The classificatory value, however, of such a peculiarity is much reduced by the results of such experiments as those of Nathusius just referred to.

MEASUREMENTS OF IFFLEY SKULL AND OF RÜTIMEYER'S 'TORFSCHWEIN,' pp. 45, 183, l. c.

	Iffley Skull. inches.	Torfschwein. inches.
Greatest frontal width between ectorbital processes .	3·2	3·7
Least width on vertex	0·8	0·9
Greatest interzygomatic width	4·6	4·7–4·9
Height of occipital from inferior border of foramen magnum	3·5	3·8–4·5
Length of vertex from level of supraorbital foramen to ridge of occiput	3·8	4·4
Horizontal distance from anterior border of orbit to posterior of temporal fossa	3·0	3·1
Length of intermaxillary along alveolar border .	2·1	1·9–2·4
Maximum length of skull from apex of intermaxillaries	10·9	—
Length of nasal bones (approximately)	5·5	—
Length from apex of intermaxillaries to inferior border of foramen magnum	10·3	—
Length from anterior external angle (apex) of frontal to middle of occipital ridge	5·9	—
Length from middle of fronto-nasal suture to middle of occipital ridge	4·8	—
Length from anterior border of orbit to posterior .	1·4	—
Greatest width of occiput	2·4	—
Breadth of nasal at commencement of naso-frontal suture, which is the point of maximum width of nasals	1·1	—
Maximum width of intermaxillaries	1·3	—
Height of lacrymal along rim of orbit	17 millims.	
Length of lacrymal along malar suture	18 millims.	

Fig. 3. Orbito-lacrymal region of *Sus andamanensis*. 1514 *b*, Oxford University Museum, adult ♀.

The skull from which this drawing was taken was that of a wild sow from the Andaman Islands, procured for me by my friend J. Wood-Mason, F.G.S., of the Calcutta Museum. It is about the same size as the two prehistoric British skulls, Plate IV. Figs. 1 and 2, and as the 'Torfschwein' of Rütimeyer—resembling this latter in the one important particular, that of its long fronto-parietal region, in which it differs from the British skull (Fig. 2). It differs from all the pigs' skulls here figured in the exaggeratedly disproportionate shortness of the malar border of its lacrymal bone, which is, as in the newly-born European pig, little more than half the length of the orbital border. In its convex frontal region we have, again, a character retained in the adult Asiatic which is transitorily represented in the European pig. This convexity of the fronto-lacrymal ridge is well marked, and, like the straight naso-frontal suture and the two other peculiarities specified, serves to differentiate *Sus andamanensis* from specimens of similar size from the Palaearctic region, and to show that it is a member of the group represented by *Sus cristatus*. Again, this skull resembles *Sus cristatus* and differs from *Sus scrofa* in being somewhat shorter, and having the paroccipitals, temporal bullae, and pterygoid processes more crowded together. The length of the nasals is a little less than that of the fronto-parietal region in the vertical aspect, as is often, though not always, the case in *Sus cristatus*, though very rarely, if ever, in *Sus scrofa*.

MEASUREMENTS OF SKULL OF 'SUS ANDAMANENSIS,' FEMALE.

	inches.		inches.
Greatest frontal width between ectorbital processes	3·3	Length from apex of frontal to middle of occipital ridge	5·95
Least width on vertex	0·8	Length from middle of fronto-nasal suture to middle of occipital ridge	5·5
Greatest interzygomatic width	4·65	Length from anterior border of orbit to posterior	1·5
Height of occiput from inferior border of foramen magnum	3·7	Greatest width of occiput	2·5
Length of vertex from level of supraorbital foramina to middle of occipital ridge	4·5	Breadth of nasals at commencement of naso-frontal suture, which is the broadest part of nasals	1·0
Horizontal distance from anterior border of orbit to temporal ridge	3·05	Maximum width of intermaxillary region	1·2
Length of intermaxillary along alveolar border	2·0	Height of lachrymal along rim of orbit	17 mm.
Maximum length of skull from apex of intermaxillaries	10·8	Length of lachrymal along malar suture	8 mm.
Length of nasals	5	Length of molar series	2·0
Length from apex of intermaxillaries to anterior edge of foramen magnum	9·1	Length of premolar series	1·3

Fig. 4. Orbito-lachrymal region of skull of wild Indian sow, *Sus cristatus*, British Museum. See 'Catalogue of Carnivorous, Pachydermatous, and Edentate Mammalia,' 1869, p. 334 (where this skull is catalogued as '716 *m*. Skull of young female. Length 10¼, height 7½ inches. Terai, Nepal; Dr. Oldham'), or Hand-List, published 1873, p. 64.

This skull has been figured to show the points in which it contrasts with the British pig's skull (Fig. 1) on the one hand, as to racial characters, and with the Indian boar's skull (Fig. 6) as to sexual characters, agreeing as it does with the one as to sex, and with the other as to age.

It is, as may be seen by constructing two triangles with the three first measurements given of it and of that of *Sus cristatus* (Fig. 6) respectively, a very much smaller skull than that; the principal points, however, enumerated in the description of that skull, as differentiating it from *Sus scrofa*, can be recognised in its smaller contours. The third molar in the upper jaw has the part posterior to its second principal lobe much smaller relatively and absolutely than is the case in boars of *Sus cristatus*; this posterior factor, however, effloresces into as many as seven cusps in the upper and five in the lower jaw of this specimen.

MEASUREMENTS OF SKULL, NO. 716 *m*, BRITISH MUSEUM COLLECTION.
'SUS CRISTATUS,' FROM TERAI, NEPAL.

	inches.		inches.
Extreme length	11·2	Length of lachrymal along malar suture	0·75
Extreme height	4·3	Interpremolar transverse diameter of palate	1·6
Base-line	10	Intermolar	1·75
Height of lachrymal along rim of orbit	1·1		

Fig. 5. Orbito-lachrymal region of skull of wild boar, *Sus scrofa*, var. *ferus*, from Germany. No. 1513*a*, Oxford University Museum.

The entire dentition was in place; the portion of the third molar which is posterior to the four primary cones is very small as compared either with the rest of the tooth or with its homologue in the Indian wild boar (fig. 6). It is, however, considerably in front at a line drawn as a perpendicular to the plane of the cutting-edges of the molars from the lachrymal canal. This skull differs markedly from the skull just named in the greater length, relatively to the height, of its lachrymal bone, in the greater length of its nasals relatively to the length of the fronto-parietal region, in the absence of any great convexity in the part of the frontal abutting upon the lachrymal, and in the convexity backwards of its fronto-nasal suture.

MEASUREMENTS OF SKULL OF GERMAN WILD BOAR, NO 1513 *a*, OXFORD UNIVERSITY MUSEUM.

	inches.		inches.
Extreme length	16·5	Width of nasals	1·3
Extreme height	5·1	Maximum interzygomatic width	5·9
Base-line from anterior margin of foramen magnum to apex of intermaxillaries . . .	14·0	Maximum frontal (interectorbital) width	4·2
Length of nasals, approximatively	8·0	Height of lachrymal along rim of orbit	1·0
Fronto-parietal length in vertical aspect . . .	6·8	Length of lachrymal along malar suture	1·2

Fig. 6. Orbito-lachrymal region of skull of Indian wild boar, *Sus cristatus*, killed at Haugul, Dharwar province; from the collection of Sir Walter Elliot, K.C.S.I. No. 72.

In this animal the third molar is only just coming into place, and its very large multicuspidate posterior lobe is not so far forward relatively as in the European wild boar (fig. 5) nor in other Indian hogs of greater age. It shows however very plainly the points, several if not always all of which have, by their presence in every specimen of *Sus cristatus* which I have examined, enabled me to distinguish it from *Sus scrofa*, var. *ferus*. It has the relatively short lachrymal—a tape, stretched as an arc across the long axis of the skull, from the anterior inferior angle of the bone on one side to the homologous point on the other, passing over the frontals, and not over any part of the nasals; the naso-frontal suture, which lies entirely in front of such an arc, is straight, and not convex backwards; the portion of the frontal which is bounded internally by the supraorbital channel, and externally by the lachrymal bone, is markedly convex. The nasals are broader, as is nearly always the case, and shorter also, which is not by any means always the case, relatively to the fronto-parietal region of the vertex, than in *Sus scrofa*, var. *ferus*.

MEASUREMENTS OF INDIAN WILD BOAR, NO. 72, SIR WALTER ELLIOT'S COLLECTION.

	inches.		inches.
Extreme length	16·9	Minimum vertical width	1·8
Extreme height	5·1	Height of lachrymal along rim of orbit	1
Base-line from anterior margin of foramen magnum to apex of intermaxillaries	13	Length of lachrymal along malar suture	0·9
Length of nasals	7·3	Interpremolar transverse diameter of palate	1·6
Length of fronto-nasal suture to middle of occipital ridge	8	Intermolar	0·25
Width of nasals at apex of frontals	2	Length of posterior upper molar	1·35
Maximum interzygomatic width	6	Breadth	0·9
Maximum frontal (interectorbital) width	4·4	Length of third lower molar	1·6
		Breadth	0·8

Fig. 7. Orbito-lachrymal region of skull of *Sus barbatus*, Borneo. 1519 *d*, Oxford University Museum.

This skull, like the other three skulls of the same species examined by me (of which two are in the British Museum), differs from those already described in large points as well as in small ones; and there can be little reason for hesitating to accept it as specifically distinct from them, and indeed from all other *Suidae*.

The contour described by the middle line of its nasal and fronto-parietal regions superiorly, the relations of the greatest width and greatest lengths both of the entire skull and of the nasal bones, the position of the plane of its greatest interzygomatic width, not posteriorly, but in the middle of the zygomatic arch, are points of large difference. The exceeding simplicity of its third molars and the persistence of the mesopterygoid as a distinct bone, are points of small difference, but yet of great morphological importance. In the shortness of its lachrymal bone it resembles the other *Suidae* without facial warts.

Its naso-frontal suture and lachrymo-frontal ridge are more like those of these pigs than those of *Sus scrofa*.

MEASUREMENTS OF SKULL OF 'SUS BARBATUS,' NO. 1519 *d*, OXFORD UNIVERSITY MUSEUM.

	inches.		inches.
Extreme length	19·7	Fronto-parietal region in vertical aspect from same point as preceding measurement to occipital ridge in straight line	9·0
Height	6·7		
Base-line	16·7		
Length of nasals from plane of postero-lateral tips	10·5		

	inches.		inches.
Width of nasals . . .	1·5	Length of posterior upper molar	1·4
Maximum (interzygomatic) width	6.8	Breadth of anterior, which is much the widest lobe . .	0·9
Maximum frontal (interector-bital) width . . .	4·5	Length of posterior lower molar	1·5
Minimum vertical. . .	1·4	Breadth of anterior lobe . .	0·8
Lachrymo-malar line . .	1·2	Intermolar space at narrowest point	1·1
Height of lachrymal . .	1·2		

Fig. 8. Left lower third molar of *Sus andamanensis*, female, 1½ times the natural size.

This tooth shows the three divisions of the third molar of the true *Suidae* in great simplicity. There are two bicuspidate lobes corresponding to the two principal lobes of molar 1 and molar 2, and, like them, enclosing a single azygos lobe in the middle line between them. This azygos lobe is developed from the second bicuspidate lobe. There is no ridge developed in *Sus andamanensis* on the anterior part of the tooth, i. e. on the face in contact with molar 2. Posteriorly to two bicuspidate lobes, which already show signs of wear, are seen five smaller cusps, occupying in all a much smaller space than the rest of the tooth. In the males of *Sus cristatus* these five smaller cusps would, as in well-fed domestic pigs, occupy a very much larger space relatively and absolutely than they do here, or even in *Sus scrofa*, var. *ferus*. In *Sus barbatus* the third lower molars are as simple as they are in *Sus andamanensis*, and, like this *Sus*, have the third molars of the upper jaw simpler still than the lower. The five posterior cusps consist of one placed mesially in the interspace between the primary cusps of the second pair, of three placed like the dots in the sign ·.·, and of a fifth, not constant, accessory cusp placed on the inner side. The four latter are represented in the upper jaw of this pig by two, and in *Sus barbatus* by one main cusp.

XXXI.

NEW POINTS IN THE ZOOLOGY OF NEW GUINEA.

THE author commenced by saying that the zoology of New Guinea has had a great deal of research bestowed upon it, and will yet have a great deal more, as a consequence of the profit which has already resulted. A point which recent zoological discoveries in New Guinea throw light upon is that there was a dry land passage at one time between Australia and New Guinea, recent discoveries in the latter country having revealed the presence there of animals similar to, or identical with, some found in Australia. This is held as proof that where Torres Strait now is there was once dry land. But against this hypothesis is urged the difference between the vegetation of the two islands. This, however, is accounted for by what Herbert Spencer calls the circumambient medium. Though people are inclined to think vegetables considerably less sensitive than animals, sometimes they are more sensitive to heat and dryness; and the author believes that it is the greater tenderness of the vegetation in those countries which accounts for the disparity observable between the vegetable growths of New Guinea and those of Australia. In the centre of New Guinea there is a high range of mountains, which attract and impart moisture to the surrounding country; while the interior of Australia consists of great barren plains, which harbour no moisture. The plants, as they have not had the means to protect themselves available to animals, have gradually altered their form to accommodate themselves to circumstances. A curious creature, covered with prickles, living on ants and other insects, and unprovided with means of militant operations, is found on both sides of the Straits. Two kinds of *Echidna* have also been discovered in New Guinea, and corresponding with them is one in Tasmania and another in Australia. These creatures could not travel over water, and so there must have been land communication at the period of their original distribution. Quite lately an *Echidna* has been found in the south-west corner of New Guinea, and sent to Professor Rolleston by the Rev. Mr. Lawes, the discoverer, accompanied by a letter, in which the statement is made that this is the first ever found. For this species the name *Echidna lawesii* is proposed. The Cassowary has also been found on both sides of Torres Straits. Proof of the existence of the Tree-Kangaroo, both in Australia and New Guinea, Professor Rolleston also considered reliable. At its conclusion the paper treated of the Admiralty-Island pig, in the first part of which Professor Rolleston pointed out the peculiarity of a glabella.

[As no description is given in this abstract of the characters which distinguish this new species of *Echidna* from other species of the same genus, I wrote to C. Robertson, Esq., Demonstrator of Anatomy, University of Oxford, to ask if any description of the species had been recorded in the Catalogue of the Museum or elsewhere. He answered that he was not aware of any description drawn up by Dr. Rolleston, but referred me to the 'Proceedings of the Linnean Society of New South Wales,' in which I find that Mr. E. P. Ramsay, Curator of the Australian Museum, Sydney, contributed March 26, 1877 ('Proc. Linnean Society N. S. W.,' vol. ii. p. 31), a note of a species of *Echidna* (*Tachyglossus*) from Port Moresby, New Guinea. Mr. Ramsay distinguishes it from the *Tachyglossus bruijnii* from the northern parts of that island. He describes its external characters on p. 32, and states in a footnote that he has 'not yet learned the name which has been given to this new species, but daily expects to hear of it from my friends in England. Should it however be still unnamed, I propose for it the name of *T. lawesii*, in honour of its discoverer.' In vol. iii. p. 244, September 30, 1878, he states that Mr. Goldie has obtained three other specimens, and he gives the measurements of each.—EDITOR.]

XXXII.

ON THE ROT IN SHEEP.

The English Lake District presents us with as simple a case for the investigation of the cause of rot in sheep as any other portion of the wide area over which that disease has spread, with, perhaps, the exception of such isolated localities as the Faroe Islands: for sheep abound upon its thousand hills, while the species of snails and slugs are but few, and the conditions of its geological formation, of its fauna and of its flora differ as widely from those of many other regions within our four seas similarly affected as it is, all things considered, possible for them to differ. Having occasion to visit the district in question last week, I used the opportunity for making a few enquiries of the farmers and shepherds there as to the natural history of this plague. These enquiries were of the simplest kind, anybody can repeat them, and I cannot but think that the answers he will receive will incline such an enquirer to think that a strong *a priori* case is made out in favour of the view put forward in 'The Times' of April 7, to the effect of identifying the black slug (*Arion ater*) or the gray slug (*Limax agrestis*) as one necessary link in the chain of causes concerned. I found the natives as intelligent and observant as I have found them to be any time during the last thirty-four years upon natural history questions; and I very rapidly got the following facts deposed to by them without any prompting on my part :—

1. The fluke disease is a disease of low grounds, and notably of pastures liable to be flooded.

2. But not exclusively of pastures liable to be flooded; for, what is of special consequence as going some way towards eliminating the pond snails (*Limnaeus pereger*) and others from the charge of sharing in the causation of fluke disease, the pasturing of sheep in a stubble rich in the 'melancholic poisonous green,' which a wet

autumn often produces, is a very sure way for producing the disease.

3. The words just given in inverted commas are not the exact words employed by my informants; those which follow are, or pretty nearly so—'a single bellyful will give the disease.'

4. 'A pasture will give the disease at the back-end of the year which won't give it after Candlemas.' This means that the winter cold and rains destroy or wash away the larval or other forms of flukes which the slugs brought with them in the autumn.

The first of these observations is, like all other observations and all experiments concerning the matter with which I am acquainted, utterly destructive of Dr. Harley's suggestion as to the self-infection of sheep from their own droppings or those of their fellow sheep. If this were possible why should it not take place on the uplands or 'commons'—as the Lake shepherds call the unenclosed mountains—where the sheep, for the sake of food on those often limited grazing spaces or 'alps,' or for the sake of shelter, often huddle together as closely as they do or can do anywhere else? It is destructive also of the often-uttered ascription of rot to damp and moisture, the mountain ridges of Westmoreland being 'many-fountained' to a degree never dreamed of by Theocritus in Mediterranean districts, or realised by our own Laureate out of these 'rainy isles.' It is, on the other hand, confirmatory, as are all accurate observations on the subject, of the view which asserts that the presence of snail or slug is a necessary factor in the causation of rot. Snails and slugs are but scantily represented, if present at all, on mountain tops; there are no slugs in Forbes' and Tschudi's lists of high Alpine ranges; and these districts, like salt marshes, owe their character for considered 'soundness,' as the phrase is, as regards the most destructive of sheep diseases, to this absence or paucity of, at least, certain mollusca.

I was first put in this pursuit upon the slimy trail of the slugs and snails specified by various well-known facts which it is here unnecessary to specify. The number of the mollusca which it is necessary to trouble about appears to me to be very distinctly and very conveniently limited by the fact pointed out by the late Willemoes-Suhm, one of the 'good company of famous knights' upon H. M. Challenger, as to the Faroe Islands. The Faroe Islands are afflicted by the rot, but they have only eight snails and slugs, all told, out of which to choose the guilty party or parties. These are, as enumerated in Siebold and Kölliker's 'Zeitschrift' for 1873 (vol. xxiii. p. 339), *Limax agrestis, Limax marginatus, Vitrina pellucida, Hyalina alliaria, Limnaeus pereger, Limnaeus truncatulus, Arion ater, Arion cinctus.* That the fluke of the Faroe Islands sheep spends a considerable part of its life as a parasite in one or more of these mollusca admits of about as much doubt as the statement that the 'giddy' disease—the 'sturdy,' as the Lake District shepherds pronounce the French word *étourdi*—of the sheep is similarly dependent upon a to-and-fro shuttlecock alternation of one animal between two others. I do not say that it is necessarily in one, and one only, of the specified eight mollusca. Undoubtedly the fluke, like other parasites, may, in its sporting tour, infest many hosts. As a matter of fact, this particular fluke (*Fasciola hepatica*) in its adult stage infests some dozen mammals beside the sheep and ourselves. But, as a matter of speculation, I incline to think that the fluke in its younger days is, like some other animals, a little, or indeed a good deal, more particular as to where it lodges than it is in later life. As a matter of practice at any rate there is no need to tell farmers to be on their guard against snails which do not infest their pastures, and of the eight just specified they need usually in England only look to the black slug and the gray slug. Willemoes-Suhm suspected the gray slug, I suspect the black slug; partly on account of its very wide distribution in space (it having been found as a 'spectre noir' in Prince Jerome Bonaparte's voyage to Jan Mayen's island), partly from the facts furnished to me by Mr. D. Gresswell, M.R.C.V.S., of Louth, to the effect that sheep which have been feeding on turnips harbouring 'black jacks' will die of rot even when shifted on to salt marshes. Any inhabitant of the Shetland Islands who will inform us that sheep-rot does infest his country, or any inhabitant of Siberia who will inform us that it does not infest his, will settle the question, in a preliminary way at least, in favour of the black slug. For 'this very common, beautiful, and exceedingly variable slug,' as Forbes calls it, is not found in Siberia, and the *Limax agrestis* is; while in the Shetlands the case of distribution is precisely reversed. Whoever will furnish us with the information required will be indeed a 'true farmers' friend.'

Anyhow, the line of prevention indicated in my previous letter of

April 7th[1] is the same whichever animal is the one to be blamed, or if both are. And in confirmation of what I there recommend let me say that rot is comparatively rare in the Lake district, while the pasturage of ducks and geese on marshy ground is a branch of agricultural industry greatly developed. It was with much pleasure that I saw two large flocks—one of ducks, the other of geese—*échelonned* some hundred yards or less apart on such ground there. They will keep down both black and gray slugs as fast as they appear, without, so far as is known, incurring any danger from this fluke.

[1] [In this letter he recommended the employment of birds, especially ducks, for the destruction of the snails. As sheep lick up snails for the sake of the relish which their salt taste imparts, he recommended that blocks of rock salt should be placed convenient for the sheep to get at.—EDITOR.]

[The publication of the above as a letter in the 'Times' of April 14, 1880, induced the Royal Agricultural Society of England to offer a grant to Dr. Rolleston for an investigation into the life-history of the liver-fluke. Dr. Rolleston was unable to undertake the research, but recommended to that Society one of his pupils and demonstrators, Mr. A. P. Thomas of Balliol College, who commenced the investigation on June 7, 1880. The results of his enquiry have been incorporated in preliminary reports to the Royal Agricultural Society ('Journal,' vol. xvii, 1881, p. 1; vol. xviii. 1882, p. 439), and in 'Nature,' vol. xxvi. p. 606. Mr. Thomas communicated more elaborate memoirs to the Royal Agricultural Society of England, October, 1882, printed in their 'Journal,' vol. xix. S. S. part 1, under the title 'The Natural History of the Liver-fluke and the prevention of Rot;' and to the 'Quarterly Journal of Microscopical Science,' January, 1883, under the title 'The life-history of the Liver-fluke (*Fasciola hepatica*).' The conclusion at which he has arrived is that the *Lymnaeus truncatulus* is the only English mollusc which can serve as intermediate host to the liver-fluke, though it is possible that in other countries some other mollusc may be the intermediary. It was at one time thought that, although *Fasciola hepatica* occurred in Australia, the genus *Lymnaeus* did not exist there. Mr. W. Hatchett Jackson has directed my attention to a paper by Mr. E. A. Smith, 'On the Fresh-water Shells of Australia,' in the 'Proc. Linnaean Society (Zoology),' vol. 16, p. 255, in which he enumerates eleven species of *Lymnaeus*, one of which, *L. victoriae*, is almost identical with *L. truncatulus*.—EDITOR.]

XXXIII.

NOTE ON THE GEOGRAPHICAL DISTRIBUTION OF *LIMAX AGRESTIS*, *ARION HORTENSIS*, AND *FASCIOLA HEPATICA*.

THAT some not inconsiderable confusion exists as to the question of the existence of *Arion ater* and *Limax agrestis* in Greenland, will be seen from the following quotation, to be found as hereinafter specified ('A Manual of the Natural History, Geology, and Physics of Greenland, together with Instructions for the use of the Arctic Expedition.' 1875. London):—

(*P.* 124.) '*Mollusca Groenlandica:*

Classis i. ANDROGYNA, Mörch.

Ordo i. GEOPHILA, Fér.

* 1. *Arion fuscus*, Müll. Probably introduced. *L. agrestis*, L., according to Wormskiold.

The species marked with an * are doubtful inhabitants of Greenland.'

Prefixed to the list whence the above passage is taken is a note to the effect that the list is the 'Prodromus Faunae Molluscorum Groenlandiae (in Rink's 'Grönland,' &c., 1857, pp. 75–100). By Dr. O. A. L. Mörch. Revised and augmented by Dr. O. A. L. Mörch, University Museum, Copenhagen. April, 1875.'

On referring, however, to the Prodromus itself, as published in Danish in 1857, I find the entry which concerns us stands simply thus:—

'*Mollusca Grönlandica:*

Order i. GEOPHILA.

Gen. i. *Limax*, L.

* 1. *L. agrestis*, L. (ifolge Wormskjold).

* Betigner at Artens Forekomst paa Grönland ikke er sikker.'

That is to say, that the line in the entry given in the Manual of 1875—'**Arion fuscus*, Müll. Probably introduced'—is altogether something fresh and new; whilst the asterisk, denoting that the animal so marked is possibly not indigenous, was removed from the Limax agrestis, and prefixed to the curious name 'Arion fuscus, Müll.'

It is difficult to understand how the late Dr. O. A. L. Mörch can have come, in 1875, to alter his previous entry in this manner. For the name 'Arion' was unknown to Müller, the author of the 'Historia Vermium,' having been introduced into malacology by Férussac, as he himself tells us[1]; and as regards the animal itself, on the supposition that Dr. Mörch, by his entry '*Arion fuscus*, Müll.,' intended to have written '*Limax fuscus*, Müll.;' and knowing that this Limax, so called by Müller, was really an *Arion* (*hortensis*), and not a slug with a posteriorly placed respiratory inlet and a continuous shell, it is still more difficult to see how he could have added the word *L. agrestis*, L., apparently as a synonym. For in the thirteenth edition of the 'Systema Naturae,' tom. i. pars vi. pp. 3101–3102, the (true) 'Limax agrestis' is distinguished from the '*Limax fuscus*' (=*Arion hortensis* hodie) of 'Müller, Hist. Verm. ii. p. ii. n. 209.'

On referring to Dr. O. A. L. Mörch's 'Faunula Molluscorum Islandiae,' communicated on the 13th April 1866, and published in 1868, in Danish, in the 'Vidensk. Medd. fra den naturhist. Forening i. Kbvn,' pp. 185–227, I find at p. 196, 3, that '*Limax agrestis*, L.' stands with a ? after its name, even though there can be no doubt from references to Olafsen, several of which are, in fact, given by Mörch, that *a* gray slug, as well as the black slug, *Arion ater*, exists in Iceland. And a suggestion at the end of the entry, to the effect that the specimens may possibly belong to the species *Limax tenellus*, appears to explain the presence at the beginning of it of a ? after the words *Limax agrestis*.

Perhaps, therefore, the true explanation of the entry in the Manual of 1875 is as follows. In the interval between 1857 and 1875 a black slug may have been proved to Dr. Mörch's satisfaction to have been found in Greenland, and he may have identified it as the *Arion fuscus* of Moquin-Tandon, which is the same as the *Arion hortensis* of Férussac, *and* as the *Limax fuscus* of Müller and Linnaeus; and he may, by a very slight slip, have entered it as '*Arion fuscus*, Müll.,' instead of '*Arion fuscus*, Moquin-Tandon,' or '*Limax fuscus*, Müll.' To his addition 'Probably introduced,'

[1] 'Hist. Nat. des Mollusques,' ii. 1820–1851, pp. 23 and 54.

some objection might be taken on the ground that there is no very strong *a priori* reason why an Arion should not exist in Greenland, considering that it exists in Iceland, the land shells of which Mörch himself[1] allows are nearly allied to those of Greenland, and is not only an acknowledged member of the circumpolar fauna[2], but the most abundant of all slugs in Finmark and Lapland. It is curious—and not only curious, but in view of the question of the distribution of *Fasciola hepatica* also important—to note what follows. In 1875 Dr. Mörch appears, after thus adding *Arion hortensis* to his former list of Greenland mollusca, to have been content to leave the entry of '*Limax agrestis*, L., according to Wormskiold,' untouched, though in smaller type, feeling probably that as the entry of the animal was overtly made only on the authority of Wormskiold, he was in no way pledged either to holding that it was *Limax agrestis*, and not *Limax tenellus*, which existed in Greenland, or indeed to holding that any Limax whatever existed there. What completes my case is the fact that in 1877, when preparing a list of the Greenland mollusca for the English translation of Dr. Rink's 'Grönland,' of 1857, Dr. Mörch omits all mention of *Limax agrestis* altogether, and his entry runs as follows (p. 436):—

'Class i. ANDROGYNA.
Order i. GEOPHILA, Fér.
1. *Arion fuscus*. Probably introduced.'

If we follow Dr. Mörch, therefore, we shall strike *Limax agrestis* out of the list of Greenland mollusca, and hold that *Arion hortensis*, which exceeds it in number in other circumpolar regions, has in Greenland displaced, or at any rate replaced, it altogether.

If, however, *Limax agrestis*, notwithstanding the advantage which

[1] See Manual, p. 135.

[2] Middendorff, indeed, in his 'Sibirische Reise,' ii. 1851, p. 419, omits the name of this small slug from his list of Circumpolar Freshwater and Land Molluscs, but five pages farther on, l. c., says in a note, 'Vielleicht ist *Limax* (Arion) *sub-fuscus*, Drap. (Drap. "Moll." p. 125, pl. ix. 8; *Limax fasciatus*, Nillsen, "Hist. Moll. Suec." 1822, p. 3) eine circumpolare Art dieses Geschlechtes;' and he proceeds to note its discovery by himself within the polar circle in Finland, feeding on sphagnum, as also in Lapland, feeding on fungi, up to 69° N. Lat. Schrenk ('Reise in Amurlande,' 1859–1867, ii. p. 692), whilst identifying the *Limax sub-fuscus* of Draparnaud with the *Arion hortensis* of Férussac, and so with the *Limax fuscus* of Müller and Linnaeus, confirms the view as to its circumpolar character, and uses it as an argument for its being indigenous in America.

its coloration might be supposed to have been likely to give it, is beaten in the struggle for existence in circumpolar districts by *Arion hortensis*, of about the same size, but of such different colour in other districts, if not in the North[1], as not only to have been called *fuscus* and *sub-fuscus*, but even to have been confounded with the true *Arion ater* (from which, indeed, it is mainly distinguished by its more mesially placed respiratory orifice and its small size), it surpasses *Arion hortensis*[2] in more southern latitudes.

Middendorff indeed expressly says, l. c.: 'In Siberien traf ich diesen *Limax* (*Arion hortensis*) nicht, sondern nur einen einzigen kleinen *Limax* in *Starowoj* Gebirge, welcher dem *Limax agrestis*, L. recht ähnlich sehe.' But this absence from Siberia, to which F. Schmidt's silence as to its presence bears some testimony, may be paralleled by the similar absence of *Paludina vivipara* (Middendorff, l. c., p. 426) and of crayfishes from the Siberian river basins[3], and, as in those two cases, when compared with the facts of a distribution elsewhere does not disprove a circumpolar character.

Gerstfeldt, 'Mém. Sav. Etrang. St. Pétersbourg,' 1859, 515 (11), refers to some few, small, ill-preserved specimens, 'einige wenige kleine und schlecht erhaltene Exemplare' of slugs from Irkutsk and Wilni and from the Amur, and speaks of them under the name *Arion ater*. Their small size may justify us in supposing them to have been *Arion hortensis;* and the bad state of preservation in which they were, and which makes Gerstfeldt himself speak doubtfully of his identification, p. 535 (31), makes this note of their presence less authoritative than it otherwise would have been, and has caused Schrenk to suggest that they were in reality specimens of *Limax agrestis*.

An illustration of the paucity and rarity of *Limax agrestis* in circumpolar regions is furnished by the entry made by Friedrich Schmidt in his list of Animals from the Region of the Lower

[1] Even in England, where the *Arion hortensis* is often of a 'deep blue-black,' and is, I suspect, the 'Black Jack' of agriculturists, it is not rarely 'yellowish,' sometimes 'gray or greenish-gray' (Lovell Reeve's 'British Land and Freshwater Molluscs, p. 11). In Amoorland it is 'graugelblich,' with three stripes, one dorsal and two lateral narrower ones; whilst its rival the *Limax agrestis* is described as 'hell-bräunlich-oder bläulich-grau.' See Schrenk, l. c.

[2] See Schrenk, 'Amurlande,' ii. 690-693, 1869; Middendorff, 'Sibirische Reise,' ii. p. 424, 1851.

[3] See Huxley, 'On Crayfishes,' p. 305.

Yenisei, 'Mém. Acad. St. Pétersbourg,' 1872, p. 48, as to this eminently social mollusc: 'In einem faulen Treibholzstamm auf den grossen Brjochow Insel (70° N. Br.) *in einem Exemplar* gefunden.' But, *per contra*, in Amoorland, Schrenk tells us, l. c., that *Limax agrestis* outnumbers *Arion hortensis*, just as *Arion hortensis* outnumbers *Limax agrestis* in Sweden, Finland, and Lapland, and that while *Limax agrestis* spreads into Spain, Portugal, Italy, Algeria, and the southern slopes of the Caucasus, *Arion hortensis* reaches no farther south than the southern slopes of the Pyrenees and Alps.

In a letter published in the 'Times,' April 14, 1880, and republished with certain omissions in the 'Zoologische Anzeiger,' May 24, p. 258–260 (Article XXXII), I suggested that *Arion ater* may be the '*Zwischenwirth*,' or one '*Zwischenwirth*,' to *Fasciola hepatica*. For, calling the small black slug upon the distribution of which I have, following Schrenk and Middendorff, just been writing, '*Arion ater*,' I have the example and authority of Forbes and Hanley, and I think that of Gerstfeldt. But now, following Schrenk more closely, I should call it *Arion hortensis*, and should wish to be understood to be of opinion that it will—as I hope, by means of experiments now being carried on in my laboratory by Mr. A. P. Thomas—be ultimately shown that the smaller of our two British Arions really is one at least of the hosts infested by the sheep-fluke, *Fasciola hepatica*[1].

As regards the distribution of the *Fasciola hepatica* in northern regions we have the authority of Leuckart, 'Die Menschlichen Parasiten,' i. p. 531, 1863, for saying that it is found in Greenland and North America; and the same excellent authority quotes (l. c., ii. p. 870, 1876) Krabbe to the effect that it is not found in Iceland. The last statement is confirmed by Jonsson in 'Deutsche Zeitschrift für Thiermedicin und vergleichende Pathologie,' Bd. v. Heft vi. 1879, p. 413, in the words 'Leberegeln kommen in Island nicht vor.' I wish to add that there is no mention of the disease which *Fasciola hepatica* causes in Olafsen's and Povelsen's two volumes of 'Travels in Iceland,' though the diseases of sheep are repeatedly treated of by those authors[2]. And a similar remark may be made

[1] [From the editorial note appended to the immediately preceding Article, it will be seen that Mr. Thomas's experiments resulted in the conclusion that *Limnaeus truncatulus* is the intermediate host.—EDITOR.]

[2] See German translation published in 1794, i. pp. 112–280; ii. pp. 46, 198, 199.

as to Siberia; neither Middendorff, nor Radde, nor the great Pallas, treating as they do so exhaustively of the natural history of that region, ever within my knowledge make any allusion to the existence there of *Fasciola hepatica* as a cause of sheep disease. As regards, however, the existence of this animal and of the sheep-rot in Greenland, as testified by Leuckart, I wish to lay alongside of it the following statement from the English translation of Rink's 'Greenland' already referred to, and edited by Dr. Robert Brown in 1877. There, p. 97, it is stated that about the year 1855 there were in the whole of Greenland only from thirty to forty cows, a hundred goats, and twenty sheep, and that this handful of cattle were located at Julianshaab, on the west coast. A statement to the same effect is given by Dr. Brown himself in the 'Manual of Arctic Instruction,' 1875, p. 27. Surely if the rot still exists in Greenland, and has not shared the fate of so many other forms of life which have finally left its inhospitable shores, we have in Julianshaab a simple case and a circumscribed area wherein to prosecute research.

If the presence of *Fasciola hepatica* in an isolated locality—that of Julianshaab, on the west coast of Greenland—is likely to prove instructive, its absence from Iceland may also throw some light upon the subject. Most or all of the mollusca which have been or can be supposed to act and suffer as *Zwischenwirth* for the *Fasciola* are to be found in Iceland, viz. *Arion ater*, *Arion hortensis*, *Limnaea truncatula* and *Limnaea peregra* [1], as well as *Planorbis rotundatus*, if not *Planorbis marginatus*. And that abundant opportunities for the introduction of *Fasciola hepatica* into Iceland have been given by the importation of sheep from abroad is learnt from what Olafsen, l. c., ii. pp. 198–199, tells us as to the ascription of another sort of sheep disease to such importation.

I incline to ascribe this immunity from rot which the sheep enjoy in Iceland to the habit which they in common with the Shetland and Orkney sheep have of feeding between high- and low-water marks upon the sea-weeds specified by Olafsen in various passages, q.v., l. c., i. 233, 279, ii. 198, and Low, 'Domestic Animals of Great Britain,' p. 59. The *Fasciola hepatica* is a freshwater animal, and would not of course be picked up in such a locality as the interval between 'Ebbe and Fluth,' to which the sheep resort

[1] See Mörch, 'Faunula Molluscorum Islandiae,' 1868, pp. 12 and 16.

even on the dark nights of winter. It is possible to speculate as to the virtues of salt as an anthelminthic, and to suggest that it may act either by enabling a better gastric juice to be secreted, and so giving the sheep a better chance of digesting the larval *fasciolae* when swallowed, or by provoking a more copious flow of bile, and so washing the young fluke out of the gall-ducts. This, perhaps, is not the place for such enquiries. But it is a pure natural history fact that localities rich in deposits of salt are favourable to the growth and health of sheep[1]. Pallas, in the wonderful eleventh Fasciculus of his 'Spicilegia Zoologica,' dwells on this in reference to the Steatopygous variety of the domestic sheep at pp. 65–67; and with reference to the Argali, the *Ovis fera Siberica,* supposed to be the parent stock of *Ovis aries*, var. *domestica*, he writes thus at p. 12: 'Omni vero tempore ubi possunt loca salsagine rorida quibus universa Siberia abundat crebro frequentant, terramque sale foetam cavant quod cervino quoque generi solemne est.'

[1] [The importance of the recommendation of salt as an anthelminthic has now been practically tested. Mr. A. P. Thomas has proved that common salt in very small proportions is fatal to the redia and sporo-cyst of *Fasciola*, as also to *Limnaeus truncatulus*, which serves as host to the intermediate forms of the liver-fluke. Mr. T. P. Heath has published an experiment in the 'Western Morning News,' Oct. 14, 1882 (cited in Mr. Thomas's paper in 'Journ. Royal Agric. Soc. of England,' vol. xix. S. S. part 1), in which sheep fed on permanent pastures with salt mixed with oats were quite sound, whilst others to which salt was not administered were affected with liver-flukes.—EDITOR.]

ARCHÆOLOGY.

XXXIV.

RESEARCHES AND EXCAVATIONS CARRIED ON IN AN ANCIENT CEMETERY AT FRILFORD.

THE paper which I have the honour of laying before the Society of Antiquaries was drawn up by me at the suggestion of J. Y. Akerman, Esq., F.S.A., and in the hope that it might serve as a continuation of his 'Report of Excavations in an ancient Cemetery at Frilford, near Abingdon, Berks,' which may be found in the Society's Proceedings for May 25, 1865.

During the years 1867 and 1868, I have from time to time, by the kindness of William Aldworth, Esq., the owner of the soil, been allowed not only to watch such quarrying operations as have been carried on upon the site of this cemetery, but also to conduct some excavations there independently of that work. The results of my observations I have arranged under two heads. First, I have given an account of the objects and discoveries of a purely archaeological character; and, secondly, I have specified the various conclusions to which my examinations of the very extensive series of human remains have seemed to me to point more or less doubtfully. Appended to this paper will be found, first, a detailed catalogue of all the very numerous objects, both of archaeological and of anatomical interest, which the liberality of Mr. Aldworth has transferred to the University Museum; secondly, a tabular catalogue, giving in one view the number, the age, the stature, and the nationalities of the human remains; and thirdly, a *catalogue raisonné* which presents a similar *coup d'œil* of the different objects of cardinal importance, which have served as fixed points for my various identifications.

Mr. Akerman's investigations had, as may be seen by referring to his paper already cited, led him to the conclusion that Roman or

Romano-British and Anglo-Saxon interments were both alike to be found in the Frilford Cemetery, but that the majority of them belonged to the latter of the two nationalities. I have, however, by the discovery of Anglo-Saxon cinerary urns placed superficially to the relicless graves of which Mr. Akerman speaks, been compelled to refer these inhumations to a period anterior to that of Pagan Saxondom, and to differ herein from the instructor from whom I have learned and to whom I owe so much. It is upon this discovery of Anglo-Saxon cremation urns, containing half-calcined human bones, and holding when discovered, relatively to relicless or all but relicless skeletons found in the ground below them, a position from 15 to 18 inches nearer the surface, that I rest almost the only conclusion to which I have ventured to come in opposition to Mr. Akerman's views. But it is hoped that a record of the somewhat extensive series of observations made in this cemetery during the last two years may serve to cast some light upon certain moot points upon which Mr. Akerman's investigations did not give him an opportunity of remarking.

The cemetery is situated in the angle intercepted between the left bank of the river Ock and the road leading from Frilford to Wantage. Frilford 'Field' is now brought under cultivation, but the tradition that this portion of it is haunted still survives in the recollections of the rustics, one of whom informed me that, though he had never seen them there himself, ghosts were supposed to be particularly likely to be seen at a single thorn-bush[1] which stood, some time back, close to the site of these graves. Great numbers of Roman coins have been and still are found by labourers engaged in ordinary agricultural work all round this spot; and fragments of very many varieties of Roman pottery are equally accessible, though, of course, much more abundant, on and in the superficial

[1] The growth of this thorn-bush may have been accidental here, but we know that thorns were purposely planted on tumuli. (See Jacob Grimm, 'Verbrennen der Leichen,' Berlin Abhandl. 1849, pp. 203, 209, 242, 244; Nillson, cit. in loc.; Max Müller, 'Zeitschrift Deutsch. Morgenländ. Gesellsch.' ix. 11; Theocritus, Idyll xxiv. 87, where Wüstemann remarks in his commentary, 'Omnibus spinarum generibus vim noxarum depellendarum inesse existimabant veteres.' See also 'Horae Ferales,' p. 69.) The neighbouring tumulus known as Barrow Hill is beset with thorn-bushes at the present day; and the British barrow of Dinnington, in South Yorkshire (p. 159), on the estate of J. C. Athorpe, Esq., was similarly clothed. The thorn may have belonged to the 'certis lignis' used, according to Tacitus, 'Germania,' xxvii, in the cremation of chiefs.

layers of the now cultivated fields. There is much more evidence to show that Roman civilisation had taken firm root in this locality, and some of this evidence will appear in the course of my account of the excavations of the cemetery. But two excavations which we made in two spots, about a couple of hundred yards distant from the cemetery, gave us a more vivid idea of the wealth and civilisation of the Roman or Romano-British inhabitants of the place, which their Saxon conquerors named Frilford, than anything which we found in the burial ground, which both races successively occupied. Mr. Aldworth had observed the greater greenness and strength of the crops upon these two patches of ground; and by his suggestion I dug into them with the result of finding [1], for a depth of ten feet or more, an aggregation of fragments of pottery of the most varied patterns and degrees of fineness mixed up with similarly fragmentary bones of the ox, sheep, pig, and dog, and with other articles, such as knives and coins, which, like the bones and shards specified, would be expected in the rubbish-heap of a great house. The site of this great house I have not found; but I strongly suspect that the quarry, whence the stones for its construction were taken, was employed for, and is now represented by, one or other, or both, of those pits of rubbish. This short history illustrates the truth of a remark recently made by the Hon. W. O. Stanley [2] as to the imperfection of 'the investigation of sites and of dwellings in the early times;' but time and opportunity may enable me to supply this deficiency. In the meantime, the discovery in the cemetery of four interments in leaden coffins, and after the Roman fashion, so fully described by the Abbé Cochet [3], furnishes additional evidence as to the character of the civilisation existing here in the times of the Later Empire, which the excavation of hypocausts and tessellated pavements might confirm, but cannot be thought necessary to complete.

Four other kinds of interment, one Romano-British and three Anglo-Saxon, have been observed and described in the following account of the excavations at Frilford. The Romano-British interments differ from those just mentioned merely in being of less expensive character; they constitute the greater part of all

[1] See Catalogue, infra, Sept. 24, 1868.
[2] 'Ancient Interments and Sepulchral Urns in Anglesea,' p. 19.
[3] 'Normandie Souterraine,' pp. 29, 30.

the interments I have examined at Frilford, and that they are Romano-British is, to omit for the present other evidence, proved by the fact that superficially to them in the soil I have found Anglo-Saxon urns containing burnt human bones, and belonging, therefore, to the first periods of Anglo-Saxondom in England. About half of the Anglo-Saxon interments discovered here were interments in the way of cremation. The other half are cases of inhumation with the well-known Anglo-Saxon relics, and, in adopting inhumation, the Anglo-Saxons either dug shallow graves without regard to the points of the compass, independently of, though often superficially to, those of their conquered predecessors; or, secondly, they dug deeper graves pointing to or towards the East, following thus Christian precedent both as to depth and as to direction, but diverging from the practice of the Romano-Britons in setting stones round the graves instead of protecting the body in a wooden or other coffin; and whilst doing this, they sometimes—all supposed scruples as to secondary interments [1] notwithstanding—displaced one body, probably that of one of their predecessors, to make room for the corpse they were interring with the same orientation. I say it is probable that where an Anglo-Saxon skeleton is found to have displaced another set of remains, the primary interment was a Romano-British one, because I think it improbable that the half-heathen custom of interring with insignia should have been combined for a sufficiently long time with the Christian method of deep and oriented interment to allow of one body thus interred being sufficiently forgotten to be safely displaced. Burial with insignia was early discontinued by Christianised populations, except in the cases of distinguished personages ecclesiastical and temporal [2], and the Anglo-Saxons I have exhumed do not appear by their insignia to have belonged to either of these classes.

[1] The Abbé Cochet, in the first edition of his 'Normandie Souterraine,' p. 185, had stated that 'l'usage d'enterrer plusieurs fois au même endroit est éminemment moderne;' but in the second edition of that work, pp. 209, 432, 436, and also in the 'Tombeau de Childeric,' p. 55, he has receded from this untenable position. Grimm, towards the conclusion of his paper, 'Ueber das Verbrennen der Leichen,' ubi supra, p. 269, quotes the words of Sidonius Apollinaris, 'Jam niger caespes ex viridi, jam supra antiquum sepulchrum glebae recentes,' to show that the practice was only too well known to the Christians of the later Roman Empire. See also Friedr. Simony, 'Die Alterthümer Halstatter Salzberg,' Wien, 1851.

[2] See 'Capitularia Regum Francorum,' ii. 852.

In all the inhumations which I have examined at Frilford, the bodies had been extended at full length, and in the cases of Romano-British burials more or less oriented. The fact that the deviation from orientation is usually towards the south may seem to indicate that the majority of deaths took place then, as now, in the winter-quarters of the year, when the point in the horizon at which the sun would rise would be south of east [1].

I. *Of the Roman Interments in Leaden Coffins discovered at Frilford.*

By a reference to Mr. Akerman's paper already quoted, it may be seen that two leaden coffins, each of which contained a skeleton, and one of which contained a coin of Constantine the Great also, were found in the Frilford cemetery in the autumn of 1864. The commencement of my researches in this cemetery dates from the discovery in it of a third and fourth coffin of similar character and contents to these, in the month of January, 1867. These interments were near to each other, ten feet only intervening between the foot of the one and the head of the other grave. The direction of the graves was 45° south of east, which, when corrected for the magnetic variation, would give E.S.E. as the true bearing. The coffins were at a depth of about five feet below the present surface of the soil, and this greater depth, as well as their greater intrinsic costliness, would seem to show that their tenants had been persons of greater wealth and consideration than the occupants of the similarly oriented graves of which we shall have to speak next. The length of the coffins is 6 feet 4 inches, and their breadth 1 foot 6 inches. Both of the coffins have undergone much mechanical change in the way of contortion and crushing, and they contrast herein to disadvantage with certain coffins of the same period in the British Museum, and in the Museum of Antiquities at York, which still retain the form which was conferred upon them at their manufacture [2]. The Frilford coffins have also undergone much

[1] Cf. Abbé Cochet, 'Normandie Souterraine,' ed. i. pp. 192, 193, 255, 265.

[2] The leaden coffins to be seen in the British Museum were dug up in Camden Gardens, Bethnal Green, in the excavations for the New Docks at Shadwell, and in Whitechapel. For the coffins in the York Museum, see Professor Phillips's 'Yorkshire,' p. 247, and 'Descriptive Catalogue of Antiquities in York Museum,' by the Rev. C. Wellbeloved, p. 77, and his 'Eburacum,' p. 112.

chemical change, the metallic lead having been changed both on their exterior and throughout their substance into the red oxide and carbonate, whereby they have suffered great loss of plasticity and flexibility. Each of them possessed a lid, which appears to have been simply laid upon the top of the rectangular coffin proper without any soldering. Large nails with square heads were found in relation with the coffin, and as woody fibre, shown by microscopic examination to be probably oaken, is still plainly enough to be detected upon the urn, even with the naked eye, it would seem that the leaden coffin had been surrounded by a wooden one[1]. An analysis of the substance of these coffins, which I owe to the kindness of Heathcote Wyndham, Esq., M.A., Fellow of Merton College, shows that it contains 3·28 per cent. of tin, and that the coffins resemble in this, as in other particulars, those described by the Abbé Cochet in his 'Normandie Souterraine,' pp. 28–31, as characteristic of the Gallo-Roman period in France. In each of these coffins was found the skeleton of a strong man, who was at the time of his death considerably past the middle period of life. Of the anatomical characters of these skeletons I shall have to speak in detail later; it is sufficient to say here that they show that the individuals to whom these bones belonged were strong men, in the possession of the means for culture and comfort which those days could afford, but who had also suffered much from the physical and other inclemencies which we know to be the natural incidents of the life of the soldier. In one of these coffins five coins were found, of which one was a coin of Constantine the Younger, another of Valens, and a third, which, like the first, was a third-brass specimen, was a coin of Gratian. By means of this last coin we are enabled to say that this interment took place, in all probability, within the short but eventful period which elapsed between the accession of Gratian and the evacuation of Britain by the

[1] This conclusion almost rises to certainty when we read the account given by Ralph Thoresby, 'Phil. Trans.' 1705, No. 296, p. 1864, of the excavation of a coffin, 'probably interred 1500 years ago,' which was seven feet long, and was 'inclosed in a prodigious strong one made of oak planks, about two inches and a half thick, which, beside the riveting, were tacked together with brags and great iron nails . . . they are four inches long, the head not diewise, as the large nails now are, but perfectly flat and an inch broad.' The length of the Frilford nails is four and a half inches, and the breadth of their heads one inch and a quarter. See also L'Abbé Cochet, 'Normandie Souterraine,' ed. i. p. 33; 'Archaeologia,' vii. 376, 381; Bloxham's 'Fragmenta Sepulchralia,' p. 39.

Legions, inasmuch as the departure of the Romans may be reasonably supposed to have entailed the collapse of the civilisation and customs which they had introduced and supported [1].

II. *Of the Roman or Romano-British Interments without leaden, but in most cases, probably, with wooden coffins, and in semi-oriented graves.*

The second and most numerous class of interments that we meet with in this cemetery are found occupying parallel, or nearly parallel, rows of trenches, running, to speak generally, from a point more or less north of west to one more or less south of east, and containing, very commonly, besides the skeletons, bones and teeth of domestic animals (though not in the great abundance noted in other Romano-British cemeteries), fragments of charcoal, oyster-shells, shards, flints, and nails, with woody fibre adhering to them. In some of these graves coins were discovered, in addition to the other objects just specified. Now, we are not justified by the presence of any, nor, indeed, by the presence of all, of these peculiarities, in concluding that any interment is Roman or Romano-British, the imitative tendencies [2] of the Teutonic races having led them somewhat slavishly into copying the customs of the world they subdued, even in points relating to such matters as the burial of the dead. Each and all of the objects have been found all but in-

[1] For a note of a discovery of leaden coffins in the neighbourhood of other Roman remains, see Schaaffhausen, 'Die Germanische Grabstätten am Rhein,' 1868, p. 131.

[2] For the imitative tendencies of the Teutonic races generally, see Coote's 'Neglected Fact in English History,' p. 44; Worsaae's 'Primeval Antiquities of Denmark,' Eng. Trans. 1849, p. 140; Engelhardt, 'Denmark in the Iron Age,' Preface, p. viii; Von Sacken, 'Leitfaden zur Kunde des Heidnischen Alterthums,' p. 158; Wylie's 'Fairford Graves,' p. 30; Merivale's 'Conversion of the Northern Nations,' p. 92; Roach Smith, 'British Assoc. Report for 1855,' p. 145. For the presence of bones of animals and their teeth in Anglo-Saxon graves, see Wylie, l. c., p. 24; Akerman, 'Pagan Saxondom,' Introd. p. xvii. For that of charcoal, Wylie, l. c., 29; Akerman, 'Further Researches at Long Wittenham,' Archaeologia, vol. xxxix. For that of shards and flints, Douglas' 'Nenia Britannica,' pp. 10 and 34; Wylie and Akermann, ll. cc. For that of the *Portorium*, Lindenschmit, 'Archiv für Anthrop.' ii. 3, 1868, in review of Wanner's work, and in his own work, 'Die Germanische Todtenlager beim Selzen,' p. 51; Von Sacken, l. c., p. 154; Akerman, 'Proc. Soc. Antiq.' 2 S. iii. 165. See also Abbé Cochet, 'Tombeau de Childeric,' passim, and 'Normandie Souterraine,' p. 31.

differently in both Anglo-Saxon and Romano-British, in Frankish, and in Gallo-Roman graves. I was first convinced that these interments, more than fifty of which have been under examination at Frilford since I first became acquainted with the cemetery, contained the remains of Romano-Britons, and not of Anglo-Saxons, by the discovery of an unmistakeable Anglo-Saxon urn, about fifteen inches above a skeleton occupying one of these graves (No. vi. Sept. 1867). Two other skeletons, one of an old woman interred with three coins (No. iv. Jan. 9, 1868), and one of an old man (No. iii. April 1, 1868), were found subsequently occupying the same position relatively to similar Anglo-Saxon urns containing similarly burnt human bones. It is possible, however, to object to this apparently satisfactory argument; first, that the deeper-lying body may have belonged to a Christianised, and the cremation urn to an apostate, Anglo-Saxon's burial; or, secondly, that the cremation urn belonged to an Anglo-Saxon funeral which took place in the heathen pre-Augustinian period, but that it was carefully replaced, after having been disturbed, to make room for one of the same race who had died after the evangelization of Berkshire by Birinus. Both these objections—the former suggested to me by Mr. Akerman, and the latter by the reading of Mr. Roach Smith's letter in the 'British Association's Report for 1855,' p. 145 [1]—are, however, fully met by the discovery, on four different occasions, of Anglo-Saxon skeletons, verifiable as such by their insignia, and with no constant relation to the points of the compass, in the same relative position to these interments as that already described as being held by the cremation urns. (See infra, Catalogue, No. xviii. Feb. 8, 1868; No ix. Sept. 25, 1868, infra.) It is possible, though not probable, that an urn, even of much fragility and elegance, may have been replaced in its entirety, heavily laden though it was with its contents; but it is impossible to conceive that a similar pious painstaking can have laid out a disturbed skeleton a second time in the full and due proportions of the unarticulated bones possessed by the skeletons found lying superficially to the 'grave-row' interments of which I am speaking as Roman or Romano-British. The variation in the direction of the two bodies lying one above the other, the deeper being always the oriented one, excludes, of course, the possibility of their having

[1] See also 'Inventorium Sepulchrale,' Introd. p. xvi. and p. 8.

been interred at the same time, as after a battle bodies are buried one above another in trenches. The funeral feast, and the visit to the burial-place of a beloved relative, will account sufficiently for the presence of the teeth and bones of the domestic ruminants, and the pig, in these graves. In the Romano-British cemetery at Helmingham in Suffolk, which I had an opportunity of examining through the kindness of the Rev. George Cardew, relics of this kind were more abundant than I have found them to be in the Frilford cemetery. Oyster-shells were found in considerable abundance in both these cemeteries, as the other indications of Roman occupation would have led us, *a priori*, to expect. I may perhaps here say, that it does not seem clear to me that any great probability attaches to an argument for the heathen character of an interment from the discovery there of such evidences of a funeral feast as the bones of domestic animals. The instinct so beautifully alluded to by Wordsworth, in his well-known poem 'We are Seven,' has in itself nothing repugnant to the spirit of Christianity, though the actual practice at the grave-side may and often did degenerate from that of the 'little Cottage Girl'

> 'Who took her little porringer
> And ate her supper there.'

Scandal arose out of the abuse of the funeral feast; but, inasmuch as the Church in all ages has acquiesced in the retention by newly-made converts of customs which, though heathen in origin, may not have been intrinsically immoral, it is easy to understand how a custom intrinsically laudable may have been tolerated when kept within due limits. As to the actual practice being rife amongst Christians[1] the numerous denunciations and inhibitions issued relating to it afford very abundant evidence.

[1] The following passages may be cited in addition to those so often referred to from the 'Capitularies' of Charlemagne. In the collection of the Canons of the Greek Synods, by Martin, Bishop of Braga in Portugal, who died in 580, we find the following words, 'Non oportet, non liceat Christianis prandia ad defunctorum sepulchra deferre et sacrificari mortuis.' See the 'Corpus Juris Canonici,' where the passage is adopted as the text of Decretum Gratiani, De Consecr. dist. i. cap. 29, § 2, under the title 'Ex Concilio Martini Papae.' Hardouin, 'Acta Conciliorum,' &c. 1611, iii. 390, has printed Martin of Braga's Collection, and, according to the margin of his edition, this particular canon comes from the third Council of Arles, and not from a Greek source. See also Gretzer, 'De Funere Christiano,' to which work I owe the foregoing quotation, lib. iii. pp. 159, 164, 166, ed. 1611, where Ambrose, Augustine, Cyprian, Gaudentius, and Faustus the Manichee, may all be found deposing to the fact of the

A few bones of the dog and some teeth of the horse were found in some of the interments, but not in such numbers or positions as to make it at all probable that the former were the relics of a favourite animal interred with its master, or that the latter were remains which, in like manner, had been buried from similar, or from superstitious notions, or which had been the leavings of the practice of eating horseflesh which we know existed in those days in spite of the efforts of the Christian priests [1].

Fragments of carbonaceous matter are to be found in Romano-British as also in Anglo-Saxon and undoubtedly Pagan interments. It is a little hazardous to pronounce quite positively as to a piece of black woody tissue that it was put into the grave as charcoal; and that its blackness is not due to the 'eremacausis,' which it has been exposed to for so many hundreds of years. If, however, such matter be in masses of considerable size, which possess on fracture the peculiar lustre of charcoal, and if it have not been impregnated

funeral feast being abused by the Christians into an occasion of great licence. I do not happen to have met with any evidence to show that food or drink was put into the graves of the *early* Christians from any influence which any pre-Christian belief may have had upon them as to its possibly being of some use to the departed in the new world. This superstition was of course operative in the case of heathens, and amongst certain of the Scandinavian races (see Lubbock's 'Prehistoric Times,' p. 89) it has lasted even down to our own times. Weinhold tells us ('Altnordisches Leben,' p. 493) that the tobacco-pipe, pocket-knife, and filled brandy flask were placed in Swedish graves (it is to be supposed only in remote districts), if not up to the present time, at all events up to the beginning of the present generation. Heathen customs, however, and customs as markedly heathen as cremation, retained their vitality to a very late period in the Baltic regions. (See for this Grimm, loc. cit.; Wylie, 'Archaeologia,' xxxvii. 467; and Lindenschmit, 'Alterth. heidnisch. Vorzeit,' Heft ii. Bd. ii. ad Taf. vi., for long persistence of heathen customs amongst the Alemanni. See also Wylie, 'Graves of Alemanni.')

[1] For the interment of favourite animals with their masters, see Von Sacken, 'Heidnisches Alterthum,' 1865, p. 155; Weinhold, 'Sitzungsberichte Phil. Hist. Klass. Akad. Wien,' Bd. 29, p. 203, 1859. The bones of a large dog were found at Long Wittenham in a Romano-British interment so near to certain human remains as to make it seem possible that the animal had been purposely so placed. For the burial of the horse (*Das Trauer-Pferd*) in Teutonic graves, and those of other races, see Keysler, 'Antiq. Select.' 1720, p. 168; Wylie, 'Graves of the Alemanni,' Archaeologia, vol. xxxvi, ibique citata; Cochet, 'Normandie Souterraine,' p. 298. For the suspension of the skull of the horse over graves, see 'Pagan Saxondom,' p. 23. For the practice of eating horse-flesh, see 'Confessional of Archbishop Ecgbert,' c. 38; the Decrees of Council held A.D. 785, under the presidency of Gregory, Bishop of Ostia; and 'Penitential of Theodore,' c. xxx. s. 17. See also Lubbock, 'Prehistoric Times,' p. 115; Keysler, l. c., pp. 322, 340; and Pearson, 'History of England,' i. 138.

with any salt of iron or other mineral so as to have been preserved by such impregnation from the decay which would otherwise have befallen it, we are justified in considering it exceedingly probable that it was put into the grave in the condition either of yet burning embers, or of charcoal. The test mentioned by the Abbé Cochet, 'Normandie Souterraine,' p. 198 ed. i. (p. 229 ed. ii.), for differentiating charcoal from decayed wood, viz. that the latter gives a sherry colour on boiling with potash, is a little unsatisfactory, inasmuch as the purest charcoal would give a similar reaction after being surcharged and sopped through and through for ages with water, more or less laden, *ex hypothesi*, with impurities. Without losing sight of the possibility that blackened woody matter may be the remnants of a coffin, it is well to consider the different explanations which may be given of the presence of true charcoal in an interment. Four such have been given, two of which refer the practice to the operation of Christian beliefs; the third refers it to the working of feelings which are neither distinctly Christian nor yet distinctly heathen; whilst the fourth explanation is applicable to heathen interments only. The two first explanations may be expressed in two separate utterances of Durandus, the first being the often quoted one, vii. c. 35, as to the placing of embers and incense, *prunae cum thure*, in the grave; and the second, a few lines further on, speaking of a Christian practice of placing charcoal in the grave to serve there as an imperishable protest against using the soil of the grave thereafter for secular purposes, 'in testimonium quod terra illa in communes usus amplius redigi non potest; plus enim durat carbo sub terra quam aliud.' The third of the four explanations refers the presence of charcoal in the graves to the holding of feasts by their side in replacement of the pagan sacrifices of former times. The fourth explanation refers us to the overt and recognised performance, or to the stealthy continuance of the eminently heathen practice of burning the body or of lighting a fire in the grave to prepare it for the reception of the corpse. Any one or all of the three first explanations are admissible in the case of the Romano-Britons; the fourth may very probably apply to the interments of the half-converted or apostatising Anglo-Saxons, to whose history we shall return [1].

[1] For the discovery of carbonaceous matter in graves, see Cochet, 'Normandie Souterraine,' ed. i. pp. 198, 255, 256, 304; Kemble, 'Horae Ferales,' pp. 98, 104;

We are, from our recollections of the classical allusions to the *naulus* or *portorium*, strongly tempted to think that the placing of coins upon the corpse must have been a distinctively heathen practice. A curious passage which I came upon in Martene's great work ('De Antiquis Ecclesiae Ritibus,' ii. 374) has caused me to attach importance to the fact that, in two of the interments I have examined here, the number of the coins interred was five. One of these interments was the first of the two in leaden coffins described already, and the other was an interment of the class of which I am now writing, and will be found in the appended catalogue under the number xxiv. of Feb. 21, 1868. Martene's words are, 'Addit anonymus Turonensis:—Quidam sortilegi contra fidem agentes ponunt *quinque solidos* super pectus mortui, et in hoc imitantur morem gentilium qui in ore mortui ponebant denarium "ut habeat quem porrigat ore trientem."' I am not aware of any explanation having been offered for the selection of *five* as the number of the coins which the *gentiles*, or those who imitated them, placed in the grave. But such a passage as the one just quoted does not, even when taken by itself, justify us in considering an interment with coins to have been always an interment without the rites of the Christian Church. Many persons act *contra fidem* and *imitantur morem gentilium*, whom, for historical purposes at least, we must consider to be Christians. As probably in the case of placing of charcoal in the grave, so, certainly, in that of the placing of coins there, the Church exercised a wise toleration, protesting, it may be, more or less directly, by the introduction of such sentences as those which our Burial Service contains, against the thought that we can take anything with us out of the world, but acquiescing in the actual repetition and continuance of the custom. Just as the custom of placing earthen vessels in tombs has survived down almost to our own time in remote districts such as La Bresse and Morvan in France (see Cochet, 'Archéologie Céramique,' p. 1, 1860), so that of placing coins on the mouth and chest of the corpse is persisted in even to the present day in parts of the country similarly

Wylie, 'Fairford Graves,' p. 29; 'Graves of Alemanni,' p. 13; Schaaffhausen, 'Germanische Grabstätten am Rhein,' 1868, p. 104; Walder, 'Anzeiger für Schweiz. Alterthum,' March 1869, p. 32. For the discovery of fragments of charcoal scattered throughout the entire mass of heathen tumuli, see Keller, 'Mittheilungen der Antiquarischen Gesellschaft in Zürich,' Bd. iii. p. 66. For the use of charcoal as being imperishable, see Augustine, 'De Civ. Dei,' xxi. 4.

remote from the great centres of life. The fact, however, that money to the amount of no less than three hundred pieces of Roman coinage was placed in the tomb of Childeric is more conclusive than any mere speculation from the analogies furnished by ancient or modern times. A kindly instinct induced persons, who probably enough had never heard of Charon, to bury with their deceased friend or relative that which they knew him or her to have valued most, and the presence of coin in a grave may convey thus to us a satire upon the departed, which it was never intended to hint at. The Abbé Cochet seems to me[1] to lay too much stress upon 'la coûtume Chrétienne de rendre à la terre les hommes nus comme ils y sont entrés.' For this principle would have prevented the burial with ornaments[2], of which, however, we are told in the 'Capitularia Regum Francorum,' ii. 852 (cf. also p. 701), 'Mos ille in vulgo obsoletus in funeribus episcoporum et presbyterorum retinetur.'

In many of these semi-oriented graves nails with woody fibre still adhering to them were found, and from their presence, as also from that of a piece of coffin-hooping (see also Dr. Thurnam, 'Catalogue, Osteological Series, Royal College of Surgeons,' ii. 881, 5712) in one of these graves, we may argue with considerable probability for the employment of coffins in some, at least, of these interments. The custom of throwing shards, and flints, and pebbles into the grave is common both to Romano-British and to Anglo-Saxon interments in England. That it was pagan and even of very early origin seems probable, and that it persisted into Christian periods is pretty certain. Shakespeare's well-known lines[3] (Hamlet, v. 1) show, however, that its pagan origin had somehow or other so strongly impressed itself upon the public mind that it was no

[1] 'Normandie Souterraine,' p. 194. See also Keysler, 'Antiq. Select.' p. 174.

[2] See also the account of the plundering of the gorgeously-arrayed corpse of Pope Adrian I. in Mabillon, 'Museum Italicum,' i. 41; Gretzet, 'De Funere Christiano,' i. 28; Chrysostom, Hom. 84; Guichard, 'Funérailles,' 1581, p. 581, where the Council of Auxerre is said to have condemned 'toutes ces bobances.'

[3] Douglas, in his 'Nenia,' appears to be the first person who drew attention to the lines of Shakespeare, referred to, see p. 10, and also p. 34. For other references to the custom, see Keller, l. c. p. 65; Wylie, 'Fairford Graves,' p. 25; Akerman, 'Pagan Saxondom,' Introd. p. xvii; Weinhold, 'Sitzungsberichte Kais. Akad. Wiss. Wien. Hist. Phil. Klasse,' 1858, bd. 29, hft. i. p. 166; Fried. Simony, 'Die Alterthümer vom Halstatter Salzberg, Sitzungsberichte Kais. Akad. Wiss. Wien. Phil. Hist. Klasse,' 1851, p. 7; Keysler, l. c. p. 106; Rev. G. R. Hall, 'Nat. Hist. Trans. Northumberland and Durham,' i. 2, 1866, p. 167.

longer practised in Christian burials. They show also that the presence of these shards cannot be explained as being due to accident. Indeed, upon several occasions, I have found fragments of pottery in such relations to the bones of skeletons, in company with which nails were found, as to make it seem highly probable that the shard, when thrown in, must have clanked upon the boards of the coffin, which the nails show us was present there. The thought that our own custom of throwing earth into the grave during the burial service may be connected with this custom, and again, that both may be connected with the classical custom referred to in Horace's line, 'Injecto ter pulvere curras,' and also Virgil, 'Æneid,' vi. 365, and in Sophocles, 'Antigone,' 256, λεπτὴ δ' ἄγος φεύγοντος ὣς ἐπῆν κόνις, will at once suggest itself; but only to be dismissed on mature consideration; for to the modern antiquary it is no paradox to say that the custom of throwing in shards was probably much older than that of scattering earth over the corpse; and I would suggest, as it is very likely others may have done before me, that the throwing in of the broken pottery may be the perfunctory representation of the deposition in the grave of the entire vase, and that the throwing of earth, for which Archytas and Palinurus begged, may in like manner represent the toilsome but unattempted process of inhumation [1]. Massillon, long before prehistoric archæology had been thought of, argued for the conclusion that a belief in a future state is a naturally implanted conviction from the fact that 'nulle part vous n'en rencontrez des peuples sans sépultures et sans vases,' and the Abbé Cochet, in his 'Archéologie Céramique,' p. 1, says that the custom of placing earthen vessels in tombs is one of the most ancient of all customs, and, as just noticed, that it still exists in secluded and remote parts of France, as in Morvan and La Bresse. This coexistence with the custom of our modern burials seems to disprove any interdependence of the two practices. Again, the fact that fragments of pottery were used in interments by cremation, as well as in interments by inhumation, seems to show that the shard and the handful of earth were not set in motion by the same impulses. In very early times earthen vessels were of great value, and it was in those days a proof of at least

[1] For the tendency of customs involving expense to assume cheaper forms, see Sir John Lubbock, 'Nat. Hist. Rev.' Oct. 1861, p. 801; 'Prehistoric Times,' p. 98 ed. i, p. 142 ed. ii.

as great affection to bury or throw on to the funeral pile an earthen vessel as it was in after ages to burn his gorgeous insignia with Pompey[1]. I take this opportunity of quoting a passage from a curious work, the only one[2] of very many old books which I have looked through in the Bodleian and elsewhere for some passage parallel to the one quoted so often from 'Hamlet' in which I have found one. This book is entitled 'Funus Parasiticum, sive L. Biberii Curculionis Parasiti Mortualium, Ad ritum prisci Funeris, Auctore Nicolao Regultio, Lubeccae,' MDCXXXVII. In describing the imaginary funeral of the parasite whom he is satirising, the author uses the following words: 'Cum quisque certatim in rogum dona cumulat, et partim trullas, cantharos, lances, alii struices patinarias, cyathos, ciboria coquinaria, omnia flammae committunt.' It is obvious, of course, that the author may be representing the throwing in of these articles as being the most natural thing to do at the funeral of a glutton, as they had been his *instrumenta artis;* and Peniculus, it may be recollected, in the 'Menaechmi' of Plautus, i. l. 25, speaks in terms of unctuous affection of his hosts' *struices patinarias*—the very words employed by Regultius. Still, I am inclined to think that Regultius may have had some recollection, or at least some tradition, of the custom considered as so distinctively heathen by the priest in 'Hamlet' when he introduced this particular feature with so much iteration into his burlesque *ad ritum prisci funeris.* Writing at Lübeck, he may well have been familiar with the Baltic provinces further eastward, which the Teutonic knights had so much difficulty in civilising and Christianising.

Roots of plants had twined themselves about and around the bones contained in these graves, and the minute mollusc[4] *Achatina acicula* was found inside the skulls in such abundance as to make it

[1] See Lucan, ix. 175.

[2] Since writing as above I have met with the following passage in Keysler's 'Antiquitates Selectae,' p. 173: 'Inde Nimischae, in pago uno miliari a Gubena distante universus adparatus culinarius erutus, cacabi, ollae, catini, phialae, patinae, urceoli, lagenulae, testante D. *Christiani Stieffii Epistola.*' This Epistola was published in 4to. in 1704, and treats of 'Lignicenses atque Pilgramsdorficenses urnas.' See Keysler, loc. cit. p. 113.

[3] See Wylie, 'Archaeologia,' xxxvii. 467.

[4] See Schaaffhausen, 'Die Germanische Grabstätten am Rhein,' p. 125; and 'Collectanea Antiqua' (vi. 201), a work with which I was not acquainted when I wrote, as above, for an account of a cemetery at Kempston.

very evident that air and moisture had very free access even to the bottom of these graves, and consequently we should not be justified in arguing from the want now in many of these graves of any traces of such perishable materials as the wood and metal-work of a coffin, to the conclusion that no coffin had been put into them 1400 years ago. The wonder, indeed, is not so much that such substances should in some instances and in such circumstances have vanished, as that they should in any have persisted to the present day. Still I am inclined to think that evidence is not wanting to show that in some cases the Romano-Britons, like other races in ancient, mediæval, and modern times, interred their dead sometimes with, sometimes without, coffins. This evidence lies mainly in the fact that in some cases a large stone has been found so near the head as to render it difficult to think any coffin, however thin its walls, can have been interposed between the stone and the body. (See Catalogue, xv[3], Sept. 26, 1868; xvii[5], Sept. 26, 1868.) But even in these interments, where coffins may not have been employed, and which consequently so far resemble the Anglo-Saxon burials by inhumation shortly to be described, three important and easily recognisable differentiating peculiarities are present. First, stones do not appear to have been placed by the Romano-Britons under the head of the corpse, as they were placed in Anglo-Saxon interments, and consequently we do not find in the former, as we do in the latter so very commonly, the cervical vertebræ impacted along the base of the skull from the occipital foramen up to the symphysis of the jaw. Neither do the Romano-Britons, at least at Frilford, appear to have set stones along the sides of their graves, as the Anglo-Saxons did. Thirdly, the Romano-British graves, when recognised as such, in contradistinction to the Anglo-Saxon interments, by the help of these external peculiarities, are found to contrast with them in a point of even greater, as it is of more intrinsic, interest, viz., in the very large proportion of aged skeletons which they contain. The male Anglo-Saxon skeletons are invariably, or all but invariably, the skeletons of young men: quite the reverse is the case with the Romano-British. To this point, as resting upon anatomical evidence, I shall have to revert in the second part of my paper; it is sufficient here to say that the difference is just what would be observed now between the cemetery of a settled civilized Christian village and that of an outlying station on the border-land between some gradually advancing empire, and the territories of some gradually receding but intermittently aggressive aborigines.

III. *Of the Anglo-Saxon Interments in the way of Cremation.*

Ten urns containing burnt bones have come into my hands during the excavations carried on at Frilford. Of these two were patterned urns, and the rest plain. A fairly perfect patterned vessel from this cemetery is to be seen in the British Museum, and two patterned fragments have been recovered by me. These three latter vessels, I incline to think, on account of their size, may have been holy-water vessels rather than cremation urns. The pattern upon them, as well as that upon the patterned urns which were found with burnt bones inside them, is the pattern now so familiar to us as the Anglo-Saxon pattern, from the memoirs of Kemble, Akerman, the Honourable R. C. Neville, and others; and the general style and conformation of all the urns, patterned and plain alike, is not much less plainly referable to the same type. Neither class of urns has been lathe-turned; in none of them is the bottom perfectly flat; they are all of a darkish colour, and, though this colour may occasionally have a tawny streaking intermingled with it, it has usually been protected from reddening by the intermixture of vegetable matter with the paste. The figured urns possess the vandykes, the punched stellate or multiradiate stamps, the circular thumb-made depressions, the encircling zones scored with a pointed stick, and the 'characteristic bumps,' so fully and accurately described by Mr. Kemble in the 'Horae Ferales,' pp. 87 and 222, as distinguishing Anglo-Saxon urns found in England as well as urns found in the North-German fatherland[1].

The Frilford urns are, with the exception of those found at Long Wittenham, the first urns of Anglo-Saxon manufacture which I have seen recorded as found in Berkshire. Mr. Wylie[2] has put on

[1] See also for figures of urns resembling those found at Frilford; Engelhardt, 'Denmark in the Iron Age,' English translation, 1866, p. 9; urn from Smedeby, Slesvig; Akerman, 'Pagan Saxondom,' Introd. p. xxviii, and pl. iv; 'Archaeologia,' vol. xxxviii. pl. 20, fig. 1; Hon. R. C. Neville, 'Saxon Obsequies,' pls. 24-33; Bloxam's 'Fragmenta Sepulchralia,' p. 59; Roach Smith, 'Inventorium Sepulchrale,' Introd. p. xv. For the discovery of a bone-punch for stamping ornaments, see Schaaffhausen, 'Die Germanische Grabstätten am Rhein,' p. 139, 1868.

[2] 'Archaeologia,' xxxvii. 473.

record similar 'finds' from some thirteen English counties, to wit, Warwickshire, Nottinghamshire, Derbyshire, Northamptonshire, Yorkshire, Lincolnshire, Gloucestershire, Oxfordshire, Norfolk, Suffolk, Bedfordshire, Cambridgeshire, and the Isle of Wight. The 'Horae Ferales,' p. 229, enable us to add a fourteenth county, Sussex, to this list. An urn in the possession of the authorities of Queen's College, Oxford, and which a short note in the catalogue existing in their magnificent library may be taken as localising with some probability to Faversham, in Kent, gives us this county[1]—in which cremation, like the paganism with which it was correlated, was earlier superseded than elsewhere by Christianity—as a fifteenth in which Anglo-Saxons established themselves whilst still heathens. Berkshire makes the tale up to sixteen. When we consider how distinctively Christianity opposed[2] itself to the practice of cremation, every fresh discovery of these distinctively Anglo-Saxon urns shows us how thoroughly overrun our England was by the 'heathen of the Northern sea[3]' in the period which elapsed between the landings

[1] For the rarity of the discovery of cremation urns, at least in an unbroken, undisturbed condition, in Kent, see 'Inventorium Sepulchrale,' xv, xlvi. 184, 186; 'British Assoc. Report,' 1855, p. 146; and Mr. Wylie, loc. cit. The Queen's College urn is indubitably of Anglo-Saxon origin. The evidence for its coming from Kent amounts only to probability, and stands thus: in Queen's College Library there is a 'List of the Collection of Egyptian, Etruscan, Greek, Roman, British, and other antiquities, formed by the late Rev. Robert Mason, D.D., from the collections of Messrs. Belzoni Salt, Burton, Millingen, and others, 1822 to 1839.' In this Catalogue there is the following entry: 'Sepulchral urns, a large and small, 2.' On the smaller of these two urns, which, however, is of Roman manufacture, there is a ticket, 'Found at Faversham, Kent.' The exteriors of the two urns have much the same colouration or discolouration, which makes it seem likely that they came from the same excavation, and they were, consequently, as we now find them, catalogued and placed together.

[2] For the opposition of the Christians to the practice of cremation, see Neander's 'Life of Julian,' English translation, p. 108; Ibid. 'Minucius Felix,' cit. p. 45; 'Acta Martyrum,' Baron, ii. p. 290; Martyrdom of S. Tharacus; Tertullian, cit. Grimm, Berlin Abhand. 1849, p. 207; 'Ep. Ecc. Vienn. et Lugundi,' fin. Euseb. H. E. v. 1, cit. Pusey, 'Minor Prophets,' Amos vi. 10; Charlemagne, 'Capit. ad. Saxon.' 789 A.D., cit. Fleury, 'Ecc. Hist.' i. 44. 45; Gruber, 'Origines Livoniae,' cit. Wylie, 'Archaeologia,' xxxvii. 467; Kemble, 'Horae Ferales,' p. 95; Schaaffhausen, 'Germanische Grabstätten am Rheine,' p. 90; 'Jahrbuch des Vereines von Alterthums-freunden im Rheinlande,' Bonn, 1868.

[3] Literary evidence for the numbers of the Saxons is furnished by such expressions as those which Claudian puts into the mouth of a personified Britannia,

'Ne litore *toto*
Prospicerem dubiis venientem Saxona ventis.'—*Laus Stilichonis*, xxii. 254.

Evidence for the sudden and continual vexations to which Britain and other regions

in it of Hengist and that of Augustine. The legend which makes Hengist land in Thanet and be buried at Conisborough, in South Yorkshire, tells obviously in the same direction, but it is always well to strengthen a conclusion based on the interpretation of such a history as this by evidence drawn from actual, tangible, and verifiable facts. And it is worth while, consequently, to put on record here certain 'finds' of Anglo-Saxon urns which have been made subsequently to, or, for other reasons, have not been enumerated among those already referred to. In the year 1859 five urns of the Anglo-Saxon type, which are now to be seen in the museum of the Philosophical Society in York, were found by F. W. Calvert, Esq., in his garden, which is about half-a-mile outside of Micklegate Bar on the right side of the road from York to Tadcaster. Several Roman urns and sarcophagi were found at the same time and place, the Anglo-Saxons having in this, as in so many other Roman stations, used the cemeteries of their predecessors. An urn with an inscription, which I have not seen, was found at the same time. Five other undoubtedly Anglo-Saxon urns are mentioned in 'The Descriptive Account of the Antiquities' of this Museum (p. 95, n. 34), as being found in tumuli on the Wolds. An urn[1] as indubitably Anglo-Saxon has been discovered at Kempston, in Bedfordshire, for a sight of which I am indebted to the kindness of Canon Greenwell of Durham. Lastly, in the 'Illustrated London News' of Jan. 25, 1868, Supplement, p. 93, some excellent figures of several urns found by Dr. Massey of Melbourne, at King's Newton in Derbyshire, may be seen; and though I have not as yet had an opportunity of personally examining these specimens, I apprehend they will be recognised as belonging to the same class as the North German urns of the 'Horae Ferales;' the South Jutland or Slesvig urns figured by Engelhardt, loc. cit. and plates 14 and 17; and those from the sixteen English counties above enumerated. If, as Mr. Kemble has said, 'wherever Christianity set foot cremation was to cease[2],' we may be doubly sure that wheresoever cremation was practised in a country which had been previously Christian, Christianity had for the time become extinct. Of the

were subjected by the Saxons may be found in 'Ammianus Marcellinus,' xxvi. 4, xxviii. 2.

[1] See 'Collectanea Antiqua,' iv. 161, vi. 166, 201 seqq.

[2] 'Horae Ferales,' p. 95.

co-existence *in place* of cremation-urns and of skeletons inhumed entire there is no doubt; and, as many authorities seem convinced that the two practices co-existed also *in time*[1], I should be slow to set against their opinion the fact of the strong feeling which the Christians entertained as to the impiety of cremation. For I read in the passages just referred to, and can believe, that a practice was not always nor immediately discontinued because it was denounced. Still, at Frilford, though in three cases urns were found above Romano-British inhumations, in no case had I any reason to think that one part of the population on this area was practising the one, at the same time that another was practising the other, of these two modes of sepulture. If it should be allowed—in dangerous opposition, it is true, to Mr. Kemble's dictum, that no pagan Saxon was buried except when burnt[2],—that the Anglo-Saxon inhumations, shortly to be described as without orientation and with relics, may have been the burials of pagans, I should be more inclined to think that the two rites may have been practised contemporaneously, as we know them to have been by several heathen nations. To the heathen the two modes of sepulture were comparatively indifferent, and very slight reasons may have determined his choice of the one or the other. With the Christian it was different, and abstinence from cremation was made to seem a corollary of some of the most sacred and cherished articles of his faith. Hence I am not disposed to think that the conquered Romano-Britons would continue to use the cemetery of their forefathers when it was constantly being, as they would think, desecrated by the deposition in it of the urns of the unbelievers. The Saxons, on the other hand, as already remarked, had no reluctance against burying in the ground which held the bones of the former lords of the soil, and as the positions of several of the urns show—

'Little they recked of those strong limbs
Which mouldered there below.'

[1] For the co-existence of cremation with inhumation, see Kemble 'Horae Ferales,' p. 918; Neville's 'Saxon Obsequies,' p. 11; Wylie, 'Archaeologia,' xxxvii. p. 456; Akerman, 'Further Researches at Brighthampton, Archaeologia,' xxxviii; 'Inventorium Sepulchrale,' pp. 165, 195; Weinhold, 'Sitzungsberichte Kais. Akad. Hist. Phil. Klasse,' bd. 29, p. 138, bd. 30, p. 176; Lindenschmit, 'Archiv Anth.' iii. 114.

[2] 'Horae Ferales,' p. 98; and, per contra, the Rev. S. Finch, 'Coll. Antiq.' vi. 220, and Thrupp, 'Anglo-Saxon Home,' p. 399.

I should add that it is possible that half-converted Saxons may have relapsed into cremation in the absence of the missionary, and under the temptation which the licence of the 'lyke-wake' created. But the practice of such a transitional period, if it ever existed, would not affect the historical argument for the overrunning of this country by heathens, which the discovery of these urns in so many parts of it furnishes.

A piece of Samian ware was found in the Roman rubbish-pit already mentioned as having been discovered within about 200 yards of the cemetery. The resemblance of its pattern to that on the Anglo-Saxon urns is very striking, though the execution and finish are as different as is the material. A pattern of vandykes, scored zones, and stellate impressions, is one which, by its simplicity, would suggest itself to the rudest nations, and I do not, of course, mean to hint that the urns found here by me were figured after the pattern of Roman ware found here by the Anglo-Saxons. Still the similarity of the two patterns is very striking, and when we consider that urns with Latin inscriptions and of Roman manufacture have been found with Anglo-Saxon patterns upon them [1], it is less difficult to imagine that the Teutonic races, years before the period we are dealing with, and while yet in their North German native country, imitated with a stick on coarse hand-fashioned clay-paste the very simple but still beautiful pattern which the Gallo-Romans imprinted on finer and lathe-turned materials. Another illustration would thus be furnished of the extreme readiness already alluded to with which the Germanic natives imitated the arts and refinements of the Romans.

Burnt human bones have been here and there met with without any urn in relation with them, but within my experience at Frilford they have been merely scattered or even single bones, the presence of which may be explained by the disinterment of an urn, and the subsequent replacing of its fragments and its contents with less care than was sometimes bestowed upon this task [2].

In none of the urns were any other contents than human bones mingled with earth and stones discovered, except in the case of the

[1] See Roach Smith, 'British Association Reports for 1855,' p. 145, and the same writer's 'Collectanea Antiqua,' v. 115, pl. x. where such an urn, bearing the inscription D.M. LAELIAE RUFINAE VIXIT. A.III. M.III.D.VII. is figured.

[2] See 'Inventorium Sepulchrale,' Introd. p. xvi, pp. 8, 9, 12, 17, 18, 19, 40, 156, 159, 175.

urn found Sept. 1867, in which a few pieces of glass were found together with the bones, and in that of the small unpatterned urn found Jan. 1867, in which the incisor of a hare or rabbit was also found in company with the human remains, and like them had been subjected to the fire.

The urns were in most instances at but a very short distance from the surface of the ground, and, shallow as the furrows are (some five inches or so) which it is usual to make in this soil, the upper rims of the urns have in several instances received injury from the plough-share. This superficial position of cremation urns enables us to understand how the many superstitions[1] as to their pullulation in the spring, &c. arose, and it is paralleled, we may remark, by the shallowness of the inhumations of the same race, to the consideration of which I now proceed.

IV. *Of Anglo-Saxon Interments in the way of inhumation without orientation, but with insignia and in shallow graves.*

The Anglo-Saxons appear to have discontinued cremation, probably at the urgent request of the Christian missionaries, without at the same time adopting the direction of the grave which the usage of their teachers, as well as of their predecessors, would have led them to adopt. The shallowness[2] of many graves containing skeletons extended at full length and adorned with Anglo-Saxon insignia may again be referred to the retention by half-converted proselytes of some of that carelessness as to the disposal of the corpse which marked many heathen races then, as, indeed, it does now. The now well-known insignia of the male and female Anglo-Saxon respectively—to wit, the umbo, the spear, the buckle, and the knife; the fibulae, the perforated beads, the similarly perforated glass ornaments; the ear and tooth picks, the scoops, the

[1] For the belief as to urns being 'natural productions pullulating from the earth like bulbous roots,' see 'Horae Ferales,' p. 86. For other superstitious relating to them, see Cochet, 'Normandie Souterraine,' p. 124; Wylie, 'Archaeologia,' xxxvii. 46.

[2] For the shallowness of Anglo-Saxon and other Teutonic interments, see Cochet, 'Tombeau de Childeric,' p. 41; Bloxam, 'Fragmenta Sepulchralia,' p. 47; Engelhardt, 'Denmark in the Early Iron Age,' p. 9; Akerman, 'Archaeologia,' vol. xxxviii, Long Wittenham; Kemble, Ibid. vol. xxxvii. 1856; Wanner, 'Alemannische Todtenfeld bei Schleitheim,' pp. 10, 20.

shroud-pin, the perforated coins, and the knife, found both with women's and men's skeletons—have been found with several skeletons at Frilford, which were interred in graves varying in depth from eighteen up to twenty-seven and thirty inches, and varying still more in their compass bearings. In four of these cases, skeletons, which must be supposed to have been Romano-British, have been found to underlie these Anglo-Saxon remains, just as similarly inhumed skeletons have been already spoken of as underlying cremation urns. In one case a large fragment of a large unpatterned urn, which resembles in style the urn found at Long Wittenham, containing human bones (figured by Mr. Akerman, 'Archaeologia,' xxxviii. 352, pl. xx. fig. 4), was discovered lying over the pelvis of an Anglo-Saxon woman, buried with disc-shaped fibulae, beads, and shards. The fragment was itself in seven pieces when discovered; but, as they have admitted of readjustment, the fragment must have been put into the grave in the condition which it is in as now restored, in accordance with the custom of carefully replacing the fragments of a disturbed funeral urn, which has been several times noted in other Saxon burials[1]. In another of these interments some Roman tiling was found set along the side of the grave, a practice which other Teutonic tribes, in their imitation of the Roman civilisation, adopted, as has been observed by Wanner[2]. In another, a spear-head with the raised ridge, which Mr. Akerman ('Pagan Saxondom,' p. x.) has observed is to be seen on the assagaye of the modern Hottentot, was found accompanying a skeleton, the sex and nationality of which were spoken to by the presence of an umbo and a buckle[3], as well as by its osteological characters. Fibulae were not found with the male skeletons; with the female skeletons the common disc-shaped fibulae were the most usual. In one case, however, the cruciform variety, such as Mr. Akerman has figured ('Archaeologia,' xxxix. pl. xi. figs. 8, 9) from Long Wittenham, or ('Pagan Saxondom,' pl. xviii. fig. 1) from the neighbourhood of Rugby, was exemplified in two fibulae found with a female skeleton, which was accom-

[1] See 'Inventorium Sepulchrale,' Introd. p. xvi, &c.

[2] 'Das Alemannische Todtenfeld bei Schleitheim,' p. 13. See also Lindenschmit, 'Archiv für Anthropologie,' ii. 3, p. 356.

[3] For the indications which the presence of a buckle furnishes as to nationality, see Akerman, 'Pagan Saxondom,' p. 58; Cochet, 'Tombeau de Childeric,' pp. 228, 234.

panied also by the earpick and toothpick and scoop so frequently found in Anglo-Saxon interments[1]. No sword has as yet been found in the cemetery at Frilford, and the general character of the Anglo-Saxon relics which have been discovered is in keeping with the absence of this mark of condition and authority, if such[2] it may be considered to be. In one case a male skeleton was reported to me to have been found lying in one of these shallow graves with its face downwards. Unfortunately I was not upon the spot when this skeleton was removed; but, though Schaaffhausen[3] has pointed out that unskilled observers may be deceived as to the position of the face in a grave, I am nevertheless of opinion that the workman who had assisted in the removal of a very large number of skeletons from their graves was right in the report he made to me. Because, in the first place, I have myself seen an instance of such a mode of interment in a Romano-British barrow; and, secondly, it is not difficult to understand how such a misplacement could occur with an uncoffined body borne to a grave, the shallowness of which bore, and bears, evidence to a carelessness which the 'lyke-wake' would be only too likely to intensify. It has often been observed[4] that the Anglo-Saxons by no means invariably employed coffins in their interments. When the head is found to have been supported upon stones placed underneath it, it is plain that the interment must have been coffinless. But I do not find in my notes of the class of shallow, non-oriented, Anglo-Saxon interments that the head had been so supported; and, inasmuch as the results of its having been so raised are ordinarily very evident, the cervical vertebrae being impacted between the *rami* of the lower jaw, and this bone being, not rarely, separated widely from the upper jaw, owing to the changes of position which the perishing of the soft parts has entailed,—it is difficult to think that this peculiar arrangement would have been left unnoticed if it had existed. A nail has occasionally been found in a grave con-

[1] See 'Pagan Saxondom,' p. 70, and pl. xxxv. fig. 4; 'Archaeologia,' vol. xxxvii, Brighthampton, No. 1; vol. xxxviii, Brighthampton, No. 16, preserved in Ashmolean Museum, Oxford; 'Fairford Graves,' pl. ix. fig. 10, object similarly preserved.

[2] See Akerman, 'Archaeologia,' vol. xxxix, Further Researches at Long Wittenham.

[3] 'Die Germanische Grabstätten am Rhein,' p. 119.

[4] Wylie, 'Graves of Alemanni,' p. 13; Bloxam, 'Fragmenta Sepulchralia,' pp. 67, 72; Akerman, 'Pagan Saxondom,' Introd. p. xvi. Compare plate xiv. with plates xxxix, lvii, lxii, and lxvi, of Strutt's 'Horda Angel-cynnan.'

taining an Anglo-Saxon skeleton, but I have never come upon nails in such numbers as to make me think it probable that they had come there otherwise than accidentally, nor have I ever found in such interments that all but infallible sign of a coffin having been employed, namely, coffin-hooping. The shallow Anglo-Saxon graves do not appear to have had stones set round their edges; and the absence of such stones is another, and complementary, illustration of the carelessness which appears to have characterised the performance of these burials. Wherever stones have been found set round a grave, the grave has had the semi-oriented bearings of the Romano-British interments, and has all but universally the same depth as these graves, and may hence be considered to belong to a distinct era of inhumation.

V. *Of Anglo-Saxon Interments in the way of inhumation in graves of the same compass-bearings, and usually of the same depth, as the Romano-British graves, but differing from them in having stones set along the edges of the grave, and in containing insignia together with the skeletons.*

I have not at Frilford come upon a grave with stones set round its edges which had not the Romano-British direction towards E.S.E., and which did not contain a skeleton with the insignia of the Anglo-Saxon race. Following the Romano-British direction, these interments have followed the same precedent ordinarily as to depth also, and the like, it may be noted, has been observed by Wanner of the Alemannian interments at Schleitheim[1]. The closeness of the stones to the sides, head, and feet of the skeleton seems to preclude the notion of coffins having been employed in these interments, and the fact that the sides of these stones, which looked towards the skeleton, were in some cases reddened in a way in which actual experiment shows that similar stones of the neighbourhood do redden under the action of fire, makes it appear all but certain that the charcoal found in these graves around and even under the skeleton must have been produced by a fire lighted in the grave before, or indeed after, the corpse was put into it[2].

[1] 'Das Alamannische Todtenfeld bei Schleitheim,' pp. 11, 18.

[2] In a note from Professor Pearson to me, in which he gives much valuable information upon other points relating to the history of this country in the times with

Mr. Kemble, in the passages already referred to [1], supposes that in the transition state from heathenism to Christianity, such practices as this may have been stealthily indulged in by the newly-made and only half-converted proselytes, and these interments lend a considerable confirmation to this view. The Abbé Cochet [2] and Professor Schaaffhausen [3] seem to incline towards supposing that the similar appearances which they have noticed are to be ascribed to the remnants of a coffin; but I am inclined to think that the absence of nails, the raised position of the head observed in some of these burials, the large size of, and the retention of a certain brilliancy by, the fragments of carbonaceous matter found in these graves, and underneath as well as around the skeletons, as well as the conditions of reddening and of position which the stones present, are points militating very strongly against the hypothesis of a coffin having been present, and in favour of a wood fire having been lighted in the grave either in preparation for, or for the partial combustion of, the dead body. No coins were found in such relations with the head or chest of any of these skeletons as to make it seem likely that they had been put in as 'portoria;' in one case, however, a coin was found perforated, for suspension, doubtless as an ornament, about the region of what had been the chest or waist of a very much water-worn skeleton. Shards and flints, and a few bones and teeth of domestic animals, were found in these as in other kinds of inhumation observed in this cemetery. In one of these interments a pair of odd fibulae, one being of the cruciform, the other of the saucer or disc pattern, was found, one upon one shoulder and the other upon the other of a female skeleton. Similarly, or somewhat similarly, 'two large cruciform and two circular fibulae of bronze,' now preserved in the York Museum, were found with a skeleton in the Danes Dale Tumulus [4]. These discoveries may seem of trifling moment, but they do go to show,

which I am concerned, he says, 'The "Anglo-Saxon Laws," vol. ii. contain several lists of superstitious practices which the Church condemns, such as burning corn upon graves. It is true that the compilations in which these ordinances occur are in one sense not authentic, that is, have been ascribed to wrong authors; but they probably represent the customary laws of the Church here and on the continent with tolerable fidelity.'

[1] 'Horae Ferales,' pp. 98-104.

[2] Opere cit., pp. 198, 255, 256, 304.

[3] Opere cit., p, 104.

[4] See Catalogue, p. 93, and 'Coll. Antiq.' vi. pl. 28.

first, that no pattern of fibulae should be considered as peculiar to any one district, except provisionally; secondly, that a very considerable uniformity may have existed in the manners and customs of the Anglo-Saxons throughout the entire length of England; and, thirdly, that, inasmuch as intercommunication between places as far apart as Frilford and Driffield must have been difficult in those days, the numbers of the invaders of these similar fashions and habits must have been considerable.

The stones were set round in the grave in but a single row from within outwards, and in height they do not seem to have extended from the bottom of the grave further upwards than a stone coffin (of which they may be supposed to have been a cheap imitation) would have done. The graves here, as at Selzen, are narrowed towards their lower ends[1]. In such interments as these the skull may or may not be found to rest upon a stone which had been put under it in the way of support, and which has caused the lower jaw to settle down upon the cervical vertebrae, and to hold them impacted between its *rami*. The Anglo-Saxon habit of thus placing stones beneath the head of the corpse may or may not be adumbrated by the mediaeval stone-pillow in monuments as suggested by the Abbé Cochet[2]; but, at all events, it goes some way towards proving that coffins were not employed in the interments in which it is noticeable.

In one of these graves a mass of what has been called a 'scoriform' lava, though it is different enough from the true *scoriae* or slag similarly found in Anglo-Saxon graves at Fairford by Mr. Wylie[3], was found at the foot of a female skeleton. The bulk it made up was about that of an orange, and, as it has separated into two co-adaptable halves, each of which resists very violent hammering, we must suppose that since it was put into the grave it must have been subjected to some disrupting agency which acted upon it with great force, and yet left it, when broken asunder, *in situ*. It is possible that the piece of lava in question may have been broken into two pieces by the action of a fire lighted in the grave, as, it

[1] See Lindenschmit, 'Archiv für Anthropologie,' ii. 3. 356, in review of Wanner's Memoir, 'Das Alamannische Todtenfeld bei Schleitheim;' Schaaffhausen, op. cit. pp. 131, 154.

[2] 'Normandie Souterraine,' ed. i. p. 192.

[3] 'Fairford Graves,' p. 24.

has been suggested, was the case with a millstone found split to pieces in a Saxon grave discovered[1] at Winster in Derbyshire, and showing, which this grave did not, signs of a fire having been lighted in it. But one of the many valuable hints which I owe to Professor Phillips has made me think that it may be to frost rather than to fire that we ought to look to account for the fractures of volcanic products such as these. A porous soil would allow the cavities of such a piece of lava to become filled with water, and a shallow grave in a severe winter might furnish the other requisite conditions. Some mortar-like matter was adherent to the exterior of the piece of lava besides and distinct from the calcareous incrustation which the water of the soil had deposited upon it. The lava itself, as containing *hauyne*, we may be justified in regarding as having, in all probability, come from Niedermennig, which is a place whence, in the time of Augustus[2], the Romans took building materials for the bridge at Trèves, and whence, as a matter of fact, millstones are now largely exported, and whence, consequently, we may think it not wholly unlikely[3] they were exported in former and Anglo-Saxon times. It is difficult, of course, to be quite sure that a sub-globular mass such as the piece of lava I found at the feet of this Anglo-Saxon female had been a piece of a quern; but fragments, of identical and closely identical mineralogical characters, found 'near a barrow in Norfolk,' and 'in a British barrow at Thetford,' respectively, have been considered as pieces of a millstone by the well-known antiquary J. Wickham Flower, Esq., to whose kindness I owe the opportunity of comparing these several sets of volcanic fragments together.

Schaaffhausen[4] has put on record several instances of Germanic interments either in coffins made out of *tufa*, or in graves with fragments of such volcanic matter set round their copes, together with other stones, and it is just possible that the Niedermennig lava may have been put, as it was in this grave, at the foot of the grave, whilst other stones were set round the sides, as a kind of reminiscence of what the 'setting' of the interment might have

[1] Extract from the 'Times,' Thursday, Oct. 23, 1856, given in 'Horae Ferales,' p. 104.

[2] Daubeny, on 'Volcanoes,' pp. 49, 64.

[3] See Bruce, 'Roman Wall,' ed. iii. 1867, p. 438, seen by me subsequently to writing as above.

[4] 'Op. cit.' pp. 122, 127; Wren, 'Parentalia,' p. 27.

been elsewhere. But I am not aware that we have any reason for thinking that the Anglo-Saxons, who, rather more than a century[1] after the first invasion, drove the 'Southern Belgae or Firbolgi' out of Berkshire into Wales and Damnonia, received any accessions to their numbers from regions so far south as Andernach and Coblentz[2], where such interments could be easily, and were frequently made; and it is more probable that a fragment of lava may have been put into a grave in its aspect of a fragment of a millstone, an implement of daily life, than in its aspect of a fragment of the same material as that out of which entire coffins or the entire 'setting' of a grave had been made elsewhere.

On the whole, I am inclined to regard these interments as belonging to a period of transition from the comparative if not total heathenism of shallower interments without orientation, and without the decent regard for the dead which the setting of stones round the graves indicates, to the more distinctively Christian mode of burial without insignia and in coffins. The greater depth and the direction of the graves I should regard as due to the teaching of the Christian missionaries; the adoption of the very graves used by the Romano-Britons may have been due merely to the imitative tendencies of the conquering races, or it may be ascribed to the influence of some remnants of the conquered Christians, who may have maintained their religion on sufferance, and their traditions as to the tombs of their fathers during the dark period which intervened between the invasion of Cerdic and the preaching of Birinus. The tricking out of a corpse with insignia of sex, or rank, or employment, seems half heathen to us who have the great truth that we can take nothing out of the world with us impressed upon us at times when we are most open to impressions; still it is just such a custom as a missionary with the proper amount of the wisdom of the serpent would acquiesce in. Time, such a teacher would know, was on his side, and he would feel that he could afford to wait.

It is possible that the differences between these two kinds of Anglo-Saxon inhumation may have been due to some social differences between the persons severally practising them, and that

[1] See Beale Poste, 'Celtic Inscriptions,' 1861, p. 71.

[2] Leo, however, in his 'Ortsnamen,' pp. 100–104, has tried to show that most of the local names near Heidelberg correspond to local names in Kent.

the deeper graves may have been dug for richer, and the shallower for poorer, persons. But the insignia in both alike are very closely similar, and I incline, therefore, to ascribe the greater care bestowed upon the latter class of interment not to any sense of the favours which a richer person had conferred in times past, but to the greater care which Christianity would teach ought to be bestowed upon the burial of the body.

The resemblance of the Anglo-Saxon manners and customs to those of the kindred but hostile race of the Franks is very familiar to the English explorer of Anglo-Saxon cemeteries, if he be acquainted either with Lindenschmit's work, 'Das Germanische Todtenlager beim Selzen,' or with the works of the Abbé Cochet so often referred to in this paper. The Merovingian[1] and the Anglo-Saxon resembled each other in their abhorrence of city life; and also in the melancholy point of their short-livedness which has already been alluded to, and which appears to be explicable by the fact that in the times we have been dealing with these races preferred a country life, it is true, to a town life, but a country life in a camp, not a country life in a village. As Temple (cit. Rapin, p. 161) and Leibnitz long ago remarked, there are other points which serve to show the community of origin of the Frank and the Saxon, such are their reckoning time by the nights, as the 'fortnight,' to say nothing of their closely allied languages. A minor point of community is furnished by their common employment of the Roman tiling to set round their graves. On the other hand, the Saxons retained the custom of cremation a century and a half longer than the Merovingians, and their urns were not lathe-turned, whilst those of the Selzen Teutons were. (See Lindenschmit, l. c., p. 15.) Holy-water vessels have not been so constantly found at Frilford as they appear to have been at Selzen, from the beautiful figures given in the monograph referred to, or as they are expressly stated to have been by the Abbé Cochet in the Merovingian interments[2].

[1] Gibbon, vi. 336, chap. xxxviii. for Merovingians; Tacitus, 'Germania,' chap. 16, for Germans generally; Coote's 'Neglected Fact in English History,' p. 123; Ammianus Marcellinus, xvi. 2-12; Pearson, op. cit. i. 264. Augustine brought Frank interpreters with him into Kent, Bede, 'H. E.' i. 25, and the Welsh poems sometimes speak of the Saxon enemy as a 'Frank;' see Skene, 'Four Ancient Books,' i. 460.

[2] See 'Archéologie Céramique,' pp. 11, 13.

VI. *Conclusions suggested by an Examination of the Human Remains found at Frilford.*

The cranial and other osteological peculiarities of the human remains which I have examined from the Frilford cemetery seem to me to throw sometimes a very unambiguous, and sometimes, it must be confessed, a more or less questionable light upon certain of the moot points in the political and natural history of the period in which their owners lived. Among those points may be specially mentioned the often-raised and very variously answered questions, as to the extent to which [1] the Anglo-Saxon Conquest was equivalent to an extirpation of the population previously in occupation of this country [2], and as to the physical and more particularly the cranial characters of the Romans and Romanized Britons. But it is worthy of note that very indubitable evidence, at least as to some of the social and moral peculiarities [3], of the conquered and the conquering races respectively, may be gathered from a careful examination of their bony remains [4].

I have subjoined in a tabular form the results of my examination of the sometimes fairly complete, sometimes exceedingly incomplete, remains of 123 burnt or buried bodies which have come into my

[1] For the question of the extent to which the Celtic population were destroyed by the Saxon Invasions, see Pearson's 'History of England during the Early and Middle Ages,' i. 99–103, 1867; Freeman's 'Norman Conquest,' i. 18, 20; Akerman, 'Archaeologia,' p. 38, Second Report, Brighthampton; Turner's 'Anglo-Saxon History,' i. 311; Wylie, 'Fairford Graves,' p. 8; Kemble's 'Saxons in England,' i. 21; D. Wilson, 'Anthropological Review,' iii. 81.

[2] For the various views which have been held as to the Roman cranium, see Ecker, 'Crania Germaniae,' p. 86, 1865; Ecker, 'Archiv für Anthropologie,' i. 2, p. 279, 1866; ii. 1, p. 110, 1867; Hölder, Ibid. ii. 1, p. 58; His, 'Crania Helvetica,' pp. 39, 40; His, 'Archiv für Anthropologie,' i. 1, p. 73, 1866; His and Vogt, Mortillet's 'Matériaux pour l'Histoire de l'Homme,' August 1866, pp. 522, 523; 'Crania Britannica,' p. 23, chap. ii. and pl. 49; Davies and Thurnam, cit. 'Indigenous Races,' p. 312; Maggiorani, cited by Ecker, 'Cran. Germ.' p. 88, and 'Arch. für Anth.' l. c.; cited by v. Baer, 'Bull, Acad. Imp. Sci.' St. Petersburg, 1860, p. 58, fig. g; Edwards, 'Des Caractères Physiologiques des Races Humaines,' p. 50; Nott and Gliddon, 'Indigenous Races,' p. 311, and Cardinal Wiseman, cit. in loco.

[3] As to the supposed degeneracy of the Britons, see Kemble, 'Saxons in England,' ii. 294, i. 6; 'Encyclopaedia Metropol.' xi. 378; Zosimus, cit. 'Mon. Hist. Brit.' lxxviii. vi. 6.

[4] As the German periodical, the 'Archiv für Anthropologie,' is conducted under the joint editorship of Ecker and Lindenschmit, and as the latter, I apprehend, is as well known among archaeologists as the former is among biologists, no apology will be needed for the constant reference which I shall have to make to its pages. It may be well to

hands from the excavations and quarrying carried on at Frilford at various times during the years 1864–68 inclusive. In spite of the ravages of fire in the cases of cremation, and the all but equally destructive working of the water containing carbonic and other acids upon inhumation in ground with the rock (coralline oolite) at an average distance of about a yard from the surface, it has been possible to identify the sex and age in all but about a sixth of the skeletons, or parts of skeletons, examined. Many skeletons, however, and many urns had been lost to science, as may be gathered from Mr. Akerman's report[1], during the various quarrying operations carried on at various times previously to his investigations, and the arithmetical results of my researches are much less valuable consequently than they otherwise might have been. But I incline to think that the tolerably exhaustive and complete collection which the great kindness of the authorities at Frilford has enabled me to make of the fruits of the excavations carried on during the last two years, may be taken as a fair sample of what the entire series was.

One of the most striking peculiarities of the series of 123 skeletons, as represented more or less fragmentarily in the University Museum at Oxford, is the very large number of old persons' remains which it presents to our view. The most superficial observer cannot fail to be impressed by this fact. A little more accurate inspection shows that the proportion of aged persons varied most surprisingly in accordance with the nationality, and that of the persons of either sex who were interred with Anglo-Saxon insignia only two could have been considered old. We are, unhappily, even now too familiar with the history of invading armies to feel it necessary to spend much time in excogitating an explanation of this fact: it is worthy, however, of mention, that a similar fact has been noted by the Abbé Cochet[2] in the burial-grounds of the kinsfolk of the Anglo-Saxons, the Merovingian Franks. The preponderance of longevity being seen to attach to the Romano-British population, the presence with these aged

add here that the English reader can find a very clear account of the classification of crania adopted by His and Rütimeyer, and alluded to very frequently by myself, as also by various writers in the periodical just mentioned, in the 'Prehistoric Remains of Caithness,' pp. 104, 105, a work written by S. Laing, Esq. M.P. and Professor Huxley, conjointly.

[1] 'Proc. Soc. Antiq.' ubi supra.

[2] 'Normandie Souterraine,' p. 183.

'frames' of coins bearing such names as those of Gratian tempts us to explain the phenomenon by the hypothesis of the young men having been taken away to fight and die in distant countries under such commanders as Magnus Maximus. Persons who some years ago had the opportunity of seeing village after village on the continent of Europe inhabited by forms like that of Tithonus, will be ready to accept this explanation as sufficient to account for the fact. Till I came to add up the various individual identifications of the two sexes which I had made from time to time, and without any reference to any historical relations which the skeletons or their owners might have possessed during life, I held this hypothesis myself. But on adding up the numbers of males and females severally, I find that I have assigned no less than 48 of the 123 bodies to the male sex, and only 34 to the female. Even if we add to the female series the 11 individuals as to whose sex I have felt myself unable to pronounce, the force of this arithmetic is but little impaired, or, indeed, not at all[1]. The fact of the great preponderance in number of aged remains may be explained by a reference to the present condition of the population on the spot. Frilford is renowned for its salubrity and the longevity of its inhabitants at the present day. The fact of the great preponderance of male skeletons is not so easy of explanation, and it is especially difficult of solution when we note that more than half of these male skeletons are aged ones. Barracks and prisons furnish an excess of male skeletons in their burial-grounds, I apprehend, but not an excess of *aged* male skeletons. I am not aware that the monks of the west had established themselves among the Atrebates before the time of Cerdic[2]. And the only hypothesis which has suggested

[1] It has been suggested to me that the soldiers, who, on the hypothesis before us, are supposed to have left their bones in foreign lands, may have taken wives with them. But it could not have been often in days of such difficulty in travelling that Lycoris

'Perque nives alium perque horrida castra secuta est.'

The soldiers of Gustavus Adolphus were, very many of them, married men, but I do not know that their wives accompanied them to his famous battle-fields. The men, too, who fought and won at Lützen had very different motives and incentives from those of the recruits who followed the standards of the various 'tyrants' and pretenders of the later Roman Empire, and it is only by means of such motives and incentives that men can be got in any large numbers to break away from family ties and join distant military expeditions.

[2] See, however, 'Hist. Mon. de Abingdon,' i. pp. 2, 3.

itself to me is that the part of the burial-ground which has fallen under my inspection may have been used by preference, though by no means exclusively, for male interments. The hypothesis of a battle is excluded by several considerations, and notably by that of the age of the skeletons.

Of the thirty-five skeletons assigned by me to the female sex, thirteen were of aged, and no less than nineteen of young, women. The great dangers of child-birth may be supposed to be indicated by these figures, and the osteophytic intracranial growths[1] so often observed in the puerperal state, and noted here in four cases, may point in the same direction. Under the head of children I have reckoned all persons below the age of thirteen or fourteen. The numbers of this class, viz. twenty-eight, which I have identified, holds a much smaller proportion to the whole number, 123, than we should expect from modern statistics. But the greater perishability of children's bones, and the lesser depths of their graves, which, if not more chemically, is yet mechanically more dangerous to their preservation, must be borne in mind in considering these figures, and should prevent us from basing any argument upon them over-hastily. Still, we may perhaps be justified in thinking that there could not have been at Frilford, even in days when glazed windows and coal were as little used as China-ware and 'China drink,' that great infantile mortality which, by weeding out all the weakly in early life, produces a population of adults with a great proportion of aged individuals.

The Anglo-Saxon remains which I have procured from Frilford have suffered much from the mechanical and chemical agencies to which the shallowness of their graves, and, secondly, the shallowness of the soil, exposed them; and the youth of their owners has still further rendered them amenable to these destructive and distorting forces. But, thanks to the reconstructive ability of Mr. Charles Robertson, I have been enabled to see that the two types of crania which have been shown by Dr. Barnard Davis to have been found with Anglo-Saxon insignia, both at Long Wittenham[2], and at Linton, in Cambridgeshire[3], co-existed side by side in the

[1] Rokitansky, 'Path. Anat.,' Sydenham Soc. Trans. iii. 208; Bock, 'Pathologie,' p. 209.

[2] See 'Archaeologia,' xxxviii. No. 107, 770 k. Oxford Univ. Museum.

[3] See 'Crania Britannica,' Dec. 4, pl. xlvi. Two other crania of this 'platycephalic'

Anglo-Saxon contingent which possessed itself of Frilford. I may remark that the two types are recognisable in specimens of both sexes, and a very fairly perfect female cranium from a grave in which a pair of fibulae and a number of beads were found, as it shows at once, and distinguishably, the tribal and the sexual characters, which have very often been confounded, and as from the surroundings with which it was found there is no doubt as to its value as a standard of reference. This skull appears to have belonged to the shorter and broader type of Anglo-Saxon crania, which was, I am inclined to think, the less cultivated of the two types. A second Anglo-Saxon female cranium found here belongs to the same type. A single female and a single male cranium of a more elongated form were also found with Anglo-Saxon insignia. The female skeleton, it may be remarked, belonged to an old person, and in this point, as also in the possession of cruciform fibulae, instead of circular ones, this skeleton differed from the two others with which we have compared it. It was chiefly from a comparison of the female Anglo-Saxon skull, with the first cranium described by me as 'cranium (male) marked A' for Mr. Akerman in his 'Report of the Proceedings of the Society of Antiquaries,' May 25, 1865, that I came to see that my assignment of this latter to the male sex had been in all probability erroneous. This cranium was reported as having been found with a fibula two feet above it, and though this by no means proves it to be an Anglo-Saxon skull from the archaeological point of view, the very close anatomical approximation of this skull to the indubitably Anglo-Saxon skull does, when coupled with this fact, lend some considerable probability to such a conclusion. In justice to myself, I may be permitted to say that the cranium and lower jaw were the only bony relics upon which I had to form my judgment as to sex, and that in my report I did draw attention to the small development of size and strength which they seemed to show that their owner must have possessed. And the authority of anatomists of no less repute than His and

type have been found in the Frilford cemetery subsequently to the writing of this paper, viz. March 22, 1869 (Nos. iv and v). Both had belonged to young men. In both the body had been buried with the head raised; and in one the grave, though semioriented, was only 18 inches deep, and the arm lay across the body, and not by the side, as in the burials of Latinised populations (see Cochet, 'Normand. Souterr.' p. 193). There were no relics, and we have not therefore more than probable evidence for their nationality.

Rütimeyer[1], Welcker[2], and Ecker[3], may be adduced to show that it is by no means always possible to decide the question of the sex of a cranium in the absence of the pelvis and other bones. It is interesting to remark that a very similar female cranium was found by the Rev. George Cardew in a Romano-British cemetery at Helmingham under circumstances such as that of having the head raised, which makes it probable that the skull may have differed as much ethnologically as it does anatomically from the skulls of the Romanized Celts, amongst whom it, as also another cranium supposed not to have been Celtic, was found. This cranium has been presented to the University Museum by the Rev. G. Cardew, and has been carefully measured and compared with other skulls supposed or known to have belonged to Anglo-Saxons. Two smallish brachycephalic or sub-brachycephalic and prognathic crania, one of which belonged to an old (No. xiv. Jan. 15, 1868) and the other to a young woman (No. x. March 17, 1868), and neither of which has any other than osteological evidence attached to it for the decision, I am inclined upon this evidence to think may have been Anglo-Saxons of the type of the two female crania just spoken of. The younger of these two women's skulls was found with the cervical vertebrae impacted between the rami of its lower jaw, and in this, as in many other particulars, resembles the female Anglo-Saxon skull from Brighthampton, to be seen catalogued as No. 5,712 D, in the College of Surgeons.

Among the entire series, besides some fourteen crania, or parts more or less fragmentary of crania, and other bones, which speak to the existence of a distinct interment without making it possible to refer the remains certainly to a distinct type, there are some four or five crania which bear a considerable resemblance to crania of what is perhaps the most common modern English type. The frontal region, without attaining any very extraordinary development, or exceeding either in vertical or transverse diameter the frontal regions of the larger specimens of brachycephalic British skulls, is, nevertheless, possessed of more equable proportions relatively to the other regions of the cranium than the great majority of ancient crania[4]. And in consequence, to some extent,

[1] 'Crania Helvetica,' pp. 8, 9.

[2] 'Archiv für Anthropologie,' i. 1, p. 127.

[3] Ibid. ii. 1, p. 110.

[4] See Broca, 'Sur la Capacité des Crânes Parisiens,' Bull. Soc. Anth. de Paris, tom. iii. 113, 1862.

of this, the entire calvaria shows a more evenly ovoidal contour than the skulls composing the rest of this series. These crania were found in graves in which no relics, except in one instance a nail, were found, and which ran in the ordinary semi-oriented Romano-British direction. And, so far as the brain case is concerned, these crania might be looked upon as embodying the result of intermarriages of the broader 'Sion' type with the narrow 'Hohberg' type, and corresponding with the 'Misch-Form' spoken of by His at p. 49 of the 'Crania Helvetica.' And they might perhaps be considered as representing the inevitable result of the settlement of a large Roman immigration in the midst of a dolichocephalic Celtic people. But, inasmuch as these crania show a not inconsiderable tendency to prognathism, and resemble herein the Anglo-Saxon, and differ from the Romano-British series, I incline to think they may have belonged to Christianised Anglo-Saxons who died before the churchyard had superseded the cemetery, but after the custom of burial with insignia had given way to the urgency with which its anti-Christian character may have been represented to the convert. The hypothesis of poverty will account for the absence of relics, but I do not incline to accept it here, partly on account of the presence of a nail, which may seem to imply the employment of a coffin in one of the interments, and partly on account of the resemblance which these skulls show to the male Anglo-Saxon cranium (No. 36, Researches at Long Wittenham, 'Archaeologia,' vol. xxxviii.) and to a female Anglo-Saxon cranium obtained for me by the kindness of the Rev. R. Taylor, from the Kemble Cemetery, described by Mr. Akerman in the 'Archaeologia,' 1856, vol. xxxvii, in neither of which cases have we reason to suspect the existence of straitened means.

The name of Magnus Maximus, the Maxen-lwedig of the Mabinogion, forbids us to think that in the days of Gratian there could have been, either in modes of life or in modes of burial, much difference between a Roman and a Romano-Briton. Tenants of leaden coffins must, from the expensive character of their interment, have been persons of distinction, such as were the 'Equites'[1] under the Roman empire; but Roman citizenship no more implied Roman blood in the days of Ambrose than it did in those of

[1] See Kemble's 'Saxons in England,' ii. 272; Pearson's 'History of England,' i. 45; Coote's 'Neglected Fact in English History,' pp. 40, 45.

St. Paul. The *Notitia*[1], indeed, informs us that races such as the Tungrians, Dacians, Moors, Cilicians, and Dalmatians, as well as Spaniards, Gauls and Germans, were employed by the imperial policy to hold Britain at the foot of Rome.

But if it is at all possible to separate and distinguish, when one is treating of the times of Maximus, between a Romano-British and a Roman interment, it may be possible to do so in such cases as those of the two interments in leaden coffins already described. The tenants of these coffins must at least have been persons of wealth, and in the enjoyment during their lifetime of all the distinctive characteristics which still remained attached to the title 'Civis Romanus.' It is true that coins were found with the one and not with the other of these two skeletons, but in all other particulars attending their sepulture they seem to have very closely resembled each other. But when we come to compare their crania we find that while that of the skeleton found with coins is of an elegantly vaulted and lofty form, that of the other is low, broad, and globose. Professor His would speak of the one as belonging to his 'Hohberg,' and of the other as belonging to his 'Sion' *typus*. The skull of the former differs but little, and that chiefly in the way of refinement, from the elongated and vaulted crania procured from British barrows of a pre-Roman period, such as the long barrow at Nether Swell, near Stow-on-the-Wold (Article xviii), calvariae from which I have side by side with that of this Roman from the leaden coffin as I write; the skull of the latter is as broad and low as another equally authentic 'Roman' cranium of about the same period, figured by Professor Ecker at pl. xx. of his 'Crania Germaniae.' So far, then, as these crania bear upon the argumentation as to whether the Roman skull was an elongated and vaulted, or an elongated and broad and flat skull, we may at first sight be tempted to rest in the conclusion that both types were equally and alike found in the imperial race. I believe, however, that it is possible to show that we should be wrong in considering with Professor His[2] that the

[1] See 'Roman City of Uriconium,' by J. Corbet Anderson, Esq., p. 129; and Hölder, 'Archiv für Anthropologie,' ii. 1, 88; Taylor, 'Words and Places,' p. 284, ibique citata.

[2] 'Arch. für Anthrop.' i. 1, p. 70; 'Crania Helvetica,' p. 38. One of the Hohberg type skulls is supposed by the authors to have come from a cemetery the graves in which were oriented, and contained swords and spear-heads as well as coins. This however does not prove that they belonged to *Roman* soldiers, but rather the contrary. See 'Cran. Helv.' p. 21, note.

former of these types, which he has also spoken of as the aristocratic type of head, is really the Roman skull *par excellence.* First, as it seems to me, the Romans themselves considered theirs to be a broad rather than a lofty-headed race. In looking at Roman monuments as reproduced for us in such works as Lindenschmit's 'Alterthümer unserer heidnischen Vorzeit,' we cannot fail to be struck by the great angle at which the ears stand out from the head; and this feature, a very striking and obvious one, is, as observation on living 'eurycephalic' persons will show, correlated with a globose and bossy rather than with a vertically-walled and narrow temporo-parietal region. The engraving of the beautiful monument to Manlius Coelius, an officer in the army of Varus, given at Heft vi. Taf. v. of Lindenschmit's work, just referred to, shows this peculiarity in the attachment of the external *concha* of the ear in each of three heads it represents; and much the same may be said of the figures given Heft iv. Taf. vi., Heft ix. Taf. iv., and especially of the uppermost of the two figures in Heft viii. Taf. vi. Busts also of the Roman emperors and of other Romans which are recognised as more or less authentic speak to the same effect. Secondly, we do find the broad and flat form of the cranium very commonly in cemeteries of undoubted Roman character in England, and the arched and centrally-ridged and narrow cranium we do find in as undoubtedly British barrows. A skull, most singularly resembling one of my globose platycephalic crania from Frilford, was recently shown me by Canon Greenwell from a cemetery at Margate, where it had been found with Roman pottery, whilst the 'Hohberg' type of skull is the very form which Retzius describes as the less common Celtic form, and calls, for the sake of distinguishing it, by the name 'Belgic[1].' Thirdly, through the kindness of Thomas Combe, Esq., M.A., of Oxford, I have had put into my hands, and into the Oxford Museum, a skull, 'found in excavating a house of the time of the Roman Republic, discovered below a vineyard, near the baths of Caracalla, on the Via Appia,' and this skull, though it belonged to a person of not more than between twelve and fourteen years of age, enables me to understand how the modern Italian anthropologist Maggiorani speaks of the ancient Roman skull as a long but broad skull, oblong and four-cornered, with broad interparietal and broad frontal regions. But

[1] 'Ethnologische Schriften,' p. 108.

I must say that the skull from the leaden coffins, of which I am speaking, as also a more or less authentic bust of Julius Caesar, and such works of art as the Roman figured in Lindenschmit's 'Alterthümer,' Heft vii. Taf. v. have convinced me that too much weight may be laid upon breadth of forehead. In these heads the broad character which they present does not depend upon the frontal but upon the parietal region, and the vertical view of the cranium presents very much such an outline from back to front as the broad side of the flint axes or celts, so familiar to antiquaries, presents from front to back. The head of the first Napoleon must have presented such a contour when viewed from above; and I believe, in spite of our tendency to connect a narrow forehead with foolishness, that a truer analysis would connect it in many cases merely with premature closure of the frontal suture, which seems hereditary in some families. This premature closing is consistent with the possession of a large cerebrum, and of great mental powers, and we cannot arrogate for it any ethnological significance, at all events in cultured races.

His and Rütimeyer ('Crania Helvetica,' p. 34) hold that their 'Sion' type of cranium, which seems to me to be represented by the broad, flattish, globosely contoured skulls, of which I have just been speaking, was the type of skull possessed by the Helvetii, their 'Celtic forefathers,' and by the inhabitants of their Pfahlbauten. And, as there is evidence to show that this same form of skull existed in pre-Roman times even in these islands, we must not suppose that the flatter and more globose skulls which we find at Frilford belonged exclusively to Roman immigrants, or to immigrants from Southern Europe, who may have been commanding as officers, or settled as upper-class *decuriones* or *equites* in the neighbourhood of this cemetery. The loftier and narrower crania, however, may with less hesitation be supposed to have belonged to men of similar station, but of British birth and blood, who had acquiesced in Roman rule, and identified themselves with Roman institutions.

Differing in the particulars specified, the osteological remains of the two occupiers of leaden coffins do nevertheless present certain important points of resemblance. Both belonged to men who were beyond the middle period of life, who were possessed of great muscular strength, but whose skulls, teeth, and jaws seem to show that they had the command of the comforts of civilisation. Whilst the skulls in both cases present the appearances of refinement, the other bones of the skeleton are much roughened by the development upon them on the one hand of ridges for the insertion and origin of muscles, and on the other of rheumatic (?) exostoses. And these same bones show, in the one case, with considerable probability, and in the other with absolute certainty, that their owner had been exposed, or exposed himself, to personal injury and violence, and had, probably, been a soldier of much service in the stormy times to which, in one case, the antiquarian relics enable us to assign his remains with perfect certainty. The left collar-bone belonging to the skull of the more globose and flatter outlines had undergone and repaired a comminuted fracture during life, and the left metatarsal of the second toe of the foot of the same side, a bone but rarely broken, had been broken, though less severely, than the collar-bone, and had, like it, been repaired during life. A fall from a horse may break a collar-bone, but injuries such as war entails are suggested to us by a history like this. The other skull, which was found with five coins, and which I have said may probably be looked upon as having been produced by the action of Roman influences upon the more roughly-hewn dolichocephalic Britons, was found in company with a left first rib, which had anchylosed with its ossified costal cartilage, which again, like the clavicle just above it, had its sternal articular end greatly enlarged. It is possible that these peculiarities may have been the result of exostotic disease, of which the other bones bear evidence, though less marked evidence than the bones of the other skeleton with which we are comparing them; but for the reason conveyed in these last words, as also because the abnormal appearances are not repeated on the opposite side, I incline to ascribe them to the working of some mechanical injury inflicted, possibly, in war, and certainly many years before death. The owner of this skull had lost, and was at the time of his death losing, teeth by caries [1], and was suffering and had suffered from exostosis in sympathy with it; the owner of the other had lost two of the molars of the right side of the lower jaw early in life, and the molars of the corresponding side in the upper jaw are little worn and suggestive of youth till we

[1] For an interesting history of dental caries, as observed in the ancient inhabitants of Britain, see a paper by J. R. Mummery, Esq. 'Trans. Odont. Society,' 1869.

look at the other side. The lower jaw in the former of the two skulls is very well formed; in the latter it is comparatively feeble, especially in the region of the chin; the teeth in both are less worn than the age testified to by the rest of their skeletons would have led us to expect. The occupiers of these coffins were both tall men; the stature of the man found with the coins must have been nearly five feet eleven inches, that of the other nearly six feet. A skeleton of an old man, the skull of which closely resembles that of the former of these (see 'Catalogue,' infra, No. xiv. May 1867), and which bears less ambiguous marks of its owner having been a warrior in the gaping, though healed, wound on its left side, belonged, as its femur of 19·5 inches length shows, to a man of fully six feet in height. The stature of each of these three warriors was much above that of the average Roman of ancient days, who spoke of the Germanic and Celtic races as possessing *immania ac procera corpora*, as it is also above that of his modern Italian representative [1], and above that of the Long-barrow British skeletons [2]. The better food of civilization may have increased the stature of the former of the two occupiers of the leaden coffins, and of the owner of the beautifully elegant and vaulted cranium (No. xiv. May 1867); whilst intercrossing would account for the increase in height in the skeleton to which the flatter skull belonged, if, with Edwards, Cardinal Wiseman [3], Sandifort, and Ecker, we should consider it to be probably Roman.

The craniography of the occupiers of the graves which I have spoken of as Romano-British or British, and which the archaeological evidence above adduced shows to have belonged to the times of the later Roman empire, is a subject of considerably greater difficulty than that of the Anglo-Saxon and of the leaden coffin interments. An examination of fifty-three of these interments, and a comparison, carried on at great cost of time, of their contents with those of several other cemeteries, has conducted me to the following conclusions as to the tribal characters of the pre-Saxon inhabitants of this district with whom I have had to deal. In the

[1] See Edwards, 'Des Caractères Physiologiques des Races Humaines,' p. 53. See Keysler, l. c. p. 220, for the stature of the ancient races under comparison, ibique citata.

[2] See Thurnam, op. cit. pp. 40–41.

[3] 'Lectures on the Connexion between Science and Revealed Religion,' p. 152, cit. Nott and Gliddon, 'Indigenous Races,' pp. 311, 312.

first place, I have not in my excavations at Frilford met with any representatives of the brachycephalic type of ancient Britons so well described by Dr. Thurnam [1], and called 'Belgic' by Professor Huxley. This is especially noteworthy, as typical examples of this form of cranium have been, through the kindness of the Duke of Marlborough, procured by me for the University Museum from the long barrow at Crawley, described by Mr. Akerman, in the 'Archaeologia,' xxxvii. 432, and supposed by him to belong to the same period in time, as it does to much the same district in space, as the Frilford cemetery. Secondly, the longer, narrower, and more vaulted skulls, supposed to have distinguished a race which in England at least took the priority in point of time of the brachycephalic and taller race just mentioned, are, in what I should consider their most typical form, all but equally absent here. That most typical form I should consider as identical with the form regarded as 'Belgic' by Retzius [2], and spoken of by him as 'a Celtic but not the common Celtic form;' and the form called 'Cumbecephalic' by Professor Daniel Wilson [3] I should regard as being but a slight modification of it. And the three skulls which I have classed in my Tabular view of results of Osteological Investigations (infra) as belonging to the 'Hohberg' *typus* of His and Rütimeyer, may be looked upon as embodying the results of the working upon that form of the Roman civilisation with which their owners were in contact. Those results are expressed by a decrease in the angularity of the external outlines, and an increase in the cubic capacity indicated in a few cases very strikingly by an open frontal suture; see p. 619, supra. Thirdly, a very large majority, viz. thirty-two out of the fifty-three, adult Romano-British interments investigated by me belong to a type which has frequently been confounded, since the time of Retzius' writings, with the dolichocephalic types just spoken of, but which that excellent ethnographer distinguished from it as 'Cimbric,' a variety of 'the common Celtic' type. Comparing this form of cranium, which I may add is by no means extinct amongst ourselves at the

[1] 'On Two principal Forms of Ancient British and Gaulish Skulls,' pp. 31, 101. Skulls of this form are considered by Sir Thomas Wilde to have belonged in Ireland to fair-headed, light-coloured, blue, or grey-eyed Celtae, or Tuatha De Danaan. See 'Beauties of the Boyne,' 2nd ed. 1850, pp. 221, 237, 239, and the figure at p. 232.

[2] See 'Ethnologische Schriften,' pp. 107, 108.

[3] 'Prehistoric Annals of Scotland,' chap. ix. 1851.

present day, with the elongated but narrow form which he supposes to have belonged to the 'Belgae,' Retzius speaks first of 'the common Celtic form,' and says it differs from the 'Belgic,' in being less narrow and compressed. The Cimbric variety, he adds, which is found in South Sweden and Denmark, is even somewhat broader still; is very like the Scandinavian Gothic form, and is of an elongated oval shape, with a greatly developed occipital region. And Retzius has, by the gift of a 'plaster cast of the cranium of an ancient aboriginal of Scandinavia regarded as the Celt' to the easily accessible and invaluable ethnological series in the London College of Surgeons [1], enabled us to understand most unambiguously what was the type of skull to which he alluded. To this type, most assuredly, the large majority of the adult Romano-British crania found in this cemetery are referable. And I may here say that a skull obtained by me, with many others, from a barrow at Dinnington, near Rotherham, in South Yorkshire (Article xiii), of which casts have been made and presented to various museums in this and other countries by Dr. Thurnam, corresponds very closely with this cast presented to the College of Surgeons by Professor Retzius, and more closely still with some of the very fine skulls obtained by me from Frilford. Professor Ecker, in writing of this cast [2], observes, apparently without having Retzius' comparison above quoted of such skulls to the Scandinavian Gothic type before his mind, that it resembles the skulls he has described as 'Grave Row,' 'Reihen-Gräber' skulls, and assigned to the ancient Germanic and modern Swedish peoples. Very similar skulls, again, I have obtained from Romano-British cemeteries of the later times of the Empire, as testified to by archaeological evidence, at Long Wittenham, in Berkshire, through the kindness of the Rev. J. C. Clutterbuck; from Helmingham, in Suffolk, through the agency of the Rev. G. Cardew, and from Towyn-y-Capel, Holyhead, by that of the Hon. W. O. Stanley. The ancient British skull from a cist at Winterborne Monkton, North Wilts, figured by Dr. Thurnam, 'Crania Britannica,' Plate 58, is closely similar in contour and proportions, as taken by measurement, to the variety of which I am here treating.

[1] See 'Catalogue,' Osteological Series, ii. 880, Prep. 5709.

[2] 'Archiv für Anthropologie,' i. 2, p. 283. As Professor Ecker considers his 'Reihengräberform' to correspond with the 'Hohberg' type of His and Rütimeyer, it would appear that he would consider this cast as belonging to that class, from which, however, its cubic capacity differentiates it.

The osteological peculiarities of this 'elongated oval Romano-British type,' as seen at Frilford, show us that we have to deal, there at least, with *times of civilisation.* For civilisation differs from heathendom in nothing more markedly to the eye of the craniologist than in the age to which persons who have lived under its influences attain; and the long skulls of which I am now speaking differ very strikingly from the long and narrow skulls described by Dr. Thurnam in this very particular, that in very many cases they belonged to very aged individuals[1]. The greater average stature of this variety of Celt (5 feet 8 inches as against 5 feet 6 inches of the older form) may perhaps be in like manner ascribed to the greater civilisation and command of the means of sustenance which we know them to have possessed. I have referred eleven female skulls to this type as against twenty-one male; the female skulls in many cases approaching very closely to the proportions of the medium-sized skulls[2]. A much greater difference, on the other hand—viz., as much as 8·5 inches, judging from the average approximatively obtained from the measurement of the long bones of ten women referred to this type—appears to have existed between the statures of either sex in this type than exists between the statures of modern[3] English men and women. It may be said that the estimation of the stature by the various methods which take one or more of the long bones as their standard, is amenable always to several sources of fallacy, and more especially in the case of female skeletons; but in savage races[4] at the present

[1] See Dr. Thurnam, op. cit. p. 60.

[2] Huschke, 'Schädel, Hirn, und Seele,' p. 48; Hölder, 'Arch. für Anthropologie,' ii. 1, p. 55.

[3] The average height of 295 adult male patients examined in the Somerset County Lunatic Asylum by Dr. Boyd, and recorded by him in the 'Philosophical Transactions' for 1861, p. 261, varied from 67·8 to 65 inches; that of 233 females from 63·2 to 61·6 inches. The average height of the modern German male is given by Vierordt in his 'Grundriss der Physiologie,' 2nd ed. p. 460, as 172 centimètres (5 feet 3½ inches); that of the German female as 164 (5 feet 2¼ inches). In the long barrow explored by Dr. Thurnam (l. c. p. 27) at Tilshead, three male skeletons varied in length from 5 feet 5 inches to 5 feet 8 inches, and three female skeletons from 4 feet 9 inches to 5 feet 3 inches. The average height of the dolichocephalic men from megalithic and other long barrows is given by the same author (l. c. pp. 40, 41) as 5 feet 5 inches as against 5 feet 9 inches for the brachycephalic men from circular barrows.

[4] Sir Andrew Smith, K.C.B., has kindly informed me that he can safely state from extensive observation made during seventeen years' residence in South Africa, that the Amakosa Kaffirs to the eastward of the Colony, average, men 5 feet 8½ inches, women 5 feet 1½ inch.

day an average difference nearly equal to that just given, as deduced from my measurements, has been observed to exist between the statures of the two sexes. And though the Romano-Britons must be considered to have been a civilised population, it must be borne in mind that the physical comfort, upon which such matters as stature depend, of their times was something very different from that of ours, when coal and glass [1] are more or less within the reach of the poorest settled inhabitants of our country. The greater relative stature of the males of this variety of the Romanised Celt may perhaps be accounted for by their having been more exposed to and invigorated by the influences of an out-of-door life; whilst the stature of the females, which is so disproportionately smaller as compared with modern ratios, may have been due to their spending their lives inside houses which, if light must have been cold, if warm must have been dark—which had no chimneys, and only in the case of the rich, hypocausts, and even in their case probably no glass.

Fourthly, a second form of cranium differing from the one just described is found with similar archæological surroundings. It resembles this form in its noble proportions and indications of culture; it equals or exceeds it in length, and is distinguished from it by its greater breadth, and, whilst considering it to correspond to the 'Sion Typus' of His and Rütimeyer, I have spoken of it in my catalogue and tables as the 'globose Romano-British' type. A very large proportion, six out of the eleven female crania, and seven out of the ten male crania, referred by me to this type, belonged to persons of considerable age. The men attained an average stature of 5 ft. 8·5 in. The crania and the other bones of this variety of men enjoying Romano-civilisation have resisted the ravages of time better than those of the other form. There is no reason, however, for supposing that this valuable peculiarity is referable to any conditions not intrinsic to the bones themselves. The mode of their sepulture is identical with that of the other form, and one of the best-marked specimens of the type in question was taken from a grave over which an Anglo-Saxon urn containing the burnt bones of an adult was found. The larger skulls in this series

[1] For introduction of panes of glass, or at least of the manufacturers of them, into England in 680 A.D., see Wylie, 'Fairford Graves,' p. 17, and per contra Corbet Anderson, 'Uriconium,' 1867, p. 69, ibique citata.

belonged in all but one instance to men of a stature little, or not at all, short of six feet, and this large stature must not be forgotten when we admire the large size of their brain-case. Only one female skull which at all approximates in size to these larger crania has come into my hands at Frilford; and this skull belonged to a woman of little, if at all, more than 5 ft. 1 in. in height. But I incline to think that the female crania, seven in number, which I have spoken of as 'the River-bed type modified by increase of size,' and which constitute in the tabular view to which I allude a third variety of the Romano-British series, with an average stature of four feet nine inches and a-half, are to be considered as the female representatives of the 'globose Romano-British type.' For, strikingly similar as the contour of these skulls is to that assigned by Professor Huxley[1] to his 'River-bed skulls,' their capacity exceeds that of those crania, and their measurements come to correspond very closely with those of the smaller male skulls belonging to individuals of smaller stature of the globose Romano-British type, whilst in their solid texture they resemble the larger skulls of that division.

Facsimiles of these female crania have been procured from many excavations in this country. I have found them in the 'Long Barrow' at Crawley, which has been already spoken of as containing skeletons with crania of the brachycephalic British type, and which, it should be added, has furnished us with evidence as to female skulls corresponding to the large brachycephalic male skulls, and differing, therefore, considerably from every variety of the River-bed type. A skull very closely similar to this Frilford variety of pre-Saxon times may be seen in the museum of the London College of Surgeons, under the number '5712 R' in the Catalogue, and with the title 'Peat skull.' And, lastly, a modern female skull obtained for me by Dyce Duckworth, Esq., M.D., from the Hinter-Rhein-Thal, near the Splügen Pass, the country of the 'Disentis' type of the Swiss anatomists, would have shown me, had other evidence been wanting, that this form of cranium has persisted into, and is abundant in, our own day.

A modification of the River-bed type is presented to us in certain small crania to which I have applied the term 'cylindrocephalic.' In this form represented by two female and undoubtedly

[1] See 'Prehistoric Remains of Caithness,' p. 120.

pre-Saxon crania (No. ii. Sept. 1867, and No. iv. Jan. 1868), the frontal and parietal tuberosities are nearly or quite obsolete, and the calvaria, elongating as if in compensation, becomes somewhat cylindroidal in its antero-posterior outline.

Only one male cranium has been found by me at Frilford, which I should class with the River-bed male skull from Muskham, and the Towyn-y-Capel skulls so intelligibly described by Professor Huxley in the 'Prehistoric Remains of Caithness,' p. 120, and frequently examined by myself in the museum of the College of Surgeons. This cranium belonged to a strong man of six feet, beyond the middle period of life, who seems, from the direction of his grave, and the copper staining upon his somewhat prognathic jaws and collar-bone, to have been acknowledged as a Romano-Briton, and to have been buried just as individuals whose osteological remains speak with some authority to their greater culture. By the possession of a slightly greater breadth, and consequently a much higher cephalic index, 78 as against 76 of the typical male River-bed skull just specified, this skull shows a tendency towards assuming the outlines of the smaller representatives of the globose Romano-British type. The fact that but one male against nine female skulls of the River-bed type has been found at Frilford amidst so many other types of head and so many marks of civilisation, is suggestive of the explanation which their having belonged to a slave population would more or less satisfactorily give. The River-bed skulls from the barrow at Crawley which have come into my hands are also all female, as I think, but this barrow has by no means been exhaustively explored. And I incline, though doubtfully, not having had the pelvis nor the long bones to aid me in forming my judgment, to refer the Towyn-y-Capel skulls in the College of Surgeons to the same sex as all the similarly-constructed crania, except the one just mentioned, found at Frilford. In the large male skulls, of which I have spoken, Professors Rütimeyer and His would, I think, recognise their 'Sion typus;' and assuredly they merit the titles *Kräftigkeit* and *Würde,* which Rütimeyer[1] bestows upon them. It may be right to hold that these crania belonged to men British in blood, though here at least Roman by citizenship; but, if we assign them to the Roman immigrants, we shall have an explanation of the enlargement of the River-bed type of skull sug-

[1] 'Jahrbuch der Schweizer Alpen' for 1864, p. 398.

gested to us at once in the very probable hypothesis of intermarriages taking place between foreigners and the, possibly aboriginal, inhabitants of the country, who may have been actually slaves, but must certainly have been in a lower state of civilisation. And in this hypothesis the paucity of male River-bed skulls would also find an explanation.

The Roman immigrants had all but certainly a preponderating proportion of males amongst them, and it would be natural to suppose that the same disproportion prevailed similarly among the swarms of the less settled, less civilised, Saxons. But I am bound to say that the craniological evidence before me leads me to think that the reverse of this very reasonable anticipation was what actually took place, at all events here; for the crania found buried with the Anglo-Saxon insignia of the female sex are most distinctly different, both as to signs of culture, and as to type and contour, from the crania which belonged to the Romano-British women exhumed here. I do not think these Rowenas with somewhat prognathic jaws, and small unhandsomely contoured calvariae, could have been 'exceedingly fair and goodly to look upon;' and I am certain that Martial, though he may not have been a physiognomist, would never have said of these Saxon females what he said of the British lady, Claudia Rufina [1], that she might have been taken by a Roman matron for one of her own country-women.

M. Serres, on the other hand, appears [2] to have convinced himself that in the Merovingian cemetery of Londinières the males belonged to the Scandinavian and the females to the Celtic race. And, upon the general considerations which have been very clearly and convincingly put forward by Professor Pearson [3] and by Mr. L. O. Pike [4], I should be inclined to think that wholesale massacres of the conquered Romano-Britons were rare, and that wholesale importations of Anglo-Saxon women were not much more frequent. Still Anderida was levelled with the ground, and its women and children, as well as its male inhabitants, were put to the sword.

[1] Claudia caeruleis quum sit Rufina Britannis
Edita, quam Latiae pectora plebis habet!
Quale decus formae! Romanam credere matres
Italides possunt.—xi. 53.

[2] Cochet, 'Normandie Souterraine,' p. 188, ed. i; 'Comptes Rendus,' xxxvii. p. 518; 'L'Athenaeum Français,' Oct. 22, 1853, p. 1013.

[3] Op. cit. p. 100.

[4] 'The English and their Origin,' pp. 59, et seqq.

And where the obstinate resistance of the inhabitants may not have provoked the invaders into cruelty, which would have been unnatural, even in the notoriously cruel Saxon (see Salvian, cit. Kingsley, 'Roman and Teuton,' p. 46), the civilisation of the former may very well have attained to such a level as to make them think a retreat into Damnonia preferable to remaining on the same spot with a race so destitute, as the Saxons were, both of the means and appliances of the arts and manufactures which make this life enjoyable, and of the beliefs which make the prospect of another comforting. At Frilford the relics of Roman manufacture, as well as other remains, show, as I have said, that a population must have existed there previously to the Saxon invasion, which was in the possession of a very considerable share of the material and other elements of the civilisation of that period. The very name of this Romanised settlement has been lost, and the Saxon name Frilford, like that of Garford, a village a few hundred yards distant, may possibly speak, as the Rev. Isaac Taylor, in his 'Names and Places,' has suggested with reference to Gateshead[1], to the destruction of a bridge by the worshippers of Frea. The name, indeed, seems to point to the same explanation as the great number of urns; and to suggest that the very real heathenism of the soldiers of Cerdic may have driven away a population who might have acquiesced in submission to such professed Christians as the soldiers of Clovis exhumed at Londinières. Such a story as that which Bede tells us[2] of the refusal of the British priests to eat in company with the Saxons, even in his time, enables us to understand in what abhorrence the Christians must have held them in the days of cremation[3]. Some Lloegrians, as the Triads tell us[4], 'became as Saxons;' but many of the Celtic tribes, as their poems show us, preferred emigration to submission and coalescence. The large Romanised towns, no doubt, made terms with the Saxons, who abhorred city life[5], and who would probably be content to leave the unwarlike burghers in a condition of heavily-taxed submissiveness. The villages would be more exposed to the violence

[1] See pp. 266, 267. Gateshead, however, may mean *Caprae Caput*. See Bede, 'H. E.'

[2] 'H. E.' ii. 4, 20.

[3] See also 'Crania Britannica,' p. 184, vol. i, and pl. xx. p. 3.

[4] Pike, op. cit. p. 46.

[5] See Pearson, op. cit. p. 264.

and lawlessness of hordes made insolent by conquest than the large towns; and I am inclined to think that where we find Roman remains succeeded by relics of the Anglo-Saxon cremation period, on a locality which now bears an Anglo-Saxon name, emigration or extirpation of a Christian population may have very often entered into the now irrecoverable history of the locality.

I further suspect that the heathenism of the Anglo-Saxon domination during the hundred and fifty years[1] which elapsed between the time of Hengist and that of Augustine is one and not an unimportant factor in the complex aggregation of conditions which has given us the Germanic language which we speak. Whilst and where heathenism reigned supreme, the performance of the Church services would doubtless cease; and in an age of few books, and those in manuscript, and in a country which, with whatever centres of civilisation and population, was, after all, but thinly peopled, it is easy to understand how the language of the vanquished succumbed in three or four generations to that of the victors, whose relics speak to their great numbers being so ubiquitously scattered over England. Even in France, where the Merovingians allowed every citizen to declare what law, Frank or Roman, he would live under, and where the priests used the Theodosian code, and so put the Germanic idiom at a disadvantage, it was still employed by the kings and nobles even in the Carlovingian period[2]. On the other hand, during my somewhat considerable practice in the way of exhuming Saxons, and my gradual familiarisation with the two facts of their great aptness at destroying and of their great slowness in elaborating material civilisation, a doubt has little by little grown up in my mind as to the extent of the debt which we are so commonly supposed to owe

[1] Professor Pearson, 'History of England,' i. 101, suggests that the long duration of the struggle may have caused the victory of the Saxon language, by allowing of the perpetual fresh arrivals of German-speaking invaders.

[2] See Gibbon, 'Decline and Fall,' ed. 1838, vi. 118, 351, 376, chap. 38, viii. 156. For an instance of the power obtained and exercised by the Christian ministers, see Fleury, 'Eccl. Hist.' viii. 34, 50, of the Council of Macon. Fleury in his small work, 'Essays on Ecclesiastical History,' tells us, p. 203, English Transl. 1721, that the Goths, Franks, and other German people dispersed into several parts of the Roman provinces, were so few in comparison with the ancient inhabitants that it was not thought necessary to change the language of the Church on their account. On the other hand, Bede tells us, that in his time God was served in five several languages in Britain, namely, Anglorum, Britonum, Scotorum, Pictorum, et Latinorum. See also Taylor, 'Words and Places,' 1864, p. 151; Lingard, 'Hist. A.-S. Church,' i. 307.

to our Anglo-Saxon conquerors. That they conquered a much divided and not very numerous Romanised population of Christians, and overran the greater part, if not the whole, of England proper whilst yet heathens, and within the comparatively short space of time during which they remained such, proves, of course, that the Saxons were superior to the Britons in the arts of war as it was then understood and carried on. But though war in our days is very intimately dependent upon the arts of peace, proficiency in the one set of accomplishments was by no means so correlated with proficiency in the other fourteen hundred years ago. And though my investigations have made me a very firm believer in the reality of the Saxon 'man and steel, the soldier and his sword,' they have not revealed to me any convincing evidence of the importation into this country by these invaders of any such distinctive civilisation as the language often held as to our 'old Teutonic constitution,' or 'the landing of Hengist in Thanet having been the birthday of English liberty,' would seem to pre-suppose. Civilisation and culture are not wholly dependent upon material conditions, but I apprehend they cannot exist without giving us some material and tangible evidence of their existence, at all events *secundum statum praesentem*, of a very different kind from what we find in pre-Augustinian Anglo-Saxon interments in England. Mr. Merivale's dictum [1] to the effect that 'it may appear that moral culture is almost altogether independent of material progress,' is too much out of keeping with the ordinarily-accepted views of the way in which the external world works upon human nature, *curis acuens mortalia corda*, to need discussion at length; and when Professor Pearson [2] says 'it would be unjust to judge the Teutonic tribes of the fifth century by the low development of the mechanical arts among them,' we expect to have evidence of some other arts and pursuits having somehow or other attained to a compensatory high development amongst these races at that time. Guizot [3], it is well known, has compared the social and political condition of the Germanic races at this period of their history to that of the Red

[1] 'Conversion of the Northern Nations,' p. 186.

[2] See, however, his 'History of England,' pp. 44, 51, 103, 112, 130, 264. The high development of the pictorial art to which Professor Westwood's magnificent work, recently (1868) published, speaks, belongs to Christianised, and therefore as little to 'unalloyed Saxondom' as do Cædmon, Bede, or Alcuin.

[3] 'Hist. Civ. Franc.' lect. vii. tom. i; cit. Merivale, ubi supra, note G, p. 185.

Indians; and when we find Sharon Turner, the historian of the Anglo-Saxons, telling us[1] that Ethelbert, after his conversion by Augustine, 'became distinguished as the author of the first written Saxon laws which have descended to us, or which are known to have been established, an important national benefit for which he may have been indebted to his Christian teachers, as there is no evidence that the Saxons wrote any compositions before,' we may be inclined to think that the views of Guizot are nearer to the truth than those of Ozanam[2], Greenwood, and Rogge.

We have historical, literary, archaeological, and anatomical evidence for saying that two or more distinct varieties of men existed both in England and France, both previously to and during the periods of the Roman and of the Teutonic invasions and dominations[3]. The earliest Welsh traditions, Professor Pearson informs me, speak 'of the social races inhabiting Britain, the Kymry, the Lloegrwys, and the Brythons,' all descended from the Kymry. The word 'Kymry' itself, however, has been supposed,

[1] 'History of the Anglo-Saxons,' i. 332. See also Taylor, 'Words and Places,' p. 339, and per contra, Kemble, on Runes, 'Archaeologia,' xxviii.

[2] Ozanam, however, cit. 'Merivale,' l. c. 187, says, 'Les lois de l'ancienne Germanie ne nous sont connues que par les témoignages incomplets des anciens, par la réduction tardive des codes barbares, par les coûtumes du moyen âge. Il y reste donc beaucoup de contradictions, d'incertitudes, et de lacunes.' Gibbon may be shown to be similarly self-contradictory by a comparison *inter se* of the following passages; vol. i. chap. ix. p. 362, ed. Milman, 1838; vol. vi. chap. xxxviii. p. 325; vol. v. chap. xxxi. p. 317. The stories told of the two Gothic Princes in the two latter passages are quite inconsistent with the statement contained in the first of the three, to the effect that 'in the rude institutions of the barbarians of the woods of Germany we may still distinguish the original principles of our present law and manners.' See Finlason's Introduction to Reeves' 'History of the English Law,' 1869, p. xl; and Professor Pearson's 'Historical Maps,' 1869, where at p. vii. the Professor speaks of the Saxon invaders as consisting of 'a few boat-loads of barbarians.' I agree as to the barbarism, but differ as to the numbers of the Anglo-Saxons. Both these valuable works came into my hands after the coming of these sheets from the printers. See per contra, B. Thorpe, 'Ancient Laws and Institutes of England,' preface, p. xxii.

[3] Gibbon, v. 351, ed. 1838, says, 'If the *princes* of Britain relapsed into barbarism whilst *the cities* studiously preserved the laws and manners of Rome, the whole island must have been gradually divided by the distinction of two national parties.' See also Pearson, l. c. pp. 99, 100; Coote's 'Neglected Fact in English History,' pp. 144, 149, 169; Skene's 'Four Ancient Books of Wales;' Gododin, 'Poems,' pp. 382, 394, 412; Broca, 'Recherches sur l'Ethnologie de la France,' Mém. Soc. Anthrop. de Paris, tom. i. 1860; Sir William R. Wilde, 'Beauties of the Boyne,' pp. 229, 232; Dr. Thurnam, 'On the two principal forms of Ancient British and Gaulish Skulls,' Memoirs of the Anthropological Society of London, vol. i. ibique citata; Huxley, 'Prehistoric Remains of Caithness,' pp. 114, seqq.

like the words 'Frank' and 'Aleman,' to denote social or confederative, rather than genealogical, community; and, though we are warned thus *in limine* against any premature attempt to harmonise the results of philological with those of craniographical enquiry, it may not be entirely hopeless to attempt to harmonise the traditions which tell us that the Romanised town populations, the 'Lloegrians,' took the side of the Saxons against their own countrymen, with the facts of our 'finds' in cemeteries. Now, these facts, as they have presented themselves to me, I have, with the help of light borrowed from many other investigators, read thus. Two varieties of capacious crania, one dolichocephalic and the other brachycephalic, have been found by me in cemeteries referable by their archaeological characters to the periods corresponding with, and immediately subsequent to the close of the Roman domination in England. These two varieties of skulls are not ordinarily found occupying one and the same tumulus, at least with the relative positions which the remains of two races inhabiting the same district peacefully usually hold to each other, and I incline, though but doubtfully, to anticipate that evidence will be ultimately produced to identify the dolichocephali in question with the Lloegrian traitors, and the brachycephali with that portion of the Kymry which preferred exile to the Saxon yoke. The fact of the dolichocephali having been found abundantly (see p. 624, supra) in the Suffolk region of the Littus Saxonicum, where the Celt and Saxon are not known to have met as enemies when East Anglia became a kingdom, is not without its significance. Their geographical distribution may indicate a greater political pliability just as their greater variety of cranial conformation indicates a greater anatomical plasticity. In the same cemeteries with both of these varieties of skulls I have found skulls which are very closely similar to Professor Huxley's 'River-bed' type of skull, and which I should be inclined to think may have belonged to a serf, or at all events to a poor, population, whose necessities may have made them as indifferent as any similar population is now to the political leanings of their masters. I should agree with Professor Huxley in considering this a very ancient form of cranium; but, though I should allow, with a knowledge of the great aptitude for modification possessed by the human cranium, that it may be connected by transitional forms with the dolichocephalic Celtic varieties[1], I am convinced that it is even more closely allied with that brachycephalic form which has been called 'Ligurian' by Professor Nicolucci, which is identified with the 'Disentis' type of Professors His and Rütimeyer, by Dr. Hölder in his excellent paper on the ethnography of Würtemberg[2], though the Swiss Professors themselves would demur to this unification[3]; and which, finally, is, I apprehend, the form considered till recently[4] by nearly all continental anthropologists as the oldest of European types. I am inclined to hold that the rough-hewn brachycephalous Briton, of whom Dr. Thurnam has written in his paper on 'the two principal forms of ancient British and Gaulish skulls[5],' was distinct from the brachycephalous 'Ligurian,' though very possibly descended from one common stock; just as I should think it very probable that the cultured brachycephalous skulls of which I have spoken were produced simply by the operation of civilising influences upon the rougher crania of similar type, but of earlier times; and as I should suppose that Roman civilisation and Roman inter-crossing elaborated the larger out of the smaller and earlier dolichocephalic skulls of this country. The five varieties which I believe may be thus distinguishable—viz., the two brachycephalous, and the two dolichocephalous, cultured and uncultured respectively, and the 'Ligurian'—will be found to be connected with each other by inosculant forms. Even under conditions of the most primitive[6] simplicity and peacefulness, the human cranium shows a great tendency to variation; and in England we must recollect that this essential liability to variation was much intensified in early times by the migrations and immigrations of the Belgae from the continent; by those of the pastoral inhabitants of the then thinly peopled, forest-covered country; and in later times by those of the Romans and Saxons. Most or all invasions entail more or less of

[1] As taught by Professor Huxley, l. c. p. 120; and 'Proc. Soc. Antiq.' April 19, 1866.

[2] 'Arch. für Anthrop.' bd. ii. hft. i. 55-57.

[3] See 'Crania Helvetica,' p. 41; 'Arch. für Anthropologie,' i. 70, 1866; Ecker, 'Cran. Germ.' pp. 76-86; Huxley, l. c. pp. 117-118.

[4] For a discussion as to the priority in point of time of the brachycephalous or the dolichocephalous form of skull, see Mortillet, 'Matériaux pour l'Histoire positive et Philosophique de l'Homme,' 1867, pp. 383-385; Ecker, 'Crania German.' p. 93.

[5] 'On two Forms,' l. c. pp. 31-44.

[6] See Bates, 'Naturalist on the Amazons,' ii. p. 129, and *per contra*, Ecker, 'Crania Germaniae Meridionalis,' p. 2; Gratiolet, 'Système Nerveux,' ii. p. 286.

intermarriage between the invaders and the invaded; and the craniographer who considers what very motley hordes passed into England under the names 'Roman' and 'Saxon' respectively, and for what long periods these immigrations continued to be made, will be cautious as to his inferences. Other disturbing conditions were introduced by the invasions specified: among them I need only mention the establishment of an antithesis between town and country life, which, in a country intersected by woods and ill-provided with roads, is equivalent to the establishment of an antithesis between civilisation and savagery. Isolation, howsoever produced, whether by social, by political, or by physical barriers, tends to exaggerate the ethnical or tribal characteristics which intercrossing tends to obliterate. But a subjective cause of much fallacy lies in the curiously corresponding psychological fact that one class of mind is as prone to overrate distinctions as another is to underrate differences.

In conclusion, I must be allowed to express my sense of the obligations I have incurred to Professor Phillips, whose advice and opinion I have very constantly sought; to Professor Pearson, whom I have consulted well nigh as frequently; to Heathcote Wyndham, Esq., M.A., Fellow of Merton College, who has given me assistance upon several chemical and mineralogical points which arose in the course of my investigations; to James Parker, Esq., for suggestions as to several archaeological matters; and to Charles Robertson, Esq., for superintending these disinterments upon several occasions when I was unable to be present.

CATALOGUE OF FRILFORD EXCAVATIONS.

October and November, 1864.

I. SET. *Cranium A.* Found with a fibula 2 ft. above it, wrongly described by me as a male skull in 'Proc. Soc. Antiq.' 2 ser. iii. 139. Probably an Anglo-Saxon woman. Middle-aged.

Cranium B. Found with a small Roman coin. Probably, from this and from anatomical characters, a Romano-British woman. Middle-aged. Elongated oval type.

Lower jaw from leaden coffin No. i. Roman man. *C.* Middle-aged. In this leaden coffin a coin of Constantine the Great was found.

Calvaria E. Asymmetrical and with a partly open frontal suture. It is possible, though this calvaria came into my hands a month later than the lower jaw *C*, that it belonged to it. Mr. Akerman says (p. 3, 'Proc. Soc. Ant.' l. c.) that the remains from the two coffins were handed over to me for examination.

Calvaria D. Incomplete. From a second leaden coffin. Middle-aged man. Capacious.

Fragments of second lower jaw, possibly belonging to calvaria *D*.

January 25, 1867.

II SET. i. *Skull* of 'Hohberg' type of His and Rütimeyer, with long and other bones, from a leaden coffin, in which were found five coins, one of which was of Constantine the younger, a second of Valens, and a third of Gratian, obiit A.D. 383. Femur 19⅛, humerus 13·9. Old man.

ii. *Skull* of 'Sion' type of His and Rütimeyer, with long and other bones, from a leaden coffin, in which no coins were found. Clavicle and second left metatarsal broken and repaired during life.

iii. *Coins* (some of which were lost in sending by post), and leaden coffins.

iv. *Plain urn*, which contained the bones of a child probably two to three years old.

2 old men, 2 middle-aged men, 2 middle-aged women, 1 child, = 7 bodies.

April 16, *May* 9, 16, 1867.

III SET. i. *Skull of old man*, of 'Sion' or globose Romano-British type. Skull circumference, 22½ in. Femur, 20¼ in. Tibia, 15¼ in. Height, 6 ft. 1 in.

ii. *Skull of young woman*, of 'enlarged River-bed type,' see p. 625, supra, with some osteophytic deposit internally, æt. 20 to 25. Humerus, 10·3 in. Tibia, 11·7 in. Stature, *circa* 4 ft. 6 in.

iii. *Lower jaw, and frontal bone, platycephalic*, said to have been found with umbo No. v. Old man.

iii¹. *Second lower jaw*, also said to have been found with umbo. Very old man.

iv.

v. *Umbo*, reported to have been found with jaws iii and iii¹, but in the grave in which the umbo was reported to have been found a secondary interment was supposed to have taken place, a tibia and femur having been observed in it lying with their relative positions reversed.

vi. *Calvaria of young woman*, of 'enlarged River-bed type,' with cephalic index 77, and some leaning towards the smaller type on the one side, as well as to the globose Romano-British male skull vi of September, 1867, on the other. It resembles skulls vi of April 1, 1868, and xi of same date, and very possibly may be the female form of the Romano-British globose type. It is mainly in length that these female skulls are inferior to the smaller male skulls of the 'Sion' type. N.B.—8 millimètres is the average excess of male length. Femur, 13 in. Humerus, 10 in. Mean stature from these two bones, 4 ft. 2·5 in. A phalanx of an ox and a piece of pottery were sent with this skull. This skull was found very near an infant's.

vi. *Child about time of birth.*

vii. *Skull of very old man*, of Romano-British elongated type. Humerus, 13·2 in. Radius, 9·2 in. Exostotic growths on humerus. Stature, 5 ft. 10 in.

viii. *Skull of strong young man*, of Romano-British elongated type. No long bones.

viii¹. *Skull of child*, first dentition, middle period of.

ix. *Bones of child*, first dentition, early period of.

ix. *Bones of child*, first dentition, early period of.

x. *Skeleton of young man*, of globose Romano-British type, very similar to skull ix of March 17, 1868. Found with fragments of Roman pottery. Femur, 17 in. Humerus, 22·5 in. Radius, 9·2 in. Ulna, 10·1 in. Stature, 5 ft. 4 in. Some carious teeth.

xi. *Skull of young woman*, of Romano-British elongated type.

xi. *Skull of young woman*, (?) of Romano-British globose type.

xi. *Skull of young man* (?). Type (?).

xii. *Calvaria of young woman*, of Romano-British elongated type, with osteophyte internally. Much water-worn, teeth all good but one, which is carious.

xiii. *Skull and long bones of young Anglo-Saxon woman*, from a shallow 18-in. grave running from west by south to east by north, the deviation from orientation being 18° north. Two fibulæ, four or five beads, and fragment of an urn. Femur, 17 in. Tibia, 14 in. Stature, 5 ft. 4 in. A piece of Anglo-Saxon and another of Roman pottery found in this grave, of small size.

4 old men, 6 young women, 1 infant, 3 children, 2 young men, 1 doubtful, = 17 bodies.

xiv. *Skull of old man*, of 'Hohberg' type, with the long bones. A large wound through skull, healed during life. Grave 3 ft. 6 in. deep, without relics, running in a direction from N.W. to S.E. Femur 20 in. long. Stature 6 ft. All the molars of upper jaw are lost. In the lower jaw the two anterior molars are left. Some teeth are carious; they are small in size. The grave was immediately on the right of that of the Anglo-Saxon woman xiii.

xv. *Skull of Romano-British woman*, with long bones, from a grave of same direction but not quite the same depth as the preceding—2 ft. 9 in. Femur, 16 in. Humerus, 11¾ in. Stature, 5 ft. A good instance, as is also xiii of Sept. 26, 1868, xxiv of Feb. 21, 1868, and cranium *A* of Oct. 1864, of the close adherence to type which female skulls, especially of the elongated Romano-British type, show.

xvi. *Skull of very old man*, $\frac{6}{10}$ inch thick, of elongated British type, with sagittal furrow posteriorly. No relics.

xvii. *Patterned cremation urn*, containing bones of child before period of second dentition.

xviii. *Plain urn*, containing the bones of a person about the period of puberty.

xix. *Skull of woman*, middle-aged, of elongated Romano-British type. Charred matter and pottery, and carious teeth from the grave. Femur, 16 in. Stature, 5 ft.

xx. *Bones of child* under 6 years.

xxi. *Skull of young person* with abnormal succession of teeth. Spongy growths in orbits and hypertrophic calvaria. Reported to have been found with two pieces of blackish pottery and a nail.

xxii. *Skull of young man*, of 'Hohberg' type, wanting jaws. Femur, 18·3 in. Stature, 5 ft. 11 in. Sent with pieces of pseudo-Samian ware.

xxiii. *Skull of male*, of 'Sion' type. Massive, weighty. Belonged to a man past middle period of life. Many teeth lost, both before and after death.

xxiv. *Skull of young person, probably woman*. Fragmentary, without history.

xxv. *Skull of young person, probably woman*. Teeth carious.

xxvi. *Lower jaw of old person*, with six teeth.

xxvii. *Bones of old woman*, Dec. 31, 1867. Elongated British type. Femur, 15 in. Stature, 4 ft. 8 in. Coffin hooping and nails found in grave.

xxvii[1]. Fragments of an urn of Anglo-Saxon ware, said to have been found with a burnt bone, Dec. 31, 1867.

1 old woman, 3 old men, 1 old person (sex ?), 3 young women, 2 young persons (sex ?), 1 young man, 2 children, 1 middle-aged woman, = 14 bodies.

September 16, 17, 18, 1867.

IV Set. i. *Calvaria and bones of old woman* from a grave running from N.N.W. by N.W. to S.S.E. by S.E. 3 ft. 2 in. deep down to the coralline oolite. A burnt flint, some pseudo-Samian ware, and some gray lathe-turned pottery, were found in the grave with her. Femur, 16⅝ in. Radius, 8½ in. Stature, 5 ft. 1½ in. Of enlarged 'River-bed' type. Lower jaw nearly destroyed by water-wear.

ii. *Similar calvaria from continuation of same trench.* 'Cylindrocephalic' female skull. Stature, 5 ft. 4 in. Femur, 17 in.

iii. *Skull and long bones of young man of elongated British type*, from continuation of same trench. A sheep's tooth close by his jaws, and two flints. Stature, 5 ft. 10 in. Right fibula a good deal curved, epiphyses not fused.

iv. *Skull of young Anglo-Saxon*, found with spear and umbo. The spear at right side of head with point upwards. This had been a secondary interment, the upper jaw of a very old man (iv[1]) having been found close to this skull, as also a manubrium sterni with articular facet for first left rib much enlarged, which could not have belonged to this skeleton. The grave was broader than the others, and had large stones set along its sides. Its direction was W.N.W. to E.S.E.

iv[1]. *Upper jaw of old person, probably male.* Sternum and large head of humerus and os calcis with it.

v. *Skull of very old man*, with skeleton, from continuation of trench whence the Anglo-Saxon No. iv came. The direction of the grave the same, but no relics nor any stones set around it. Skull like iv and xi of March 17, and i of March 23, 1868. A mixed form combining the Hohberg with the Sion type. Can these skulls have belonged to Christian Anglo-Saxons? See p. 616, supra. Femur, 18 in. Stature, 5 ft. 8 in.

vi. *Skull of old man* of 'Sion' type, found with skeleton at a depth of 15 inches below the patterned urn ix, the urn occupying a space corresponding with the top of the sacrum of the skeleton below. Skull bones a little roughened by water-wear, but also strongly made, and indicating, as do the other bones also, both age and great strength by their various outstanding processes. Ceph. index, 78. Stature, 5 ft. 8 in. Found with several pieces of flint and with pieces of pottery.

vii. *Skeleton of very old man*, of elongated Romano-British type. A typical skull such as No. 5709 in Royal College of Surgeons, which belonged to an 'ancient aboriginal inhabitant of Scandinavia regarded as the Celt;' and called 'dolichocephalic by the donor, Professor Retzius.' Found in a trench between the trench with skeleton No. vi in it to the south, and the one with the Anglo-Saxon No. iv in it to the north. Femur, 19¼ in. Stature, 5 ft. 11 in. Osseous upgrowth of acetabulum, and hypertrophy of head of femur to correspond. Cephalic index, 72.

viii. *Young woman*, æt. 17 to 19, from trench in same direction, but to south of others, dug September, 1867, and to north of trench containing Romano-British woman, xv of May, 1867. Found with flints and shards.

ix. *Urn* found above skeleton vi, containing bones of child under 8.

x. Fragment of probably a *holy-water vessel.*

About 4 ft. of ground had fallen in to the right of the pit, looking towards the River Ock.

2 old women, 1 young woman, 4 old men, 2 young men, 1 child, = 10 bodies.

January 9, 1868.

i. *Child's bones*, between 9 and 10 æt. Many fragments of scoriform lava, probably Niedermennig (Daubeny, 'Volcanos,' p. 50); no other relics brought with it; compare Wylie, 'Fairford Graves,' p. 24, and account of Anglo-Saxon woman, xxii, Jan. 6, 1869. (Cf. Schaaffhausen, 'Die Germanische Grabstätten am Rhein,' 1868, p. 122, and p. 608, supra.)

ii. and iii. *Young women* (20, 23), placed close, side by side, in the same trench; ii a little shallower than iii, and a little further forward, and with the left humerus across the cervical region of iii. Roman pottery and nails. The legs of the two skeletons were wide apart. The iron relics (nails) were found on the pelvis. No nails near the feet. Probably buried at same time; coffins in same trench.

iii[a]. *Delicate unpatterned urn* with child's bones, about 6 in. below the surface, and 18 in. above skull No. iv; the place about a yard to the right of the place where the patterned urn of September 17, 1867, was found. Child's age towards the end of 5–6th year.

iv. *Old woman's skull*, much senile atrophy, found 18 in. below urn iii[a] with face upwards. No soil had fallen into the skull; one coin, the largest, was found on lower jaw; two smaller ones on atlas and axis, which are stained in consequence. Coins not identified. One nail was found on the right side of the head, but none on the left, nor at the feet. All the bones are very light. Humerus, 10·710 in. Femur, 14·5 in., gives stature 4 ft. 6 in.; humerus (say 11 in.) gives stature 4 ft. 10·6 in.; mean 4 ft. 8·3 in., small skull, cylindrocephalic.

v. *Skull of a child*, removed by the men. Close by was found a piece of lead, possibly from a leaden coffin, and with the bones an ulna, which had belonged to a very powerful man, which had been part of a fractured segment repaired during life. Compare account given p. 620, supra, of skeletons from leaden coffins. First dentition complete; second not begun.

3 children, 3 women, 1 old, 2 young, = 6 bodies.

January 15, 1868.

vi. *Strong urn*, not patterned, containing child's bones. Removed by men from earth a little to right of No. i of Jan. 9. Aged probably about 9–10. Premolars not displaced; milk molars.

vii. *Child's bones*, a little to right of urn vi. Early period of first dentition.

viii. *Child's bones* under 6, at extreme left of 'fall,' i.e. of mass of earth thrown down in quarrying operations. Removed by men.

ix. *Young man*, with nails and Roman pottery, nails at head and feet. Elongated British type.

x. *Child*, much decayed. Early period of first dentition.

xi. *Calvaria, man*, middle age, with pot and flint. Long bones much water-worn. Elongated British type. No lower nor upper jaw.

xii. *Old man*. Face upwards, and left arm across body. About middle of 'fall' and to right of viii and xi. Elongated British type, but vertically carinate like the preceding specimen. The crossing of the arms may point to his being an Anglo-Saxon. See xxii, Jan. 6, 1869.

xiii. *Old woman*. Calvaria and femora, 16·7 in. Osteophytes and pacchionian pits. Water-worn. 'Sion' type. Compare vi of April 1, 1868; xi of April 1, 1868; vi of May, 1867.

xiv. *Old woman*. Skeleton sent by carrier. Right humerus, 11·7; left, 11·2. Femur, 13·4. Anglo-Saxon woman.

4 children, 1 young man, 1 middle-aged man, 1 old man, 2 old women, = 9 bodies.

January 20, 1868.

xv. *Knife*, with much rust and (?) woody fibre adhering to it. Found close to xii in the loose earth which had fallen to the bottom of the pit. Could this knife have belonged to the little old woman, xiv (?); see skull, which is much more like that of the Anglo-Saxon woman xiii of May, 1867?

xvi. *Urn*, a little to the left of the knee of xiii of January 15. *The urn unopened.*

xvii. *Skeleton of woman* (*young*). Her tibia was 1 ft. beneath the humerus of an Anglo-Saxon, xviii, who was lying in a direction from S. S.S.W. to N. N.N.W. and at right angles to her grave, which was in the ordinary Romano-British direction from W.N.W. to E.S.E. In the intersection of the graves a beautiful coin of Constans was found. Left radius injured during life and repaired. Both humeri malformed. With this came part of upper jaw of old person, with three teeth from canines inclusive backwards, of elongated Romano-British type. Stature, 4 ft. 11 in.

1 young woman, 1 age and sex uncertain, = 2 bodies.

February 8, 1868.

xviii. *Anglo-Saxon young man*, with umbo, spear, knife (no buckle), Roman tiles, stones round grave; coin of Constans in intersection of his grave with that of xvii; some animal's (sheep ?) bones in grave (see Akerman, 'Pagan Saxondom,' Introd. p. xvii); grave from head S. S.S.W. to N. N.N.E.; foot at right angles to grave xvii 2 ft. deep; xvii 3 ft. A tooth of ox between head of Anglo-Saxon and feet of Romano-British woman. Femora only partially recovered, a large stone over their lower ends having crushed them very much. Fragments of great size. Clavicles long and curved. Humerus, 13·1; radius, 9·3. For Roman tilings, see 'Archiv für Anthropologie,' i. 3, 356; (see xxii of Jan. 6, 1869).

xix. *Man, strong, beyond middle life, probably*. Head of River-bed type, parietal protuberance. Ribs broken and repaired in life; abscess at root of one molar. A good deal of exostosis on left humerus. Copper staining on jaw, and clavicle. No nails found with body. Romano-British direction of grave. See p. 628, supra.

xx. *Skull of middle-aged man*, with Roman tile. Very elongated, with long bones. Bones loose. Skull peculiarly elongated.

xxi. *Skull of young man*, found with vertebra of ruminant in grave, and a fragment of pottery. Diseased hip. This cranium has some approximation to the modern form of English crania, and resembles herein cranium No. v of Sept. 1867, crania xi and iv of March 17, and cranium No. i of March 23, 1868.

2 young men, 1 aged, 1 middle-aged, =4 bodies.

February 21, 1868.

xxii. *Old Romano-British man*, large skull with long bones. Copper stain on left ulna, immediately to N.E. of grave xviii, so that the Roman tiles found in taking out the skeleton were supposed to belong to this skeleton. A good deal of charcoal and decayed wood was found near the head, but not near the legs nor pelvis. Of elongated type. Femur, 18·4; ulna, 10·3. The man, however, who took out the tiles supposed them to belong to Anglo-Saxon No. xviii. In the cases described by Wanner ('Das Alemannische Todtenfeld bei Schleitheim,' 1867, pp. 13, 16; 'Archiv für Anthrop.' ii. 3, p. 356), Roman tiling was similarly employed. In some cases the graves were, as here, so close as to have only a single tile as a wall between them. See xxii of Jan. 6, 1869.

xxiii. *Old woman* (? old man). Skull and long bones. Femur, 17; humerus, 12·2; radius, 8·9; ulna, 9·5; stature, 5·4. This is a very old skeleton, and I think the sex may be doubtful, but it is probably, from lower jaw's muscular markings, a male. The forehead is vertical, but perhaps abnormally; the vertex is carinate. Globose type.

xxiv. *Skull of old woman*, with five coins; one of Valens, and one of Constans I. Flint, ball-shaped, chipped. Flat flint and Roman pottery. Femur, 15·6; tibia, 12·4; radius, 8·4. Lower jaw nearly destroyed by water-wear. Elongated type.

xxv. *Man*, prime of life. Frontal suture patent.

xxvi. *Fragmentary cranium of old person;* bones of young person of eighteen to twenty wrongly assigned to it.

xxvii. *Long calvaria*, man prime of life. Lower jaw a good deal water-worn, and the long bones lost, perhaps destroyed by decay.

xxviii. *Infant.*

xxix. *Child*, first dentition complete.

1 child, first dentition complete, 1 infant, 2 men, prime of life, 2 old men, 1 old woman, 1 old, doubtful of which sex, prob. male, 1 young person, 18–20, =9 bodies, counting xxvi as two.

March 4, 1868.

i. *Skeleton of Romano-British woman*, adult, of globose type, like No. xi of April 1, 1868. Skull larger and more strongly made than most female skulls, and a nearer approximation to male skulls of same type. Teeth considerably worn; no wisdom teeth developed. Orthognathous, with posterior sagittal furrow. Femur, 16·2; tibia, 13·4; humerus, 11·2; stature, 5 ft. 8·10 in.

1 adult woman.

March 17, 1868.

i. *Skull*, with long bones and patellae, of a *very strong young man*, buried with fragments of Roman pottery, black and red, and nail, with wood adhering to it, from coffin. Femur, 18·5; humerus, 13·3. Skull, flat and broad, to be reconstructed. Hyoid fully ossified. Elongated type.

ii. *Skull of old man*, with femur, and tibiae, and nails near head. Of elongated, flat type. Large. Very large bones. Femur, 18·9.

iii. *Skull of young woman*, with long leg-bones and patellae; short stature; teeth carious; and abscess in alveolar processes. Elongated type. Lower jaw all but destroyed by water-wear.

iv. *Skull of adult man*. No femora; no lower jaw; carious teeth. Skull high and long, but not delicate, though possessing transverse post-coronal depression. (Compare skull v of Sept. 1867, and skull i of March 23, 1868.) No femora were found with it; the skeleton having been thrown down in a 'fall' during the quarrying operations.

v. *Skull of young Anglo-Saxon woman*, very much contorted and distorted by post-mortem pressure, found in a grave 2 ft. 4 in. deep, with six beads, some near head, some over chest, perforated, of various sizes, of blue spongy glass, striated concentrically; fibula on either shoulder of flat shape, circumference gimped, and immediately within a circle of stamped round depressions, diameter, 1·3, of much the same pattern, but not identical, nor of quite same weight as another fibula of uncertain date and place from this cemetery; of quite different pattern from the two other sorts of fibulae found here, though of same general shape, flat, as fibulae of xiii, May, 1867. A skewer-shaped bronze pin, 4 in. on the left breast; a knife, 3 in. long, near the waist. For pin fastening shroud, see 'Pagan Saxondom,' p. 71, pl. XXXV. fig. 5; 'Archaeologia,' xxxv. 477. The direction of the grave was not quite that of the Romano-British, viz. W.N.W. to E.S.E., but was very nearly this, running, as it did, from a little to the north of W.N.W. to a little to the south of E S.E. There was some Roman pottery in the grave, animals' bones, an ox tooth, an oyster-shell, and a flint. The skull and the other bones are much water-worn. But we can see that the skull is small and short, that the nasals rose from a level with the glabella, which was little prominent, though underlaid, as also the similarly low superciliary ridges, by sinuses. The parietal tubera are fairly marked, the minimum frontal diameter apparently very small, 3·7 in., though it may have been diminished by compression, the same minimum frontal being 3·9 in. in each of the two other Anglo-Saxon women from Frilford. The interior of the skull has the smooth polished appearance characteristic of youth. The wisdom-tooth in the right half of the lower jaw is very small, and not at all worn. The premolars are also little used in comparison with the two true molars, though more than the third molar. The chin seems to have been emarginated unusually below, but to have been fairly pronounced. The upper jaw, judging from a small portion of the right side, must have been slightly prognathous. None of the teeth are carious. There is copper staining on some of the ribs, the clavicle of the left, and the humerus of the right side.

vi. *Patterned urn*. Probably a holy-water vessel, with characteristic bosses. Found a little to the south of the grave of Anglo-Saxon woman No. v. It was about 4 in. from the surface of the ground with its top edge, which had escaped the plough; its bottom was about 6½ in. Close to this urn or holy-water vessel was a mass of *infant skull bones*, the child having been about (before or after) the time of natural birth. It is possible that the diggers of the Anglo-Saxon woman's grave may have disturbed this urn in the process, and having broken the urn may have reinterred it in fragments, and its contents apart from the fragments. A plain fragment, which does not appear to have belonged to the patterned fragment, was also found at some little distance from the patterned urn and the baby bones. And it is again possible that the child may have been deposited in the urn of which this latter fragment was a part. But I incline to think this was not the case, as the child's bones do not bear marks of fire; and though the Roman rule expressed in the words 'minor igne rogi' (Juv. xv. 140; Plin.

vii. 16) may not apply to an Anglo-Saxon interment, these bones may have belonged to a still-born child, for which no urn would probably have been used.

vii. *Skull and some long bones, imperfect, of young woman*, wisdom teeth not through. A piece of grey spongy pottery; no other relics. Romano-British direction, W.N.W. to E.S.E.

vii[1]. *Skull and long bones of child* of 8 years, with two pieces of Roman pottery.

viii. *Skull and some long bones of old man.* Skull both globose and elongated. Humerus roughened at point of origin and insertion of muscles.

ix. *Skull and long bones of old woman* (? very old man), with Roman pottery. Skull of type of x, May, 1867. Femur, 16·3; tibia, 13·2; humerus, 11·4; stature, 5 ft. 1 in. It is doubtful, I think, whether this skull may not be a very old man's. The lower jaw shows great marks of old age. The straight clavicles point the other way. Of 'Sion' type.

x. *Skull with long bones, of young woman*, possibly Christian Anglo-Saxon. This skull was sent by the men, but without relics. The type seems to be that of Anglo-Saxon woman xiii, May, 1867, and of woman, 771 m, Oxford University Museum (see p. 616, supra), from Helmingham. The wisdom teeth are, though little worn, very small in upper jaw. The jaw prognathic. Some little doubt as to sex from slope of forehead and parietes and large mastoids, but, nearly certainly, female. Femur, 15 in.; stature, 4 ft. 8 in. (1–51) by 4 = 56; tibia, 11·5; fibula, 11·2; ulna, 8·3; radius, 7·5; humerus, 10·4. The cervical vertebrae, from 7th onwards, having been impacted into the interior periphery of the lower jaw, it is probable the head was raised when the body was buried, and hence that this may have been an Anglo-Saxon interment. With this skull compare skull 5712 D, Royal College of Surgeons, which belonged to an Anglo-Saxon woman from Brighthampton, and No. xiii, of Sept. 26, 1868, infra.

xi. *Strong young man*, with long bones. Protuberance on right parietal. Buried with nail. Wisdom teeth either not coming or retarded. Second molars little worn. Compare skull ix, supra, and vii, Sept. 1867. Taken out of grave by the workmen, as also No. x.

1 infant, 1 child of eight, 4 young women, 3 old men, 1 adult man, 2 young men, = 12 skulls.

March 23, 1868.

i. *Young woman*, aet. 17–18, no relics. Buried in grave running W.N.W. to E.S.E. Good skull of modern well-developed European type. Ceph. index, 78. Height, 5 ft. 4 in.; humerus, 11·9 in.; femur, 17 in.; skull, 7 in. long, 5·4 in. broad, circumference 19·6 in. Compare skulls vii of Sept. 1867, and xi and iv of March 17, 1868, for somewhat similar conformation. Can these skulls have belonged to Christian Anglo-Saxons? See p. 616, supra.

ii. *Old woman*, skull and long bones. Romano-British direction. No relics. Sutures much obliterated. Exostoses in antrum maxillare. Extreme length of skull, 7 in. The roots had reached into its interior. Vertical forehead. Elongated type. Femur, 15·5 in.

ii[1]. *Child*, with first permanent molar not through, at a short distance from ii; a fragment of pottery, Romano-British, with it.

iii. *Skull of strong adult man*, with no long bones. Of broad platycephalic type. Teeth small, considerably worn, one carious. A nail found with bones. The skull was full of the small mollusc *Achatina acicula*.

iv. *Skull with long bones, very perfect, of very strong adult man*, found with Roman tile and Romano-British pottery. Femur, 18·8 in.; humerus, 13·8 in.; radius, 9·9 in.; ulna, 11 in.; stature, 5 ft. 11 in. Of globose type.

1 young woman, 1 old woman, 1 child, 2 adult men, = 5 skulls.

April 1, 1868.

i. *Young Anglo-Saxon man*, lying with head at N.N.E. and foot at S.S.W., the very reverse of the compass-points held by the head and foot respectively of Anglo-Saxon xviii of Feb. 8, 1868, and of Anglo-Saxon women of May, 1867. The body was thrown down in the 'fall' of the quarry, and was described as 'not being in a grave, but lying above and at right angles to the other graves.' There was a buckle 1⅛ in. long on the pelvis (cf. 'Pagan Saxondom,' p. 58; 'Tombeau de Childeric,' p. 234), and adhering to it some coarse flax fabric, as proved by the microscope. This skeleton has the left radius and ulna bronze-stained, and in the neighbourhood into which the bones were thrown a spear-head with a central raised ridge (like the assagaye of the Hottentots, 'Pagan Saxondom,' p. 10), an umbo, and a knife were found. These latter the workmen thought belonged to a child, i[a], which occupied a grave in the ordinary Romano-British bearing; but it is much more probable that they belonged to this skeleton, which had an Anglo-Saxon buckle upon its pelvis, and from which, in the wrench and jerk of the fall, the umbo and knife may very readily have been dislocated. The diameter of the umbo was 6.2 in.; height, 2.8 in.; lesser circumference, 14.4 in. There were four broad-headed rivets on the broad periphery, with three eyelet-holes between each pair. Its type was therefore the ordinary one found here. This umbo was exchanged for one in the possession of the Aldworth family. The skull appears after reconstruction to have been of the platycephalic ovoidal Anglo-Saxon type. Cf. 'Crania Britannica,' Plate XLVI, and plate added in description of Plate IX.

i[a]. *Child*, probably boy of about 12.

ii. *Skull of old woman*, with Anglo-Saxon ornaments, such as are described p. 70 of 'Pagan Saxondom,' and figured Plate XXXV. fig. 4; and 'Fairford Graves,' Plate IX. fig. 10; 'Cran. Brit.' Plate XX. p. 5; and Brighthampton, 'Archaeolog.' xxxvii. No. i. 38; No. xvi. in Ashmolean Museum; and with fibulae such as are figured at fig. i. in Plate XVIII. 'Pagan Saxondom,' as found near Rugby. The body lay in a grave running from W.S.W. to E.N.E:, not an unusual bearing for an Anglo-Saxon here. The grave was 27 in. deep. Stains of green on left clavicle and right rib i., the pins on the ring having been on the left shoulder, and the fibulae one on each shoulder. The fibulae are similar, also, to the two figured by Mr. Akerman in the 'Archaeologia,' vol. xxxix. Plate XI. figs. 8 and 9, as found at Long Wittenham.

iii. *Skull and femur and patella of old man*, dug out of a grave with Romano-British bearings, and from under an urn, iii[a], containing burnt bones of an adult. A nail was found in the grave with this old man. Skull eminently globose. Femur, 18.8 in.; stature, 5 ft. 11 in. Lower jaw nearly destroyed by water-wear.

iii[a]. *Urn* not reconstructed, plain, containing *adult bones*. It had a flat stone on the top of it.

iii[b]. *Plain urn*, with two bosses each on opposite sides, not pushed out from the inside, but stuck on to the outside, containing burnt bones of *an adult*.

iv. *Man past middle-life*. Skull of globose type, with some of long bones. In a very much deeper grave than usual, 40 in. deep, in usual Romano-British direction, from W.N.W. to E.S.E. Many nails in grave with the bones, with wood, probably oak, adherent. A fragment of old Roman pottery, the bottom of an urn, in grave. Femur, 18.18; tibia, 14.8; stature, 5.11.

v. *Skull of child*, 6½ years old, with two amber beads. Came from last grave but one on right side, as did also the skeleton No. ix, which had a coin with it, and also osteophytes internally in skull.

vi. *Skull with a few broken long bones of old woman,* very like a modern Swiss skull, 768 B, in Oxford Museum, with a cephalic index of 82. The oblique dip away of the posterior half of the parietal makes its distinctive character from the Anglo-Saxon xiii of May, 1867, and the skull xiv of Jan. 16, 1868. It is shorter and broader than the River-bed type, but its longitudinal arc has the same contour. Again, No. vi of 1867, with cephalic index 79, resembles it very much, and by vi of May, 1867, we pass to vi of Sept. 17, 1867, with cephalic index 78, and to the female skull or calvaria xi of April 1, 1868, xiii of Jan. 15, 1868.

vii. *A child's skull,* removed by the men. First dentition only.

viii. *Young woman 25 to 30, skull and long bones,* found near child with two amber beads. Skull of Romano-British, elongated, coronally-constricted type. See 'Crania Britannica,' Plate LVIII. Wisdom-teeth not through the gum, though the crista is anchylosed to the ilium.

ix. *Skull of woman* 25 to 30, found with a coin which is lost, and in last grave but one on right side, whence came the child with the two amber beads. Osteophytes on inner surface of skull. Femur, 16·5 in.; tibia, 13·6 in.; stature, 5 ft. 2 in.

x. *Fragments of skull of old person, probably female.* No history; found in 'fall' with fragments of Roman pottery; skull of platycephalic type with the posterior sagittal 'rainure,' supposed to characterize Celts and Scandinavians. See 'Bull. Soc. Anthrop. de Paris,' 1863, p. 319; 1864, p. 283. Internally, in correspondence with this, is a very deep furrow for the longitudinal sinus; showing of course that the bottom of the two furrows outside and inside the skull corresponds to a line of arrested growth, and that the skull has grown out on either side in lines of the parietal tubera, to fit itself to the growing brain. In other skulls, as for example skull No. ii of March 23, 1868, this parietal vallecula on the exterior corresponds with raised ridge along the line of the longitudinal sinus. See 'Jour. of Anatomy and Phys.' iii. 253, 1868, also Article xiii, p. 159.

xi. *Skull and long bones of old woman?* from extreme right of quarry; of globose Romano-British type, resembling skull No. vi of Sept. 1867, and vi of April 1, 1868, and vi of April, 1867. Femur, 14·5 in.; stature, 4 ft. 6 in.

xii. *Child* with first set of teeth—removed by me. Romano-British direction of grave.

2 old men, 2 adults ♀ ♂, 1 young man, 1 boy 12, 3 children, 4 old women, 2 young women, = 15 bodies.

September 24, 1868.

i. Key of Roman type . . . .
ii. Stag's-horn hair pin . . . .
iii. Bronze ring . . . .
iv. Two knives . . . .
v. Spoon . . . .
vi. Coin-shaped Kimmeridge shale

The excavations on this day were carried on upon two patches of ground which Mr. Aldworth had observed to have stronger and greener corn growing upon them than was to be seen elsewhere. Great quantities of the bones of the domestic animals, exclusive of the horse but including the dog, were found, together with the articles specified and numbered. No human remains were observed however. These spots appear to have been the rubbish-pits of some house of a person of considerable wealth, an 'eques.' See Pearson, 'History of England,' i. 45; and Coote, 'Neglected Fact in English History,' pp. 40–45, cit. in loco.

vii. Coin. One of the many coins imitated from Roman originals in 5th and 6th centuries. Very common in England.

viii. Pottery of very many patterns and degrees of fineness, from very fair and fine Samian down to very coarse ware. Some of both bestudded interiorly with particles of silex; some with pattern very like that of the Anglo-Saxon urn. See Bruce, 'Roman Wall,' p. 438; A. Corbet's 'Uriconium,' p. 63.

September 25, 1868.

ix[1]. *Anglo-Saxon girl*, with two plain bronze fibulae, in a grave from 18 in. to 24 in. deep, lying over Romano-British woman (xvii[5] of Sept. 26, infra p. 648). The skeleton's upper half ran from W.N.W. to E.S.E.; but the lower half of the body was twisted at an obtuse angle to the upper half, and lay from N.N.E. to S.S.E. This distortion probably accounts for the displacement of one of the fibulae from the right shoulders on to the manubrium sterni. Towards the lower end of this grave a beautiful coin was found, Byzantine, 4th century. Decentius. Many bones brought of a child of 9 aet. For the view that men had only one fibula and women two, see L'Abbé Cochet, 'Tombeau de Childeric,' éd. 2de. 1859, p. 228. Can this distorted position correspond to the 'contraction from the hips' described by Canon Greenwell, at Kirby-under-Dale, *Times*, 1841?

x[2]. *Skeleton of a child*, 12 to 14 months, from a grave running from W.N.W. head to E.S.E. foot. The depth of skeleton was 23 in.

xi[3]. *Skeleton of child*, 6 to 7, found lying immediately above xii[4] in a grave with bearings W. for head, E. for feet. There was a large stone at its head, and in the grave were three pieces of Roman pottery, one beautifully patterned, and a horse's tooth. First true molar just coming into place. Could this child have been a Christian Anglo-Saxon?

xii[4]. *Skeleton of adult male*, probably 25 to 30 aet., lying underneath preceding skeleton, head at W.N.W., feet at E.S.E. Femur, 19·2 in.; humerus, 13·7 in. Globose Romano-British type. Stature, 5 ft. 10 in.

September 26, 1868.

xiii[1]. *Skeleton of adult woman.* Femur, 16 in.; stature, 5 ft. In a grave running from W.N.W. by N.W. to E.S.E. by S.E., its depth being 35 in. to stone which was under the back of her head. The vertex of the head was horizontal, the frontal norma looking E.S.E. and the vertebrae of the neck being underneath the base of the skull. The head of the humerus was 2 in. from the skull. No relics nor traces of nails in this grave. Possibly a coffinless one. Elongated type. A number of shards were with this skeleton, but I think it may have been an Anglo-Saxon woman, such as No x of March 17, 1868.

xiv[2]. *Skeleton of young man.* Femur, 17·6 in.; stature, 5 ft. 4·4 in. The skull lay on its right side, in a grave running from W.N.W. to E.S.E. of 36 in. deep, without pottery or nails. Elongated British type.

xv[3]. *Skeleton of old man.* Femur, 17·19 in.; stature, 5 ft. 7·6 in. From a grave running from W.N.W. to E.S.E. 32 in. to top of skull, which was lying on its right side, not raised. The lower jaw a little on one side, not, however, so much as the head. A stone 9 in. long, 5 in. across, and 3½ thick, was so close to the forehead as to render it difficult to think a coffin could have been present. Elongated British type. Very fine skull, nearly of same size as the largest skull of the Dinnington series. See 'Journal of Anatomy and Physiology,' vol. iii. p. 253, 1868, also Article xiii, p. 160.

xvi[4]. *Skeleton of old man.* Femur, 16·3 in.; stature, 5 ft. 1½ in. From grave of same direction and depth as others; no nails, but some fragments of pottery and 'marks of burnings.' Elongated British type.

xvii[3]. *Skeleton of young woman*, enlarged River-bed type. Femur, 16 2 in.; stature, 5 ft. $\frac{7}{10}$ in. From a grave running N.N.W. by N.W. at 36 in. deep, one foot deeper than the Anglo-Saxon girl's grave No. ix[1] of Sept. 25, under which it ran. There was a large stone close to the forehead. The head was on its left side, looking slightly upwards. A large fragment of the rim of an urn was found between the left os innominatum and sacrum of this skeleton. A small nail was also found in this grave.

September 28, 1868.

xviii. *Skeleton*, reported by men by whom it was taken out as having been discovered in levelling the ground and smoothing the inequalities caused by the excavations of Friday and Saturday, Sept. 25 and 26, and as having been in a grave of same direction as, but of much less depth (viz. only 18-19 in.) than, the other graves. It was 'lying with its face downwards, as also its leg bones; and was found with two pieces of iron, and also a knife. One of the pieces of iron reached from its right elbow to its shoulder; the other was between the hip bone and the bottom of the grave. The knife was underneath the frame, about the middle of the body. The piece by the arm was a long piece all joined in one.' Probably buried when bearers drunk. A coin was sent with this skeleton. Not verified. (? Postumus) Young (? middle aged) Anglo-Saxon man of broader head type, many carious teeth, bones much water-worn. Femur, 17·8 in.; stature, 5 ft. 7·2 in. Had received and repaired during life a severe injury on left frontal and both parietals.

xix. *Skull with one long bone*, the humerus considerably worn, the rest decayed or water-worn, as also the lower jaw. No relics nor iron. In a grave of same direction as preceding, and as the Romano-British grave, but deeper. Of elongated Romano-British type. Old man.

xx. *Child.* The long bones not brought, having been beneath the growing crop of turnips. First dentition in place.

xxi. *Bones of infant about time of birth*, taken out of a grave about 18 in. deep, and of ordinary direction, W.N.W. to E.S.E.

TABULAR VIEW OF RESULTS OF OSTEOLOGICAL INVESTIGATIONS.

I. *Table of Skulls and Skeletons illustrating the several Types and Nationalities.*

The Celtic or Romano-British cranium of the 'Cimbric' type of Retzius is illustrated by—

Male crania, with average stature of 12 skeletons, 5 ft. 8·3 in.
- vii. May, 1867.
- xvi. " "
- iii. Sept. 1867.
- vii. " "
- ix. Jan. 15, 1868.
- xi. " "
- xii. " "
- xx. Jan. 20, 1868.
- xxi. Feb. 8, 1868.
- xxii. " "
- xxv. Feb. 21, 1868.
- xxvii. " "
- i. March 17, 1868.
- ii. " " ? A. S.
- iv. " " ? A. S.
- v. Sept. 1867 ? A. S.
- iii. March 23, 1868.
- xx. Sept. 26, 1868.
- xi. Mar. 17, 1868 ? A. S.
- xix. Sept. 28, 1868.

Female crania, with average stature of 9 skeletons, 4 ft. 11·5 in.
- Cranium B. Oct. 1864.
- xi. May, 1867.
- xii. " "
- xv. " "
- xix. " "
- xxvii. Dec. 31, 1867.
- xvii. Jan. 20, 1868.
- xxiv. Feb. 21, 1868.
- iii. Mar. 17, 1868.
- i. Mar. 23, 1868, ? A. S.
- ii. " "
- viii. Apr. 1, 1868.
- xiii. Sept. 26, 1868.

The globose Romano-British type, the 'Sion typus' of His and Rütimeyer, is illustrated by—

Male crania, with average stature of 11 skeletons, 5 ft. 8·5 in.:
- i. April, 1867.
- x. May, 1867.
- ii. Leaden coffin, 1867.
- xxiii. May, 1867.
- xxiii. Feb. 21, 1868.
- iv. March 23, 1868.
- xii. Sept. 25, 1868.
- viii. March 17, 1868.
- ix. March 14, 1868.
- vi. Sept. 1867.
- iii. April 1, 1868.

1. *Female cranium*, of size corresponding to male variety of globose type. — March 4, 1868.

8. *Female crania*, of smaller size, and of a type which may be called *the enlarged river-bed type*, with 7 skeletons averaging 4 feet 9·5 inches, are represented by—
- vi. April, 1867.
- ii. " "
- xi. April 1, 1868.
- vi. " "
- i. Sept. 1867.
- xvii. Sept. 1868.
- xiii. Jan. 15 1868.
- xi. May, 1867.

One *male cranium* which belonged to a man of 6 feet may be looked upon as furnishing a *form transitional from* an enlarged river-bed type to the globose Romano-British form. — xix. Feb. 8, 1868.

2. *Female crania*, of a small size and a type which may be spoken of as the *cylindrocephalic river-bed type*, with a mean stature of 5 ft., are represented by—
- ii. Sept. 1867.
- iv. Jan. 1868.

Of skulls, which, though not found with, are, from other causes, conjectured to have belonged to, *Anglo-Saxons* . . there are
- 3 *Male crania* . . . . v. Sept. 1867. iv. March 17, 1868. xi. " "
- 3 *Female crania* . . . . x. March 17, 1868. xiv. Jan. 15. 1868. i. March 23, 1868.

Of Roman, or Romano-British, skulls of the 'Hohberg' type and stature 5 ft. 10·5 in. . there are 3 *Male crania* . . . . . i. Leaden coffin, 1867. xiv. May, 1867. xxii. " "

Anglo Saxons with relics . .
- *Female crania* . . . . . . A. Nov. 1864. xiii. May, 1867. v. March 17, 1867. ii. April 1, 1867. ix. Sept. 25, 1868.
- *Male crania* . . . . xviii. Feb. 8, 1868. i. April 1, 1868. xviii. Sept. 28, 1868.
- *Males—Bones imperfect* . . . . v. May, 1867. iv. Sept. 1867.
- *Child* . . . . v. April 1, 1868.

II. *Numerical Table.*

Men from Leaden coffins, in 1864 and 1867	Old men	2	4
	Middle aged	2	
Men of the Hohberg type, besides one from leaden coffin, i, 1867	Old	1	2
	Young	1	
Anglo-Saxons with relics or in urns	Young or middle-aged men with relics	4	21
	Old man	1	
	Young women	2	
	Middle-aged	1	
	Old	1	
	Children	2	
	Adults in urns (one about puberty)	3	
	Urn unopened	1	
	Children in urns	5	

Category	Sex	Age	No.	Subtotal	Total
Skeletons supposed to have belonged to *Anglo-Saxons*, though found without relics	Men	Old	1	3	6
		Middle-aged	1		
		Young	1		
	Women	Old	1	3	
		Young	2		
Romano-Britons of elongated oval capacious type, called 'Cimbric' by Retzius, 'Ethnologische Schriften,' p. 108	Men	Old	11	21	32
		Middle-aged	3		
		Young	7		
	Women	Old	4	11	
		Middle-aged	1		
		Young	6		
Romano-Britons of globose or 'Sion' type, the male crania of great size occasionally, and the female in only one instance approaching the larger male crania in dimensions	Men	Old	7	10	21
		Middle-aged	2		
		Young	1		
	Women	Old	6	11	
		Middle-aged	1		
		Young	4		
Male skeleton of enlarged River-bed type (Old)					1
Skeletons the type of which has not been determined, the bones having been too much injured by water-wear or otherwise	Men	Old	1	2	14
		Young	1		
	Women	Old	1	6	
		Young	5		
	Sex undetermined	Old	2	6	
		Young	4		
Children found without relics and in graves	Infants			4	23
	Within period of first dentition			14	
	From period of commencement of second dentition to that of puberty			5	
				Total	123

Of which 123 there are:

- 30 children
- 25 old men
- 13 old women
- 19 young women
- 15 young men
- 8 middle-aged men
- 3 middle-aged women
- 2 old persons of undetermined sex
- 4 young persons of undetermined sex
- 3 adults from urns
- 1 urn unopened

Of which 123—
48 are men.
35 are women.

FIXED POINTS FOR ARGUING AS TO DATE AND NATIONALITY OF THE SKELETONS FOUND AT FRILFORD.

I. *Coins.*

In leaden coffin No. i, Jan. 1867, five coins, of which one was a coin of Constantine the younger, one a coin of Valens, one a coin of Gratian.

In the leaden coffin opened by J. Y. Akerman, Esq. F.S.A., Oct. 1864, and also in one of the graves opened by him at the same time, a coin of Constantine I. was found.

In the point where *graves* xvii *and* xviii of Jan. 20 and Feb. 8, 1868, intersected, a coin of Constans was found. Possibly accidentally fallen in.

With *skeleton* xxiv of Feb. 21, 1868, an old woman, of the elongated oval Romano-British or Celtic type, five coins were found, of which one belonged to Valens, and another to Constans I.

With *skull* ix of April 1, 1868, a coin was found, which is lost.

With *skeleton* iv Jan. 9, 1868, of a very old woman, of small cylindrocephalic type, three coins were found, which could not be identified.

In the grave (but towards the lower end of it, whither it may have found its way accidentally) of the Anglo-Saxon girl No. ix of Sept. 25, 1868, a Byzantine fourth-century Decentius.

In the Roman rubbish-heap, examined Sept. 24, 1868, a coin was found, one of the many imitated from Roman originals in fifth and sixth centuries. Very common in England.

With the skeleton No. xviii of Sept. 28, 1868, a coin was sent, Postumus (?).

II. Relics.

Arms and ornaments. *Lower jaws* iv of April and May, 1867. Reported to have been found with an umbo No. v.

Skeleton No. x ii of May, 67. Anglo-Saxon woman. Was found with fibulae and beads.

Skeleton No. v of March 17, 1868. Anglo-Saxon woman. Was found with fibulae, beads, and pin.

Skeleton No. ii of April 1, 1868. Anglo-Saxon woman. Was found with fibulae of Midland counties type, with scoops and pickers on ring, and with a knife. But see 'Further Researches, Long Wittenham,' 'Archaeologia,' xxxix. Pl. XI. p. 142.

Skeleton No. iv of Sept. 1867. Anglo-Saxon man. Was found with an umbo and a spear-head.

Skeleton No. xviii of Feb. 8, 1868. Anglo-Saxon man. Was found with an umbo, a spear-head, a knife, and some Roman tiles set round his grave.

Skeleton No. i of April 1, 1868. Was found with an umbo, a spear-head with a central raised ridge, a buckle, and a knife.

Skeleton No. v. of a *child*, April 1, 1868. Was found with two beads, not spherical, and therefore probably Anglo-Saxon.

Skeleton No. ix of Sept. 25, 1868. Anglo-Saxon girl. Was found with two fibulae. A coin was also found towards the lower end of her grave, but may have fallen or worked its way into the grave without any intention on the part of the burying persons. The coin was a fourth-century Byzantine coin of Decentius.

Skeleton No. xviii of Sept. 28, 1868. Anglo-Saxon man. Was found with the face downwards, and with two pieces of iron, probably remnants of a crushed umbo, a knife, and a coin, which was considered as probably of Postumus.

III. Urns.

i. Anglo-Saxon . . . . . .
- *Plain urn*, iv of Jan. 25, 1867, containing bones of a child, 2 to 3 years old.
- *Patterned urn*, xvii of May, 1867, containing bones of child before second set of teeth.

1. Cremation	*Plain urn*, xviii of May, 1867, containing bones of person before age of puberty. *Patterned urn*, ix of Sept. 1867, containing bones of child under 8 years of age. *Plain urn*, iii[a] of Jan. 9, 1868, containing bones of child from 5 to 6 years of age. *Plain urn*, vi of Jan. 15, 1868, containing bones of child from 9 to 10 years of age. *Plain urn*, xvi of Jan. 20, 1868, unopened. *Plain urn*, iii[a] of April 1, 1868, containing bones of adult. *Plain urn*, with two bosses, iii[b] of April 1, 1868, containing bones of adult. Fragments, with bones, were found Sept. 1867, and Dec. 31, 1867.
2. Holy Water Vessels (?)	Patterned vessel in British Museum, of date 1864. Patterned fragment, figured Plate i, found Sept. 1867. Patterned fragment, figured Plate ii, found March 17, 1868, No. vi.
ii. Roman	A Roman vessel was found perfect at bottom of one of the walls.
iv. Skeletons found under urns	vi of Sept. 1867, under urn No. ix. i[v] of Jan. 9, 1868, under urn No. iii[a] of Jan. 8, 1868. iii of April 1, 1868, under urn No. iii[a] of April 1, 1868, not reconstructed.
v. Skeletons found under other skeletons, which are identifiable by their relics . .	xvii of Jan. 20, 1868, was under Anglo-Saxon man, xviii of Feb. 8. xvii of Sept. 26, 1868, was under Anglo-Saxon girl, ix of Sept. 25, 1868.

See also No. 3 (xxxvii), of Jan. 20, 1869, found under Anglo-Saxon woman, xxxviii with fibulae, No. 2 (xxiii) of Jan. 6, 1869, found under Anglo-Saxon woman, xxii with fibulae, both of which are in the Oxford University Museum, as also No. xii[4] 4 of September 25, 1868, p. 647, supra.

XXXV.

FURTHER RESEARCHES IN AN ANGLO-SAXON CEMETERY AT FRILFORD, WITH REMARKS ON THE NORTHERN LIMIT OF ANGLO-SAXON CREMATION IN ENGLAND.

THE first discovery in the cemetery at Frilford, subsequent to those already recorded in the 'Archæologia,' xlii. pp. 417–485 (Article xxxiv), was made on March 22, 1869, when a leaden coffin was found, containing the bones of a young woman, with a toilet comb [1] at the right of the back of her head. This brings the number of leaden coffins found at Frilford up to five; one of them has already been figured in 'Archæologia,' xlii. pl. XXIV. figures 7 and 8.

The second was the discovery of some fragments, which, when fitted to the three fragments found in September, 1867, one of which is figured in the 'Archæologia,' xlii. pl. XXIII. fig. 2, p. 423, make up the larger portion of what is often called a 'holy-water vessel.' The fragments of September, 1867, were to my eyes so distinctly Saxon that I had one of them figured, and the unexpected discovery of the remaining fragments enabled us to build up the urn shown in the annexed woodcut. I imagine that a plough's coulter had knocked out the first discovered fragments. No burnt bones were found quite close to the urn, but one fragment was found a little way off [2].

This reconstructed vessel may be compared with vessels of somewhat similar shape, and possibly similar purpose, found in Roman

[1] For difference between toilet and other combs, see Anderson, 'Proc. Soc. Ant. Scot.' June 10, 1872, p. 551, and woodcut in loco.

[2] The fragment, which with a triangular apex pointing upwards, occupies about the middle point in the front upper border of the urn figured above, is the same fragment which is figured with its apex pointing downwards, pl. XXIII. fig. 2, 'Archaeologia,' xlii.

cemeteries, for instance, at Hardham, Sussex, as figured by Professor W. Boyd Dawkins[1]. But urns more similar still have been found in many Teutonic cemeteries in England, as well as in France and Germany.

May 3, 1870.—An old Anglo-Saxon woman, with tweezers[2], knife, metallic button, and small metallic ornament at head of humerus; large stones set by the sides of the graves, as described in 'Archæologia,' l. c. p. 438; but no nails. Depth of grave 2 feet 6 in., direction north-west to south-east. Abundance of charcoal in the grave; arms extended, patellæ *in situ*. Tibiæ platycnemic.

Anglo-Saxon urn, Frilford. Scale ⅓ linear.

May 3, 1870.—Fragment giving about three-sevenths of the circumference of an Anglo-Saxon 'holy-water vessel,' or, perhaps, rather of a rudimentary representation of cremation urn; found near the bones of a young person. This vessel has the characteristic German angular projection round its body, the vandyking and the stamped pattern, &c., which we are familiar with in urns of larger size intended for the reception of burnt bones. Its small size, as well as the fact that many such vessels have been found with buried bodies and without any bony contents, shows that this

[1] 'Sussex Archaeological Collections,' vol. xvi. p. 58.

[2] For figures of similar tweezers, see Lindenschmit, 'Alterthümer,' Bd. ii. Hft. v. Taf. vi, where they are said to be found usually in men's graves, but sometimes in women's. Neville, 'Saxon Obsequies,' pl. II; Cochet, 'Normandie Souterraine,' p. 219, pl. VII. fig. 35.

vessel cannot be considered as a cinerary vessel[1]. Cochet, in his 'Arch. Céramique,' p. 13, explains what he calls the mystery of the custom by the often-quoted passage as to holy water from 'Durandus,' vii. 35, 37. I think this passage of little weight[2], considering that Durandus lived in the thirteenth century. I incline to consider these vases, another example of which, from Haslingfield, is herewith figured, and which sometimes have been, as at Selzen, found to contain combs, shears, beads, fibulæ, flint and steel, and bronze rings, in fact everything that an ordinary cremation urn does contain except the bones, to be rudimentary representations of such cremation urns. Solemn occasions are tenacious of their symbols, and will hold to them or keep hold of them in miniature when they can no longer maintain them in full proportions. The

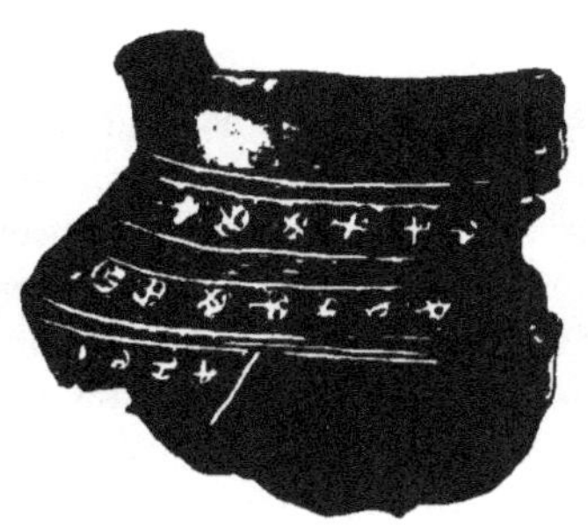

Fragment of Saxon urn, Frilford.
Scale ½ linear.

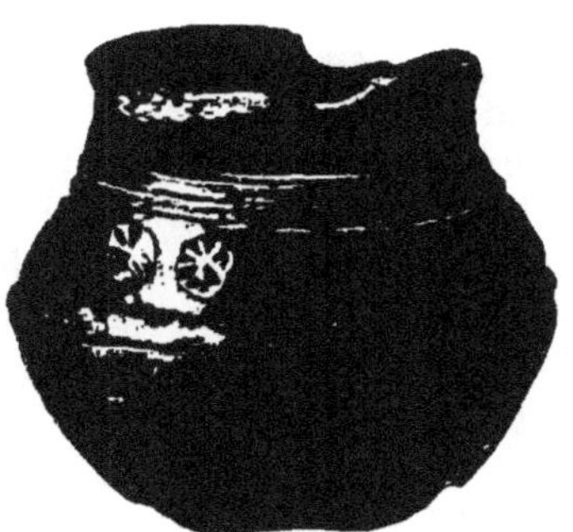

Saxon urn, Haslingfield.
Scale ½ linear.

wide range over which this diminutive representation of the larger Germanic urn has been found is another argument in favour of my

[1] For the greater fineness of workmanship in these smaller vessels, see Kemble, 'Horae Ferales,' p. 225; Roach Smith, 'Collect. Antiq.' iv. 161-196.

[2] It was thus as given by Cochet l. c.: 'Corpus ponitur in speluncâ in quâ . . . ponitur aqua benedicta . . . Aqua benedicta ne daemones qui multum eam timent ad corpus accedant: solent namque desaevire in corpora mortuorum, ut quod nequiverunt in vita saltem post mortem agant.' Cochet's own words are: 'Tous les cimetières Mérovingiens et même Carlovingiens que nous retrouvons . . . montrent toujours aux pieds du mort un vase vide dont les hommes d'aujourdhui nous demandent le sens et le mystère. Nous croyons l'avoir trouvé dans la piété naïve et grossière, peut-être même matérielle et superstitieuse, de nos pères. Nous supposons donc, non sans fondement, qu'ils auront mis dans ce vase une eau sacrée préservatrice des obsessions et des possessions démoniaques si fréquentes chez les vivants et dont les morts ne leur paraissaient ni exempts ni affranchis.'

view, which is based upon the recognition of an acknowledged tendency of the human mind as opposed to a view which can only appeal to a superstition of probably much more limited geographical range[1]. A somewhat similar vessel, both as to size and contour, from the Oberpfalz of Bavaria, may be found figured in 'Die Sammlungen des Germanischen Museums,' Nürnberg, 1868, p. 67.

May 23, 1870, iii; May 23, 1870, vii.—Skeleton of old Romano-Briton lying in a grave such as described in 'Archæologia,' xlii. p. 422, undisturbed 18 inches below skeleton of a young Anglo-Saxon, æt. about 17, with umbo, spear, and knife.

May 23, 1870, iv b; May 23, 1870, iv a.—Skeleton of old Romano-Briton, buried with coffin, lying from 4 ft. 7 in. below skeleton of old Anglo-Saxon woman, lying in the contracted position without any relics, and, indeed, with disproof of any coffin, with two cruciform fibulæ, a shroud-pin, an iron ring, and a knife[2].

In both cases the long axis of the upper grave formed more or less of a right angle with the long axis of the semi-oriented lower one. This shows that the burials could scarcely have been simultaneous: for the other conclusions which can be based upon the finding of two skeletons, verifiable as Saxon and Romano-British respectively in the relation specified, see my previous

[1] For the general literature, see Cochet, 'Arch. Cér.' p. 14, ibique citata; 'Normandie Souterraine,' pp. 199, 267; 'La Seine Inférieure,' p. 530; 'Tombeau de Childeric,' p. 391, ibique citata; Akerman, 'Researches at Long Wittenham,' Archaeologia, xxxviii. pp. 342, 346, 352 (note), 330, 333, 342, 352; pl. XX. fig. 2; 'Pagan Saxondom,' pl. XXII, where an urn 8 in. high is described as containing tweezers, shears, comb, and knife, though it is not stated whether any bones were found in it or not. See also 'Inventorium Sepulchrale,' 1856, Introd. p. xxvi, and Neville's 'Saxon Obsequies,' p. 9, where vessels like these are said to have been very frequently, as regards the entire number (viz. three or four times out of twelve), found with infant skeletons, and to have been found either at head or foot, 'though in the grave of an adult two small vases were found, one on each side of the former.' This difference in placing seems to me to favour my view as above stated. The Selzen vases were, it is true, or nearly always, at the feet, and those found in the French interments of the same period, always, according to Cochet, l. c. But at Hallstatt (see V. Sacken, 'Das Grabfeld von Hallstatt,' 1868, p. 107) the position of these vases was most variable: 'bald standen sie zur rechten, bald zur linken Seite des Skelettes, neben dem Kopfe, bei dem Hüften oder zu den Füssen, bei Verbrennungen in der Regel neben den Brandresten, selten auf denselben.'

[2] These four skeletons, with the relics accompanying them, were presented to the Cornell University, Ithaca, United States. A more detailed account of these objects than that given above may be found in the 'Register' of that University for 1870-1871, pp. 50-54.

Memoir on Frilford, Article XXXIV. It is worth while noticing that this Anglo-Saxon woman was 5 ft. 5 in. in height, an instance of what is said to be usual, but what I have found to be by no means invariable, viz., an equality, or an approach to equality, in the stature of the German women and men[1]; next, that her skull was found five inches above her sternum, three stones having been placed underneath it; and, lastly, that the knees were at a higher level by several inches than either the ankles or the hips, besides being, as the statement of the body having been in the contracted position implies, out of the line of the long axis of the skeleton. These points are not ordinarily found in Christian burials. The arms were, however, crossed, and the hands folded inwards, as was often done in such interments; with which, again, on the other hand, this Anglo-Saxon burial appears to have contrasted in the body's being turned somewhat on to the left side, a point which, from the crushed condition of the skeleton, lying only about 2 feet from the surface of the ground, it was difficult to make out.

The question now arises, Were such non-oriented, contracted, shallowly interred, but relic-provided, bodies, the bodies of heathen or of Christianised Saxons? Mr. Kemble's dictum, 'Horae Ferales,' p. 98, to the effect, that, 'if there is any equivocation in the matter, it lies the other way; a few half-converted Christians may for a while have clung to the rite of burning, but no Pagan Saxon was buried without it,' is well known; but I am of opinion that this is one of the few mistakes which Mr. Kemble made. This one mistake of Mr. Kemble led him logically to a conclusion to be found at p. 230 of the same valuable work, the 'Horae Ferales,' in a remark printed from the MSS. left behind him. Speaking of the rarity of Saxon urns in Scottish Museums, one from Buchan, to be seen in the Museum of the Antiquarian Society, being specified as the only one he knew of, Mr. Kemble remarks, 'If they (Anglo-Saxon urns) should turn out to be very rare there (in Scotland not merely in Scottish museums), it would be evidence that no very important settlement was made there by the Saxons before their conversion to Christianity; a result which history seems to bear out. It was, in fact, Christianity which united the Saxons sufficiently to make them capable of acting *en masse* against their

[1] For figure of a skeleton with skull similarly raised, see 'Grabfeld von Hallstatt,' tab. iii. fig. 4.

neighbours.' Without raising any objection to the view which would assign the tendency to attack one's neighbours *en masse* to the religion which is ordinarily said, and by members of the Society of Friends believed, to teach lessons of peace, I would remark, that history does not seem to me to bear out Mr. Kemble's view, and that the finds in many unmistakeably Teutonic burials by interment seem to me to suggest the idea of heathendom by their shallowness, their want of orientation, their possession of secular relics, and by the frequency, especially in the north, with which the skeleton is discovered to be in the contracted position. In the case of Kent, the great salient facts recorded by the historians as to the conversion of Æthelberht are almost or even quite as indisputable as the facts of the 'Inventorium Sepulchrale' with regard to the comparative rarity of cremation urns in that earliest to be founded of Saxon kingdoms. It is true, as Mr. Kemble himself has shown ('Horae Ferales,' p. 91), that cremation urns are not entirely unknown in Teutonic cemeteries in Kent; but no one can doubt that this comparative rarity in that locality, when coupled with the facts that Kent was sufficiently powerful and thickly peopled for the Frankish King Charibert to give his daughter to the King of Kent, and that this King Æthelberht, and, by consequence, most of his Court, were nevertheless heathen, shows that a Saxon population, at all events when firmly established in a country, could give up cremation before taking up with the teaching of the missionaries.

The drawings which I lay before the Society[1] represent a number of urns from a Saxon cemetery at Sancton, co. York, a village a little south of Market Weighton, and the once better known Goodmanham. These urns, the acquisition of which I owe to the kindness of Charles Langdale, Esq., of Houghton Hall, mark, as I believe, and as far as is known, the northern limit of cremation as practised to any considerable extent by Teutons in the north of England. But, little[2] as we do know of the history of the Conquest of Northumbria, we have some reason for believing that Æthilfrith was an unbeliever, and that by his great victory at Dægsastan in 603 a Pagan Saxondom was established under his

[1] These drawings were reproduced in the 'Archaeologia,' plate xxxiii, 1879, vol. xlv.

[2] For statements as to this littleness, see Stubbs, 'Constitutional History,' p. 61; Freeman, 'Norman Conquest,' i. 25, 26.

rule from the Humber to the Forth. If Æthilfrith was a heathen, such no doubt were his followers; and, if the whole of Northumbria was heathen in 603, its two component sub-kingdoms of Bernicia and Deira were, it cannot be doubted, at least as pagan for the period little short of a couple of generations which intervened between the date of the battle of Dægsastan and that of the landing, before A.D. 547, of Ida the Flame-bearer at Flamborough Head. The bones, however, of the unsung heroes of these wars have not previously been found in cremation urns, at least in any abundance, though contracted Teutonic burials are common enough between the two latitudes mentioned.

XXXVI.

ON THE THREE PERIODS KNOWN AS THE IRON, THE BRONZE, AND THE STONE AGES.

I HOPE that in the observations I am about to submit I shall make plain the differences which have enabled antiquaries to divide pre-historic times into two principal divisions, namely, the Stone and the Bronze ages—and to draw a tolerably sharp line of demarcation between these periods—and the Iron age, in which we are now living, and to which the interment examined this day in Oakley Park belonged.

It has been said very truly that 'Les divisions des êtres, des objets, des sciences sont la source la plus commune des erreurs de l'esprit humain;' and naturalists regard the aphorism 'Nature is not so strict a classifier as man' as being one of their commonplaces. And I do not say that no one of the three ages has been overlapped at either end by another, nor can I accept all the minute subdivisions of these periods which some specialists have urged upon us. But gradations at both ends of any series should not prevent us from seeing, acknowledging, and holding, that it may be distinct enough in its middle; and if old forms of implements and weapons are enabled by isolation as regards locality to live into contemporaneity with newer ones, that is only what happens with older forms of animal and vegetable life which isolation of the same kind often preserve as living fossils, but without for a moment making us doubt the propriety of referring them to an age distinct from ours.

Let us begin with the Iron Age, with the age of which we know most, and so work our way gradually upwards and backwards through the Bronze into the Stone periods. The drawings I exhibit represent the implements and weapons used by the Romans and by the Anglo-Saxons, which were of iron. The Roman weapons are, some of them, such as we see them to be on monuments and some other works of art; but in some instances they were drawings of the actual weapons themselves as found on battle-fields and elsewhere. The *hastae* and *pila* of the Saxon and Frank, for such were in reality their spear and the *angon*, were, on the other hand, found where I have never found the Roman weapons, viz., in the graves of those warriors. This difference, as to the fitting out of their deceased, depended, I would suggest, upon a difference of views as to the future state of the dead; and this difference did not consist, I apprehend, in that the one race held that the future state would be such a peaceful one that weapons of war would be superfluous in it, whilst the other believed that it would be more or less a continuation of the life of assault and battery they had so richly enjoyed here; but in a very different opposition of beliefs. The 'sunt aliquid Manes' of Propertius was as beautiful a piece of poetry to the Roman as any other of the beautiful poetry of that sweet poet, but it was nothing more. The Teuton, on the other hand, held firmly on to the belief in another world; and this belief accounts for the deposition of weapons in the graves of their dead.

I am inclined to think that the Iron Age would be better spoken of as the 'Steel Age.' For there is no reason why we should not suppose that iron, as distinct from steel, may have been in the hands of many tribes before they came into the possession of Bronze; and if the iron was soft iron merely, bronze would be much more useful and trustworthy for the purposes of war and the chase, for which so many ancient and modern races have mainly lived. A very striking instance from Roman history of the comparative uselessness of untempered iron tools for such purposes is given us by Polybius, Hist. ii. c. 33. There we read, in a probably somewhat unjustly unfavourable account of that somewhat rare animal, a liberal military commander, that his colonels saved him, as colonels have in later times saved other generals, from disasters, by the following tactics. The Gauls came to the fight armed with long pointless soft iron broadswords. These, the Roman tribunes had observed, bent after each blow delivered on to a sufficiently resistent body. Such a body they sought and found in the *pilum* —that best of pikes or bayonets, with which a man could parry or thrust, but with which he could not strike or slash. The brave barbarian came up *ferox viribus,* brandishing his broadsword, its downward strokes were parried, and the malleable iron, glancing downwards, bent as malleable iron will do, and left its gigantic owner at the mercy of an Italian, some five feet six inches in stature, who then brought into this sword-play a weapon which he had been taught to use *punctim, non caesim.* The same tactics succeeded at Culloden, as the tactic of thrusting and giving point always will succeed when masses of men in rows, not isolated individuals merely, are pitted against each other on the thrusting *versus* the slashing plan, though the slashing sword at Culloden was of good steel enough. The point for our present purpose in this story of the victory of Flaminius over the Insubrian Gauls lies in the proof it gives us of the existence, so lately in the world's history as B.C. 224, of a warrior-race fighting with soft iron instead of steel swords.

The red kidney iron ore, which we know so well from the fact of such large fortunes having been made out of it in the country lying north of Ulverston, and that other hematite known as 'hematite brun,' 'Brauneisenerz,' 'Sumpferz,' and the specular iron ore of Elba and Norway, are widely 'distributed,' very rich in their percentage of iron, and allow of its being easily (even if wastefully, as we should consider it) extracted. In the metallic state, iron is seldom met with naturally; as the 'Dictionary of Chemistry,' *sub voc.*, p. 335, tells us, *telluric* iron is very rare; and *meteoric* iron, the other variety of native metallic iron, now that the common Greek name for iron is known not to have any relation to any *sidereal* origin, but to express simply the dew-like out-sweating of the metal as reduced in the primitive 'bloomeries' of those early Dactyli, Elfins, and Dwarfs, whom we know as 'Tubal Cains,' 'Vulcans,' 'Sindris,' and 'Wayland Smiths,' has lost any claim which it may have been supposed to have had to being considered the primitive source of iron weapons. Hence it is of the utmost consequence to keep in mind the fact that certain widely diffused ores of iron are very easy of reduction, as the examples furnished by the metallurgy of certain African negroes, and of wild tribes in India and in Borneo, abundantly prove. And it is possible enough that in pre-historic times one of the more easily reduced ores of iron may have been reduced, and even found to be malleable, before not only bronze but even the mode of reducing a copper ore was discovered. Still, this would not prove that bronze must, as has been maintained, have, as being a more complex invention, been a later discovery than that particular modification of iron known as steel. The rigid resistent bronze would make a better weapon, especially for that most efficient process of sword-play, 'giving point,' the thrusting *punctim*, as opposed to mere slashing *caesim*, of the Roman military writers, than would such soft iron as in the absence of the knowledge of converting iron into steel could only have been available to the savages who reduced it. A pike may 'bend bravely,' even when made of good steel, and *a fortiori*, as the quotation from Polybius shows, when made only of untempered iron. When, however, once the art of making steel out of iron was discovered, and soft iron took on 'the ice-brook's temper,' the wider diffusion of the material, and the greater facilities of this process of manufacturing an equally useful article out of it, caused the displacement of bronze just as many a similar discovery has caused the displacement of many another product of toilsome elaboration by the introduction of another and simpler one. It should not, however, be forgotten that 'cementation' is a preliminary process to that of hardening and tempering in the manufacture of steel, that it is a process requiring several days as well as the combination of several other conditions if it is to be successful, and that the improbability of pre-historic men stumbling early and easily into the knowledge of a process consisting of a considerable number of heterogeneous operations is, in spite of the now apparent simplicity of those operations, as great, perhaps, as the improbability of their similarly stumbling into the discovery of bronze.

Coming in the second place to the Bronze Age, and facing the discussions hereinafter to be bibliographised, I have to say that, as against all quotations from old and modern authors, and as against all records, by whomsoever recorded, of the discovery of iron weapons in Bronze Age Tumuli, I am as confident as I can be of anything in Anthropology that no iron will be, though bronze not rarely is, found in Tumuli of the Bronze Period and Round Shape in this country at least. I have been present and assisted in more ways than one at the examination of many 'Round,' 'Bell,' 'Bowl,' 'Cone' shaped Barrows, and in the *primary* interment in such Barrows I have never seen any other metal than bronze. It is common enough to find a Saxon burial with iron and bronze both in the superficial layers of such barrows, but the superficial position of such burials shows their posteriority in point of date. The central interment at the bottom of the barrow on or sunk into the natural soil, may or may not contain a bronze dagger, may or may not contain weapons of flint, survivals from, and religious or ceremonial reminiscences of, the Stone Age, may or may not contain vessels of pottery, may contain a skeleton in the contracted position, or burnt bones either in an urn, or in a case of bark, or simply naked in the earth (I have seen all these cases); but it has never contained any shred of iron within my experience, nor, as I believe, within that of any person who can be trusted to distinguish between a primary and a secondary interment.

It will be said by some in answer to this that iron is oxidizable and perishable in an eminent degree, and that it would disappear whilst the bronze would remain. This suggestion I will not characterize as one of the study as opposed to one of the barrow, but as one of the laboratory, and the laboratory with its strong reagents supports it in a way that the slow and weak or wholly inert chemistry of the deep sand, or rubble, or gravel-filled grave does not. Of course, if you conceive a stream of water, acidulated even slightly with nitric acid, to pass constantly over an iron spear-head, there is no difficulty in estimating the time which will be necessary for the entire disappearance of an implement so tested. But no such agent is available in many, I might say most, Bronze Period graves. In some such graves you may find the objects they contain encrusted with a deposit of carbonate of lime, which would have protected an iron weapon of the Bronze Period if there had been any to protect; or you may find, as I am happy often to have seen, the bones in a capital state of preservation, and contrasting to great advantage with the corroded and 'perished' bones of Saxons *whose iron weapons were, nevertheless, very present with them;* or the grave itself may contain a considerable quantity of free carbonic acid, as other sunk wells do, and yet may be so dry from conditions of superjacent and subjacent rubble and soil as to have afforded no means for the removal of any results of any slight erosion which its contents might have suffered. The phenomena disclosed by the spade must be compared with those disclosed by the test tube; there is here a *makro-* as well as a *mikro*-chemistry.

One of many other convincing arguments for the conclusion that iron came into general use later than bronze, is, to my mind, the fact that in Switzerland you find, as in the museum at Berne, bronze bracelets ornamented with beads of iron, and, as in a knife from Mörigen, the blade of a cutting instrument made mainly of bronze but similarly inlaid with strips of steel. On this, Désor and Favre ('Le bel Age du Bronze Lacustre,' 1874, p. 16) remark, '*Or pour qu'on ait employé le fer en guise d'ornement il fallait bien qu'on le tînt en grande estime et qui il ne fût pas très commun.*' The larger use of iron when a thin blade of it was carried on a handle of bronze, the retention by such iron blades of the leaf-shape of the bronze blades which they displaced from their bronze pommels, and finally the exceedingly rich ornamentation of the pommels of the iron swords found at that most instructive discovery at Hallstatt, are all similarly indications that iron was of later introduction than bronze; that at first it was the scarcer of the two materials. If, afterwards, iron was made a servant of all work, and bronze was retained simply for the manufacture of ornaments, as by our Anglo-Saxon forefathers, this is but a history which can be paralleled by that of many other household goods!

Copper, as distinguished from bronze, is, on *a priori* grounds, likely to have been discovered and used long before metallic iron. For it is much more abundant in the metallic state in nature, as for example in Siberia, in the Faroe Islands, in many Cornish and in some Welsh mines, in Brazil, Chili, and Peru, and, above all, in large masses near Lake Superior, in North America. And, in addition to being there available and obvious as a red metal—or, indeed, *the* red metal—copper is malleable and ductile immediately after fusion, and acquires considerable hardness *when mixed with other metals.* These last five words from the 'Dictionary of Chemistry,' *sub voc.* 'Copper,' bring us face to face with the question: Where was made the discovery of the advantages to be gained from alloying copper with tin, and so obtaining bronze? It was of course likely to be made in some district in which the ores of these two metals were to be found in proximity. There are three such areas. Firstly, Cornwall: but as against the claims of our westernmost county are to be set, not only the ratiocinatively weighty words of Caesar as to pigs or ingots of bronze, *Aere utuntur importato*, but many materially ponderable arguments in the shape of bronze celts themselves demonstrably also imported as manufactured. Secondly, Khorasan: as to the existence of tin in which district we have hints from Strabo and from Burnes, but have lately had full and authentic information furnished to us in a paper written by the late illustrious Von Baer but twelve days before his death on November 28th, 1876, and published in the 'Archiv für Anthropologie,' ix. 4, p. 262. It appears that the Vice-President of the Imperial Geographical Society of St. Petersburg, Herr Semenow, at the instance of Von Baer, procured the following report as to the existence of tin and copper and other minerals in the neighbourhood of the places tolerably familiar to us, as Herat, Merv, and the Bamyan Pass. Here is the Report:—

'Ein Bewohner der Stadt Meschhed, Aga Mamed Kasym Ragim, Arrendator eines der vielen Kupferbergwerke in Chorassan, theilte mir mit, das 20 Farsangen (1 Farsange = annähernd 7 Werst) von der Stadt Utschan Miot Abot sich die reichsten Lager von Zinn, Eisen, Kupfer, Schwefel und Blei befinden und 2)6 Farsanger von Meschhed ein Zinnbergwerk das sogenannte Rabotge Alokaband ist. Die Genauigkeit dieser Angaben ist bekräftigt durch den Vorsteher der russischen Kaufmannschaft in Chorassan, den Bucharen Hadschi Ibrahim der wohl bekannt ist mit der hiesigen Gegend und mit vielen Personen die sich mit Bergwerksarbeiten beschäftigen: ausserdem habe ich mich factisch von dem Vorkommen des Zinns hier überzeugt durch Ueberfluss von zinnernen Waschkrügen und grossen Schüsseln alter einheimischer Arbeit, welche aus dem Zinn des Ortes gefertigt sind wie mir die Besitzer sagten.

'Nach den Aussagen der Kaufleute, die durch Handelsinteressen mit Merw in Verbindung stehen, sind die bergigen Theile Turkmeniens das von Stamme Teke eingenommen wird, überhaupt reich an verschiedenen Erzen unter welchen sich auch Zinn vorfindet. Genauere Nachrichten jedoch über diesen Gegenstand werde ich geben in der Ausarbeitung der Tagebücher meiner Reisen im nordöstlichen Persien.

'Hierzu schreibt der Geheimrath Semenow: "Diese Nachrichten sind nach meiner Bestellung gesammelt und mitgetheilt von einem Reisenden, der im Auftrage der Geographischen Gesellschaft und des Herrn Gluchowskoi eine Reise nach Ost-Persien (Meschhildo) zu Stande gebracht hat. Er heisst Ogorodnikow."—P. Semenow.

'Diese Nachrichten machen es höchst wahrscheinlich, dass zu der vielen Bronze, die man in den Ruinen von Assyrien und Babylonien gefunden hat, das Zinn aus der Gegend von Chorassan kam, wo man die Drangianer zu suchen haben wird. Wie weit hin das Vorkommen des Zinns sich erstreckt, ob bis zum Bamyan Passe der das natürliche Thor im Hindukusch aus Afghanistan und Indien in das Flachland des Orus bildet, bleibt noch künftigen Untersuchungen vorbehalten. Dass aber bis zur Entdeckung der Zinngruben in Cornwallis alles Zinn zu den vielen Bronzen, die in allen Ländern des Mittelländischen Meeres und in Skandinavien gefunden sind nur aus dieser Gegend kam, möchte ich doch bezweifeln.'

There is yet a third locality in which copper and tin are found in a condition of proximity, which may well have led to the combination of them into bronze. This locality is no less extensive, and no less ancient a seat of human history, than the region lying between Birmah and Banca, inclusively. This is what Mortillet writes, 'Revue d'Anthrop.' i. v. 1875, p. 653:—

'Reste le groupe de l'extrême Orient Asiatique. C'est là évidemment où il faut chercher l'origine du bronze. Les principaux gisements sont dans la presqu'île de Malacca et surtout dans l'île de Banca, mais ils s'étendent dans d'autres iles de la Sonde et remontent jusque dans l'empire Birman où l'étain est encore exploité actuellement dans le district de Merguy. Ce minéral dans tous ces gisements se recueille de la manière la plus simple et plus facile dans les alluvions. Ce sont bien certainement les alluvions les plus riches du monde en étain et celles qui occupent la plus grande étendue. Il est donc tout naturel que ce soit celles qui les premières aient attiré l'attention de l'homme. La cuivre se rencontre dans les mêmes régions. Tout le monde connaît les gisements de cuivre des îles de la Sonde, Timor, Macassar, Borneo. La Birmanie anglaise présente des mines de cuivre à côté des ses exploitations d'étain. Le pays se trouve donc dans les meilleures conditions pour avoir vu naître l'industrie du bronze.'

Von Baer himself, *l. c.*, thinks that Ceylon may very probably have been one source whence tin was procured by the Phoenicians trading for it with the Malays, as they traded doubtless with them, at second, if not at first hand, for the cinnamon which still retains its Malay name, little altered though it has passed through the mouths of so many Western races. I a little doubt the correctness of the introduction of the Malays into the picture; for if the Malays brought bronze, or even only the ores of either or both of the metals forming it, to meet the Phoenicians at Ophir, it is difficult to understand how they should have failed to carry the knowledge they could not thus have failed to gain, with them on their colonising expeditions over Polynesia. Yet Polynesia was in the Stone Age till quite recently, though the common fowl and the pig had been carried to some of the most remote of its islands, and the dog even to New Zealand, in times beyond the memory, if not beyond the traditions of the natives, and long before they came into *rapport* with Europeans; and we learn from Mr. J. Crawfurd, 'Trans. Ethn. Soc.,' iii. 1865, p. 353, that it was the Gentoo traders of the Coromandel coast who brought tin from Malaysia to India, when the Europeans first came into relation with Malaysia in the early years of the sixteenth century.

There can be no doubt that many of the bronze weapons now found in this country were imported as made up; if my memory does not deceive me, hollow bronze weapons have been found upon our south coasts, containing still the cores on which they had been moulded for the use of our natives, who were balked of them by the shipwreck of the vessel laden for them. On the other hand, masses of bronze in the rough, pigs, that is, or even ingots, have also been found in this country, together with smelting apparatus and moulds; so that bronze must also have been worked up here, as there is abundant evidence (see Klemm, 'German. Alterthumskunde,' 1836, p. 151) to show it was also in Germany. Every nation, the most refined, perhaps, not more than the most barbarous, has its own fancies as to the patterns of its own weapons, as much as its own clothes, its own architecture, and its own ceremonials; and this feeling of independence would shortly evoke a demand for the raw material and a production of moulding apparatus. A very instructive story, bearing upon the possible working of this desire for variation, is told by Major-General Lane Fox, F.R.S., in a paper on 'Primitive Warfare,' read by him June 5, 1868, at the Royal United Service Institution. As the paper in question was printed only and not published, it may be allowable here to reproduce it. It runs thus:—

'The next principle which we shall have to consider is that of variation. Amongst all the products of the most primitive races of man, we find endless variations in the forms of their implements, all of the most trivial characters. A Sheffield manufacturer informed me that he had lately received a wooden model of a dagger-blade from Mogadore, made by an Arab who desired to have one of steel made exactly like it; accordingly my informant, thinking he had found a convenient market for the sale of such weapons, constructed some hundreds of blades of exactly the same pattern; on arriving at their destination, however, they were found to be unsaleable. Although precisely of the same type as those in general use about Mogadore, all of which to the European eye would be considered alike, their uniformity rendered them unsuited to the requirements of the inhabitants, each of whom prided himself upon possessing his own particular pattern, the peculiarity of which consisted in having some almost imperceptible difference in the curve or breadth of the blade.'

Persons who, like myself, incline to the belief that the regions round the Bay of Bengal were probably the seat, not only of the discovery of the stream-works oxide of tin, but also of that of its alloy with copper, will be tempted to assign more weight than is due to the fact, or supposed fact, of the bronze-swords having such small handles, as it may be thought Hindoos or people like them would have. I am not quite clear that this bronze sword, leaf-shaped or other, has always a very small hilt; certainly in some cases, if we imagine the hilt to be wrapped round with leather or other material suitable for the purpose, it will not turn out to be at all too small for the grip of an ordinary English hand of the present day. At any rate there can be no doubt that in this country the skeletons of the Bronze Period belonged to much larger, and stronger, and taller men than did the skeletons of the Long Barrow stone-using folk who procured them. In some parts of England the contrast in this matter of size between the men of the Bronze and those of the Stone Age is as great as that now existing between the Maori and the gentle Hindoo; and in some, though not in all, parts, the Bronze-users appear to have as entirely extirpated the Stone-users, as the Maoris, in their cannibal days, would have extirpated any similarly weaker race. The facts as seen by me, when in company with Canon Greenwell, and upon other occasions, appear to me to justify some such statement as this, as to the introduction of bronze into this country. The stone-using inhabitants of Great Britain, if not also of Ireland, may have had their first introduction to a knowledge of bronze in the way of peaceful barter and commerce. Some probability is given to such a view as this by the fact that some of the earliest bronze axes are evidently moulded upon the pattern furnished by stone weapons, just as in North America, where there was a Copper Age, the copper arrow-heads are modelled (see Lubbock, 'Pre-historic Times,' sec. ed., 1869, p. 245) on the type of their stone ones. But with improved and advanced Bronze weapons in this country, we find, invariably within my experience, an improved and advanced race of men, so far as powerful limbs, tallness of stature, and capacious crania, do make one race of men superior to another. This race of men, besides their physical, present us with many ceremonial and other differences; their burial mounds are round; their pottery is of another kind, or kinds rather, as they have funeral as well as other wares, the former of which the stone men had not; the ornaments they buried with their dead are of a different kind, type, and material; finally, the numbers of dead interred in round barrows, and the numbers of round barrows themselves, are very much greater than those of the dead interred in long barrows, and than those of the round barrows themselves. All this seems to me to point to a conquest of this country having been effected by Bronze-using invaders, who came in great numbers, probably as has been elsewhere suggested, from the Cimbric peninsula, which was once again in the Iron Age, viz. in the Iron Age of Swegen and Cnut, an *officina gentium victricum*. If the Danes in a recent war had been as much in advance of their enemies in the adoption of improved weapons of war, as *ex hypothesi* they were in the Bronze Age, and *de facto* in the Iron Age, of the inhabitants of these islands, Schleswig-Holstein might still have been an appanage of Denmark.

I take this opportunity of remarking that anybody who will take the trouble of reading the few lines which come in Hesiod's 'Works and Days,' 144–148, just before the often-quoted line as to the sequence of the Bronze and Iron Ages, will find that he had somehow become as much impressed with the vast size and brute strength of the bronze-using people as I, in spite of the currently accepted statements as to the small hands of the men of that era, have become from actual handling of the bones. Bronze Age tumuli, however, may have been excavated, indeed, as the history of the examination of the Tomb of Theseus, at Scyros, shows, they actually were excavated in the days of the very early bards, such as the one just referred to.

Virgil's line

> 'Grandiaque effossis mirabitur ossa sepulcris'

expresses the tendency to magnify the size of such *trouvailles;* still there was solid fact for what Hesiod wrote, *l.c.*, and Ovid might have given more space to insisting upon this very distinctive characteristic of the Bronze Age, than he has in his reproduction of Hesiod, Metamorph. i. 125–127:—

> 'Tertia post illam successit ahenea proles,
> Saevior ingeniis, et ad horrida promptior arma;
> Non scelerata tamen: de duro est ultima ferro.'

In modern Europe we have but some half-dozen millions of men under arms at the present moment, and we have lost by war in the last twenty-five years something under a couple of millions only, by the accidents inseparable from modern fighting; and it is difficult for us, consequently, to realise, even approximatively, the terrible conditions prevalent in the 'bella, horrida bella' of the Bronze Age. Hesiod appears to have been much impressed by what tradition told him of it; he does not, however, appear to have thought his own time so very much better, as we have such good reason for thinking ours is.

I have sometimes thought that the comparison (for which see Max Müller, Lectures, ii. p. 256, 8th edition, 1875) by the Sanskrit writers of copper to the muscles or flesh of an animal, may really have been a comparison of greenish bronze to muscles taking on a greenish hue from decomposition, and that we should thus save ourselves from supposing that copper, which, as a metal, is eminently 'red,' should have been contrasted by our forefathers with something, iron, to wit, which they compared to blood. At any rate, leaving both mythology and pathology, I may say that Sir John Lubbock has given us excellent reasons for doubting whether Europe, or at least the western part of it, ever went through a pure copper stage, as America, so rich in native copper, did. And as regards metallic tin being used for weapons at least, I have come upon only a single statement which could bear such a meaning; this statement is given by Klemm, 'Germanische Alterthumskunde,' p. 19, in the following words: 'Ein Stück aus reinem Zinn fand Kortum in der Ruhenthal Grabstatte,' s. 8. 105. It is plain, however, that this may have been a 'find' of an ornament as distinguished from an implement made of unalloyed metallic tin.

There are two Greek words standing at the end of line 612 of the 'Agamemnon' of Aeschylus, which mean 'baths for copper,' but which are usually translated 'dyeing of copper' and are supposed to be a proverbial mode of indicating an impossibility, or, as the Germans put it, an 'Unding.' I strongly suspect that these words have attained this secondary signification, not from any reference to colouring, but simply to 'tempering,' and that the mode of tempering bronze having been a secret, it has to be considered something *supra-* and ultimately *contra-*natural. If this suggestion is true, we have in it a fresh argument for the view which teaches that the discovery of alloying copper with tin was extra-European in origin. There is another new argument for the same conclusion, and for the corollary to it, that bronze, like the Jade, Jadeit, and Nephrit of the preceding or Stone Period, and like all imported articles in such times, must have been scarce and highly valued; and this argument lies in the fact that the use of stone weapons survived so long after the introduction of this alloy. This was forced upon me in the examination this year of certain barrows in Somersetshire, proved to be of the Bronze Period by the discovery in them of bronze weapons, with burnt human bones, in which worked flints were in such abundance, that had it not been for the discovery of the bronze implements, we might almost have supposed that we were dealing with interments of the Stone Age. These barrows were in a district (that of Castle-Cary) the surface strata of which are low down in the lower secondary formations, yet the worked flints, and they not only 'strike-a-lights' or 'thumb-flints,' but scrapers, were as abundant as they might have been in a tumulus upon a chalk down. Their varied quality and great quantity render it impossible to think that they are in such a district merely thrown in ceremonially, and are evidence to the effect that, though tin and copper were available enough, and side by side, at no greater distance than Cornwall, those particular deposits had not then been utilised for the manufacture in question.

Let us now pass to the Stone Age. I have not the knowledge requisite for subdividing the Bronze Age into distinct periods; and looking at the question in the light which played over the Somersetshire hills, when I was employed, as just now stated, upon them, I doubt whether any subdivision of it, as it was in England, can be justified. A Copper Age, no doubt, must have existed, and did exist, in America, antecedently to the Age of copper alloyed with tin; but there is no evidence that it ever existed in England, at least. More may be said, on the authority of Polybius and on other evidence, for the subdivision of the Iron Age into two periods, one of which, the earlier of course, had not learnt the art of tempering iron, whilst to it a second, 'the age of steel, succeeded then.' But as regards the Stone Age we have no need to have recourse to mere probable arguments and *a priori* evidence. There is no doubt whatever that the Stone Age is divisible into two great periods upon several principles, which, however, make their several sections in the same plane. We can look at a stone weapon and ask ourselves one or other of these three questions; firstly, was it intended to be used in the hand, or used as hafted? Secondly, has it been polished and ground up, or has it been left simply chipped over with conchoidal fractures? Thirdly, was it found in company with pottery, however rude, or was it found in some river gravel-bed, in company with no other evidence of human handiwork, but with the bones of mammoth and rhinoceros? If a stone weapon is so fashioned that we can see that it was intended to be stuck into a handle or haft, and if it is polished, we may be sure that it belonged to a later than the mammoth period in this country, and that it may be spoken of as Neolithic in contradistinction to the Palæolithic weapons. It is true that in the great factory for flint weapons, which has been described by Major-General Lane Fox ('Journal Anth. Inst.,' v. 3, 1876), at Cissbury, an implement, or implements, which could only be used as held in the naked hand, came out during the period of the excavations carried on there, and amongst multitudes of 'celts,' which were as obviously intended to be used in handles. But survivals were not unknown in the great Stone Age any more than in our great Steel Age; and for the very various manipulative processes connected with the working of a *Flint-mine*, with its tortuous galleries, necessitating an amount of 'body-bending toil' no way inferior to that necessitated by the galleries of the modern coal-pit, a pointed stone weapon which had a blunt end fitted for a hand grasp would not rarely have its advantage. The fact that at Cissbury, as also at Grimes Graves, in Norfolk (for which see 'Journal Ethn. Soc.,' N.S., ii. p. 214), and at Spienne, in Belgium (for which see 'Mém. Soc. Sci. et Arts du Hainaut,' 1866–7, p. 355), it was found worth while to undertake and execute such extensive works as are those flint-mines, enables us to realise the meaning of the words 'Stone Age' very vividly. The demand for these weapons was so great that it was found profitable to go through all this toil to supply it; the margin of advantage which made it profitable, lying in the mineralogical fact that a flint taken freshly out of its chalky matrix, and retaining its normal hygrometric properties, is more workable and plastic than a flint which has been rolled about the world in floods *per mare per terram*. A modern workman will break flints fresh from the chalk for a shilling, whilst for an equal amount of results for gravel pebbles he will charge you eighteen-pence. It may seem something of a contradiction to the principles of the identity of the period of handled, with that of polished, as opposed to chipped flints, to say that the flints manufactured at Cissbury were, with the few exceptions alluded to, all intended to be fitted with handles, and yet that they were all left unpolished; but the process of polishing a flint, when *finely* chipped, as these are, is a very easy one, involving only the use of a little sand and water to rub the broad chipped cutting edge into smoothness, on a stone such as modern savages use for the purpose; and the modern manufactory of metal weapons shows us that weapons and implements of all kinds are, from certain considerations of expediency, stored and stacked in an unfinished state, before being sent out on, or for, sale.

One point I should wish here to put upon record, relatively to the excavations at Cissbury. In my paper on 'The Animal Remains found at Cissbury,' published in the Journal of the Anthrop. Inst. for July, 1876, vi. p. 22, as also in 'British Barrows,' p. 742 (this volume, Articles XVII, XIX), I expressed myself as having been much impressed by what I had seen to the effect that the pitfall, especially as eked out with certain accessories, had counted for a great deal in the economy (if this be not to profane the word) of the Stone Age. In the earlier of the two places referred to, I say: 'Hurdles of gorse probably were arranged on the principle of the wicker hoops in a decoy, and it is easy to see how, by such a plan, eked out, perhaps, by the firing of heaps of the same useful material, a wild bull, or a herd, might be driven over a pitfall.' In the latter, I say: 'It requires a greater effort of imagination on our part to imagine a pack of wild dogs co-operating with priscan men in driving a herd of wild cattle or wild pigs (both of which were represented in the Cissbury Pits) along a track in which a pitfall had been dug and covered over. Still what we know justifies us,' &c. When I wrote these words, I was very distinctly of opinion that the suggestion they contain was, however obvious, yet entirely an original and novel one; I was rudely, yet not unpleasantly, undeceived a few days ago, when verifying, as it is always well to do by often-quoted lines, the lines of Lucretius, v. 1285. I 'tried back,' as I have heard it expressed elsewhere, to the preceding context, which greatly fascinated me, not only by its grand roll and flow, but also by the singularly clear insight which it gave me into the way in which its author had faced the great problem of 'Kulturgeschichte.' In that context I came, to my great surprise, upon two lines, 1249–1250, which contain a suggestion at once half-coinciding with, and half-contradicting my own as just quoted. Lucretius, undoubtedly, can claim priority as to the part of his hypothesis in which he and I agree; and perhaps I had better not claim originality as to either part of mine; but this question is of little consequence. Here are the lines of Lucretius, v. 1249–1250:—

'Nam fovea atque igni prius est venarier ortum
Quam sepire plagis saltum canibusque ciere.'

To this disquisition on the several Ages of Iron, Bronze, and Stone, I will here append an account of the disinterment of a skeleton of the Iron (Roman) Period, which took place by the permission of the Earl Bathurst, on August 27, in Oakley Park, during heavy rain, in the presence of the Hon. Mrs. Lennox, the Hon. Miss M. Ponsonby, Professor A. H. Church, Christopher Bowley, Esq., R. A. Anderson, Esq., E. C. Sewell, Esq., and myself. The skeleton was contained in a stone coffin, covered by a flat stone slab, much of the same character as the undoubted Roman coffins found at York and elsewhere in England, though, unlike many of them, it contained no relic besides the skeleton itself, and a Roman nail, of a type known at Cirencester. The dimensions of the coffin were :—

External	Length	7′ 2″
	Width at N.E. end	2′ 8″5
	Width at S.W. end	2′ 4″
Internal	Length	5′ 8″
	Width at N.E. end	1′ 5″5
	Width at S.W. end	1′ 5″.

Its bearings were from N.E. by E. to S.W. by W., a rather unusual orientation, it being more common to find the feet at a point a little south, than at a point a little north, of the rising sun, deaths being more numerous in the winter than in the summer quarters of the year. The head, however, was at the northeastward end, and this appears to make it probable that this coffin dates as far back as the time when the Romans had relinquished the practice of cremation, without accepting the religion, or, at least, the religious practices of Christianity; to a time, that is, between the death of Severus, in the first decade of the third, and the accession of Constantine, in the first decade of the fourth century, A.D. The skeleton was in good preservation; the only disturbance to which it had been subjected, of a violent kind, previous to our exploration, having been quite recently inflicted by some gay young anthropologists from the day school, who, in defiance of the school-board's inspector, had, in their zeal for science, been poking sticks through a chink at the north-east end of the coffin, and had slightly displaced the skull inwards, besides damaging its outer table and exposing the diploe. The lower jaw, however, had not been displaced. On the back of the skull, and also around the first cervical vertebra, there is a considerable deposit of lime, probably the remains of quicklime which the Romans often put into their coffins, as may be seen to great advantage in the museum at York. I did not observe this till the bones were cleaned in the museum here; and I did not note whether there was any hole in the bottom of the coffin, whereby an exit would be possible for this lime as dissolved by carbonated water passing down into the coffin. Some of the other bones were blackened in places by carbonaceous deposit from the leaves and other vegetable matters, such as, if my memory serves me, beech-nuts, which had found their way into the coffin and decayed there, and also from the decay of the soft parts of the body and the wrappings of it. The nail found in the coffin may, indeed, appear to indicate that some sort of coffin of wood was used, as well as the coffin of stone; there would have been plenty of room for one, as the length of the Roman body was but five feet one inch, whilst the internal length of the stone coffin was 5′ 8″; but I think this nail may have worked its way in from without, through the same chinks which gave inlet to the other foreign bodies already mentioned.

The left arm lay alongside the body, and the left hand rested on the pelvis; the right arm was stretched upwards with the hand at the face; the left leg was drawn up to the centre of the body, or thereabouts. The distance from the end of the coffin to the pelvis was 2′ 11″, leaving a space of about 4″ between the sole of the foot and the end of the coffin.

My thanks are eminently due to Professor A. H. Church and to E. C. Sewell, Esq., for their help before, after, and during this disinterment. For the measurements and descriptions following, I am more entirely responsible than for what has preceded.

MEASUREMENTS AND DESCRIPTION OF SKELETON, FROM STONE COFFIN IN OAKLEY PARK,

On the Estate of the EARL BATHURST, August 27, 1877.

Measurements of Skull.

External length	7·3″	Frontal arc	5·2″
Fronto-inial length	7″	Parietal arc	4·5″
Extreme breadth	5·8″	Occipital arc	4·5″
Upright height	5·5″	Greatest frontal width	4·9″
Absolute height	5·2″	Greatest occipital width	4·8″
Circumference	21·3″	Basicranial axis	3·75″

Measurements of Face.

Length of face	2·25″	Width of lower jaws, ramus	1·3′
Breadth of face		Interangular diameter	3·6″
Basio-subnasal line	3·4″	Cephalic index	79″
Basio-alveolar line	3·5″	Antero-posterior index	55″
Height of orbit	1·5″	Femur	17″
Width of orbit	1·6″	Tibia	12·2″
Length of nose	2·0″	Humerus	11·5″
Width of nose	1·0″	Stature	5·1″
Depth of lower jaw, at symphysis	1·1″		

Age, about 30.

A well-filled-out skull, on the whole, though the parietal tubera are still distinguishable, as is often the case in female skulls; the forehead is vertical, but the parietal region slopes with considerable obliquity in its posterior two-fifths; the plane of the superior occipital squama lies distinctly behind that of the posterior part of the parietals, so that a very marked undulation is formed at the line of meeting of the two bones. The relation of inferiority held by the height to the breadth of the skull is probably merely a sexual character; the vertical contour being eminently that of the dolichocephalic type of skulls, whilst the smallness of the mastoid, the slightness of the supra-orbital ridges, and the feebleness of the lower jaw, show, what the characters of the limb and trunk bones also show, viz., that the owner of this skeleton was a woman.

This woman had lost the second molar of the left half of her upper jaw some time before the evolution of the wisdom teeth of the lower jaw of the same side, and probably not very long after the evolution of the second molar of the same side of the lower jaw. The first molar of the right half of the upper jaw had been similarly lost early in life; the second molar next to it was largely excavated, and the wisdom tooth, on to which that carious cavity opened, had an abscess at its fangs. The lower jaw teeth, though all sound except the left second premolar, are much crowded together. It is not clear that the wisdom tooth of the left upper jaw was ever developed. Six abnormalities is a large proportion in the dental series of a woman who was not much beyond thirty years of age.

The slightness and straightness of the collar bones, the horizontal direction of the neck of the femur, the characters of the *os innominatum* and other bones, show the skeleton to have belonged to a woman of about thirty years of age, or a little over. The suture between the first and second vertebrae of the sacrum is widely, but not symmetrically open, and its patency with a greater width on the left than on the right side must be considered as due to some morbid process. All the sutures and epiphyses of the limbs are closed and anchylosed; so also are, to a great extent, the sagittal and lambdoid sutures in the skull. The characters of the facial bones, such as those of the elevation of the nasal bones, and the proportions indicated by the measurements, show that this Romano-British lady may have deserved the praise of Martial as expressed in the following lines:—

> 'Claudia caeruleis quam sit Rufina Britannis
> Edita, quam Latiae pectora plebis habet!
> Quale decus formae Romanam credere matres
> Italides possunt.'

BIBLIOGRAPHY OF THE BRONZE CONTROVERSY.

Crawfurd, John, 'Supposed Stone, Bronze, and Iron Ages,' Trans. Ethn. Soc., N.S., iv. 1–13. 1866.

Wright, Thomas, 'On the true assignation of the Bronze weapons,' ibid., 176-196.

Lubbock, Sir John and Frederick, 'On the true assignation of the Bronze weapons,' ibid., v. 105–115. 1867.

Lindenschmit, 'Zur Beurtheilung der alten Bronzefunde diesseits der Alpen, und der Annahme einer nordischen Bronze-cultur,' 'Archiv für Anthropologie,' viii. 3, 161, 177. 1876.

Hostmann, Review of Hans Hildebrand's 'Das Heidnische Zeitalter in Schweden,' translated by Mestorf, 1873, 8vo., A. A. viii. 278–314. 1876.

Mestorff reviews Engelhardt in the following pages, 315–320, of the same Journal.

Sophus Müller, 'Dr. Hostmann und das nordische Bronzealter,' ibid., ix. 2 and 3, 127–141.

Hostmann, 'Zur Kritik der Kulturperioden,' A. A., ibid, 185, 219. 1876.

C. E. V. Baer, 'Das Zinn von wo?' A. A., ix. 4, 187, 7, pp. 263–269.

Sophus Müller, 'Zur Bronzealters Frage,' 27, 41. }
Hostmann, 'Zur Technik der Bronze,' 41–63. } A. A., x. 2. 1877.
Lindenschmit, 'Zu den Vorstehenden,' 63–73. }

Hostmann, 'Hoher Alter des Eisen-verarbeiten,' A. A., x. 4, 418.

See also for an excellent, though not distinctly controversial, account of the Bronze Age in North Europe, a memoir, in an octavo of less than 140 pages, by Sophus Müller. 'Die nordische Bronzezeit und deren Periodentheilung' von Sophus Müller, aus dem Dänischen von J. Mestorff, Jena, 1878.

The beautifully illustrated work, 'Le bel Age du Bronze Lacustre,' by E. Desor and L. Favre, 1874, has been referred to above.

The first volume of the Newcastle Archaeologia, 1822, contains a valuable article by the Secretary, the Rev. John Hodgson, under the title, 'An Enquiry into the Era when Brass was used in purposes to which Iron is now applied.'—pp. 17–99.

XXXVII.

ON THE STRUCTURE OF ROUND AND LONG BARROWS.

PROFESSOR ROLLESTON read papers to the British Association, September, 1880, *On the Structure of Round and Long Barrows,* his remarks being illustrated by a number of diagrams. Premising that one of his objects was to preserve barrows from being spoilt, and thus to prevent the destruction of certain links in the history of our species, he described the construction of barrows which he had explored, and urged the absolute necessity of very great care being exercised in such exploration. Speaking of urn burials in round barrows, he briefly referred to the question of the cremation of bodies, and the idea of it. Why did the people burn their dead? He believed the idea was this—that all savage races, when they had to deal with an enemy, were exceedingly prone to wreak certain ignominies on dead bodies. Burning the bodies put it right out of the power of the enemy to do this, and the urn enabled people to carry away their friends who were so burnt. In time of pestilence it became actually necessary for sanitary considerations to burn the dead, and it was only in time of plague or war that we found that cremation or burning became the order of the day, and that was readily explicable by the fact that men always did what they could on the principle of least action, because burning was a troublesome process. Any universality of burning was explained by the fact that ancient history was simply one great catalogue of plague, and pestilence, and war, and the like. Of course he was an enemy to cremation, because it did a great deal of harm, preventing us from knowing what sort of people our predecessors were. Professor Rolleston chronicled the finding in a barrow of the Bronze period of a man laid out at full length, the general rule being that of burial in a contracted position. As regarded the date to be assigned to these things, he might give it as his opinion that no Roman ever used a bronze sword, nor crossed swords with an enemy using a sword of that material. As regarded the long barrows, that mode of burial stretched all the way from Wales to the Orkneys, and in them was found not a scrap of metal. His opinion was that the idea of the construction of these barrows was taken from limestone mountain headlands projecting into the sea, such as might be seen by a little trip in their immediate locality[1]. The men lived in caves, and the idea for the place of burial was taken from the place of living, it being often found that a man made the house in which he lived his burial-place.

[1] [The Meeting of the British Association at which this paper was read, was held at Swansea.—EDITOR.]

XXXVIII.

ON THE CHARACTER AND INFLUENCE OF THE ANGLO-SAXON CONQUEST OF ENGLAND, AS ILLUSTRATED BY ARCHAEOLOGICAL RESEARCH.

THERE are numerous points of general and living interest relating to the Anglo-Saxon conquest of this country which are very largely dependent upon archaeological research for their elucidation. Amongst these may be mentioned the question of the extent to which the Romano-British population previously in occupation was extirpated; the question of the relative position, in the scale of civilisation, held by victors and vanquished; and the question of the extent of our indebtedness as to language and laws to one or other of the two nationalities. Light is thrown, even upon points apparently of the most purely archaeological character, from such literary sources as histories of the nomenclature of localities, as the records of monasteries, as illustrations in manuscripts, and as laws. But the graves of the Anglo-Saxons and their contents have been for the present investigator the primary, and such literary works as those alluded to, and such as many of those published under the direction of the Master of the Rolls and by the Early English Text Society, have been only a secondary source of information. They have, however, been by no means neglected by him.

It may be well to begin by stating how an Anglo-Saxon is to be distinguished from a Romano-British interment. Anglo-Saxons, during the period of their heathendom, which may be spoken of roughly as corresponding in England to a period of some 200 or 300 years onwards from their first invasion of the country in force, were interred in the way of cremation, and in urns of the pattern so common in the parts of North Germany and of Denmark whence they are supposed on all hands to have come. A reference to any manual of archaeology, or an inspection of any such series as that figured by Mr. Kemble in the *Horae Ferales* from the museum in Hanover, will show the unmistakable identity of the pattern, fashion, and moulding of such urns as these, and those which I have had figured after digging them up in Berkshire. The Romans and Romano-Britons had given up the practice of burning the dead long before the time of Hengist and Horsa. When they practised it in England, their urns were of a very different kind, being well burnt and lathe-turned. All the Romano-Britons I have exhumed in the cemetery at Frilford, which has furnished me with the tolerably wide basis of something approaching to 200 interments of all kinds, were interred much as we inter our dead now. They were oriented, though by the aid of the sun and not by that of a compass; and, dying in greater numbers in the winter quarters of the year, had the bearings of their graves, as has been observed by the Abbé Cochet, pointing a little south of east. Now a Romano-British interment in this way of burial has to be distinguished from an Anglo-Saxon one in the same way of non-cremation, and this may be done thus:—the Romano-Britons never buried arms nor any other implements which could be of use in this, and might be supposed to be of similar use in the next world, together with a corpse. Funeral ware, such as lachrymatories, I have not found in company with coins of the Christian Emperors; but such articles stand in relation to quite a different idea from that which caused the Teuton to inter the dead with spear, shield, and knife; to say nothing of the common *situla* and sword. The Anglo-Saxons are supposed by Kemble to have relinquished cremation only when they assumed Christianity. It is a little difficult to be quite sure of this: at any rate, when we find, as we often do, an Anglo-Saxon in a very shallow grave, which may point to any one point of the compass, and in the arms and other insignia which it contains, such clear proof that its tenant thought that whatever he may *not* have brought with him into the world, at all events he could carry *something* out, we are tempted to differ even from such authority as Mr. Kemble's. But I am inclined to think that in some cases it is possible to identify the tenant of a properly oriented grave as having been an Anglo-Saxon. In many such graves Anglo-Saxons are to be recognised by virtue of their insignia; and mixed up with their bones may be found the bones of the Romano-Briton who occupied the grave before them. But further, in some such cases it is possible to be nearly sure that we have to deal with an Anglo-Saxon, even though there be no arms or insignia in the grave. These cases are those in which we have evidence from the presence of stones under the skull that no coffin was employed in the burial, and in which stones are set alongside of the grave as if vicariously. In many such cases the craniological character of the occupant of such a grave lends some colour to this supposition. But upon such identifications as had been come to in the absence of arms and insignia I have based no statistics. The results of the statistics of the cemetery which I have explored, as stated above, when brought to bear upon the large questions alluded to at the beginning of the paper, would lead us to think that the Anglo-Saxons conquered, firstly and most forcibly on account of the shorter lives they led. An old Anglo-Saxon male skeleton was a rarity, an old Romano-British one a very common 'find' in my excavations. Nothing however in this life is, from the natural history point of view, more characteristic of real civilisation or real savagery than this point of the duration of life. The Merovingian Franks had, like the followers of Cerdic, been observed to have led short lives, merry—as the Capitularies of Charlemagne teach us of their kinsmen—with those kinds of mirth the end of which is heaviness. The next question which suggests itself upon the mastery of these facts and figures is, Were not these men merely soldiers encamped? are not these statistics just such as a cemetery similarly explored now-a-days, say at Peshawur or Samarcand, would yield? Not altogether such; for, however improbable it may seem, it is nevertheless true that the Anglo-Saxons, at all events in Berkshire, appear to have brought their own wives with them, and not to have provided themselves with wives from the families of the conquered previous inhabitants. The figures of the crania of females interred with Anglo-Saxon insignia, when compared with figures of the crania of Romano-British women, show a very great difference, to the disadvantage of the former of the two sets of females. The soldiers of Cerdic, who conquered this part of Berkshire about half a century or so after the time of the first invasion, resembled the soldiers of Gustavus Adolphus in very little else, but they appear to have resembled them in being accompanied by their wives. Whether this was the case elsewhere in England, I do not know. I am inclined to think that savagery was no great recommendation, nor heathendom either, to a Christianised female population in those days; and that the reluctance which would on these grounds interpose itself to prevent intermarriage between Romano-Britons and Saxons, sets up as great an *a priori* improbability against the theory which assumes that such intermarriages did take place, as the difficulty of bringing wives over in the ships of those days sets up in its favour.

Indeed, on the hypothesis of much intermarriage, the actuality of our Anglo-Saxon language is a very great difficulty. We do speak a language which, though containing much Celtic and a good deal of Norman-French, is nevertheless 'English.' Now we know, from finding cremation urns of the Anglo-Saxon type all over England nearly, that the whole of the country was overrun by a heathen population; to thus overrun it, this population must have been (relatively at least) numerous: add to the two conditions of heathendom and multitude, which may be considered as proved, the third condition of isolation, which may be considered as matter for dispute, and then the fourth, of this heathendom and isolation lasting from the time of Hengist to that of Augustine,—and the present fact of our language being what it is is explained.

For proving anything as to the period of which I have been speaking, a period which is rendered prehistoric not so much by conditions of time as by conditions of space—the absence of contemporary historians having been entailed by geographical and political isolation—arguments of two kinds, literary arguments and natural history arguments, must be employed. Neither the one kind nor the other is sufficient by itself. The empires of the natural sciences and of literature touch at many isolated points, and here and there they lie alongside of each other along lengthy boundary lines. But empires need not be hostile though they be conterminous; and that the empires of which we have just spoken may be united happily and in a most efficient alliance from wants in common, may be seen from the title-page of that most excellent German periodical, the 'Archiv für Anthropologie,' where we have the name of the physiologist Ecker coupled in editorship with that of the antiquarian Lindenschmit. The necessity for a combination of the two lines of evidence and argument is as obvious when we have to controvert, as when we have to establish a conclusion. If you have to attack or resist a force comprising both cavalry and infantry, you must have both cavalry and infantry of your own; otherwise some day or other, either in a country intersected by woods, or in some open plain furrowed into deep undulations, one of the two arms in which you are deficient will take you in one or both flanks, and you will be surprised, broken, and routed.

XXXIX.

JADE TOOLS IN SWITZERLAND.

[The Geneva Correspondent of the 'Times' newspaper, Dec. 17, 1879, having referred to the 'find' of a jade instrument in the bed of the Rhone in the course of excavations then being conducted, and having raised the question of the region from which the mineral Jade had been derived, a correspondence extending over some weeks took place, in which Professor Max Müller, Mr. Hodder M. Westropp, Dr. Rolleston, Professor Story-Maskelyne, Major Raverty, and A. B. M. took part. Dr. Rolleston's letters were principally directed to show that the source of jade is Oriental, and in this he was supported by Professor Story-Maskelyne.—EDITOR.]

LETTER 1.

Dec. 25, 1879.

YOUR correspondent Mr. H. M. Westropp calls the supposition that the jade tools found in the Swiss lake-dwellings came from the East 'a wild hypothesis.' Now, we must judge of the difficulties of the past by what we can see of the possibilities of the present, and I make bold to say that some thousands of men who read those words this morning must have had in their pockets implements which had travelled with the single individuals carrying them, in their single lifetimes, over much greater distances than this 'wild hypothesis' supposes the westward migrating Stone-Age men to have traversed in many generations. Hence I submit there is no *a priori* improbability attaching to the view in question.

This view is, indeed, one of the best-established facts which recent prehistoric research has brought to light.

So long ago as 1865, M. Fellenberg, *père*, analysed four implements of nephrite from Meilen and Concise, and one of jadeite from Mooseedorff. This nephrite was found to have precisely the same chemical composition as the well-known New Zealand nephrite, analysed by Scherer; and the jadeite to be similarly identical with the mineral known under that name from China, and analysed by

Damours. Now, of nephrite and jadeite, it is known that, with the single exception constituted by the discovery of an unworked fragment at Schwemmsal in Saxony, no unworked specimen of either has been found nearer to Switzerland than, for nephrite, Turkestan and the environs of Lake Baikal; and, for jadeite, China.

Mr. Max Müller says it was a harder business for the westward emigrants of the Stone Age to carry the ponderous tools, which their Aryan language represents, than the lighter ones which the Messrs. Fellenberg, *père et fils*, have taught us so much about. It is not for me to defend the Professor; but, once more to illustrate the past from the present, I may say that many parents, if there are any who feel sceptical as to this ponderation, can have the opportunity of convincing themselves of its correctness by taking stock of the amount of Greek and Latin their sons have been successful in transporting home with them through the perils and trials of a railway journey this Christmas.

But there are some things harder to transport with you than pocket-knives, whether of jadeite or nephrite, and harder to carry with you even than Aryan or any other languages. These are domestic animals. The sheep and the goat are both found in the Stone-Age pile-dwellings along with these weapons. Does Mr. H. M. Westropp think a goat would be easier of transport than a celt (of jade)? Let him try. Yet it would be 'a wild hypothesis' indeed which should aver that these animals came otherwise into the Switzerland pile-dwellings than by slow transportation from the East.

I am anxious to hear how your correspondent will meet this difficulty. But before I write again I should be glad to be assured that any antagonist who may reply to my suggestions had consulted the following references:—

Keller—'Lake Dwellings;' English translation by J. E. Lee, F.G.S.; I. p. 195, 2nd ed. 1878; and Fischer and Fellenberg, *citt. in loc.* Edmund von Fellenberg—'Bericht an die tit. Direction der Entsumpfungen über die Ausbeutung der Pfahlbauten,' Bern, 1875. L. R. von Fellenberg—'Neues Jahrbuch für Mineralogie,' p. 619, 1866.

LETTER 2.

Jan. 2, 1880.

YOUR correspondent Mr. H. M. Westropp is slow to take a well-meant hint such as the one with which I ended my letter published by you on December 25.

1. The views which he ascribes to M. Desor, it is true M. Desor did once hold, but he repudiated them at the International Congress of Anthropologists and Archaeologists which met at Brussels in 1872. Here are his old and his new views, as given in the 'Comptes Rendus' of that Congress, published in 1873. Speaking of 'the wild hypothesis' of the importation of jade into Switzerland, M. Desor, p. 352, says:—'Pour ma part, je n'ai pu admettre une hypothèse aussi peu vraisemblable, et pendant longtemps j'ai pensé avec M. de Mortillet que ces roches devaient se trouver dans les Alpes mêmes. Mais voici tantôt vingt ans que nous cherchons sans rien découvrir.' A page and a little more follows; and on p. 353, M. Desor sums up his altered view thus:—'Ma conclusion que j'émets à défaut d'autres, est donc celle-ci; ces roches, vu leur petit nombre, vu leur admirable état de conservation, sont des reliques des temps les plus anciens; elles ont été apportées d'Orient par les premiers colons qui ont succédé aux peuplades de la pierre taillée.'

2. Mr. Westropp asks me to account for the presence of jade axes in Ireland; but I do not know that any have ever been found there, though there is no reason why they should not, except the reason which explains their absence from Denmark and Sweden and which I leave your readers to divine. I say I do not know of any jade hatchets having been found in Ireland; neither did Sir Thomas Wilde, if we may judge from his 'Catalogue of the Antiquities of the Royal Irish Academy,' p. 39, 1857.

3. Sir John Lubbock's words in 'Prehistoric Times,' 4th ed. p. 82, are:—'Though, perhaps, it would not yet be safe to conclude that these jade axes were introduced from the East, no European locality for jade or jadeite has yet been discovered; and it is perfectly possible they may have passed from hand to hand and from tribe to tribe by a sort of barter.'

4. I cannot verify the second reference made to Sir John

Lubbock, writing, as I am, in haste for an afternoon post. But if the sheep and goat were known at all in the Stone Age—and every collection of Swiss-lake antiquities shows they were—my argument stands.

Let me say in conclusion that archaeologists in England will be glad to have a notification of the various kinds of animals which were found in such abundance in the Bienne Lake pile-dwelling at Schaffis. In that station—one of the oldest in the Stone Age—both jadeite and nephrite implements were found; but whether sheep and goat bones were found in company with them I cannot quite positively say. Of course, it has been put on record that the two sets of objects were so found in the Stone-Age pile-dwellings of Mooseedorff, Locras, and Wanwyl.

ADDRESSES

AND

MISCELLANEOUS PAPERS.

XL.

ADDRESS ON PHYSIOLOGY IN RELATION TO MEDICINE IN MODERN TIMES,

DELIVERED BEFORE THE BRITISH MEDICAL ASSOCIATION, AT OXFORD.

The fact that my connection with the University of Oxford has put me into the honourable office of giving this address will, I hope, justify in your eyes my adoption for it of an arrangement, and my choice for it of a set of topics, which that local position has suggested to me. If I were to say that I had chosen for the subject of this address the bearings of the studies which it is the business of my life to teach here, upon the interests of the medical profession, I should be giving it too ambitious a title; it is but with some and with few of these bearings that I propose or feel myself competent to deal. I shall limit myself, firstly, by selecting only such topics as, having been pressed forcibly upon my own attention in my own peculiar course of labour, have come to assume, in my own eyes at least, a considerable importance, and have seemed, in consequence, not unlikely to prove possessed of interest for others also; and I shall limit myself, secondly, by abstaining from going over ground which has, within my own knowledge, been occupied by persons who have on previous occasions stood in the position which I now occupy before you.

Let me throw the heads of my address into a few short phrases, and say that I propose, with your permission, to speak firstly of the bearing of certain portions of the very extensive range of subjects, comprised under the titles Anatomy and Physiology, upon certain points and problems which come before the attention of the medical practitioner in the course of his actual duties; and secondly, of the illustration which some of the conclusions recently come to in Biological Science cast upon the validity of certain principles which are ordinarily looked upon as authoritative canons for the regulation of the reason in medical, and, indeed, in other investigations. Under the term Biological Science are included, besides pure Physiology, Human and Microscopic Anatomy, Comparative Anatomy also; and in this place, as your visit to the Museum will have convinced you, we give considerable, but as we hope, not undue prominence to this latter branch of study. I propose to speak of the bearings of Biology on Medicine in each, but, owing to our local speciality just alluded to, specially in the last mentioned of these four departments. And I must ask you to bear in mind that the very constant reference which I shall make, if not in my address, at all events in my notes, to the works and writings of others, is in like manner to be explained by my wish to have a distinctive colouring given to this address by the local peculiarities of the great educational centre in which we are assembled. For one of the most distinctive peculiarities of this ancient University is the formation within its precincts of such a library of modern science as will shortly have no superior, and but few rivals, in the world. This we owe to the well-advised administration of the funds of that famous physician, Dr. Radcliffe; and it is from a wish to make a sort of acknowledgment of the obligation which medical and other sciences owe to him and his trustees, that I shall so constantly, at least in print, refer to the chapters and pages of the innumerable books which their enlightened munificence has put here at the disposal of the student. It is not, I can assure myself, from any irritable anxiety to impugn or depreciate the work of others, that I have so constantly consulted and specified their pages; nor, I trust, have I allowed myself to be tempted into the unpardonable fault of using, or rather abusing, a great opportunity by making upon it a petty personal display. Rather have I felt it to be my duty to occupy this hour as you occupy your lives, in doing what it may be possible to do for the good of humanity. Your presence and your example make me feel that any other course would be but impertinence; and I have therefore kept constantly before my eyes Bacon's sentence in condemnation of all empty parade of useless erudition—*Vana est omnis eruditionis ostentatio nisi utilem operam secum ducat.*

The title of Niemeyer's work on medicine, the seventh edition of which has recently appeared and come into my hands, will furnish me with an excellent text for my first head—The Connection and

Interdependence of Medicine and Physiology. That title runs thus:—'Lehrbuch der speciellen Pathologie und Therapie, mit besonderer Rücksicht auf Physiologie und pathologische Anatomie, von Dr. Felix von Niemeyer. Siebente vielfach vermehrte und verbesserte Auflage,' Berlin, 1868. From this work I will take my first illustration of the nexus and connection of which I have to speak; and I believe that, though I am obliged to dissent from the explanation therein given of the facts I shall refer to from its pages, the explanation which they seem to me to bear, or rather demand, shows even more clearly than the one there given the intimacy of the alliance which is now becoming so close between the experimentalist and the practitioner. In writing (vol. ii. p. 334) of a form of neuralgia, the pain of which those who have suffered from it themselves, or, indeed, have seen others suffering from it, will allow is not exaggerated by the application to it of the words 'fast unerträgliche,' Niemeyer remarks[1] that the physiologists Dubois-Reymond and Dr. Möllendorff refer its origin to the existence of a dilated state of the *arteria carotis cerebralis*. This state of dilatation these authorities explain by a reference to certain facts in the physiology of the cervical sympathetic as discovered now some seventeen years ago by Bernard, and elucidated still further by Waller, Budge, and Brown-Séquard. And in like spirit, or at least by a reference to certain anatomical facts in the arrangement of blood-vessels, which he supposes to become dilated and distended, the great Göttingen anatomist, Henle (cited by Niemeyer, vol. ii. pp. 319–339), explains the causation of certain neuralgiae. The neuralgia which is apt to haunt the sixth, seventh, and eighth intercostal spaces of the left side, he has suggested may be explained by the peculiar arrangement of the left or smaller azygos vein in that region; and the greater relative frequency of neuralgia of the first than of the second and third divisions of the fifth cerebral nerve, he refers to the greater quantity of dilatable veins with which it is beset in passing from the inner to the outer surface of the sphenoid. Now I venture, though with some diffidence, as I find myself in opposition to such names and authorities, to dissent from the explanations thus given of these facts, which I

[1] Incorrectly however; Dubois-Reymond himself explaining hemicrania as due to a tetanus of the muscular coat of the arteries. See Reichert u. Dubois-Reymond's Archiv, 1860; Brown-Séquard, 'Journal de la Physiologie,' iv. 13, 1861.

suppose must be acknowledged, with perhaps an exception as to the relative statistical frequency of neuralgia of the first division of the fifth, to be only too real facts. By laying the physiology of the cervical sympathetic alongside of the natural history of an attack of neuralgia, we shall be enabled, I believe, to see that there are stages in each corresponding with stages in the other, but that it is a stage of spasm in the one, and not a stage of relaxation and congestion, which corresponds with the stage of pain in the other. Stimulation of the upper cervical sympathetic produces, more or less immediately, contraction of the blood-vessels of the head and dilatation of the pupil, and diminution of the temperature. This is the first line of operation, resulting in what Brown-Séquard ('Lectures on the Physiology and Pathology of the Nervous System,' 1860, p. 142) calls 'Decrease of Vital Properties.' But after a while the reverse of all this takes place, and the vessels dilate. Now whether this be so in consequence of exhaustion, as is ordinarily said (e.g. Funke, 'Physiologie,' vol. ii. p. 772), or not, as Dr. Lovén (who says in Ludwig's 'Arbeiten Phys. Anst.,' Leipzig, 1866, p. 11, that the sympathetic is not so easily tired) thinks, is of no consequence, or of little consequence, as the fact of the sequence of events is accepted as I have stated it on all hands. Indeed, similar alternations of alteration in the calibre of vessels take place, as is well known, spontaneously, as the phrase goes—whether rhythmically or not, still chronometrically in relation to the needs of the animal and its tissues; in the arteries of the rabbit's ear (Funke, loc. cit. ii. p. 771, citing Schiff and Callenfels), in the veins of the bat's wing (Wharton Jones, 'Phil. Trans.' 1852), in the arteries of the frog's web (Lister, 'Phil. Trans.' 1858, p. 653); and the occurrence of these latter alternations makes the occurrence of the former more intelligible. Now, a similar alternation from a stage of contraction of blood-vessels, of coldness of skin, of shivering, of total absence of heat, redness, swelling, or tenderness, to one of increased circulation, swelling, heat, and tenderness, constitutes two stages in an attack of neuralgia, homologous with the two described as occurring in irritation of the sympathetic. It is rare, I believe and Dr. Anstie teaches ('Stimulants and Narcotics,' 1864), for pain, as opposed to tenderness, to persist after congestion; and pain in tissues differs as much from tenderness as remorse in a conscience differs from tenderness in that organisation; and the two things are well-nigh equally exclusive the one of the other in both cases. Perhaps the mere apposition of the two sets of occurrences side by side is sufficient to justify my conclusion that the congestive stage of the physiological experiment is not the homologue of the painful one of the morbid history, mere apposition being sometimes sufficient to decide us on more difficult homologies, at least in the negative. But I may add, that the argument from the *juvantia ac laedentia*, as the older physicians and physiologists phrased it, gives some confirmation to the view which teaches that spasm and starvation go in company with pain; relaxation and congestion only with tenderness. I will put the facts before you as premises; you will piece them together into an argument for yourselves. Chloroform[1] is the greatest of *juvantia* in neuralgia; chloroform, indeed, and ether, in equal parts, as recommended[2] by Mr. R. Ellis, may be safely entrusted to a safe person for self-administration; and, if taken persistently as well as prudently, may keep the attack in suspense until the enemy, from weariness or chronometric obligations, retreats or withdraws[3]. But this great reducer of neuralgia, this great and blessed producer of 'indolence,' as Locke called it, is also the great reducer of muscular spasm, as we know from its action and our employment of it in cases of hernia, and, indeed, of tetanus. Now, if it relaxes muscles which we can see in the limbs and trunk without the aid of a microscope, we may think it not improbable that it will do the like by muscles which we cannot see without the aid of that instrument, in the arterioles. Chloroform, secondly, has the reverse action to that of the sympathetic, in dilating the blood-vessels of the head[4]; as, indeed, also has alcohol, itself too a producer, though

[1] For action of chloroform, see further, 'Chloroform, its Action and Administration,' by A. E. Sansom, M.B., 1865; 'Asthma,' by Hyde Salter, M.D., 1868, p. 216.

[2] See 'Lancet,' February, May, June, 1866, and pamphlets published by Hardwicke and Brettell.

[3] The anaesthetic effect of bisulphide of carbon in various kinds of headache, as pointed out by the late Dr. Kennion ('Medical Times and Gazette,' July 18, 1869, p. 77), may, perhaps, be similarly explained.

[4] I find that both Mr. Durham ('Guy's Hospital Reports,' iii. vi. 1860, p. 153) and Hammond ('On Wakefulness,' p. 25) are agreed that, in animals under chloroform, the veins of the brain become distended. I do not, however, lay any great stress upon this fact; firstly, because the veins may become distended under the influence, not so much of the chloroform, as of the more or less partial hindrance to respiration which its inhalation implies; and, secondly, because we have, as yet (Funke, vol. ii. pp. 769–773), no very distinct evidence for the production of effects on the veins by the

less directly, of the 'indolence' we desire, as well as of much that we deprecate. And chloroform, thirdly, antagonises the sympathetic in its very obvious action on the pupil.

It may be bold in me to venture further in this direction; yet, as a member of the British Medical Association, and as a reader of our admirable Journal, I may perhaps be allowed to say, before I return to my own more immediate subjects, that the account given by Dr. George Johnson in that periodical for March 21st, 1868[1], seems to me to indicate that epilepsy itself is but a frightful caricature of neuralgia, and of the results of vaso-motor irritation and contraction. The presence of dilatation of the pupil in all those sets of cases may be thought perhaps but a slight indication in the direction of identity of cause. High spirits and great vivacity are not rarely, in both diseases alike, precursors of an attack; while counter-irritation, which both Schiff and Setschenow[2] are agreed in considering a strong and universal reflex depressant, is not rarely, in both diseases alike, both, *ex hypothesi*, dependent on reflex vascular constriction, a preventive[3].

sympathetic. But, taking the facts as given, we must allow that venous fulness, though inferior doubtless to arterial replenishment, is still, as the growing prostate of the aged, the rank hairs shooting up round old ulcers, and the cock's spur transplanted to the cock's comb, show, a more or less favourable condition for growth and nutrition; whereas pain is correlated always with malnutrition and ordinarily with atrophy, and is now always spoken of as a 'depression' rather than as 'an exaltation of function.'

[1] See Brown-Séquard, 'Lectures,' l. c. p. 179, ibique citata.

[2] Setschenow's words are these ('Neue Versuche,' p. 23, 1864): 'Es giebt endlich bei meinen Gegnern einen Versuch, an dessen Richtigheit ich keinen Grund zu zweifeln habe, an welchem die einseitige Trigeminus-Reizung eine starke allgemeine Reflex-Depression hervorrief.' These words seem to furnish something like a rationale of the picking of the nose in helminthiasis, as also of much of that counter-irritation of the fifth nerve at its periphery which so-called 'nervous irritable' persons practise on themselves in the way of 'tricks.' Malgaigne practised similarly on his patients, as certain savage races do upon themselves with their labrets, ear- and nose-rings.

[3] These views I came to entertain without any knowledge—or perhaps I should rather say without any conscious recollection—of those which Dr. Radcliffe had put before the world in his lectures delivered at the College of Physicians, and published in 1864 in his work on Epilepsy, Pain, and Paralysis. I have not altered what I had written in consequence of my consultation of this most valuable work, to which I resorted after seeing a reference to it in Dr. Anstie's book. This latter work I have already quoted in the text, and I found it most useful and suggestive to me. I believe, indeed I hope, that what I have written is more or less in accordance with Dr. Radcliffe's views. But it is as much inferior for purposes of consultation, and indeed from other points of view also, to what Dr. Radcliffe has written, and I have read of his, on the same subject, as a skull when just removed, and that in a somewhat

I will now proceed to give, in the second place, an account of a physiological experiment worked out for us, from time to time, in the laboratory of Nature, which throws not a little light on a question which, I learn from Dr. Wilson Fox's work on 'Dyspepsia' (p. 141), is still a matter of debate among pathologists—the question, to wit, 'of the influence of perverted innervation in causing inflammatory, or sometimes even still severer morbid changes[1].' It is well known that stags, after injury to the testes, have corresponding changes wrought out in the corresponding horns. Such a specimen I can show you from the Christ Church Museum, founded by Dr. Matthew Lee. In a curious old work dedicated to him, in company with two others of the King's Physicians, by Dr. Richard Russell[2], and styled 'The Economy of Nature in Acute

fragmentary condition, from a barrow, is to the same skull when pieced together and reconstructed, as you may see many such skulls in the Museum, by Mr. Robertson.

[1] The following references to authorities on this vexed question I herewith append:—

Lister, 'Philosophical Transactions,' 1858, p. 627.
Beale, 'Philosophical Transactions,' 1865, part i. p. 447.
Virchow, 'Archiv,' vol. xvi. 1859, p. 428; 'Cellular Pathology,' Chance's translation, pp. 311, 312.
Paget's 'Surgical Pathology,' Turner's edition, p. 237, 1863.
'British Medical Journal,' 1866, p. 402.
Anstie, 'Lancet,' 1866, vol. ii. p. 548.
Simon, 'Holmes's System of Surgery,' vol. i. p. 62.
Bernard, 'Leçons Physiol. Pathol. Syst. Nerv.,' vol. ii. p. 518.
Brown-Séquard, 'Lectures on the Physiology and Pathology of the Nervous System,' p. 143.
Budge, 'Handbuch der Physiologie,' p. 794.
Funke, 'Handbuch der Physiologie,' vol. ii. p. 776.
Donders, 'Spec. Physiologie,' p. 140.
Billroth, 'Die Allgemeine Chirurgische Pathologie und Therapie,' 1868, p. 72.
Niemeyer, 'Lehrbuch der Pathologie und Therapie,' 1868, ii. pp. 320, 340, 428.
Handfield Jones, 'Functional Nervous Disorders,' p. 11; Lectures in 'Medical Times and Gazette,' 1865.
Samuel, 'Moleschott's Untersuchungen,' Band ix. p. 18.

[2] It may be interesting to record here, in passing, that Dr. Richard Russell lived at Reading, that he was a friend of Dr. Chapman and of Dr. Frewen of this very place, and that the copy of his work to which I have referred, and which exists in the Christ Church Scientific Library attached to the Christ Church Museum, and deposited with it in the University Museum Buildings, did in 1760, eleven years before Dr. Russell's death, belong to Dr. Chapman. Now these facts and dates render it not improbable that this very specimen may have been given by Dr. Russell to Dr. Chapman, possibly together with this copy of his book. It is unfortunate that so much should be left to speculation; but this digression may be justified by the moral which it conveys, to the effect that we are bound, when receiving a specimen into a museum, to put on record forthwith, for the benefit of our successors, a note of its history and donor. Dr. Russell

and Chronical Diseases of the Glands' (pp. 21–24), five cases of injury to the testicular gland in stags are recorded. In such cases as these after the injury to the testis, the horn may or may not be shed annually, and it may never thenceforward lose its 'velvet;' but it never becomes the dry lowly vascular weapon of offence which, in a fortnight or three weeks from the present time, we shall see the bucks polishing their hard leather-coated horns into against shrubs and trees. It remains vascular and spongy within, and coated outside with a hairy skin, which may be prolonged into pendulous outgrowths. Being sensitive and fragile, and bleeding easily, it acts as a second sexual disqualification; and, as I am speaking of this correlation of growth, I may be allowed to add, that its reality is further testified to by its absence after similar lesions in reindeer, where both bucks and does are alike horned. Now, I submit that the unilateral correspondence of malnutrition, such as we have here, is as good an instance and exemplification of Pflüger's first law of reflex action, the law of unilateral (*gleichseitig*) transmission of stimulus, as any unilateral or homolateral twitching of any muscle can be in response to any one-sided stimulus. Only the reflex action shows itself in the way of nutrition—a sort of reversed hemiplegic nutrition, it is true—and not in that of movement nor in that of secretion [1].

Let me, as in the former case, lay alongside of the physiological experiment a parallel to it from pathology. This I will do by the help of Budge, who, at pages 794–795 of his 'Handbook of Physiology,' gives us the two following short histories, which have

was the author of several other works besides the one I have quoted. Their existence has escaped the notice of Dr. Munk, in his interesting volumes, 'The Roll of the Royal College of Physicians,' vol. ii. p. 132. Their titles are: 1. 'De Tabe Glandulari sive de usu Aquae Marinae in Morbis Glandularum Dissertatio;' in 1 vol. 8vo.; pret. 5s. 2. 'A Dissertation concerning the Use of Sea-water in Diseases of the Glands;' to which is added an Epistolary Dissertation to R. Frewen, M.D.; in one volume, 8vo.; price 5s.

[1] See Otto, 'Neue seltene Beobachtungen Samml.,' vol. ii. p. 10; Elsaesser, 'Diff. Sex. Mamm. praeter partes sexuales,' p. 36. Since writing the above, I have seen a note to p. 22 of Mr. Paget's 'Surgical Pathology,' edited by Professor Turner, in which the fact that no disturbance of nutrition is effected by mere transplantation of the testis *in cocks*, is brought forward to show that no mere nervous disturbance can account for these alterations of nutrition. I do not think that these negative results, obtained from experiments on half a dozen *birds*, can outweigh the positive facts of unilateral correspondence in malnutrition which have been so frequently observed in mammals. (See Hunterian Catalogue, Osteological Series, vol. ii. p. 591.)

come under his own observation. His words run thus in translation: 'After a long-continuing stagnation of blood at the end of the small intestine and the beginning of the large, in consequence of which exudations and adhesions of the peritoneum ensued, the entire right half of the body became weaker than the left, was tired sooner by exertion; the right foot became cold sooner, under the same circumstances, than the left; the right ear became much more rapidly the seat of vascular dilatation than the left; and other similar phaenomena developed themselves. After a great abscess under the right gluteus maximus, and an immense loss of pus, the right hand and the entire right arm became not only evidently thinner to the eye than the left, but also actually smaller.' These cases are decisive as to the interference of the nervous system in the process of nutrition; and, though organs and structures, such as the epithelial and the cartilaginous, both physiological and, I suppose, morbid, may and do exist and grow in animal bodies being as devoid of blood-vessels and nerves as though they were found in vegetables, still any arguments based upon these undoubted facts can be met at once, if so we care to meet them, with the more or less accepted physiological axioms, 'The interdependence of parts augments with their development; the solidarity of organs increases and is more intimate with each superaddition of a fresh factor to the entire economy.' But these cases do not, of course, touch the question of the way in which this nervous influence comes to act on the tissues, whether mediately through the blood-vessels, or immediately on the tissues with which they are supposed, in the case of the salivary glands by Pflüger, though not, as I apprehend, by Dr. Beale ('Phil. Trans.' 1865, p. 447), to become continuous. Nor does such an experiment as that of Bidder, in which a salivary gland, under nerve-stimulation, picked out two-thirds of the entire quantity of iodide of potassium in the circulation, to one-third picked out by the substance of its fellow, not so stimulated. For a more innervated gland is also a more vascular gland; and of the two antecedents, greater nerve-current and greater blood-current, we have no right from this experiment to say that the one rather than the other is the cause of this particular consequent. And much probability will come to attach itself to Virchow's views, according to which innervation is not proven to increase nutrition directly, but works only mediately by its influence on the blood-vessels, in the minds of persons who may be averse to multiplying laws by cases such as these. We go to a case, as I suppose most of us may, like myself, have gone, and we frequently find one side of the body hot, and the other cold. This latter, the friends will tell us, is the paralysed part; we find that it is not; and Bernard's experiments, and Brown-Séquard's (l. c. p. 146), enable us to understand why this is so. An excellent case to the same effect, showing how increase of vital properties may take place in the entire absence of any connection with the upper part of the cord or brain, may be given from a paper of the late Sir B. C. Brodie's, in the twentieth volume (1837) of the 'Medico-Chirurgical Transactions.' 'A man was admitted into St. George's Hospital, in whom there was a forcible separation of the fifth and sixth cervical vertebrae, attended with an effusion of blood within the theca vertebralis, and laceration of the lower part of the cervical portion of the spinal cord. Respiration was performed by the diaphragm only—of course, in a very imperfect manner. The patient died at the end of twenty-two hours; and, for some time previously to his death, he breathed at long intervals; the pulse being weak, and the countenance livid. At length, there were not more than five or six respirations in a minute. Nevertheless, when the ball of a thermometer was placed between the scrotum and the thigh, the quicksilver rose to 111° of Fahrenheit's scale. Immediately after death, the temperature was examined in the same manner, and found to be still the same.'

The larger size of a horse's hoof, the nerves of which had been divided, should probably be similarly explained by the greater afflux of blood which would set in thither temporarily until the continuity of the nerve was re-established. (Ogle, 'Med. Times and Gazette,' Nov. 3, 1866.) And, finally, such an occurrence as the inflammation of skin, cartilage, or cornea, after its own sweet will, and not in the line of an irritated nerve passing through it or near it (Virchow's 'Cell. Path.,' Chance's translation, p. 299), seems to speak plainly enough to the self-sufficiency of animal cells to respond to what Niemeyer calls 'Insulte,' without appealing to any higher powers for assistance; just, in fact, as though they were as little animal, as truly vegetable, and as independent of any cranio-spinal centre as the gall-producing oak or willow.

But, in spite of all this, I am inclined to think that the direct action of nerves on cells is a *vera causa*; and, even if our highest microscopic powers do succeed in proving that nerve-tissues are never continuous with any other tissues in any part of their distribution, it must still be recollected that such intervals as may be demonstrated will be, if not insensible, at all events infinitesimal; and nerve-force may well be sufficient to act across such gaps as these. (See Dr. Radcliffe's 'Lectures on Epilepsy,' 1864, pp. 13 and 330.) I can appeal for my justification to Professor Lister's experiment, recorded in his paper on the Cutaneous Pigmentary System in the Frog ('Phil. Trans.' 1858, pp. 636–639), in which certainly the nerve-system is shown to have some control over the molecular movements of concentration and diffusion quite independently of the blood-vascular system. The cessation of the circulation in a frog's web entails the concentration of the pigment; therefore Professor Lister took a pale frog—i. e. one in which the pigment was already concentrated; and, tying a ligature above the ankle, so as to eliminate the condition of cessation of the blood's circulation, he then eliminated the condition of nerve-influence from the cranio-spinal axis by amputation above the ligature. *Cessante causa, cessat et effectus;* the nerve-force is removed; and the pigmentary diffusion which it had held in check is set up and continues, until superseded by the *post mortem* concentration which ordinarily takes place, and produces that lightening of the dark hue usually seen in the frog after death. This experiment, which I have not given in full, nor in Professor Lister's own words, is a very striking one; and I hope I may remark, without offence to any representatives of the German Fatherland, to which physiology owes so much, that much that has been recently written and worked at there might have been spared, had Mr. Lister's papers been as well known to them as they will be to their successors. They seem to me to mark an era in the literature and in our knowledge of the essence of inflammation.

Here, if I may be allowed to digress somewhat, I would remark that Professor Lister's suggestion made in 1858 (loc. cit. pp. 619 and 640) as to the probability of the existence in the limbs of a ganglionic apparatus co-ordinating molecular and other movements at the periphery, sometimes independently, sometimes subordinately to the cranio-spinally placed nerve-centres, may seem to have found a justification in Professor Beale's demonstration in 1865 of the ending of the muscular nerves in nucleate reticular plexuses. Assuredly, the discovery of these net-works bearing nuclei does away with the necessity for any further carrying on of the apparently interminable discussions as to the existence of an 'idio-muscular' as opposed to a 'neuro-muscular' contractility. But I will take this opportunity of saying, that there are not wanting purely physiological considerations, which, though not by any means amounting to demonstration, do nevertheless lend some little probability to the 'neuro-muscular' explanation of those movements which take place in muscles separated from all connection with central nerve-organs. Firstly, these movements are, within my experience, more marked and frequent in the muscular tissues of young animals; and the history of the development of nerves would lead us to expect to find a greater degree of independence in the peripheral nerve-system, than we should look for in the adult organism; for nerves do not grow from cells in the direction of what we know in the adult state and under low powers of the microscope as their branches; but, as Von Hensen has shown (Quain's 'Elements of Anatomy,' 7th edit. p. clxiv), two nerve-cells are connected by a fibre, and it is by the withdrawal of the one cell from the other, and the elongation, so to say, of the interconnecting fibre, that the peripheral and central ganglionic systems respectively assume their adult relations. And just so, I may add, in certain annelids and lamellibranchiata, we have, as we not rarely do have in the lower animals, a stereotyped though but partial adumbration of what is but a single scene in the moving diorama of the development of the higher; and we find the peripheral nerve-system studded with eyes or other sensory organs, and possessed of a prominence and importance relatively to the central nerve-ganglia which is only temporarily seen in the development of more perfect creatures.

Secondly, many of the cases of death in adults, in which this irritability is found to exist most commonly and markedly, are cases in which, from very various reasons, the functions of the intracranial nerve-centres are put into abeyance at a very early stage in the process deathward. Such are (see Nysten, cited by Brown-Séquard, 'Proceedings of Royal Society,' 1862, p. 211) cases of decapitation, of asphyxia, and of sudden haemorrhage from a large artery. Now we know that movements do continue in a portion of intestine which has been deprived of its mesentery, and we ascribe the production of these movements to the presence in the walls of the intestine of the plexuses demonstrated to us by Meissner and Auerbach; and, if we may ascribe like effects to like causes, we may ascribe the *post mortem* twitchings of muscles to Professor Beale's neuro-muscular apparatus. In like manner we should expect from similar reasons, and we do find, as a matter of fact, this same neuro-muscular irritability greatly prominent in the small-brained cold-blooded vertebrata, and in hibernating mammals. In all of these animals alike, the central nerve-system is small relatively to the entire mass of their bodies; whilst in birds, or at least in the more highly organised of the class—for birds, like other bipeds, differ as to the mass and use of the brain (see Parker, 'Zool. Soc. Trans.' v. 1862, p. 207)—the brain may hold a more favourable relation to the entire mass of their bodies than in any other class of animals; and in birds, as is well known, with some few reptile-like exceptions, such as the peewit, muscular irritability ceases almost with their last act of expiration.

Whilst speaking of this condition occasionally found in the muscles after death, I am tempted to say a few words of the empty condition of the arteries which is almost constantly found after death. I observe that Von Bezold has explained this well-known phaenomenon as being due to a last nervous impulse communicated to the small peripheral vessels *from the brain*. His words are ('Untersuchungen aus dem Physiologischen Laboratorium in Würzburg,' Heft ii. pp. 358, 359, 1867): 'Sicher ist aber ein ungemein wichtiges Moment hierbei die Innervation der Muskeln in den kleinen Gefässen des Körpers. Man stelle sich vor, dass in der Agonie, in Todes-Kämpfe, das vasomotorische central Organ im Gehirn noch in Krampf-zustande versetzt wird, welche mit Pausen der Erschöpfung abwechseln . . . Ausserdem ist gezeigt worden, dass jenes letzte Ueberpumpen des Blutes aus den Arterien in die Venen, bei den Säugethieren wenigstens, unter dem Einfluss einer letzter Thätigkeit des Gehirns geschieht.' Surely all these 'Vorstellungen' would have been spared, if Professor Von Bezold had been acquainted with Mr. Lister's papers, or even with those points in them to which I have referred. Indeed, that his view is untenable, is clear from a consideration of the fact that the circulation can be kept up, and will, like the muscular irritability, persist in a decapitated animal for a long while after death, if artificial respiration be put in play. The empty state of the arteries *post mortem* is most probably to be explained by the action of the peripheral nerve-system on the arterioles; though Dr. Alison would have explained it by the attraction *a fronte* force of the tissues around the capillaries; but Von Bezold's view of the source of the nervo-muscular action of the peripheral vessels is, I apprehend, more untenable and less plausible than most theories which have 'had their day and ceased to be.'

The following short history seems to me to be a good instance of the action, or rather of the want of action, of the peripheral nerve-system upon the arterioles. A man, who came some years ago under my own care, had had a bullet pass through his arm just above the elbow, so as to sever the musculo-spiral nerve. The scars of exit and entrance were in the lower third of the arm. Under ordinary circumstances, the soft parts of the lower arm maintained their normal consistence; but their power of resisting changes of temperature was greatly impaired, as well of course as the sensibility of the parts supplied by the injured nerve. I recollect seeing the swollen state of the inner side of the hand one cold raw morning when the man was on sentry duty, and had his hand chilled down by the musket he had to carry. Now, I apprehend that this turgescence is to be explained by saying, that the local or peripheral nerve-system of the affected parts was competent under ordinary circumstances to regulate the calibre of the arteries; but that its activity was liable to be depressed, as under the circumstances related, into actual abeyance, in the absence of any possibility of any assistance being supplied to it from the cranio-spinal nerve-axis. Thus, under the depressing effect of cold, which seems to work here much as it does in checking the regeneration of artificially amputated parts in snails and in salamanders ('Müller's Physiology,' by Baly, 2nd edit., i. p. 444; Bonnet, 'Œuvres,' tom. v. i. pp. 328, 329), the peripherally placed ganglionic system was put into abeyance; and turgescence of the vessels it ordinarily supplied with 'tone' ensued. Just similarly in mammals the skin of which has been covered with an impermeable varnish, and in which death is as much due to the chilling down which the destruction of the non-conducting power of their hairy integument entails as to the penning-in of its various acrid and volatile and other secretions, oedema and vascular congestion are to be observed in the skin, as well as in other organs (Ranke, 'Physiologie,' p. 456). The flame of mammalian life, like the flame of inorganic combustion of carburetted hydrogen, can only be sustained at a high temperature; a certain reduction is as fatal to the one as it is to the other in the Davy lamp, and the vitality of the more exposed peripheral is more easily depressed than that of the more protected central nerve-system.

I should be paying but a poor compliment to the judgment which has provided a microscopic exhibition for the instruction and entertainment of this evening, if I were to dwell at any length upon the relations borne by Histology to Medicine and Surgery. And, secondly, if I were to dwell in the least adequately upon the importance of a knowledge of Microscopic Zoology to the diagnosis, and what is better than the diagnosis, and even than the therapeutics, the prophylaxis of diseases of all kinds, from those which are considered trifling or contemptible by most men, except those who suffer from them, up to those which excite world-wide anxieties, such as trichiniasis or cholera, I should have to extend my address to a length you would shudder to think of. Upon one single point I will make a few remarks; and the purport of these will be to show how the manipulation of such an instrument as a catheter may find, if we are to do justice to our patients, its regulative condition in the manipulation and revelations of the microscope. I had myself recently come to suspect that the determining condition of the triple phosphatic alkalescence of the urine was to be looked for and found in the presence of some of those organisms which Pasteur has proved and hygienists have believed to be the real causative agents of fermentations and putrefactions. One accepted view of the causation of this most mischievous metamorphosis is, that the coats of the bladder, in consequence of altered innervation, as after spinal injuries, act upon the urine as so much dead matter acts on blood in causing its coagulation, or as the tissues round about the capillaries act when they are in an abnormal condition upon the rows of blood-corpuscles within those canals; and by this 'catalytic' agency break up the urea and throw down the ammoniaco-magnesian phosphates. Another view ascribes the like effect to the 'fermentative' working of the abundant catarrhal mucus, which is in some cases flaked off from the inner walls of the bladder. Now, neither of these views suggests a *vera causa* for the effect for which they profess to account. Blood coagulates when in contact with non-vitalised matter, and blood-corpuscles arrange themselves in *rouleaux* under similar circumstances. But Mr. Lister, who has shown us so much which bears on this matter, has shown us also, and that in the last number but two (July 18, 1868) of our Journal, that urine will remain for an indefinite period undecomposed in a properly constructed, which happens to mean a properly contorted, receptacle, even though that receptacle be as little vitalised as glass. And Niemeyer has shown, what I dare say many who are now honouring me with their presence have observed, but, I think, not recorded, that urine often retains its acidity through protracted cases of vesical catarrh, and in spite of cumuli of clouds of 'fermentative' mucus; which are, therefore, as little of *verae causae* as is 'catalysis' itself. But the presence of vibrios in the urine, and that before it leaves the bladder, is a *vera causa*, i.e. a present condition, and therefore possibly a cause, or connected with the cause, of the phaenomena to be investigated (see Beale, 'On the Urine,' p. 196); and the idea that the alkalescence in question depended upon them, an idea which I had not the time to find an opportunity of verifying for myself, I find has been verified for the benefit of others by Niemeyer, with the assistance of Traube and Teuffel. 'In the course of last year,' says Niemeyer (l.c. ii. 66, 1868), 'I arrived, partly by means of an observation of Traube's, partly by means of experiments and observations of my own, which have been published by Teuffel in the "Berlin Klinische Wochenschrift," at the conviction that it was not the vesical mucus, but lower organisms, which probably get into the bladder by means of the introduction of badly cleaned catheters, and excite there the alkaline fermentation of the urine.' Now, whether the vibrios find their way into the bladder exclusively on dirty catheters or not, I apprehend that the addition of some carbolic acid to the oil used for lubricating these instruments, whether they be guilty or not of what is here laid to their charge, will be a piece of practice calculated to prevent the alkalescence which the vibrios cause by preventing these vibrios themselves from entering on their evil activity. Mr. Lister's paper, just alluded to, will show that this is an experiment which may very safely be tried; if carbolic acid can be safely introduced into a wounded pleural cavity, assuredly we need not hesitate about the passing of it into a bladder. Thus many scientific researches, undertaken in the first instance for the elucidation of speculative truth, and for the rectificatiou not of unsound organs and functions but of unsound theories and explanations, and prosecuted throughout by the aid of the most refined methods and instruments, come ultimately to bear upon such matters as catheterisation and alkalescent urine. I would not, however, be thought to undervalue the worth of researches carried on at whatever cost with the sole object of procuring correct notions as to the way in which processes, even wholly beyond our power of modification, have been and are being carried on. It is a great and positive gain when we get rid of one false hypothesis, one single false formula which by frequent repetition has attained to the dignity of a philosophic axiom, and acquired a sort of prescriptive right to 'warp us from the living truth.' The Chemists, as I am informed, are conspiring to effect what the old Greeks would have called a 'Catalysis' of the kingdom of 'Catalysis' itself, and its banishment to the Limbo of Vanity, there to herd with Phlogiston and many other and younger as well as older unsubstantial Idola Theatri; and though these alterations of theory may not as yet have affected the oxygen we breathe, nor even have enabled us as yet to regulate with any greater precision the processes of fermentation with which we have for so many ages had an empirical familiarity, they have given us at least a warning as to maintaining always that proper diffidence as to the all-sufficient validity of our theories, by whomsoever promulgated or endorsed, which is so constantly of avail in actual practical work. The phaenomena, let me add, to account for which the hypothesis of Pangenesis has been recently (Darwin, 'Animals and Plants under Domestication,' 1868, ii. p. 403) put forward provisionally, are, and will, we may believe, always remain, beyond our control; but there will be no one, I suppose, who will not feel an interest in observing how the revelations in the all but infinite divisibility of 'germinal matter,' which we owe to Professor Beale, may come to bear upon the explanation of the marvellous phaenomena of reproduction and hereditary transmission. Nor can I leave this subject without remarking that it is in great probability upon the self-multiplication of such infinitesimal particles as this hypothesis of Pangenesis postulates that processes, to the naked eye the very reverse of Genesis, have been found to depend; and the Blue Book on the Cattle-Plague ('Third Report of the Commissioners,' &c., p. 151) will show you that here, too, we are dependent on the employment of the very highest powers of the microscope; and it is scarcely necessary to add that its employer was in this case Professor Beale.

If I have been short in speaking of the advantages which the histology of modern days has conferred upon its therapeutics, I might be shorter still in dealing with my third head—the dependence, namely, of the healing art upon the facts of Anthropotomy—that is to say, upon the naked-eye knowledge of the structure with which it has to concern itself. Some little, however, I must say with your permission. Some persons are inclined to think that there is some sort of antagonism between the interests of microscopic and those of naked-eye anatomy; and hints more or less obscurely expressed may be found to this effect here and there in writings even of the present day. It is in much the same spirit that persons are found to say that the sending of missionaries to the heathen abroad entails so much curtailment of similar work at home, and others will say that the starting of any fresh charitable institution necessitates the subduction of so much from the funds available for those already on foot; and that others again will say that the encouragement of natural science is 'inimical' to the progress and cultivation of literary and classical studies. Now, all these views depend upon the radically false assumption that intellectual and moral activities are limitable and measurable by certain quantitative conditions, just as a man's expenditure is or ought to be limited and measurable by his balance in the bank. This analogy is a wholly fallacious one, but it has nipped many an excellent project in the bud. A truer analogy is furnished us by the history of those infinitesimal scraps of germinal matter of which I was just now speaking, which are hard to destroy even with floods of carbolic acid and copperas, and which possess a faculty of self-multiplication wholly unparalleled within my experience in the history of the metallic objects of which we were just now speaking. Activity and earnestness, in fact, which are some of the best things, resemble some of the worst in being eminently contagious. The example of a strenuous labourer in one field, spreads into the weedy acres of his slothful brethren on the right and left; and the improvement of the microscope has but been accompanied by a more thorough and accurate working out of human dissection. Let us leave metaphor and general statements, and come to facts. I have in my possession a work written for the use of anatomical students in the University of Edinburgh —a place then, as now, at least on a level with the most advanced centres of such education elsewhere in Great Britain. Its title is, 'The Anatomy of the Human Bones and Nerves, with a Description of the Human Lacteal Sac and Duct, by Alexander Monro, M.D., late Professor of Anatomy in the University of Edinburgh. A new edition, carefully revised, with aditional Notes and Illustrations, by Jeremiah Kirby, M.D., author of "Tables of the Materia Medica." 1810.' The date of its appearance takes precedence, therefore, by a dozen years at least, of the first appearance of an achromatic combination; and if the development of microscopic zeal had really been injurious to the diffusion of thorough anthropotomical knowledge, we should find here in perfection that precision and fulness which *ex hypothesi* are the exclusive fruits of individual attention and undistracted concentration. Now, a few weeks ago, I was pursuing some anatomical researches into the homologies of the shoulder-joint muscles, and by the suggestion of one or two of my friends, amongst whom I may mention Dr. Boycott, I took up the line of argument for homological identity which innervation furnishes. Being deep in the country, I was reduced to consult, in the absence for the moment of other books, the work I have just mentioned, for a small matter in the composition and decomposition of the brachial plexus. This is what I found to satisfy my enquiry in a book expressly treating, you will please to recollect, of the nerves, and written by one of those 'famous old anthropotomists' who were not distracted by 'microscopische Spielereien.' 'The fourth cervical nerve, after sending off that branch which joins with the third to form the phrenic, and bestowing twigs on the muscles and glands of the neck, runs to the armpit, where it meets with the fifth, sixth, and seventh cervicals, and first dorsal, that escape in the interstices of the *musculi scaleni*, to come at the armpit, where they join, separate, and rejoin in a way scarcely to be rightly expressed in words; and, after giving several considerable nerves to the muscles and teguments which cover the *thorax*, they divide into several branches, to be distributed to all parts of the superior extremity. Seven of these branches I shall describe under particular names' (p. 291). These seven branches have the particular names of Scapularis, Articularis, Cutaneus, Musculo-Cutaneus, Muscularis, Ulnaris, and Radialis. Of such little trifles as the connection of the second and third cervical sympathetic ganglia which gives off heart-nerves with the arm-nerves, upon which connection the pain down the inside of the arm in heart-disease Niemeyer (ii. 338) supposes may, and the older anatomists would have said *must* depend; as the subclavius nerve and its connection with the phrenic, and so with the shoulder-tip pain in liver-disease, we have just as little mention made as we have of the nerves supplying those small muscles, the pectorales. Surely knowledge is not like a volcanic archipelago, where the upheaval of one mass of solid ground entails the submergence of another; rather it resembles some vast table-land which is rising, and now and then at accelerated rates of progress, out of the waters, and has, in these days of the subdivision of property and of labour, its broad and continuous surface seized upon, partitioned out by enclosures, and put under cultivation by various occupants so soon as ever its outlines are recognisable.

My last topic in this division of my address is the connection which Comparative Anatomy has with Medicine and Surgery, and the bearing which a cultivation of this department of Biology has, or is likely to have, upon the interests of the profession. Of the benefits which Comparative Anatomy receives at the hands of medical practitioners there is little occasion to speak; or rather Mr. Parker's volume on 'The Shoulder Girdle,' just published by the Ray Society, may speak for me; it is only less vast than valuable, and will constitute the commencement of a new epoch in the science. But what I have to speak to is, not the benefits which Comparative Anatomy receives, but those which it can confer. And I believe that the educational working of this study is perhaps the particular line along which the best fruits for the profession, and for the public, may reasonably be looked for. Any study which forces its students into that most valuable knowledge —the knowledge of when a thing is proved, and when it is not— is *ipso facto* an ally of real medicine, and a deadly enemy of quackery. A person who has in any way become acquainted with what reasoning and reasons are in one subject, will be apt to look for similar reasons and similar reasoning when he has to deal with another, and especially and rightly if that other be a closely allied subject. And when natural knowledge shall have become more widely and generally diffused, an end which we may hope to help towards accomplishing by means of our School of Physical Science, quackery, with its painful spectacles of reputation and confidence unfairly withheld and more unfairly bestowed, will cease to flourish in its present rank exuberance. A worker in Biology gains reputation accordingly as he is acute enough to observe and generalise for himself, and accordingly as he is conscientious enough to make himself master of and duly acknowledge the labours of others. It cannot be said that learning, talent, and labour are equally certain to secure prominence either for a medical doctrine or a medical practitioner. The medical doctrine obtains currency, acceptance, and popularity, and the confidence of an ill-educated public, by virtue ordinarily either of the effect on the imagination which its paradoxical character secures for it, or of the effect on the ear of the alliterative ring of the phraseology in which it is embodied. The success of persons, again, in the medical profession, and in some other walks of life too, may depend on personal qualities quite other than any connected with diligence, attainment, or ability—upon, say, certain peculiarities of manner, either in the way of polish or in that of roughness. The greatness of the stake, his own health, for which a man is playing when he adopts a particular doctrine or doctor of medicine, no doubt disturbs the balance of such powers of judgment as he may have, much as in Gessler's hopes the placing of an apple on the head of Tell's son disturbed the steadiness of the father's hand and eye. But habits of thought, as of other things, may be acquired by a proper course of education, and habits, like drill, steady a man under emergencies; and a scientific training enables a man to set about forming a right judgment in a right way and upon proper and legitimate grounds, even when nothing less than life itself is at stake. If a knowledge of such a subject as Comparative Anatomy, and of its external aspect, scientific Zoology, is a knowledge which will give the layman more power of forming right decisions, it is perhaps needless to labour long at showing that this self-same knowledge may be of the like service to the professional man. A sort of practical proof of its value is furnished us by the fact, that in Edinburgh as in Germany a dissertation on some subject of Comparative Anatomy is often accepted as a thesis for the degree of Doctor in Medicine. By such a regulation we have the obvious fact recognised, that the same sort of skill in the employment of methods, the same familiarity with organs, tissues, and functions, the same reasoning powers, are employed in investigating the problems of life wheresoever existing. The second aphorism of the 'Novum Organon' applies to the one as to the other line of investigation, that of Human, and that of Brute Biology:—*Instrumentis et auxiliis res perficitur;* and alike in both, *nec manus nuda nec intellectus sibi permissus multum valet.* Comparative Anatomy, finally, besides thus benefiting the public firstly, and the profession secondly, is of use to Human Biology and Medicine, as such, inasmuch as it casts so much light upon the problems which the more highly evolved organs, functions, and other relations of our own species render in a much higher, and, indeed, sometimes in the highest degree, difficult or impossible to investigate. Answers to what are riddles in Human Anatomy and Physiology are often to be found given in very simple language in the structures and functions of the lower and lowest animals. Of such hints furnished by the brute creation towards the proper solution of certain problems which concern each and all of us in dealing with our own species, I will herewith, by your permission, give a few. Of the use of rest towards the repair of injuries I presume there is little doubt, but the best established teaching is all the better for the support of a few concrete examples. Now, in what animals do we find the greatest capacity for repair of injuries, and for the reproduction of lost parts and limbs? Precisely in those in which the whole of life is carried on at the lowest rate, and in the nearest approximation to rest which is compatible with animality,—in those animals, to wit, which breathe water, and have but its scanty percentage of dissolved oxygen to sustain their animal functions. The metamorphoses which an animal may have undergone, or may have to undergo, have very little directly to do with its power of recovery from injury, or of regenerating a lost limb. No animals go through more complex metamorphoses than do many of the Crustacea, and nearly all the Echinodermata; yet assuredly no other class has a larger capacity for the reproduction of lost fragments of their bodies.

Now the latter of these classes is exclusively, and the former all but exclusively, aquatic. The more perfect, again, an insect's metamorphosis, i.e. its power of building up tissues and organs, the more perfect ordinarily, or rather the more profound, has been its quiescence as a pupa. Indeed, the very exception here proves the rule, and proves it to be a good one; for such hemimetabolous insects as, like May- and dragon-flies, come, in their imago state, to differ almost as much from their larval forms as the imagos of many holometabolous insects do from their larvae, are during those preparatory stages as completely aquatic as any crustacean (Westwood, 'Introduction to Entomology,' vol. ii. pp. 29, 38; Carus and Gerstaecker, 'Handbuch der Zoologie,' p. 29). I am aware that there is such authority as Mr. Paget's ('Surgical Pathology,' ed. Turner, p. 123) and Mr. Darwin's ('Animals and Plants under Domestication,' vol. ii. p. 15) in favour of regarding the power of repairing injuries as standing in an inverse ratio to the amount of metamorphotic change through which an animal has gone; and I must therefore take the more pains to show that my explanation, to the effect that this happy power depends mainly upon the peacefulness and quiet with which the various processes of life are carried on ordinarily, and after the mutilation, is the truer one. My opponents' case would rest on such facts as these which follow. I will give them first, and then show how they really support my views. The larvae or tadpoles of the tailless Batrachia, but not the adults, says Dr. Günther (Darwin, loc. cit., and Owen, 'Comparative Anatomy of the Vertebrate Animals,' vol. i. p. 567), are capable of reproducing lost limbs. So with insects, says Mr. Darwin, l.c., 'the larvae reproduce lost limbs, but, except in one order' (the Orthoptera, and amongst them the Phasmidae[1]), 'the mature insect has no such power.' There is, however, one common property which lies at the bottom of the power of repair both in the larval forms and in the perfect adult animal, both in the invertebrata and in the vertebrata specified. This common property is the *comparative insignificance of the apparatus for aërial respiration:* in all alike —in the larva of the anurous amphibia, in the larva of the butterfly,

[1] There seems to be some little doubt whether even a Phasma can regenerate lost parts after its last moult, and some authorities would not consider it adult till after such ecdysis. The crustacea, however, moult many times after attaining the adult state, i.e. a state in which they can reproduce *the species*.

and in the orthopterous insect—the lungs or the tracheae, as the case may be, contrast to disadvantage with those of their congeners, or adult representatives, which have come to differ from them in having lost the power of reproducing lost parts. But active respiration is a prerequisite for activity of function and rapidity of rate of vital processes: and the absence of this is, according to my argument, the cause of the presence of the reparative power. The lungs are of course all but wholly in abeyance in the tadpole, and the tracheae have no vesicular dilatations developed upon them in the caterpillar forms of any insect, nor in the adults of the non-volant Orthoptera. In the Phasmidae, the curious 'walking-stick' insects, we observe just the same sluggishness, combined with great tenacity of life, which we observe among mammals in the Bruta. Let me add some more facts in further illustration of my position. The Myriapoda, which Mr. Newport has shown to possess this power of repair up to the time of their final moult, are so little like the more typical insects, as to have been classed with the Crustacea, by no less an authority than Von Siebold. Any one, again, who will compare the simple non-cellular lung of the adult Batrachian newt *Salamandra aquatica*, which possesses an unlimited power of repair *as an adult*, but *not in its young stages* (Bonnet, 'Œuvres Hist. Nat.' v. Pt. i. p. 294), with the lung of the adult frog, will have little difficulty in understanding how their power of repair differs out of all proportion more than the amount of the metamorphotic changes they severally go through. The land Salamander, *Salamandra terrestris*, has, so far as I know, escaped the hands of Spallanzani and Bonnet; its adult lung being little inferior in extent and development of spongy matter to that of the adult anura, I should expect the power of regeneration to be reduced to zero as in them. If the teaching of Comparative Anatomy has forced me to differ from the teaching of Mr. Paget, there are other facts in the same region of research which, as it seems to me, put one of his other many valuable doctrines in a clearer light than even his own clear enunciation of it. 'Each man's capacity,' says Mr. Paget ('Lancet,' Aug. 24th, 1867), 'for bearing a surgical operation may best be measured by the power of his excretory organs in the circumstances in which the operation will place him.' Now, I am inclined to ascribe the very considerable, and indeed, on my views, somewhat exceptional powers of reproduction which two sets of air-breathing terrestrial animals, the pulmonate snails and the earth-worms, possess, to the great development of their excretory apparatus. Living, as they do very ordinarily, in atmospheres laden with carbonic acid from decaying vegetable matters, they must get rid of the products of their waste and wear in the shape of fluid solutions; and the alkaline secretion with which the bodies of both are so abundantly slimy, furnishes just the required medium. When injured or mutilated, these animals can withdraw themselves pretty completely from the atmospheric oxygen by shedding out this secretion, and it at the same time disembarrasses their system from any excess of carbonic acid which may be generated within it. Thus they can attain the most perfect possible condition for repair and regeneration, the minimum of activity of all save the excretory organs; and I submit that it is possible that these two conditions may be connected as cause and effect, just as in the reverse direction a defeat of surgical skill may be connected with the presence of a fatty kidney or liver, or the excitability of a nervous system. It is going perhaps too far to attempt to explain the much greater power of repair which Amphibia possess as compared with either Pisces below or Reptilia above them to the larger size, and consequent smaller aggregate surface and less perfect aërating power of their blood-cells, and to the transpirability of their naked skins, which execute such important depuratory work for them, and are so closely connected and correlated with their lungs, livers, and kidneys. It is curious, however, and interesting to remark that the older anatomists, in commenting on the very obvious solidarity of these latter organs, went on, in their ignorance, I imagine, to a great extent of the nature of amyloid and other degenerative changes in such cases, to observe that it was illustrated by the 'fact' that, as the lungs grew smaller, so the kidneys grew larger in phthisis[1]. (See Funk, 'De Salamandrae Terrestris Vitâ,' 1827, and Meckel, 'Pathol. Anat.' vol. i. 613, 646.)

Verloren, as quoted by Donders ('On the Constituents of Food,'

[1] For accounts of experiments as to regeneration of lost or destroyed parts, see Darwin, 'Animals and Plants under Domestication,' vol. ii. p. 15, ibique citata; Owen, 'Comp. Anat. of the Vertebrata,' vol. i. p. 567; Newport, 'Phil. Trans.' vol. cxxxiv. 1844, ibique citata; Paget, 'Surg. Path.,' ed. Turner, p. 123; Spence Bate, 'Ann. and Mag. Nat. Hist.,' August, 1868, citing Mr. Lloyd of Hamburg, p. 118; McIntosh, 'Experiments on Carcinus Mœnas,' p. 28.

translated by W. Daniel Moore, M.D., p. 24), has shown how the history of insects bears on the question which we are about to have expounded to us of the 'Relation of Food to Force;' and the very title of Bischoff's and Voit's work, 'The Laws of the Nutrition of the Carnivora,' shows how this subject, to which I shall no further allude, but leave it to the able handling to which it has been entrusted, is dependent on the life, and the modes of life, of the lower creation. But I would say that it was from a study of the structures of the class of animals last mentioned—viz. the Carnivora—that I first came myself to be convinced that the uterine mucous membrane would, if properly looked for, be found in all animals alike to stretch after delivery over the area previously occupied by the placenta; and assuredly there is no one of the many complex and hard to be investigated problems of human physiology to which we are more bound to be thankful for light whencesoever obtained; and this, though the light come, as, in justice to Dr. Matthews Duncan and Dr. Priestley, I must say it did, in the way of illustration and confirmation rather than in that of discovery. (See 'Zoological Society's Transactions,' V. 1863, p. 289, Article X; Dr. Duncan's 'Researches in Obstetrics,' p. 206.)

These facts of the structural and functional arrangements of the lower animals have been used recently to illustrate some other points of uterine pathology and therapeutics in our own species. In a work by Dr. F. A. Kehrer, of Giessen, in two parts, the former of which was published in 1864, and treats of 'Die Zusammenziehungen des weiblichen Genital-canals,' and the second of which was published in the year 1868, and treats of 'Die Vergleichende Physiologie der Geburt des Menschen und der Säugethiere,' I find no little light thrown upon the question of relative position, whether as cause or effect, in which early and late abortions respectively stand to imperfect involution of the uterus. And I find also *in loco* a very distinct admonition as to the inexpediency of allowing fear of decomposition to terrify us into what is called 'meddlesome midwifery.' These extracts I think you may be interested to hear; I will simply quote them, and leave you, who are so well able to do it, to make the application of them for yourselves. 'Finally, let it be remarked that in rabbits in the earlier stages of gestation I saw the foetus with its foetal envelopes protrude entirely out of the os uteri, whilst the placenta still remained firmly attached in the uterus; a phenomenon which indicates either that in the earlier stages of pregnancy the placenta materna is less lacerable, or that the motor power of the uterus is a relatively smaller one, and one which finds its analogy in the occurrences which take place in abortions and premature deliveries in the human subject.' Heft i. p. 52. In his second Heft, p. 166, Dr. Kehrer, in speaking of the retention of the placenta being sometimes followed by symptoms like septic poisoning and sometimes not, has wise words to the following effect:—'What chemical changes may be set up in the retained placenta is clearly dependent hereon, whether the after-birth is shut off from contact with the air by the genitalia or not; for, if air find access to it, the membranes of the ovum putrefy; if the air be excluded, a process of decomposition, probably identical with one of maceration of the foetus, but wanting further chemical investigation, is set up. The occurrence of the one or the other eventually is so far of importance, as thereupon hangs the after-effect of a retention of the placenta upon the general health. In fact, we observe in women, just as in the animals mentioned, sometimes only insignificant symptoms, sometimes emaciation and sickness; sometimes, as after the absorption of putrilage from the decomposing membranes, a violent, even a fatal pyaemic fever. In the face of these facts, it seems to me to be rational in ruminants, in which the cotyledon can only be detached from the uterus by considerable violence, and scarcely even then, completely to *avoid all introduction of the hand into the cavity of the uterus after delivery*, just with the object of keeping it free from the ingress of air, and to leave the separation of the placenta rather to the natural forces. We shall thus best avoid putrefaction being set up in the cavity of the uterus, and so expose the animal the less to the risk of pyaemia.' I have not quoted from the recently published works of Dr. Matthews Duncan and of Dr. Graily Hewitt, though I have specified in my notes the pages of those works which bear upon what I have just quoted from a foreign source. I have forborne to do this, not because I think those works less valuable, but because I think them more so, and I presume they will be in your hands as they have passed through mine [1].

[1] Dr. Graily Hewitt, 'The Diagnosis, Pathology, and Treatment of Diseases of Women,' second edition, 1868, pp. 32, 342, 393; Dr. Matthews Duncan, 'Researches in Obstetrics,' pp. 276, 284, 285; Cazeaux, 'Traité des Accouchemens,' pp. 334, 349.

The human anatomist who has once seen in the lower animals the structures which represent, as it were, in exaggeration or caricature, the human costocoracoid membrane; the tuberculum pubis, with the homologue of the clavicle which is attached to it as Poupart's ligament; or the supracondyloid process of the humerus—is not likely to forget their existence when, with either scalpel or bistoury in hand, either for the ligature of an artery or the setting free of a hernia, he has to deal with their representatives in the human frame. But, if I am right in thinking that the ciliary muscle in the eye would not have secured for itself the notice which it has done of late years, had it not been for the much more obvious manifestation of a similar, if not homologous, muscular apparatus in the eye of a bird, I apprehend that I am justified in saying that every surgeon who performs Mr. Hancock's operation of cylicotomy for glaucoma is under obligations, whilst so doing, to Comparative Anatomy and Sir Philip Crampton. I need not speak of the bearing of the discovery of this muscle in the human eye by Mr. Bowman upon the physiology of its adjustment to clear vision at varying distances. Pure Physiology, again, unassisted by Comparative Anatomy, has made out much of pure function; but, much as has been attempted in the way of experiment with infusions of pancreatic substance, and with the introduction of cannulae into the duct of the gland, I am inclined to think that a comparison of the relative size of the gland in the carnivora and the herbivora respectively, in a dog, say, and in a rabbit, points as unmistakably as any of the lines of experiment just referred to—which, by the very nature of the case, are greatly beset with several sources of fallacy —to the fact that this salivary gland is concerned as much with the digestion of albumen and fat, as with that of starchy substances. With the remark that Hyrtl's discovery ('Wien. Zool.-Bot. Ges.,' 1861, cit. Henle, 'Handb. der Anat.' ii. 310) of the diverticular character of the glomeruli in the kidney of the selachians and amphibia bears not a little upon the existence of a similar arrangement between the vasa recta and the renal arterioles in the human kidney, whereby, as by the direct communication shown by Schrœder van der Kolk to exist between the arteries and veins of the pia mater, the capillary circulation may be skipped, and the tissues in relation with it left at rest, I leave this part of my subject, and begin the concluding portion of my address.

In this part of my address I propose to consider, mainly by the light of recently attained biological results, the value of two great rules for the conduct of the understanding, each of which has a legitimate sphere of application, but the former of which enjoys, it seems to me, more and the latter less than its deserved prominence. The first of these two regulative principles has received the endorsement of Newton, and it stands as his first 'Regula Philosophandi,' at the commencement of the Third Book of the Principia. It was known in the days of the schoolmen as the 'Razor of Occam,' and in later days it has been styled the 'Law of Parsimony' or 'Economy.' Newton enunciates it as follows:—'Causas rerum naturalium non plures admitti debere, quam quae et verae sint et earum phaenomenis explicandis sufficiant. Dicunt utique philosophi: "Natura nihil agit frustra;" et "Frustra fit per plura quod fieri potest per pauciora." Natura enim simplex est et rerum causis superfluis non luxuriat.' I know that this Regula has great influence on the minds of many biologists, and I believe that this its influence is by no means always for good. This is not a subject in which authorities ought to count for much; but I may say that, while the names of Aristotle, Malebranche, Maupertuis, St. Hilaire, Goethe, Bichat, and Dugald Stewart may be quoted in approval of this rule, the names of Bacon, Mill, and De Candolle may be brought forward by those who repudiate it or curtail its application. Our motto, however, is, 'Nullius addictus jurare in verba magistri;' and our business is to ask, not what men have laid down, but how Nature operates. Can a phaenomenon have more than one cause, or can it not? Is it possible, for example, and to put the question in a concrete and most practically interesting point of view at once, that a fever which we know can spread by infection or contagion, can also originate spontaneously? I rather incline, though but doubtfully, and after an imperfect examination of imperfect data, to anticipate that a negative answer to this latter question will turn out some day to be the true one; but I do not know that there is anything in the analogy of Nature to compel us to incline towards this negative conclusion *a priori*. Such a phaenomenon, at all events, as a living animal, is often enough produced by two or more distinct processes, within the limits of the same species: as, for example, from ova of different character, summer ova or winter ova, impregnated or unimpregnated ova; by fission or gemmation; through two different series of metamorphotic changes. And such a phaenomenon as the production of a particular tissue may depend—in the case of adipose tissue, for example—upon the employment in Nature's laboratory of one or the other of two different chemical compounds. Pain may be, as Dr. Handfield Jones has shown paralysis is, produced in one case by the impact of shock upon nerve-centres, in another by the curtailment of their supply of blood. In each and all of these cases, the maxim which has many a time been sonorously enunciated in these Schools, 'Entia non sunt multiplicanda praeter necessitatem,' would, if listened to, have closed our ears and eyes to at least one-half of the truth. That Bacon would have classed this maxim with his 'Idola Theatri,' I think I am justified in saying, for that in the very next section (Section xlv) of the 'Novum Organon' to that in which he treats of these delusive notions, I find these words:—'Intellectus humanus ex proprietate suâ facile supponit majorem ordinem et aequalitatem in rebus quam invenit;' and if I am told, as by Mr. Mill ('Logic,' vol. ii. p. 379, ed. 1846), that Bacon, in the actual practice of investigation, acted as though there were no such thing as Plurality of Causes, I need only answer that herein his practice did not differ from his precepts at all more widely than does the practice of many other writers, of many practising, of many teaching doctors, differ from theirs. I have a satisfaction in quoting the living De Candolle, who enjoys one of the first and best deserved scientific reputations of the day, in repudiation of Maupertuis' famous principle of 'least action.' De Candolle writes thus in his 'Géographie Botanique,' vol. ii. p. 115: 'Nous aimons à croire aux moyens simples, peut-être uniquement à cause du peu de portée de notre esprit.'

What, then, is the legitimate application? where does Nature really bind herself to the observance of a 'Law of Parsimony'? In, as I think, three distinct lines of her operations.

Where an organ can be diverted from one and set to discharge another function, there Nature will spare herself the expense of forming a new organ, and will adapt the old one to a new use. She is prodigal in the variety of her adaptations, she is niggard in the invention of new structures (Milne-Edwards, cit. in Darwin's 'Origin of Species,' p. 232). The complicated arrangement of co-operating muscles whereby the bird's third eyelid is drawn across to moisten and wipe its eyeball without undue pressure on the optic nerve, is manufactured, if so we may express ourselves, out of the *suspensorius* muscle, which in other animals has but the simple function of slinging up the eye. The scarcely less complex and beautiful arrangement of the bird's *levator humeri* is the result of a modification of a *subclavius* muscle. (See 'Trans. Linn. Soc.' vol. xxvi. 1868, Article XI.)

Secondly, where, by availing herself of the inorganic forces always at work and available in the circumambient medium, whatever that medium may be, or where, by the employment, as in what is called 'Histological Substitution,' of a lowly organised or vitalised tissue, such as elastic tissue, she can spare herself the manufacture of such expensive structures as muscle, there Nature adopts a line of practice which we call a Law of Parsimony. Where a suspensory muscle for the eye can be dispensed with altogether, as where there is a more or less closed bony orbit, as in ourselves, and an air-tight cavity formed by it together with the soft tissues lining it, there atmospheric pressure is trusted to steady the eye in the socket, as it refixes the tooth loosened by inflammation, and holds the head of the femur in the acetabulum. The eye of the burrowing mole, on the other hand, loses its *recti* and *obliqui* before it verges itself into total extinction; but this very *suspensorius* it retains after the wreck of its other property, as its guardian in the undivided undifferentiated temporo-orbital fossa.

Thirdly, where matter that would otherwise be wholly refuse, and to be rejected, can be utilised, there Nature exemplifies this law by her 'utilisation of waste substances.' The transverse colon, with its various contents, aids and ekes out the elastic recoil of the lungs in expiration; and by its near approximation to the stomach has, as Duverney long ago pointed out, the shock of the ingestion of fresh food propagated directly to it as a warning against sluggishness in the discharge of its own function. The air we use in speech, as Mr. Paget has pointed out, we could not use for breathing.

Many other instances of the 'Law of Parsimony' might be given; but I know not of any which cannot be reduced under one or other of these three heads; I know of none, that is, which can be in any way held to negative the tenability of a law of 'Plurality of Causes.'

The second great principle for the regulation of the understanding of which I wish to speak, is one which, I believe, possesses less currency and notoriety, and is less observed than it deserves. This canon bids us, in considering a complex phaenomenon, to be most careful that we omit none of the circumstances which can by any possibility be of the essence of the case. And as the possibilities of Nature are all but infinite—as, for example, the investigator of problems of Geographical Distribution knows well that a 'secret bond' may bind up together, and that inextricably, the interests of organisms removed as far as possible to all appearance from each other in the scale of life; as a fly or a plant may, by its increasing and multiplying, make half a continent uninhabitable or inhabitable by the highest mammals; I apprehend that in biological and medical problems, by the phrase 'all the circumstances which can by any possibility be of the essence of the case,' we mean practically, '*all* the circumstances of the case,' without any qualifying limitation. But we will let Descartes, to whom the enunciation of this rule is usually and, so far as I know, rightly assigned, enunciate it for us in his own words. These run thus (Œuvres, tom. xi. 1826, ed. V. Cousin, p. 23): 'Règle Septième pour la Direction de l'Esprit. Pour compléter la science, il faut que la pensée parcoure d'un mouvement non interrompu et suivi tous les objets qui appartiennent au but qu'elle veut atteindre et qu'ensuite elle les résume dans une énumération méthodique et suffisante.' Some of the very greatest advances which have been made of late in practical diagnosis have been made in the spirit of this recommendation. The application of a chemical test to the urine for information as to the expediency of giving or withholding wine in the case of a sinking life, would have seemed to Swift, could he have had any idea either of such a procedure or of the employment of a sphygmograph for the same object, more absurd than any of the follies he ascribed to the philosophers of Laputa. But as Archbishop Whately—a name to be greatly honoured here, and, indeed, wherever else liberality, and fearlessness, and ability are held in respect—has well pointed out, the absurdities of Laputan aspirations are less wonderful than the actual attainments of modern science. And to these results science has attained, because her votaries have known that what may seem to Swift, and such as Swift, to be but curious and dilettante, otiose, or even disgusting, may turn out ultimately to be essential elements in problems, the solution of which promotes directly and greatly the interests of man, and the glory of Him to whom nothing is common or unclean. Could anything have seemed at first sight to be more impertinent, more otiosely curious and trifling, than to enquire during an epidemic of cholera what was the nature of the subsoil in the area it was ravaging, to what depth it was porous, and at what level the water was, and had been previously, standing in it? Yet, as I think, Von Pettenkofer has at last fought out and won his battle on these points (see 'Zeitschrift für Biologie,' Bd. I, Heft iii. und iv. 1865; Bd. II. Heft i. 1866; Supplemental Heft, 1867, p. 54; Bd. IV. Heft iv. p. 400); and the distinguished President of our 'Public Medicine' section, Mr. Simon, who is as little prone as most men of science to take up over-readily with any new wind of doctrine, has told us ('Report of the Privy Council for 1866,' pp. 366 and 457) that certain of his carefully observed cases of the distribution of this disease seem to illustrate and find their explanation in Von Pettenkofer's theory. I have pleasure in adding that I see, by papers published by the illustrious Professor of Munich in the 'Allgemeine Zeitung'[1] for June last, that he has been able to show that, amongst all the other circumstances of the case at Gibraltar and at Malta, there were still to be found, all guessing and objections notwithstanding, the porous subsoil and the retreating ground-water, as factors in the complex constituting an area or arena for cholera. Let us attend to and note always all the circumstances in every complex phaenomenon which we have to investigate, but let us not betake ourselves over-hastily to the process of eliminating antecedents, until we are quite sure that they do not enter as factors into its causation.

I may say, in conclusion, that attention to this seventh rule of Descartes might have saved such students of Natural Science as have fallen into materialism from falling into it. The Physiologist, as such, has nothing to do with the data of Psychology which do not admit of being weighed or measured, nor of having their force expressed in inches or ounces. This language, which I long ago employed myself ('Nat. Hist. Rev.,' April 1861; 'Med. T. and Gaz.'

[1] See also 'Allgemeine Zeitung,' December 8, 1868, No. 343. 'Ueber das Verhältniss der "amtlichen Choleraberichte" zum Boden und Grundwasser.' Von Dr. Max von Pettenkofer.

March 15, 1862, Articles I, II), coincides with an utterance which I am glad to see in Mr. Herbert Spencer's recently issued first number of his new edition of 'Principles of Psychology.' There (part i. chap. i. p. 48) Mr. Spencer says, 'It may safely be affirmed that Physiology, which is an interpretation of the physical processes which go on in organisms in terms known to natural science, ceases to be Physiology when it imports into its interpretations any psychical factor, a factor which no physical research whatever can disclose or identify, or get the remotest glimpse of.' But, I apprehend[1], if the Physiologist wishes to become an Anthropologist, he must qualify himself to judge of both sets of factors. There is other science besides Physical Science, there are other data besides quantifiable data. Schleiden, a naturalist well known by name to all of us, compares the Physical Philosopher ('Materialismus der neueren deutschen Naturwissenschaft,' p. 48), who is not content with ignoring, without also denying the existence of a science based on the consciousness, to a man who, on looking into his purse and finding no gold there, should not be content with saying, 'I find no gold here,' but should go further and say, 'there is no such thing as gold either here or anywhere else.' It is interesting to note that here in Oxford, till within a few years of the present time, we narrowed the application of the word 'Science' in what seems now to be a curiously perverted fashion. For, ignoring all the physical world as entirely as though we had been already disembodied, we used the word to denote and connote only Logic, Metaphysics, and Ethics. By a 'student of science' in my undergraduate days was meant a student of the works of Aristotle, Kant, and Sir William Hamilton. The wheel has since made somewhat of a circle; our nomenclature, like much else belonging to us, is altering itself into a closer correspondence with the usages and needs of the larger world outside; the so-called 'student of science' of the year 1850 is now said to go into the 'School of Philosophy;' and 'the student of science,' as our terminology runs in the year 1868, will be found at the Museum studying the works of Helmholtz, Miller, and Huxley. I do not say this by way of triumph, but rather in that of regret, little disposed or used though I am,

[1] In the 'Anatomical Memoirs of John Goodsir,' edited by his successor, Professor Turner, and published subsequently to the delivery of this Address, some remarks to the same effect as these may be found; vol. i. pp. 268 and 292.

and hope always to remain, to regret or deprecate change as such. For there is a philosophy of both subjects, and a science also in both; and I would hope that both the one and the other might still retain a lien on the two words and the two things, nor suffer its rival to establish a claim for sole possession by its own default in exercising a right of usage.

Advocates of the dignity of man are wont to regard, or to profess to regard, with something like horror, doctrines which would hint that either his bodily structures or his mental faculties, his 'more pure and nobler part,' may have attained their perfection in the way of gradational evolution. But it is not clear to me that the horror expressed for these conclusions is much more legitimate than the arguments with which they have been assailed by Prime Ministers and others, in the Sheldonian Theatre close by, and elsewhere within our precincts. For dignity rests upon responsibility —a man is worthy or unworthy, accordingly as he can or cannot make a good answer when called upon by a voice, either from within or without, to account for his conduct or for his character. And just as a man is responsible for the employment of the wealth he possesses to the Government under which he is suffered to enjoy that wealth, no matter in what way he may have become possessed of it, whether by the hereditary transmission of a family estate, or in any other of several feasible and conceivable ways, so is a man responsible for the employment of his corporeal and mental faculties, howsoever he may have been allowed to become seized of them, to that larger and largest Government under which he has his being. I believe, however, that, if men would take as much and the same care in these psychological questions as the physiologist does in his experiments and observations, to overlook none of the conditions and circumstances of the entire complex of phaenomena upon which they undertake to decide, they would come to see that above, and often behind, but always beside and beyond the whirl of his emotions and the smoothly fitting and rapidly playing machinery of his ratiocinative and other mental faculties, there stands for each man a single undecomposable something—to wit, himself. This something lives in his consciousness, moves in his will, and knows that for the employment and working of the entire apparatus of feelings and reasoning it is individually and indivisibly responsible. Its utterances have but a still small voice, and the turmoil and noise of its own machinery may, even while working healthily, entirely mask and overwhelm them. But if we withdraw ourselves from time to time out of the smoke and tarnish of the furnace, we can hear plainly enough that, howsoever the engine may have come together, and into its present being, the engineer, at all events, is no result of any processes of accretion and agglomeration. Science, business, and pleasure are but correlatives of the machinery in its different applications and activities; *we* are something besides all this, manifesting ourselves to others in the decisions of our will, and manifesting ourselves to ourselves in our aspirations and consciousness of responsibility.

> 'And e'en as these are well and firmly fixed,
> In dignity of being we ascend.'

I have heard this line of argumentation likened to an attempt to defend Sebastopol by balloons. 'Whilst you are in the clouds, your city will be taken beneath your feet.' But a position, though airy, may yet be impregnable. There are those present who will recollect how the highest placed forts of that town were never taken, but continued to the last to answer shot for shot, and shell for shell, to the Allies. The attacking forces knew not the strength of those north forts till they entered them, but when they entered them, they entered them as friends.

XLI.

THE HARVEIAN ORATION, 1873.

Τί τόδ' ἄχνυμαι,
φθόνον ἀμειβόμενον τὰ καλὰ ἔργα;
φαντί γε μὰν οὕτω κεν ἀνδρὶ παρμονίμαν
θάλλοισαν εὐδαιμονίαν
τὰ καὶ τὰ φέρεσθαι.

PINDAR, *Pyth.* vii. 18 sqq.

TO

GEORGE BURROWS, M.D., F.R.S., D.C.L.,

PRESIDENT OF THE

ROYAL COLLEGE OF PHYSICIANS OF ENGLAND,

WHOSE OWN ATTAINMENTS,

WHOSE SYMPATHY WITH THE INTELLECTUAL ACTIVITY

AND PROGRESS OF OTHERS,

WHOSE UNSWERVING DISCHARGE OF GREAT, AND WHOSE

CONSCIENTIOUS FULFILMENT OF DETAIL DUTIES,

FURNISHED FOR MANY YEARS A VALUABLE EXAMPLE

TO THE

STUDENTS OF ST. BARTHOLOMEW'S HOSPITAL,

THIS HARVEIAN ORATION

IS GRATEFULLY DEDICATED

BY ONE OF HIS FORMER PUPILS.

OXFORD, 1873.

MR. PRESIDENT AND FELLOWS OF THE COLLEGE OF PHYSICIANS:—A man whose lot it is to live away from London may well feel some diffidence in accepting an invitation to lecture before a metropolitan audience; and, Sir, when you honoured me by requesting me to deliver this year's Harveian Oration, I felt and expressed this natural hesitation. I wish to record that you pointed out to me that my function in Oxford was to pursue and lecture publicly upon the very subjects with which Harvey occupied himself; and I suggested to myself that what could with any propriety form the substance of a course of lectures in the one place, could, *mutatis mutandis*, furnish materials for an address in the other. I felt besides, that, as the President of the College of Physicians is by virtue of his office one of the five electors to the Linacre Professorship, the Linacre Professor might seem scarcely justified in declining an invitation to appear before the learned body to which in part he owed his position; and, though I mention it last, I felt first of all that a wish expressed to me, not so much by the official whom I am now addressing, as by the individual who now more than twenty years ago introduced me to Harvey's hospital, and has persistently befriended me ever since, was a wish which I ought not lightly to disregard. If now, Sir, I follow an example which you have often set me, and, without needless preface or further personal allusions, address myself at once to the business before me, I shall thereby pay you the best of all compliments, by showing you that your teaching has not been wholly thrown away upon your former pupil.

The time allotted to me I propose to occupy, firstly, in expounding with all possible brevity certain advances recently made in our knowledge of the anatomy and physiology of the circulatory organs; and, secondly, in giving the as yet unrecorded history of one of the many attempts to rob Harvey of his rightful rank in the noble army of discoverers, which were made in the latter half of the seventeenth century.

Some of the last, if not the very last, of the many fruitful experiments which Harvey performed in the way of interrogating Nature as to the circulation, were experiments in the way of injection. If the writer of a work which appeared but some forty-three years ago, 'On the Diseases and Injuries of Arteries[1],' had taken the pains to repeat those experiments which Harvey performed more than two hundred and twenty years ago, and when in

[1] 'On the Diseases and Injuries of Arteries, with the Operations required for their Cure; being the substance of the Lectures delivered in the Theatre of the Royal College of Surgeons in the spring of MDCCCXXIX.' By G. J. Guthrie, F.R.S. London, 1830.

his seventy-fourth year, we should not have had the following statement at page 9 of his book: 'I have conceived that the arteries contained air in an uncombined state, which may assist in keeping them distended, and in facilitating the circulation; but I have not been able to prove it.' The fact that Harvey performed experiments in the way of injection may be unknown to many persons who are too well informed to conceive that the arteries may or can, compatibly with the carrying on of any circulation, contain air in an uncombined state; for these experiments are not to be found recorded either in the treatise 'De Motu Cordis' or in either of the two letters to Riolanus; which two compositions were, in the older editions of Harvey's works, printed as three parts of a single treatise, under the names of 'Exercitatio Anatomica i. De Motu Cordis,' etc., 'Exercitatio Anatomica ii. De Circulatione Sanguinis,' and 'Exercitatio Anatomica iii. De Circulatione Sanguinis;' and were, till the appearance of the College of Physicians' edition in 1766, the only published[1], as they are still the best

[1] The statement made (by Dr. Akenside; see Pettigrew, 'Medical Portrait Gallery,' Preface, p. 7, citing Dr. F. Hawkins) in the Praefatio to the College of Physicians' edition of Harvey's works, to the effect that only two of Harvey's Letters had been published prior to the year 1766, is not correct. Horstius, as Harvey's words in the 'Epistola Sexta,' p. 631 ('Harveii Opera,' ed. 1766), show, when read in connexion with the Epistola immediately preceding it, received three letters from Harvey. By consulting Horstius' work referred to by Dr. Akenside, *l. c.*, I found at pp. 61–65 the letter, which appears in our edition as 'Epistola Tertia responsoria Morisono,' published by Horstius in 1656 with the omission of the first six and a half, and also of the last three and a half lines. These lines Harvey had doubtless ordered his amanuensis —a functionary of great importance to one who wrote so bad a hand (see p. 165, ed. 1766, or Harvey's own autograph MS., No. 486, Sloane Coll., British Museum)—to omit when he bade him copy and send to Horstius, 'eadem quae antea medico cuidam Parisiensi (sc. Morisono) responderat.' Horstius does not publish Harvey's letter (the 'Epistola Quinta' of our edition) of date Feb. 1, 1654–5, but appends the last letter of the three (the 'Epistola Sexta' of our edition) to his own answer to Harvey's earlier communication. I shall henceforward refer to the College of Physicians' edition of Harvey's works as 'ed. 1766,' and to Dr. Willis' most valuable translation of them, published by the Sydenham Society in 1847, as 'ed. Willis.' I throw out as a topic for future discussion the question whether Dr. Willis is right in following the editions of Harvey's writings of an earlier date than 1766, in retaining the negative in the sentence (at p. 131 in both his edition and in that of 1766) in the second epistle to Riolanus which refers to the Critias of Plato. I think Dr. Willis is right, and that Dr. Lawrence was wrong; but to do this it is necessary to sacrifice Harvey's credit for knowledge of Plato whilst vindicating the consecutiveness of his reasoning. Harvey himself would probably have accepted this alternative. It is right to add, however, that so far as my reading of the edition of 1766 has carried me, I have come upon no other case where I have been forced to think that Dr. Lawrence may have blundered.

known, records of Harvey's work and labour upon the circulation of the blood. The experiments to which I refer are put upon record in a letter of Harvey's to P. M. Slegel, of date 1651 (see 'Harveii Opera,' ed. 1766, p. 613; ed. Willis, p. 597). They were undertaken with the object of giving a final and happy despatch to all the quibbling objections of Riolanus, 'omnes Riolani circa hanc rem altercationes jugulare;' and they consisted, firstly, in forcing water from the cava into the right ventricle whilst the pulmonary artery, the 'vena arteriosa' of those days, was ligatured—whereby Riolanus' suggestion as to the permeability or porosity of the interventricular septum was shown to be untenable; and, secondly, in forcing water from the pulmonary artery round into the opened left ventricle, whereby the lesser circulation was demonstrated, to use Harvey's own favourite word, αὐτοψίᾳ; or, to use the very words employed by him upon this very occasion, by an 'experimentum ἄφυκτον a me' (in his seventy-fourth year) 'nuper et collegis aliquot praesentibus exploratum.' Simple as this experiment may seem to us now, I do not think that any apology is required for the drawing of attention to it; for it is only twenty-eight years ago (see 'Edinburgh Medical and Surgical Journal,' vol. lxiii. p. 20) that Dr. Sharpey, to whom our Baly Medal has been so recently and so fitly assigned, had to perform the very closely similar experiment of injecting defibrinated blood into the thoracic aorta, with the very closely similar object of showing that the force of the heart was sufficient to account for the passage of blood through the intestinal and hepatic vascular systems—nay, to perform an all but identical experiment, adding on to it but the means for estimating and reproducing the force put out by the ventricle concerned. If such experiments as these were necessary in 1845, how much more necessary must have been the still simpler experiments of Harvey in 1651! At that time, the prestige of Riolanus the younger 'pressed heavily upon mankind.' Harvey himself had called that individual 'anatomicorum coryphaeum' in 1649: and, in the very year and letter we are dealing with, he calls him 'celebrem anatomicum.' And Pecquet, the discoverer of the thoracic duct, in his work, also of this selfsame year 1651, the 'Experimenta Nova Anatomica,' a work spoken of by Haller ('Bibliotheca Anatomica,' i. p. 443) as 'nobile opus et inter praecipua saeculi decora,' has the following remarkable passage: 'Ita sentiunt non vulgaris peritiae medici Harveius, Veslingius, Conringius, Bartholinus, aliique complures; nec melior ipse Joannes Riolanus (quod mirari subit pro eximiâ viri, quâ in rebus anatomicis caeteros anteivit, sagacitate). Audi hanc in rem illius sententiam.' p. 4, l. c. This, I think, I will spare you; but I will remark that, after this singular—or perhaps, alas! not singular—instance of the blundering judgments which contemporary writers may pass upon each other, no young man, nor indeed any old one—for Harvey was in his seventy-fifth year when he first read Pecquet's work (see 'Ep. Tert.' p. 620, ed. 1766; p. 604, ed. Willis)—should overmuch fret if his own age, in his own estimation, do him scanty justice. Posterity ordinarily—I do not say always—rectifies these false judgments; it has done so, at all events, in the cases of the men so grotesquely grouped together by Pecquet[1]. Haller, for example, writing in 1774 ('Bibliotheca Anatomica,' i. p. 301), speaks of Riolanus as 'vir asper et in nuperos suosque coaevos immitis ac nemini parcens, nimis avidus suarum laudum praeco, et se ipso fatente anatomicorum princeps.' The duty of attacking and abolishing such a man may, or indeed must, have been a disagreeable one to his contemporaries. They appear to have shirked it: it was their duty to have faced it, notwithstanding it might have been disagreeable.

Harvey used for these experiments a somewhat rough injecting apparatus, 'quemadmodum in clysteribus injiciendis fieri solet' (p. 614, ed. 1766; p. 597, ed. Willis). The modern experiment which I wish first to introduce to your attention rests for its accomplishment upon the employment of the delicate injection-syringe (for *Einstichung*) of Ludwig, and of the fine soluble Berlin blue for the substance to be injected. Here, as in many other instances, our superiority to our forefathers rests mainly or wholly upon our possession of more delicate, or upon our command of more powerful agents; and the delicate syringe and the penetrating soluble injection-mass help us to discoveries and demonstrations impossible in default of such means; just as the superior lenses of Malpighi and Leeuwenhoeck helped them to the discovery and

[1] See also, I would add, Gregorius Horst, the father of Harvey's correspondent of the same name, in his 'Opera Medica,' i. p. 83 (1661), where Riolanus is spoken of as 'anatomicorum hujus saeculi fere primum;' and consult Bartholinus himself, who, in his work 'De Lacteis Dubia' (1654), refers to 'multis Riolani observationibus quibus rem anatomicam immortali nominis celebritate auxit.'

demonstration of the capillary circulation, unknown to the discoverer of the makroscopic circulation. The experiment to which I refer has its results fairly represented in the accompanying drawing (Fig. 1) of a specimen prepared by myself at a class-demonstration.

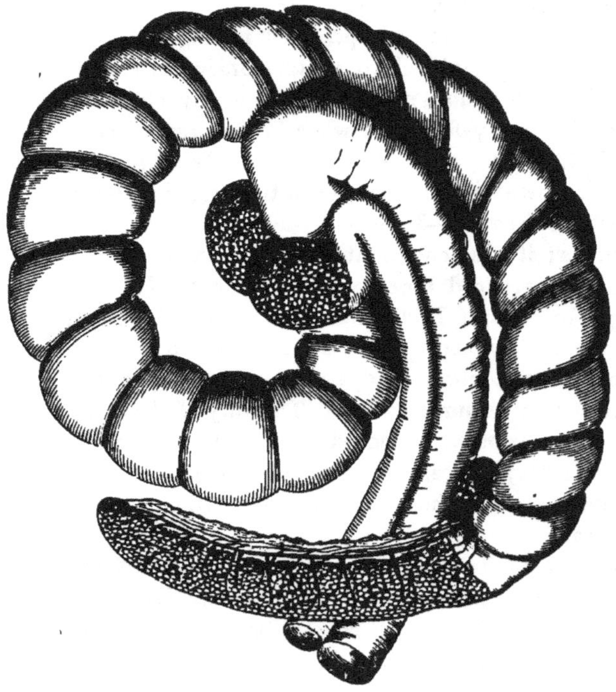

FIG. 1.—Vermiform Appendix, Caecum, and parts of Small Intestine and Colon of Rabbit (*Lepus cuniculus*), with the Peyerian and also some of the mesenteric glands in connexion with them injected. The ileum and colon lie side by side, the former describing a siphon-shaped curve before ending in a dilatation, known as the '*sacculus rotundus*' and homologous with the aggregation of Peyerian follicles situated in man just anteriorly to the ileo-caecal valve. The *sacculus rotundus* occupies the centre of the figure; above and a little to the left of it another somewhat similar aggregation of Peyerian follicles is seen just beyond the ileocolic valve. The colon curves backwards externally to and concentrically with the ileum. The caecum consists of two portions: one larger in calibre, thinner in walls, and sacculated spirally; the other, the homologue of the human 'appendix vermiformis,' smaller in calibre, but with much thicker walls. It has been injected as described in the text, and the injection is seen to have passed into some mesenteric glands situated close to the junction of the two segments of the caecum, and between it and the colon.

It gives a figure of the lacteals injected, by the means just specified, as they exist upon the terminal segment, here widely globular, of the ileum, upon a single disciform patch upon the commencing colon, and, finally and chiefly, upon and all around the walls of the colossal vermiform appendix of the rabbit. In this latter place it is but what the Germans call, and have called (Frey, 'Das Mikroskop und die mikroskopische Technik,' 4th ed. 1871, p. 255), a Kinderspiel, to insert the point of the fine *Einstichung* syringe charged with the soluble blue injection just beneath the peritonal coat at the caecal end or elsewhere, when, upon pressing the piston, a reticulation of blue will spread itself over the surface of the tube, enclosing as islands the solid substance of the Peyerian follicles. It needs but a little perseverance in the way of gentle pressure to cause superficial tubular lymphatics to arise into view, and to declare their true character by their contrast with and distinctness from the blood-vessels, as well as by their moniliform character speaking of their richness in internally placed valves. Passing over the convex walls of the appendix, they join larger trunks which run along its mesenteric border; these larger trunks in their turn enter the mesenteric glands, and form in their substance reticulations strikingly like those formed previously in the walls of the intestine around the solid substance of the Peyerian follicle—suggesting thus to the naked eye the similarity, and by consequence the homology, which a microscopic examination enables us to prove to exist between the lymph-sinuses and the solid masses they surround in the Peyerian follicles and in the mesenteric glands respectively [1].

It is the demonstration of the relation of the lymphatic or lacteal vessels, or sinuses, as the case may be, in different animals, to the solid ampulla-like mass in the Peyerian follicles which the modern method of puncture can claim as being eminently its own

[1] I take this opportunity of expressing my surprise that Henle has not seen his way towards accepting this view of the real nature or *Bedeutung* of the Peyerian follicles. In his 'Gefässlehre' of 1868 (p. 404) he refers us back to his 'Eingeweidelehre,' p. 57, of 1862, where, as also in the second edition, 1873, p. 62, the absorbent character of these structures is denied, just as it was by Hyrtl in his 'Handbuch der Topographischen Anatomie,' 1860, p. 646, and by Teichmann, 'Das Saugadersystem,' 1861, pp. 86–91. The view which I have adopted was accepted by a distinguished Fellow of this College, Dr. Burdon-Sanderson, in the Eleventh Report of the Medical Officer of the Privy Council for 1868, p. 96.

attainment; for many years ago—in 1784, in fact—and three years before the appearance of Mascagni's splendid work, with similar figures and histories of similar experiments ('Vasorum Lymphaticorum Historia et Iconographia,' 1787), the continuity of the lacteal radicles upon the walls of the intestine with the 'lymphpaths'—to borrow a word of later coinage—in the mesenteric glands, and finally, after passing through successive lines of these apparently solid structures, with the thoracic duct itself, had been demonstrated by Sheldon, then Professor of Anatomy in our Royal Academy of Arts. These are his words (from p. 49 of his work, 'Of the absorbent System,' 1784), describing his plate No. 5: 'In the fifth plate of this work, upon a portion of human jejunum from an adult female subject, seventeen lacteal vessels are injected with quicksilver, by inserting pipes into them upon the intestine. They were remarkably large and varicose in this subject, and as the quicksilver was poured into the lymphatic injecting-tube to fill these vessels, it frequently ran out in a full stream by the jugular vein, which was opened. This circumstance rendered it evident that the mercury had passed through the whole course of the lacteals and thoracic duct, and had penetrated even into the venous system. It is, I believe, the only instance in which the thoracic duct has been injected from the lacteals on the intestines[1].'

Sheldon's first plate, I may add, when compared with his letterpress on p. 37, appears to show that what he calls 'ampullulae' were really Peyerian glands, and that he had repeatedly seen these glands

[1] I have some pleasure in pointing out that by making a reference to the plates of the venerable Professor Arnold, fasc. i. tab. i. fig. 2, 1838, it may be seen that the quicksilver injection could sometimes give as correct results as the 'silver method' of modern microscopy for the detection of lymphatics by their epithelium. The figure I refer to shows the fourth ventricle plexus without, the velum interpositum on the contrary with, lymphatics injected with quicksilver. The use of the silver method has enabled me to prove that this representation is correct: abundance of choroidal villi can be procured—and very beautiful objects they are when treated with 0.25 per cent. solution of nitrate of silver—from the plexus in the fourth ventricle, but no lymphatic vessels. These can be shown from the velum interpositum by the use of the same reagent. The use of quicksilver as an injection-substance has not always led to as happy results as in the instance just given. Not to specify other cases, it is curious to note that the penetration of this material into the parieto-splanchnic ganglion of the Lamellibranchiata when thus employed by the skilful Italian anatomist, Poli, and its distributing itself thence into the nerves given off from it by displacement of their granular neurine, seduced him into supposing these structures to constitute a lymphatic system, 'cisternam lacteam et vasorum lactiferorum surculos.' See 'Testacea Utriusque Siciliae,' tom. i. p. 39, 1791.

distended in the way of natural injection with chyle, as it is easy enough to see them distended in an animal, such as a rat, which can be got to feed on fatty food, and can be killed at a proper interval of time afterwards. He appears to have had very serious as well as reasonable doubts as to the existence of any foramen in the apices of these 'ampullulae'; but the authority of Lieberkühn, whose 'Dissertatio Anatomica' (p. 18) he had himself edited, appears to have weighed with him more than his αὐτοψία. Near, therefore, as Sheldon came to seeing the whole truth, he just failed of doing so entirely and completely; and the views which Lieberkühn had put forward (p. 10, loc. cit.) as to the great number of the Peyerian glands in the lower segment of the small intestines, being a proof that they held relation to secretion or excretion rather than to absorption, prevailed and have prevailed, even into our own day. These are Lieberkühn's words: 'Quare ad finem ilei plures quam in integro intestino positi erunt? Nonne propter faeces jamjudum exsuccas et indurescentes ut lubricatae valvulam facile transeant nec laedant?' In Henle's ordinarily and marvellously excellent treatise on Anatomy, of date 1841, I find (p. 895) the excretory character of the Peyerian follicles taken as something certain; the only thing left uncertain being the question as to whether their contents found their way into the cavity of the intestine by a constantly patent, however small, duct, or by dehiscence, as ova from an ovary. In 1850 the real meaning, the true physiological import, of these glands was proved by Brücke. The method of injection, of which I have spoken, enables us to demonstrate or exhibit what was then proved, and that with the greatest ease. It is difficult to understand how any one can now doubt that the Peyerian glands are really but the pileorrhizae of the roots, the glands the tubera, and the thoracic duct the trunk or stem of the absorbent tree.

If any apology be needed for my dwelling so long upon a point of anatomy which has not merely much historical, but also much practical, interest—the Peyerian glands being the part of the organism especially affected by the poison of typhoid fever, which I see has, amongst other aliases, that of 'Peyerian fever' (Walshe, 'On Diseases of the Heart,' 3rd ed. p. 208)—I would add that I was till recently under the impression that the actual demonstration, the doing, of that which my Figure 1, p. 734, represents as done, might have been a fitting exhibition for me to go through upon the present occasion, following herein the example of Harvey, 'viliora animalia in scenam adducentis.' I have, however, learned that this very demonstration on the appendix vermiformis of the rabbit has been often performed in Germany, and, indeed, also in England; and I judged, consequently, that it might be superfluous, as it would not be novel, to exhibit it here and now.

Having been thus disappointed in my intention of demonstrating something new in this direction, I cast about in another for something of the same character. And in the heart of a bird, the Australian Cassowary (*Casuarius australis*), killed at Rockingham Bay, lat. 18 deg., on the east coast of that continent, and sent me by my former pupil, J. E. Davidson, Esq., I came upon a structure which I am well assured has never been either described or figured before. It possesses upon this ground some claim upon our attention; but it possesses stronger claims than any which mere rarity could give it, being a structure which, though it has never been seen in any other member of the class Aves, is largely developed, and, indeed, exactly reproduced in the hearts of certain mammals, and does not fail to be represented, at least rudimentarily, in our own. The structure in question is a 'moderator' band, holding precisely the same relations to the other parts of the right ventricle in this bird which the band so named by Mr. T. W. King in the 'Guy's Hospital Reports,' vol. ii. p. 122, 1837, holds in many, if not in all, Ungulate mammals. This, I presume, is made plain by a comparison of the two diagrams (Figures 2 and 3), showing, one of them the heart of this bird, the other the heart of a sheep, with the right ventricle similarly laid open in each case. The advantage, which in the struggle for existence, and specially in that very common phase of it which takes the form of a race for food or from an eater, which an animal with such a muscular band passing directly across the cavity of its right ventricle from its fixed to its movable wall must possess, is not a difficult thing for any man to understand who has ever either watched in another or experienced in himself the distress caused by the over-distension of any muscular sac[1]. A band of similar function—I do not say definitely of

[1] Since writing as above I have been reminded of what I ought not to have forgotten, viz. that my friend Dr. Milner Fothergill has discussed this very subject in his work, 'The Heart and its Diseases, with their Treatment,' London, 1872, p. 6.

precisely the same morphological importance—has often been figured as existing in the hearts of most or all Reptilia below Crocodilina; and it serves in them to close up and expel the blood from the pulmonary compartment of their imperfectly divided ventricle.

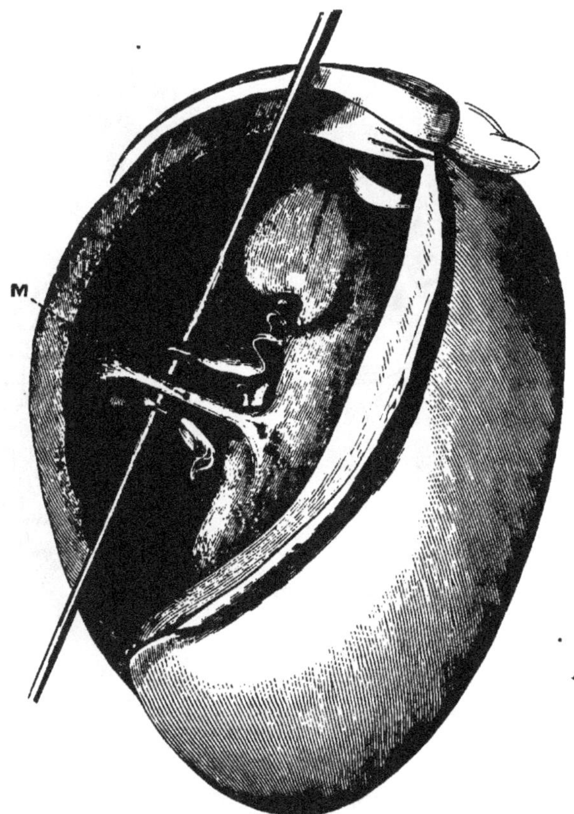

FIG. 2.—Heart of Sheep. The right ventricle is laid open. The letter M indicates the moderator band. A probe has been passed between the chordae tendineae, passing from a musculus papillaris arising from the movable wall of the ventricle to the two main segments of the tricuspid valves and the outer wall of the ventricle.

Such being the function of this moderator band, what is its morphological bearing, and what traces can we find of it in ourselves, tempting us to speculate as to the nature of the secret bond which brings us into relations of affinity not only with the mammalian class, but with an older stock, the many-sided potentialities of which embraced not only mammals, but all warm-blooded animals, and not only all warm-blooded creatures, but warm-blooded animals and

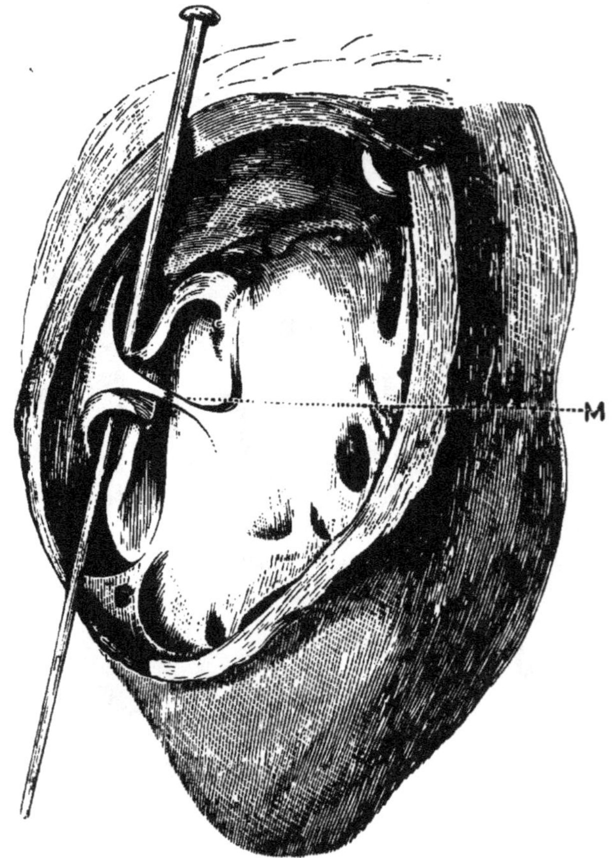

FIG. 3.—Heart of Australian Cassowary (*Casuarius australis*), prepared similarly to the heart of sheep, figured in Figure 2, and showing at M the moderator band. A line of papillary upgrowths is indicated at the upper end of the upper segment of the muscular auriculo-ventricular valve, passing downwards towards the point of origin of the moderator band from the septum. They extend in the specimen over a space of about three-eighths of an inch, and are indicated in the woodcut by dark transverse lines.

reptiles also? The valves of the heart in the higher vertebrata, when regarded from this point of view of development—the safest if not the sole criterion of homology—may be spoken of as being but trabeculae flaked off from the inner surface of the wall of a muscular sac, and subsequently made more or less membranous in the way of specialisation and its correlative economy. Thus, as Gegenbaur (Vergleichende Anatomie, 2nd edition, p. 836) has remarked, the intervalvular space in these animals corresponds to the entire cavity of the spongy-walled heart of fishes and amphibia; and the sinuous intertrabecular cavities in the spongy walls of these latter animals correspond with the chief part—viz. the extravalvular part of the ventricular space—in mammals, birds, and Crocodilina. Now, the musculi papillares represent the disposal or destination of the innermost layer of the right ventricle, according to Dr. Pettigrew (see his paper, Phil. Trans. 1864, p. 479); and I would submit that the moderator band is but a specialisation of the next layer in order from within outwards—to wit, Dr. Pettigrew's sixth layer, which he has figured (Plate XIV. fig. 33) as proceeding in a spiral direction from right to left, much as the fibres of the moderator bands I have figured do. A study of the heart of the rabbit will put this matter in a very clear light, and further open our eyes to see and recognise the rudimentary representation of this moderator band in our own hearts. If we look at the outer aspect of that very constant musculus papillaris, which passes in man from the outer and movable wall of the right ventricle to distribute its chordae tendineae to the two more anteriorly placed of the three segments of its auricular valve, we shall frequently see that its longitudinal fibres are crossed nearly or quite at right angles by a slender fibrous band, so that we have before us an appearance not wholly nor essentially unlike that presented by the striae longitudinales of Lancisi and the fibres of the corpus callosum when viewed in their mutual connection. This band of fibres can sometimes be traced up towards the conus arteriosus, and be seen not to die away until close upon the point of origin of the most anteriorly or upwardly placed of the chordae tendineae which arise from the septum to pass to the hindermost of the three segments of the tricuspid. The points between which this line of fibres lies may be observed to be the very same as those between which the moderator bands in the cassowary and the sheep stretch as free columns in the diagrams before you. It is not altogether rare to see this band raise itself from the position of fusion, like the ventricular wall, and assume the character of a cylindrical band for a lesser distance, but with no less distinctness as a column, than in the Ungulata. Such a case I had actually before me whilst writing this, and you have it now figured before you (Fig. 4)[1].

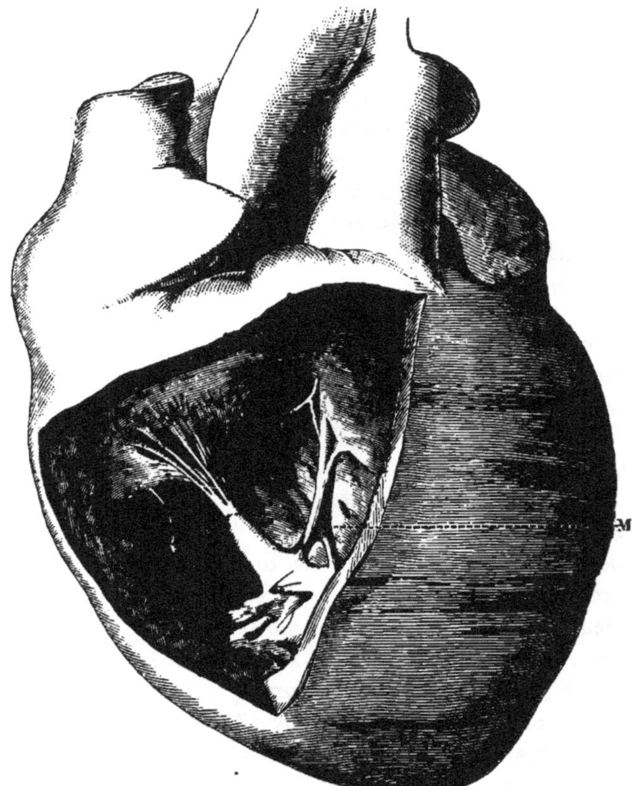

Fig. 4.—Human Heart. The right ventricle has its cavity exposed. The moderator band M is seen to pass up from the base of a musculus papillaris arising from the outer or movable wall of the ventricle into the neighbourhood of the conus arteriosus. Close to this, its upper end, a chorda tendinea arises by two roots, and passes up to the most posteriorly placed of the three cusps of the tricuspid valve.

[1] Since this oration was delivered I have received two communications relating to the presence of a moderator band in the human heart. One of these, from my former pupil, J. C. Galton, Esq., F.L.S., was accompanied by a sketch in which a moderator band was drawn as passing in a human heart from the insertion into the movable wall of the ventricle and the very constant musculus papillaris supplying

Every gradation, in fact, exists between the entire obsolescence of the moderator band, which we sometimes see in the human heart, through the typical, and I should anticipate, constant, but not functionally important, representation of it in the rabbit, up to the important and structurally prominent development attained to by it in the ungulate mammal, and this solitary instance for the class of birds, and the sub-class, with such generalised affinities, of struthiones.

And, speaking of the method of gradations, I take this opportunity of saying that its application in the case of the muscular right auriculo-ventricular valve of birds will, in my judgment, put an end to the disputes which have taken place as to its homology with one or other of the two valves in the crocodiles. The two portions of the valve in the *Casuarius australis* are so nearly equal—the larger being 1·7 inch, as against 1·4 of the smaller—as to do away with the difficulty which might be felt in holding that both crocodilian valves are represented here. There are other reasons for this view, which I reserve for another occasion. But whilst speaking of the heart of the bird, I cannot forbear pointing out how the structural arrangements of its auricle, differing as they do strikingly from those of the same compartment in the mammalian heart, help us by that contrast to get a true idea of the working of this latter. Firstly, the walls of the bird's right auricle are relatively thicker, not only as compared with the walls of its own ventricle, but also as compared with the walls of the corresponding auricle in the mammal, the musculi pectinati standing out in as sharp relief as the similarly working muscular ridges in a hypertrophied bladder, and inclosing anfractuosities and recesses almost as deep. But, secondly, and what is of more importance, the bird's auricle is furnished with a large and functionally active valve, protecting the entrance of the great veins, and preventing regurgitation into those vessels just as the auriculo-ventricular valves prevent regurgitation from the ventricles. It is fair to argue *a priori* that if the

what I would call the '*conad*' and '*dextrad*' cusps of the tricuspid, to an origin on the interventricular septum, sending a root up to the point of origin of one of the chordae tendineae of the third cusp, called '*septal*' by Mr. Galton. See also Mr. Galton's Letter to 'British Medical Journal,' July 26, 1873, p. 83. I have to thank Dr. Headlam Greenhow for a reference to another notice of the presence of a moderator band in a human heart. It will be found in an interesting paper of his in the 'Transactions of the Pathological Society,' vol. xxi. 1870, p. 88.

[It should be stated that Mr. T. W. King had figured and described a moderator band in the human heart and in several mammals other than the Ungulates.—EDITOR.]

mammalian auricle had counted for as much in the action of the heart as the bird's, its force would have been economised by the placing of a large and functionally useful valve in the site of the rudimentary Eustachian—a structure altogether absent in many mammals, and variable, as rudimentary structures very often are, in ourselves. The *a priori* argument of Comparative Anatomy is abundantly borne out by the appeal to experiment. Marey, in his 'Physiologie Médicale de la Circulation du Sang,' 1863, whilst referring (p. 36) to other evidence from Comparative Anatomy than that which I have adduced, cites, in support of the view that the auricle has but an accessory and subordinate *rôle* in the functions of the heart, an experiment of Chauveau's, in which the auricle of a horse, being exposed and irritated, lost its contractile power for a time, during which, nevertheless, the ventricles continued to contract and the circulation to be maintained. Colin, again ('Traité de la Physiologie Comparée,' vol. ii. p. 257, 1856), found that the left ventricle continued to be filled with blood even when the corresponding auricle was prevented from contracting by the insertion into it of a finger. And further, Magendie had long ago noted, in experimentation, what many here present may have noted in pathological or clinical observation—viz. that the auricles may remain extremely distended for hours, and, like other muscular sacs similarly conditioned, unable to contract and empty themselves, without the circulation for all that being brought to a standstill. It was Dr. Pavy's paper, treating (in the 'Medical Times and Gazette' of November 21, 1857) of the case of a man (E. Groux) with a congenital fissure of the sternum, which first drew my attention to these points; and his summary of what takes place in the dog is so clear that I herewith reproduce it:—

'In the dog, the contraction of the ventricles is sharp and rapid, instead of prolonged, as in the reptile, and does not appear to occupy nearly so much time as half the period of the heart's action. The ventricular contraction communicates a sudden impulse to the auricles, occasioning in them a distinct pulsation, which is instantly followed by a peculiar thrill, wave, or vermicular movement, running through the auricular parietes down towards the ventricle. This thrill or wave is coincident with the passage of the blood from the auricle into the ventricle, and takes place so instantaneously after the ventricular contraction, that the one movement appears to run on, to continue itself into the other. There is then a pause, which seems comparatively of considerable duration, and which is succeeded by a recommencement of the heart's action, beginning with the ventricular contraction.'

Dr. Pavy has very kindly gone to the trouble of repeating the experiment upon which these statements are based; and from a letter with which he has favoured me, I gather that the auricular contraction detectable by the cardiographic tracing, as immediately preceding the ventricular contraction, is also detectable, of course during the pause just mentioned, by the eye, unassisted by the cardiograph, and turned simply upon the exposed heart, in which the auricular appendix is seen to become redder or more flesh-coloured at the moment in question. And he further remarks that this auricular contraction, difficult[1] though it be to be observed under physiological conditions, may be exaggerated into considerable prominence in disease entailing contraction of the auriculo-ventricular orifices, and may then make itself known by a presystolic murmur.

I should now be glad to draw attention shortly to a few memoirs which have appeared comparatively recently, and which treat of matters of considerable interest, not merely as scientific problems, but also as practical questions. First among these I would name the paper which appears in the third volume of Professor Ludwig's 'Arbeiten,' 1868 (having previously appeared in vol. xx. of 'Bericht Math.-Phys.-Klass. K. S. Gesellsch. Wissensch.,' Leipzig), by Professor Ludwig himself and Dr. Dogiel. In this paper we have a number of experiments recorded as performed with the hearts of dogs removed from the body, and as nearly as possible emptied of blood; and the conclusion which the authors come to is that the heart of the dog, when removed from the body and emptied of blood, still produces a sound during the systole of the ventricles which is not essentially different from that which is recognised as the normal first sound of the heart. The authors add, however (p. 85), that they do not think these experiments entirely exclude the possibility of the tension of the auriculo-ventricular valves

[1] I apprehend that Dr. Walshe's account of the auscultatory phenomena as occurring under normal conditions will be accepted as correct. It runs thus ('Diseases of the Heart and Great Vessels,' 3rd ed. 1862, p. 65): 'In the normal state the blood enters the ventricles from the auricles with a current so calm as to prevent audible sound from being thereby produced in the former cavities.'

entering as a factor into the production of the first sound; and hereby they would be guarded from coming into contradiction with most English authorities—as for example, Dr. Walshe ('Diseases of the Heart,' 3rd ed. 1862, p. 62). Dr. Guttmann, however, in a paper of no great length, but of considerable merit, published subsequently to the one just mentioned, and in Virchow's Archiv for 1869, points out with much acuteness what, when once pointed out, is ever thereafter obvious—viz. that it is, in the nature of things, impossible, with all possible precautions in the way of emptying the heart of blood, to empty the complex phenomenon made up by a systole of the heart of the condition of tension of the auriculo-ventricular valves. Surely the musculi papillares will contract with the rest of the ventricular walls, and, contracting, will they not stretch the chordae tendineae and the valves? For myself, I would say that we are more likely to overrate the share taken by the valves than to underrate that taken by the muscular walls. I need not say to this audience that the fact with which we are all familiar, of the alteration in the first sound produced by disease of the auriculo-ventricular valves, does not absolutely prove that they produce any part of it during health; and, finally, to my own ear at least, a modification of Wollaston's experiments, which anybody can try for himself by making his temporal and masseter muscles contract at any time of perfect stillness, appears to produce a sound which is scarcely, if at all, different in quality from the first sound of the heart. A judgment, however, upon the nature of a sound, or, indeed, an aggregation of sounds, as in music, is one upon which two observers may very well differ, as neither of them can lay his proof of supposed identity or difference alongside of that which the other may possess, or may suppose he does.

It is with much pleasure that I refer to Dr. Rutherford's paper on the Influence of the Vagus on the Vascular System, which appears in the Edinburgh Royal Society Transactions for 1870, vol. xxvi. In that year, having to deliver an address to the Biological Section of the British Association at Liverpool, I made bold to say that the results to which Dr. Rutherford had come, and which were then only known to me in an abstract in the 'Journal of Anatomy and Physiology' (May 1869, p. 402), would prove to be of the highest value and importance. His memoir now published *in extenso*, and extending over forty-two pages, as fully justifies my prediction as it will fully repay any one who will take the pleasant trouble of reading it. The most important result in a practical point of view is the demonstration which Dr. Rutherford has given of the nerve-circle, whereby, in the way of reflex action, the all-important secretion of gastric juice is called forth. The sensory impulse caused by the ingestion of food into the stomach is propagated upwards by the vagi to the medulla oblongata, where it throws into abeyance the vaso-motor nerve-cells, which, whilst the stomach is empty, keep the blood-vessels of the gastric mucous membrane constricted, but which, when their activity is inhibited, allow the zonular fibre-cells of these blood-vessels to dilate, and allow the increased afflux of blood thus called for. That relief will result to some of the countless martyrs to dyspepsia out of the demonstration of this physiological relation of vagus, sympathetic, and peptic glands, I do not doubt. Possibly, I would add, Owsjannikow's observations as to the working of hydrate of chloral as a depressor of arterial tension (Ludwig's 'Arbeiten,' 1872, p. 32) may prove valuable to persons engaged in practice, by pointing out, in however shadowy a fashion, the road to a more rational and systematised, even if less general use of this drug than that which I am told is now made of it. It may seem a paradox, but it is none the less true for all that, to say that, for the activity of many organs, a paralysing and inactivity of certain nerve-centres in connexion with them is a prerequisite. The activity of such, indeed of most, organs is but intermittent and occasional, being but intermittently and occasionally called for, whilst the constringing activity of the sympathetic has to be constantly at work to prevent waste of force[1].

Owsjannikow's paper (also to be found in Ludwig's 'Arbeiten,' 6th year, 1871, and in the 'Bericht Math.-Phys.-Klass. K. S. Gesellsch. Wissensch.,' Leipzig) just referred to, and published two years subsequently to Dr. Rutherford's, gives, as the result of a number of experiments performed in Professor Ludwig's laboratory at Leipzig on rabbits, and independently at St. Petersburg on cats, the con-

[1] The phenomenon of the distension of the corpora cavernosa, a phenomenon used by Harvey himself in the way of illustration (p. 129 of the 'Epistola Secunda ad Riolanum'), I may adduce in the way of illustration also, being, as it is, dependent upon a similar nervous mechanism; and being shown so unmistakeably, in cases where it follows lesions in the nuchal region, to result from paralysis of nerve-centres situated there or thereabouts.

clusion that the ganglionic centres of innervation for the entire sympathetic system occupy but a small space at the base of the brain, two strips to wit, one on each side of the median fissure in the floor of the fourth ventricle; of, in the rabbit, a length of about four millimeters, beginning about four or five millimeters anteriorly to the calamus scriptorius, and ending about one or two millimeters behind the level of the corpora quadrigemina. The title of such a book as Eulenburg and Guttmann's 'Die Pathologie des Sympathicus auf Physiologische Grundlage' (Berlin, 1873) is an encouragement to those who hope to see fruit arise from such researches as these in the way of additions to our means for meeting, or at least understanding, human disease and suffering.

It has long been known (Budge, 1855) that the sympathetic nerves which supply the vessels of the head and iris do not pass directly or by the shortest possible route to this their distribution, but pass down the spinal chord for a greater or lesser distance, and then turn outwards, and pass from the anterior nerve-branches to bend upwards, much as the recurrent laryngeal nerve does. That other vascular regions receive their vaso-motor supply by this apparently circuitous and, till the history of development is taken into consideration, paradoxical route, is from time to time being demonstrated. Dr. Pavy, to whom I have already referred, many years ago identified and mapped out one segment of the road along which nerve-force passes to the liver and prevents or allows the occurrence of diabetes. Further exploration of this route we owe to Cyon and Aladoff ('Bulletin de l'Académie Impériale des Sciences de St. Pétersbourg,' tom. xvi. p. 307; 'British Medical Journal,' December 23, 1871); and this same investigator, working still in the same line of investigation, as it is in these days usually necessary for an investigator to work if he will make himself a name as a discoverer, has also shown us (Ludwig's 'Arbeiten,' 3rd year, 1868) the track along which the vaso-motor nerves of the anterior limbs pass, proving that these nerves pass down in the spinal chord as low as the mid-dorsal region before leaving it to turn upwards in the sympathetic chain to join the brachial plexus.

Of all the results, however, which have been attained to in the line of experimentation now under consideration, those come to by Brown-Séquard and demonstrated by him at the meeting of the British Association held at Liverpool in 1870, and subsequently published in the 'Lancet' of January 7, 1871, seem to me to be certainly the most striking and possibly the most important. Could anything have been more surprising to him whose memory we here this day commemorate, than to have been told that an injury to a particular part of the brain, the pons, called after the excellent anatomist whose life ended in the very year in which his had begun, would produce haemorrhage in certain parts of the lungs, and anaemia, oedema, and emphysema in others? This is an easy experiment to repeat; it is one which might have been done in the days of Harvey as easily as in those of Bernard, of Budge, of Ludwig, and of Brown-Séquard. But easy though it would have been to perform, I am bold to say it was well for Harvey that he never happened to perform it. For considering that, like Haller, he knew nothing of the contractility of arteries; considering that Hunter had not performed his now well-known experiments with the umbilical arteries; considering, Sir, that in that excellent work on Physiology by Johannes Müller, the translation of which in 1838, by our late and never sufficiently to be lamented friend Dr. Baly, we owe to your suggestion, I find several pages (vol. i. pp. 202–206, 214–219, ed. 1840) devoted to *disproving* the muscular contractility of arteries; considering, that it was not till three years later, in 1841, that Henle's work, already referred to, appeared with its still unsuperseded figures, Plate III, figures 8, 9, and 10 of the arteries with their circular muscular coat, and with its excellent summary in letterpress of the whole subject, pp. 518–526, and especially pp. 524, 525; when I consider that nothing of all this had been done, to leave unmentioned other advances connected with names of men yet living to speak for themselves and for us—I say it may have been well that Harvey never came upon the facts relating to the alterations of lung-substance being entailed by destruction of brain-substance, not difficult to be observed and reproduced, which we owe to Brown-Séquard. For if he had come upon them, how could he have explained them in the absence of the entire chain of connecting facts, in the forging of which chain so many successive workers—Purkinje, Valentin, Weber, Burdach, Stilling, and others—have all contributed links? Might not even Harvey, often as he withstood such temptations, have, nevertheless, in default of power to assign the real causes of such a phenomenon, been driven back upon some of those explanations which he himself so forcibly denounces in the words ('Epistola Secunda ad Riolanum,' p. 116), 'Vulgo scioli cum causas assignare haud norunt dicunt statim a spiritibus hoc fieri et omnium opifices spiritus introducunt, et ut mali poetae ad fabulae explicationem et catastrophen θεὸν ἀπὸ μηχανῆς advocant in scenam.' It is a hard thing for any man to abstain from speculating as to the cause of any well-established phenomenon, especially if it be of striking interest and importance; it is a hard thing for any man to do more than keep pace with his own generation; and those who have spent any time in reading the works of Harvey's contemporaries, will best appreciate the difficulty he must have had in setting himself free from the influence of the *idola theatri* referred to.

I pass from this reflection to an exposition of the claims which have been put forward on behalf of Walter Warner, the editor in 1631 of 'Harriott's Algebra,' to the discovery of the circulation of the blood; and I do this by a natural transition, Walter Warner having been a man in whose mind, all his mathematics notwithstanding, the *idola* in question greatly abounded. Warner's claims are alluded to by Dr. Willis in a note to his excellent 'Life of Harvey' (see p. lxiv). They are put forward by Anthony Wood, upon the authority of Dr. Pell, a man distinguished as one of Oliver Cromwell's diplomatists, and afterwards as an assiduous supporter of the then young Royal Society; and upon that of Dr. Morley, some time Dean of Christ Church, and afterwards Bishop of Winchester (see Wood, 'Athenae Oxonienses,' vol. i. p. 461, 2nd ed. 1721; vol. ii. p. 302, ed. Bliss). Aubrey, a contemporary of Wood's, appears, from a note at p. 417 of the second volume of his 'Lives of Eminent Persons,' to have had the same story from Izaak Walton, who gave Dr. Morley again as his authority; and Aubrey repeats the tale with certain additions, and notably with that of Dr. Pell's authority, at p. 577 of the same volume. The same story was pointed out to me by one of the officials in the Bodleian Library as being given in an anonymous biographical Miscellany to be found in the 'Rawlinsonian Collection,' B 158, pp. 152–153. This MS. appears to be of the latter half of the seventeenth century, and its legend runs to the following effect. A certain Henry, Earl of Northumberland, being imprisoned in the Tower, did, for the better passing of his time, get several learned persons to live and converse with him; one of these men (whom, Aubrey tells us,

l. c. p. 368, the world called the Earl of Northumberland's magi) was 'Mr. Warrener.' And the MS. proceeds, 'He was the inventor, probably, of the circulation of the blood, of which subject he made a treatise, consisting of two books, which he sent to Dr. Harvey, who epitomised and printed them in his own name; he usually said that Dr. Harvey did not understand the motion of the heart, which was a perfect hydraulik. . . . . . Dr. Pain, that very ingenious and learned canon of Christ Church, told me that he had seen and perused this book of Warrener's.' Finally, the excellent 'Biographia Britannica' has embalmed Wood's and Aubrey's story, in the articles 'Harriott' and 'Harvey,' pp. 2542 and 2550, ed. 1757. Many *a priori* improbabilities will at once be seen to attach to this story, and it is easy enough to discredit more than one of the witnesses. But I have better than indirect evidence to bring forward, and I will have the agreeable mental exercise of excogitating it to the ingenuity of my hearers, which ingenuity will be sharpened, no doubt, by their regard for their own Harvey, and strengthened by the belief that

> 'Whatever records spring to light,
> He never shall be shamed.'

I may be asked, after this quotation, why I should have thought it worth while to investigate Walter Warner's claims at all. I will shelter myself, in the first instance, behind the example of Sir George Ent, who, feeling and acting by Harvey as Launcelot in his better days felt and acted by Arthur, took similar pains to set aside the similar fable as to Harvey's indebtedness to Father Sarpi. And, in the second place, I will remind my hearers that it was but as recently as 1838 that an article appeared in the 'London and Westminster Review,' in which the claims of the Italian monk just mentioned were once again brought forward with surprising confidence, plausibility, and ignorance.

It was possible, I thought, that the same paltry but evil spirit which animated Dutens in writing his 'Inquiry into the Origin of the Discoveries attributed to the Moderns' (1767 [1]), and in coming

[1] Dutens was as well acquainted with the excellent work of William Wotton, 'Reflections upon Ancient and Modern Learning,' published in 1694, on the other side of the question, as a little bitter mind can ever be with a work or the working of a noble and generous one. His repeated references to it show this, as also the unimprovable character of his shallow poverty-stricken spirit.

to the conclusion that every great man in modern times had been anticipated by somebody or other in ancient ones, might still be going about in dry places, and might wholly enter into and entirely fill up the soul of some small antiquary, who, coming under such inspiration and guidance upon the passages which I have collected, might proceed to instruct the literary public as to Warner's claims.

Whilst considering what indirect evidence might be brought together to rebut this possible attempt at detraction, I came upon what led me to the discovery of the direct evidence I have promised to lay before you, in the shape of a clue which brought me, after a somewhat tortuous course, upon Walter Warner's actual autograph MS. I found, whilst following up Dr. Pell's history, scattered through Dr. Birch's unindexed 'History of the Royal Society,' that Dr. Birch had procured a number of MSS. of Mr. Walter Warner's for that Society mixed up with Dr. Pell's (see vol. ii. p. 342; vol. iv. p. 447). Coupling this statement with the voucher for Warner's claims, ascribed by Wood and Aubrey to Dr. Pell (who, however, is never reported in Dr. Birch's History, so far as I found, to have given currency to this statement), I thought that by these MSS. I should be able to test the truth of these statements. But the librarian of the Royal Society knew nothing of any MSS., either of Pell's or of Warner's; and, as the result will show, it would have been odd if he had—at least, in his official capacity. I then made inquiry of the Duke of Northumberland, in whose library the MS. of Warner, once a pensioner of his house, might possibly be preserved; but Mr. J. E. Martin informed me that this hope was a vain one. I found that Sion College had once possessed one MS. of Warner's; but I learnt from the Rev. W. W. Milman that they had lost it, and much besides, in the great fire of London in 1666. Finally, when taking the register of Merton College up to the British Museum for the purpose of comparing the entries made in that volume during Harvey's wardenship with his one authentic autograph MS. now in the national collection, I bethought me of making, at the same time, some inquiries as to Warner and Pell; and at last, when I least expected it, and had nearly ceased to hope for it, I came upon Walter Warner's MS., contained in Dr. Birch's collection (which, according to him, had been made over to the Royal Society), under the title, '4394, Birch Collection,' numbered on in continuation of the Sloane Collection.

Mr. E. Maunde Thompson, by the employment of various scientific methods, the observation of which went some way to compensate me for the tedious labour entailed upon me by the result to which they brought him, identified the MS. as being really Warner's, and brought its date down to a year close upon 1610, half-a-dozen years or so, therefore, before Harvey first lectured at the College of Physicians. The MS. being thus identified I set myself down to look through its 416 folio pages, the average number of lines in a page being thirty-three or thirty-four; the average of words, many of them idle ones, being eight or nine in a line. I do not think it is very likely that I have missed any clearer exposition of Warner's views than the one which I am about to read from page 138; nor do I think that, by choosing it, I can in any way misrepresent them, for they are stated elsewhere in the treatise in very much the same words, e.g. page 137. These, then, were his views:—'By this spontaneall pulsatory motion the bloud is continually extracted from the vaines (propter fugam vacui) as well originally exsuctory as secondarily circulatory and propelled into the arteries (propter fugam penetrationis), but with some diversity in the distribution, some part thereof being propelled up into the head by the internal jugular arteries, ad plexum choroideum for spirito-faction, the rest into all the rest of the arteries in universum corpus for organo-faction. Out of that part of the blood that is propelled by the jugular arteries up to the head, the spiritus confusus or immersus thereof being expressed and segregated in plexu choroidi, either by excussion or exhalation, and animal spirits, thereof made by the self-operation of the prae-existent in somno, it is again distributed as before, one portion thereof being still derived and transmitted to the heart, ad motum spontaneum pulsationis ciendum, and so about again, perpetua circulatione durante fabricâ corporeâ, and all violent destructions or impediments abstracted.'

It is, perhaps, needless to dwell further upon Warner's claims—certainly I do not propose to trouble you with reading to you any more of his speculations and conclusions. I have, however, had a copy made of pages 140, 141, 142, 194, and 195, and, though the gift may not seem a very valuable one, it will enable any fellow of Harvey's College to satisfy himself abundantly, and within our own walls, as to the real merits of the claimant before us, if the College will allow it to find a place in their library. In the words of Harvey's favourite poet,

> 'His saltem accumulem donis et fungar inani
> Munere.'

In all seriousness it is something to know what a contemporary of Harvey, and he a mathematician of some eminence, could write less than ten years before the actual demonstration of the circulation of the blood was given to the world.

Let me say, however, that I do not think it by any means impossible that Harvey may have read this treatise of Warner's, hard though the labour of gathering hints, or rather warnings, from its many guesses must have been to him. For in many parts of Harvey's treatise, 'De Motu Cordis,' we meet with phrases which seem as if they had been used with a special reference to Warner's views; and his dissertation has at least this claim upon my gratitude, that it has made me think that I understand Harvey's meaning the better for having read it. I fancy, in fact, that I recognise such phrases in Harvey's words ('De Motu,' pp. 58, 61, ed. 1766; p. 56, ed. Willis), 'Absque dolore vel calore *vel fugâ vacui*,' and in such words as 'longè plus est quàm partium nutritioni congruens est,' p. 64; 'αὐτοψία, non mentis agitatio,' p. 133. He might have been alluding to almost any page of Warner's MS. in his repudiation (p. 116, see 'Epistola Secunda ad Riolanum') of the hypothesis of various sorts of spirits. But there is one of Harvey's many noble and candid, whilst measured and well-balanced utterances, which seems to me to be admirably suited to serve as a text for an exposition which perhaps some future Harveian orator may undertake, of the exact relation which his discoveries held to the knowledge and the ignorance, not only of Walter Warner, but of all others of his contemporaries or predecessors. These words run thus ('De Motu,' p. 34, ed. 1766; p. 33, ed. Willis):—'Sed et hoc' (viz. the transmission of the blood by the action of the heart, from the veins into the arteries, through the ventricles of the heart into the whole body), 'omnes aliquo modo concedunt et ex cordis fabricâ et valvularum artificio positione et usu colligunt. Verum tanquam in loco obscuro titubantes caecutire videntur et varia subcontraria et non cohaerentia componunt et ex conjecturâ plurima pronunciant ut ante demonstratum est.' This may be translated thus:—'But it may be said, that all competent persons accept these views in a more or less modified form, and have been convinced of the truth of them from the structure of the heart, and the contrivance, position, and use of the valves. But they seem to me to make as little use of their eyes as men do who are stumbling about in a dark place, and their account of the matter is made up of heterogeneous, contradictory, and incompatible statements, and very much of it is pure guesswork, as I have already shown.' These words, the Latin ones, not my translation of them, were published, if not written, nine years (see p. 5, 'Harveii Opera,' ed. 1766, Dedicatio) and more after Harvey had first proved the facts of the circulation, and from them we gather that his discovery had, even so early as that date, got out of the stage in which a discovery is considered to be untrue, and got into that in which it is said that everybody knew it before. In no subject could it have been easier to make out a plausible case than in this of the circulation of the blood. Piccolhomini (an acquaintance with whom I owe to Mr. Walter Warner, see his treatise, pp. 194, 200, 201) had given a diagram, it is there before you [1], copied from the copy of his work in our library, of the junction of the portal and hepatic twigs, incorrect enough, no doubt, and obtained by a false method (see Harvey, 'Epistola Prima ad Riolanum,' p. 105, ed. 1766), but still something in the way of a working hypothesis (see Piccolhomini, 'Anat. Praelect.,' Romae, 1586, p. 117, and Warner, MS. p. 194). Servetus had speculated, but rightly, as to the lesser circulation; so had Caesalpinus; and on Harvey's own showing (p. 15, and ed. Willis, ed. 1766), Realdus Columbus; and Walter Warner, p. 132 (4394 Birch Coll. MS.) had spoken of the heart, in 1610, as being 'a mere muscle, very strongly and artificially woven, and contrived with omnimodal nerveous fibres, direct, transverse, and oblike, as it were of purpose, for dilatation and contraction, according to the fashion of other muscles.' And of the action of the auriculo-ventricular and arterial valves, Harvey himself, *niveâ animâ*, with untarnished sincerity, repeatedly (see 'De Motu,' pp. 14, 51, 53, 67, 81) speaks as of something known to all men, 'id omnes norunt' (p. 44). What then, it might have been triumphantly asked, was there left for Harvey to discover, when the action of the valves of the heart, its

[1] I have not thought it necessary to reproduce it in a woodcut.

muscular character, and so much else, was already to be found in the writings or teachings of his predecessors? To all this we can answer, as indeed, it seems probable, was practically answered even within Harvey's lifetime, what was left for Harvey to discover was nothing less than *the circulation itself*. His predecessors had but impinged, and that by guesswork, upon different segments of the circle, and then gone off at a tangent into outer darkness, whilst he worked and proved and demonstrated round its entire periphery. His demonstrations and direct proofs were all new, and his indirect arguments nearly all new. Whenever he made use of anything already known, he most punctually acknowledged it. Of his demonstration in the way of injection I have already spoken; of his demonstration of the use of the valves in the veins, and his proof that they are similar in function to the arterial, a fact previously unsuspected (see p. 65, l. c.), the thirteenth chapter of the treatise 'De Motu' speaks with figures; of his indirect, but irrefragable argument, in the eighth chapter, from the quantity of blood thrown out by the heart at each pulsation, an argument which a mathematician such as Harriott, or Warner, might have hit upon, but, so far as I have found, did not, he speaks himself as being 'adeo novum et inauditum ut verear ne habeam inimicos omnes homines;' and finally, the argument, which though it be indirect, every morphologist will allow to be not only most exquisite, but also most convincing, for the circulation in the adult warm-blooded animal, drawn from the relations held by the venae cavae to the efferent arteries in the embryo, and in all animals with but a single or an imperfectly divided ventricle, 'unus duntaxat ventriculus vel quasi unus,' and of which I would recommend every one who is not already acquainted with it, to gain a knowledge from the seventh chapter of the same book, was his, and his alone. With regard to all these points, with regard, that is, to the circulation as a whole; with regard to the actual demonstration and exhibition of it as opposed to mere guessing about it; with regard to all, or nearly all, sound reasoning as to any large portion of it, Harvey might have said with Lucretius,

> 'Avia Pieridum peragro loca nullius ante
> Trita solo';

or in the words of a poet of another country, and a later age,

> 'We were the first that ever burst
> Into that silent sea.'

I do not wish to assert that Harvey was wholly independent of the works of his predecessors; he himself would, as his repeated references to them show, have been the very last man to make any such claim for himself; nor would I say that he owed nothing to the times—

> 'The spacious times of great Elizabeth'—

in which he lived. It is true, I think, in science, as it is also true in morals and politics, that the times make great men as much as great men make the times. Many metaphors have been used to express this latter half-truth. Such is the metaphor, an acquaintance with which I owe to Mr. Picton's new and striking work, 'The Mystery of Matter,' p. 265, used by St. Augustine, in which great men are compared to great mountains, dwelling apart in loneliness, and sending floods of blessings down upon the little hills and plains at their feet. Such, again, is the metaphor used by Wordsworth in apostrophising Milton:

> 'Thy soul was like a star, and dwelt apart.'

Such is the metaphor used by Sir Coutts Lindsay, in his poem on the Black Prince, where a hero

> 'Stands like a beacon, throwing light far out
> Over the rippling tides of centuries.'

Now all these metaphors strive, and profess, to express but half a truth, and they are imperfect even for this imperfect purpose, as they are borrowed from inorganic nature and the arts, and are unfit to be used as illustrations of the complexities of life and thought. I would venture to suggest a metaphor which has struck me, during this investigation, as being more appropriate and close-fitting, even if less beautiful, than those which I have quoted. A group of horsemen are attempting to cross an arm of the sea, up which the tide has been running, and obscuring the ridge, or spit of sand, by which it is fordable. They form themselves into a line, and advance slowly: rider after rider flounders off into deep water, and, if wise, retires towards the rear of the cavalcade of his companions, who still feel and advance upon the bottom beneath them. The line by degrees narrows into a column, and the column, after a longer interval, narrows into a single file. To the foremost horseman courage is necessary, as imagination is to the discoverer, and, impelled by this feeling, he may put a wide interval between himself and his companions, and reaching the opposite bank long before them, may have leisure to look down upon them, may be looked up to by them and by the rest of the world, whilst for some time in solitary occupation of that vantage-ground. Such I conceive to be a fair representation, in the way of metaphor, the best and shortest way, perhaps, of representing such complex relations, of the relations held by Harvey, and indeed by most or all discoverers, to their contemporaries, to their compeers, and to the conditions whereby they are surrounded.

It may be expected, perhaps, that, coming from Oxford, and having been recently elected a Fellow of the College the Wardenship of which Harvey held for something more than a year (April 1645 to Midsummer 1646), I should have made search for whatever records there may be left of him unpublished in Oxford, and especially in Merton College. After diligent search, I have to report that there is but little to be learned of Harvey's history from any unpublished document which I have been able to find in Oxford. The Merton College Register gives the following account of his election to the Wardenship. In 1645 King Charles I, after the execution of Archbishop Laud, took upon himself the functions of Visitor, and, having removed Sir Nathaniel Brent from the office of Warden, for having joined 'the Rebells now in armes against' him, he directed the Fellows to take the customary steps for the election of a successor. This course consisted in giving in, after due inquiry, three names to the Visitor, in order that one of the three, the one we may suppose it would be understood who was named first, should be appointed by the Visitor. Harvey was so named by five out of the seven Fellows voting; and, after a dispute of which it is unnecessary to give an account, he was duly elected on receipt of a second letter from the King. A couple of days after his admission to the office, on April 11, 1645, Harvey summoned the Fellows into the hall and made a speech to them, to the effect that it was likely enough that some of his predecessors had sought the office of Warden to enrich themselves therefrom, but that his intentions were quite of another kind, wishing as he did to increase the wealth and prosperity of the College[1]. He finished his address to the assembled Fellows

[1] I would here remark that it was well perhaps for the College of Physicians that Harvey was, by the success of the Parliament, forced to vacate the office of Warden,

with an earnest appeal to them to cherish that mutual concord and amity amongst themselves, which recent occurrences, we may suppose, had tended to weaken. In the other pages of the Register for the period between April 1645 and the Midsummer of 1646, I find the name of Charles Scarborough, the *protégé* of Harvey, and afterwards frequently an office-bearer in this College; but there is little or nothing of special interest to us in the rest of the record, beyond the fact that Harvey appears to have attended the College meetings and so to have discharged his duties, amongst which the providing for the contingency of a siege and famine was one. Mr. Pettigrew ('Medical Portrait Gallery,' 1840) has put on record the fact that Harvey's signature is to be found in the Liber Computorum of Merton. The College Register, however, is not so enriched, as I can state upon the authority of Mr. E. Maunde Thompson, who compared the pages relating to Harvey's wardenship with the autograph MS. in the British Museum, when I took the Register up to London for that purpose[1].

otherwise he would, no doubt, have kept his word, and Merton College would have gained what the College of Physicians, or some others of his legatees, would have lost.

[1] Mr. E. M. Thompson has made another search for Harvey's missing MS. 'De Anatomiâ Universâ,' which Dr. Lawrence mentions at p. xxxi of his 'Harveii Vita' (ed. 1766), and which Dr. Willis tells us at p. vii of his Preface (ed. Willis) had then (1847) been twice looked for in vain. Mr. Thompson's search has also been equally fruitless; he writes to me thus, under date June 3, 1873:—'Harvey's "Anatomy" *was* once upon a time in the British Museum. In the first volume of the old MS. Catalogue of the Sloane MSS. (now marked Sloane MS. 3972 A), there is this entry on p. 57:—

"C. $\frac{(6)}{230}$.—Praelectiones anatomicae universales per me, Gulielmum Harveum medicum Londinensem Anatom. et Chirurg. Professor. Anno Dom. 1616, aetatis 37, praelect. April 1, 1617."

To which is added, "This is the author's foundation and first Lecture of the circulation in his own handwriting," and opposite to it is this note by Sir F. Madden, "In the place of 230 (which seems missing) Ayscough substituted the bracketed no. (6)." So you will see from this that the MS. was missing in Ayscough's time. I have ransacked our MSS. without finding any clue; so I think you may make up your mind that it was borrowed, and has gone the way of borrowed books in general.—E. Maunde Thompson.' Wood says ('Fasti Oxonienses,' ii. 6) of Harvey, 'But more in MSS. hath he left behind him, the titles of which you may see in the Epist. Dedicat. before a Historical Account of the Colleges Proceedings against Empyricks' (1684, London, Ch. Goodall). Moved by this authority, though Goodall only says that Harvey *designed* these treatises, I looked over a large number of medical MSS., assisted herein by Mr. Walter de Gray Birch, in the Sloane Collection of the British Museum, without the desired result. Subsequently I found that Harvey himself in 1650 (p.

Of Harvey's, as of Berkeley's sojourn in Oxford, we know little; little, indeed, has been recorded, with the exception of the somewhat uncertain gossip of the gossiping Aubrey. But what we do know of the place during those years which elapsed between the battle of Edgehill and 1646, makes us certain that scientific, and indeed any other work, must have been carried on in it under great disadvantages. We read of the plague, and of the 'morbus campestris,' described by a former Harveian orator and Linacre lecturer as desolating the town and driving people out of residence. It was, besides, a centre for military operations; and military life has been shown, by the experience of all ages (though this experience appears to have been lost upon the heedlessness and ignorance of this), to be out of harmony with the habits of men, old as was Harvey then (aet. 64–68), young as our undergraduates are now, who are, or who ought to be devoted to study. Whatever else of Aubrey's tales of Harvey I may disbelieve, I can believe that the words addressed to Charles Scarborough, 'Prithee leave off thy gunning and stay here,' are his.

If, however, we wish to have a real and truthful picture and image of Harvey before us, we must do by him as we have to do by Shakespeare, by Aristotle, by Butler, and several other great writers: we must lay our minds alongside of his, as it is revealed to us in his works. It is only the writings of great men which will bear or repay such treatment: no commentary nor any biography can give us the real and vivid sensation of having the men before us which we get from a perusal and reperusal of their books. Having used for this purpose what Mr. Tom Taylor has recently spoken of[1] as 'the invaluable three hours before breakfast,' I have come to persuade myself that I have obtained something like a trustworthy idea of what Harvey really was. Previously, however, to doing this, I gave Christian burial to much of what

163 and p. 502, ed. 1766; p. 148 and pp. 481, 482, ed. Willis) had recorded the loss of his 'adversaria multorum annorum laboribus parta,' and especially of his work 'De Generatione Insectorum,' when his house was plundered in the Civil War. Later again I came upon the following passage in Lower's work, 'Tractatus de Corde,' ed. 1669: 'Quid quod et Harveius si per aetatem et otium licuisset plura polliceri videtur ipse, Lib. de Circulat. Sanguinis, cap. 9. . . . . . *Sed quod maxime dolendum est et ille voto suo et nos spe nostra excidimus.*' Hence I fear there is now little hope either of recovering or of discovering the lost MS.

[1] See speech at Eighty-fourth Anniversary Dinner of the Royal Literary Fund, 'Times,' Thursday, May 29, p. 12.

Aubrey has left on record about him, feeling more and more strongly as I grew better aquainted with Harvey that—

'These were slanders: never yet
Was noble man but made ignoble talk.'

I will speak first of his scientific *character*, though it may seem strange to speak of scientific *character*, as *character* implies, perhaps, a moral element; and science, so far as it is really science, and based exclusively upon sound reasoning, has no moral element in it; reasoning, so long as it is sound, being of one kind always, and devoid therefore of all distinctive or personal factors. It is necessary for me to say that I do not forget that Harvey was but eighteen years junior to Bacon,

'Whom a wise king and Nature chose
Lord Chancellor of both their laws.'

But neither do I forget that the Novum Organon was published in 1620, subsequently to the discovery and actual demonstration of the circulation (see Dedicatio to the treatise 'De Motu Cordis'), if not to the publication of the treatise on the 'Motion of the Heart;' and that the Royal Society, with its motto, 'Nullius addictus jurare in verba magistri,' was a foundation of a much later date. And consequently, I think, we may feel justified in saying that, so far as the purely scientific factor of a man's nature can be said to have any distinctive or personal character at all, independence, or robustness, or manliness, whichever word we may like to choose, as shown in superiority to mere authority and the weight of great names, was a distinctive character of Harvey as a man of science. With Riolanus in full vigour, and Van der Linden growing towards maturity, as champions of antiquity, it required not a little manliness to assert, 'contra receptas vias per tot saecula annorum ab innumeris iisque clarissimis doctissimisque viris' (Riolanus was often thus spoken of), 'tritam atque illustratam' ('Dedicatio,' p. 5), the claims of simple Nature 'quâ nihil antiquius majorisve auctoritatis' ('Epistola Secunda ad Riolanum,' p. 123). This element of real manliness shows itself again, I think, in Harvey's power of abstaining from suggesting a rationale of what he felt he did not understand; as, for example, in what is known (out of England, at least) as the 'Problem of Harvey' (see 'De Partu,' pp. 132, 549, ed. 1766; p. 530, ed. Willis)—a problem which, I think, could not have been answered till the 'works and days' of Bernard[1]; and in the cases of several other problems instanced by himself (p. 132, 'Epistola Secunda ad Riolanum'), and hidden then, to use his own metaphor (p. 630, ed. 1766; p. 613, ed. Willis, 'Epistola Prima ad Horstium'), in the well of Democritus.

For the culture which Harvey had bestowed upon his literary faculties we have better evidence than Aubrey's, better even than that of two more trustworthy witnesses than Aubrey—Bishop Pearson, to wit, and Sir William Temple: we have the evidence of his own writings as to his familiarity with one of the greatest writers of antiquity. Bishop Pearson, as Dr. George Paget has reminded us (see p. 15 of his 'Notice of an Unpublished Manuscript of Harvey,' 1850), writing in 1664, but seven years after Harvey's death, and Aubrey (see p. lxxxii of 'Life,' by Dr. Willis, prefixed to the Sydenham Society's edition of his works, 1847), have told us of Harvey's high appreciation of Aristotle's writings; but in his own writings he refers to the Stagirite more frequently, I think, than to any other individual. And, as regards Vergil (the Latin author whom probably, if but one Latin classical writer could be saved from destruction, most men would choose to be that one, as Aristotle probably would be the similarly to be chosen Greek), Sir William Temple ('Miscellanies,' Part ii, On Poetry, p. 314) has told us that 'the famous Dr. Harvey, when he was reading Virgil, would sometimes throw him down upon the table and say he had a

[1] I refer to Claude Bernard's experiments on the influence of vitiated air ('Des Effets des Substances Toxiques et médicamenteuses,' 1857, p. 125), which show so plainly that organisms can attain a power of tolerance as against morbific agencies if time is allowed them to become gradually adjusted to such environment. The principle demonstrated in these experiments has been brought into greater prominence by Sir James Paget in his striking account ('Lancet,' June 3, 1871, p. 734), so interesting to all of us for other than purely scientific reasons, of his serious illness in 1871. As regards the 'Problem of Harvey,' the *foetus in utero* has been habituated to lowly arterialised blood; the blood of the umbilical vein is not scarlet in colour, and hence, I submit, may be explained the tolerance by a child which has come into the world but has not yet breathed in it, of conditions which entail death by suffocation in a child which, *having breathed air*, is exposed to them. This physiological principle has, among many other practical bearings, the practical value of furnishing an answer to the Philistine argument so often brought forward by Antisanitarians in favour of the retention of abuses, in the words 'See to what a good old age people live in the middle of it all!' The answer is, 'They have become habituated, and are living *in spite of*, not *because of* these surroundings: immigrants die in the process of acclimatisation.' Such persons, and indeed all persons, may read with profit Mr. G. H. Lewes' 'Physiology of Common Life,' vol. i. pp. 372–377, upon this subject.

devil.' It was a similar spirit which dwelt in Sir Philip Sidney, who never heard the famous ballad of 'Percy and Douglas' without feeling his 'heart moved more than with a trumpet.'

It may seem to some but a small matter to vindicate for our great discoverer claims to a familiarity with Greek; still, any one who will look at such passages as the one in the 'Exercitatio de Partu,' p. 553, where he speaks of the mischief done by meddlesome midwives, or other passages (pp. 116, 129, and 133, 'Epistola Secunda ad Riolanum;' p. 613, 'Ep. ad Slegelium'), will see, I think, that he had Greek in abundance at his command, and used it just when it helped him to express his thoughts more clearly and concisely than any other words at hand at the moment. He used it, in fact, like a man of sense and real learning, when the use of it would save him time or trouble—two things, of one of which he had all too little, whilst of the other he had all too much for his and our good. Let me add that, in the one authentic MS. which we now possess of Harvey's (No. 486, Sloane Coll., British Museum), a MS. never intended for publication, and consisting but of rough notes for lectures to be delivered, I find that he employs Greek words in several places (e. g. pp. 65, 66 and 87)[1].

His style has been spoken of as being more or less inelegant and unadorned; and the Latin tongue which he used lends itself but grudgingly and awkwardly to the purposes of science, being strictly a political language, habituated and framed to describe the march of the legions, the disputes of the forum, or the denunciations of the moralist. Still, Harvey's style has always an impressiveness and solidity of its own; and sometimes, as for example in the glorious eighth chapter, 'De Motu Cordis,' it rises into real eloquence where a great occasion justifies the use of repetitions, of antitheses, and abundance of metaphors. But, though the use of stilted phraseology was common enough among Harvey's contemporaries, and though

[1] I have no sympathy with the eagerness which scientific men sometimes (see Fritz Müller, 'Für Darwin,' p. 28; Dallas' Engl. Trans. p. 42) show in repudiating a knowledge of Greek, but on the other hand I should be sorry to be thought to overrate its value. I am so far from doing this that I incline to thinking that, when through want of leisure or of means, or through some other deficiency, a young man cannot add on more than a second foreign language to his acquirement of Latin (which I presuppose), that second foreign language should, in the case of Englishmen, be, for linguistic and educational, as well as for more lowly practical reasons, not Greek but German.

his imagination was vivid and active enough, his study (for to this perhaps we may ascribe it) of the excellent models mentioned saved him, as such a study can save a man, from falling into the use of false or extravagant imagery.

Harvey, besides the advantages accruing from acquaintance with the great minds of the past, enjoyed also those which may be gotten from familiar intercourse with great contemporary minds. These advantages constitute in themselves a second education; and they were at Harvey's command for the period of more than forty years during which he was prominently before the public. It is recorded as one of the many distinctions of John Greaves (see 'Life,' by T. Smith, 1699, p. 44), the once celebrated astronomer and antiquary, and a man whom we can well believe to have done more, as a Fellow of Merton, than give a silent vote for Harvey when he was chosen Warden, that he was one of the friends of Harvey as well as of Archbishops Laud and Usher. It is indeed in a letter to this latter dignitary, and in answer, we may suppose, to an appeal from him on behalf of Harvey, that we find John Greaves pledging himself in a postscript, under date Sept. 19, 1644, the year before Harvey's election as Warden of Merton, to the following effect: 'If I may serve Dr. Harvy (*sic*) I shall be most ready either here or at Leyden to do it.' (See 'Life of James Usher,' by Richard Parr, D.D., 1686, p. 510)[1]. His well-known connection

[1] I owe this last reference to the 'Biographia Britannica,' *sub voc.* Greaves. For a further account see Wood's 'Athenae Oxonienses,' vol. iii. ed. Bliss, 1817. To the former of these sources I owe a second and more interesting reference, viz. to Birch's edition, 1737, of the 'Miscellaneous Works of John Greaves,' where, at the end of Greaves' 'Treatise on the Pyramids' (pp. 136, 137), we have given us an account of a conversation between him and Harvey. It runs thus: 'That I and my company should have continued so many hours in the Pyramid and live (whereas we found no inconvenience) was much wondered at by Dr. Harvey, his majesty's learned physician. For, said he, seeing we never breathe the same air twice, but still new air is required to a new respiration (the *succus olibilis* of it being spent in every expiration), it could not be but by long breathing we should have spent the aliment of that small stock of air within, and have been stifled; unless there were some secret tunnels conveying it to the top of the Pyramid whereby it might pass out and make way for fresh air to come in at the entrance below.' The Fellow of Merton was not wanting in an answer to the future Warden, assuring him, amongst much else not wholly correct, that 'as for any tubuli to let out the fuliginous air at the top of the Pyramid none could be discovered within or without.' Harvey replied, 'they might be so small as that they could not easily be discerned, and yet might be sufficient to make way for the air, being a thin and subtil body.' It has, indeed, been left to our own times and to v. Pettenkofer to demonstrate and exhibit the action of the capillary pores in the

with the court must have constantly brought him into relation with the statesmen of those stormy times. His legacy to his 'good friend Mr. Thomas Hobbs, to buy something to keepe in remembrance' of him, is touching, even if trifling, evidence in the same direction.

Travel, which even in our day confers a kind of culture peculiar to itself, must have been doubly necessary in days when, in the absence of the steamship and the railway, an insular position must have kept its inhabitants very nearly as inaccessible to 'the thoughts that move mankind,' as it had happily kept them to the Armada. Sir George Ent's interesting and entirely trustworthy account of the interview with Harvey which resulted in the publication of the treatise 'De Generatione,' will show any one who will consult it that Harvey had drawn from his opportunities an insight into what might be expected, and what since his time to some extent has been realised, from enlarged opportunities of observing not only 'men, manners, cities, climates, governments,' but also the wonderful facts of the unequal allotment, in the various parts of the earth, of useful inorganic products, and of that mystery of mysteries, the distribution of organic life. (See 'Works,' ed. 1766, p. 162; ed. Dr. Willis, p. 146.)

Having been thus fortunate in securing for himself all the advantages which the various educational agencies of his age would furnish, he added on to all that they had effected, or could effect, the yet more elevating and glorious discipline of long sustained and finally successful labour. He attained a position of mental dignity in which he could feel neither unduly anxious for the applause of his compeers, nor unduly moved by the reproaches and misrepresentations of his enemies (see 'Dedicatio,' p. 164; 'Epistola Secunda

constituents of a mass of 'solid' masonry (see his 'Beziehungen der Luft zu Kleidung, Wohnung und Boden,' 1872, pp. 41–45, and especially the figures p. 42). What Leeuwenhoek and Malpighi did for the capillaries of the animal body in supplementation of Harvey's work, and in correction of one of his few errors, that v. Pettenkofer has done in supplementation of Harvey's suggestion as to 'tubuli so small as that they could not easily be discerned' in structures like the Pyramids. It is, perhaps, not more than curious to note that Harvey was equally right in suggesting the existence of larger 'secret tunnels': an account of the discovery and opening of them may be found in Colonel Howard Vyse's 'Operations carried on at the Pyramids of Gizeh,' 1837, i. pp. 3, 263, 285–288; ii. pp. 160, 161; and an amusing history of the inconveniences endured in the interior of the Pyramids previously to the discovery of these 'air-channels' is given by Colonel Coutelle in 'Description de l'Egypte, Antiquités, Mémoires,' ii. p. 46, 1809.

ad Riolanum,' p. 109); the impact of these opposite forces resulting, however, in much benefit for mankind, as without them Harvey might, it is likely enough, have delayed the publication of his works indefinitely. Being self-contained without being self-conscious, he was yet, like all men of real genius, large-hearted and sympathetic. Whilst he could, in a spirit of perhaps a little overstrained charity, make excuses (see p. 614, 'Epistola ad Slegelium') for the pestilent and irrepressible Riolanus, he would, we may be also sure, have felt an emotion of gratitude upon each of the many instances in which his own true-hearted adherents, Sir George Ent and other Fellows of this College, fought his battles for him, and vindicated for him successfully and during his own lifetime his own irrefragable claims. And I can believe that, answering to the character of the dignified, stately, and high-minded man so well drawn by the author whom he often quotes (Aristotle, 'Eth. Nic.' iv. 3 (7)), and considering himself worthy of great respect, being worthy of it, he would not have looked disapprovingly upon our attempt to show him respect by the Tercentenary Memorial to which you, Sir, have lent the sanction of your name. I can further conceive of Harvey as entirely sympathising with the men who have now in their hands the torch of knowledge which once passed through his, of applauding without any shadow of jealousy the work of the many workers who in these days are going over the ground trodden by him under far less favourable circumstances and with far less assistance from ancillary [1] sciences and their various and still novel instruments and methods. The same spirit which caused him repeatedly to say (as, for example, to Sir George Ent, p. 163; to Horstius, p. 630), 'haec cum mirâ, ut solet, promptitudine effundens,' that he doubted not that much now hidden in darkness would be brought to light by the indefatigable industry of the coming age; the same spirit which dictated the provision in his will bidding 'his lo. friend Mr. Doctor Ent' sell certain of his 'books, papers, or rare collections,' and, 'with the money buy better,' would have caused him, could he have been

[1] Such an experiment, for example, as that put on record by Professor Haughton ('Principles of Animal Mechanics,' 1873, p. 151), as performed by Professor Macnamara with his assistance, and as showing that the time occupied by absorption, circulation, and secretion occupies less than four minutes, requires the employment of iodine; and iodine has been discovered and isolated but some sixty-two years.

amongst us, to point out, as a matter for congratulation, in how many directions his discoveries have been extended and added to, and how well replaced had been the many works the loss of which had been so 'crucifying' to him.

There was not in Harvey's mind that defect in the way of a deficiency of interest in theological questions which constitutes in the minds of some eminent scientific, and some eminent literary, men such a lamentable void. He has, on the contrary, in several places taken pains to state his views upon this highest of subjects. To one of these passages (from the work 'De Generatione,' Exercit. Quinquagesima, p. 385, ed. 1766; p. 370, ed. Dr. Willis), as Mr. E. B. Tylor has pointed out to me, Professor His of Leipzig, a worker whom Harvey would have hailed as a colleague, has referred in one of his always excellent papers, published in the 'Archiv für Anthropologie,' Bd. iv. 1870, p. 220, on 'Die Theorien der geschlechtlichen Zeugung.' It is just in the investigation of the problems indicated in these last words that, as has often been remarked, the question of the existence of other than purely material forces presses itself most closely upon the mind; and hence, perhaps, the repetition by Harvey of his views regarding it, more than once or even twice, in his treatise just referred to (see Exercit. 49, p. 730; Ex. 50, p. 385; Ex. 54, pp. 419, 420). These statements are all to the same purpose. I have chosen one of them—the last one of the three just cited (not the one quoted by Professor His)—to repeat here, because, besides its philosophical and other interest, it has some literary claims upon our attention, it being not quite impossible, considering its line of thought and arrangement of words, that Pope, who borrowed on all sides, and made acknowledgments on none, may have had it before him when he composed his Universal Prayer. It runs thus:—

'Nempe agnoscimus Deum, Creatorem summum atque omnipotentem, in cunctorum animalium fabricâ ubique praesentem esse, et in operibus suis quasi digito monstrari; Cujus in procreatione pulli instrumenta sint gallus et gallina. Constat quippe in generatione pulli ex ovo omnia singulari providentiâ, sapientiâ divinâ, artificioque admirabili et incomprehensibili exstructa ac efformata esse. Nec cuiquam sane haec attributa conveniunt, nisi omnipotenti rerum principio; quocunque demum nomine id ipsum appellare libuerit: sive mentem divinam cum Aristotele; sive cum

Platone Animam mundi; aut cum aliis Naturam naturantem; vel cum Ethnicis Saturnum aut Jovem; vel potius, ut nos decet, Creatorem ac patrem omnium quae in coelis et terris; a quo animalia eorumque origines dependent, cujusque nutu sive effato fiunt et generantur omnia.' ('De Generatione Animalium,' Ex. 54, pp. 419, 420, ed. 1766; p. 402, ed. Willis.)

I have detained you far too long; but, feeling that my praise of Harvey has been all too feeble, I am anxious, in ending, to employ in honour of Harvey certain lines of singular beauty and force which, though composed in commemoration not of him, but of another famous Englishman, may nevertheless be applied to him with a singular appropriateness:—

'Remember all
He spoke among you, and the man who spoke;
Who never sold the truth to serve the hour,
Nor paltered with Eternal God for power;
Who let the turbid streams of rumour flow
Thro' either babbling world of high and low;
Whose life was work, whose language rife
With rugged maxims hewn from life;
Who never spoke against a foe.

. . . .

Whatever record leap to light,
He never shall be shamed.'

XLII.

THE MODIFICATIONS OF THE EXTERNAL ASPECTS OF ORGANIC NATURE PRODUCED BY MAN'S INTERFERENCE,

A LECTURE DELIVERED BEFORE THE ROYAL GEOGRAPHICAL SOCIETY, MAY 12, 1879.

THE modifications of the external aspects of organic nature produced by man's interference form so large a part of the results of all human activities whatever, that the very first thing to be said in a single evening's lecture on the subject should consist in a specification of the particular spots in that vast area which the speaker proposes to touch upon. I propose, then, with your permission, firstly, to glance at certain of the alterations, positive and negative, in the landscape of our own country, which we ourselves and our fathers before us have intentionally or unintentionally produced; secondly, to notice a few of the many alterations produced by disforesting in our own and other countries; and thirdly, to show what our knowledge as to the localities to which the parent stocks of the majority of our domestic animals and of our cultivated plants may be assigned, implies, as to the modifications of other regions of the world's surface which man has produced by his processes of importation and acclimatisation. Room may perhaps be found for a few speculations as to the future after these details as to the past and present.

I do not propose to enter into the large question of the extent to which man may, with any propriety, be spoken of, as he has been, as a 'geological agency,' a 'telluric' or a 'cosmic' agent; and I will at this very outset of my lecture profess that I think man's power of modifying the climate of the earth upon which he lives must be considered [1], when all the facts of the case are taken

[1] Upon this large question, one only of many large questions which the various details of this subject suggest, and by which, even when most in the concrete, they

into account, to be confined within much narrower bounds than some writers are willing to admit. It is possible to overstate the extent to which man can go in the direction of exhausting the soil by wasteful or neglectful agriculture, and to fall over-easily, not to say over-willingly, into despair as to the restoration to fertility and political consideration of countries so mismanaged. And if it is possible to overstate man's influence upon the dry land and its inhabitants, it is necessary to be very cautious as to asserting for him any power of altering, except infinitesimally, the vast area of marine life. Now, as the surface of the sea is to that of the land as four to one, and as I feel somewhat desirous of showing that the extent of the subject I have chosen is not quite so disproportionately large in relation to your time and my abilities as the mere words in which it is announced might seem to indicate, I should like to dwell a little upon this delimitation of it before entering upon the subject itself.

For one of those striking suggestions 'qui font penser si elles

excite general interest, it is well to hear Mr. Robert Rawlinson as he spoke in a lecture on Meteorology, delivered November, 1868, before the Royal Engineers at Chatham (p. 7):—

'It is certainly true that man modifies climate over tracts that have been cultivated; but it is asserted, further, that in various parts of the world, through cutting down forests, and in consequence of other operations, the works of man, climate has been so far modified as to have had its character absolutely changed. "The Thames is not now frozen over as in times past," one place has more rain than formerly, another place less, and so on. If by assertions such as these it is intended to be implied that any works of human hands have actually altered the current course of nature, I must meet such allegation with a positive denial. The most stupendous of human works can affect only the comparatively small and narrow space of the earth's surface upon which they may have been executed. Evaporation has only an indirect and incidental reference to the land—its real dependence being on the great ocean and the greater sun. And so, while man may exert an influence upon climate over the little area of his operations, his works can avail nothing to affect the grand features of nature even over that small area, or to disturb the majestic scale on which she accomplishes her purposes. Cosmical meteorology is unaffected, and must continue to be unaffected by human agency. The powers of man can never seriously modify the heat of the sun, cloud, rain, or climate, as these have reference to the world at large; all statements, therefore, which would assign cosmical atmospheric effect to the cutting down of forests, to land drainage, land cultivation and such-like agencies, must be treated with practical disregard.'

For other discussions on the same subject, see Reclus, 'The Ocean,' sect. ii. pp. 93-95, *ibique citata*: Unger, as regards Egypt, 'Sitzungsberichte Akad. Wiss. Wien,' xxxviii. pp. 89-93, 1859; De Candolle, 'Hist. des Sciences,' 1873, p. 412; Link, 'Urwelt und Alterthum,' ii. pp. 128-160, 1822.

ne font pas croire,' has been made to the effect that man's interference has been potent, even over the sea, to an extent which men of science have not usually claimed, and poets have denied to be possible. Mr. G. P. Marsh, the author of a well-known work on 'The Origin and History of the English Language,' 1862, as well as of the highly interesting work on physical geography which appeared in 1864, under the title of 'Man and Nature, or Physical Geography as modified by Human Action,' and as a second edition, ten years later, under the title of 'The Earth as modified by Human Action: a new edition of Man and Nature,' suggests in this latter work that the phosphorescence of the Mediterranean, unknown to, or at any rate scarcely noticed by the ancient writers, may have been greatly increased since their days through human action in the way of extirpating the whale. 'Is it not possible,' writes Mr. Marsh [1], 'that in modern times the animalcula which produce it (the phosphorescence of the Mediterranean, the most beautiful and striking of maritime wonders) may have immensely multiplied, from the destruction of their natural enemies by man, and hence that the gleam shot forth by their decomposition or by their living processes, is both more frequent and more brilliant than in the days of classical antiquity?' In a more utilitarian spirit Middendorff, in his 'Sibirische Reise [2],' points out that a continuance of the wasteful destruction of the whalebone whale in the northern seas will render it impossible to utilise for man's profit the innumerable small crustacea and mollusca of the Polar seas which that whale converts into train oil! The profligate inconsiderate slaughter again by the Kolushes of the sea-cow, *Rytina Stelleri*, a sirenian 'whale' of the region of Behring's Straits, which lived upon sea-weed, has reduced these savages to the necessity of using this self-same sea-weed for manuring their potatoes, which useful vegetable, however, gives them a much less savoury and sustaining food than was manufactured, so to say, for their forefathers in the organism of the sea-cow they extirpated. It is perhaps a little ungracious to point out that the most elegant of these three correlations and interdependencies is not so definitely demonstrable as the other two. In the first place, it may be objected as regards Mr. Marsh's suggestion, that the Mediter-

[1] *Loc. cit.*, 1st ed. p. 114; 2nd ed. p. 104.
[2] Band iv. t. 2. p. 848, 1867.

ranean whales[1], not comprehending in their number the right whale, *Balæna mysticetus*, are not whales which would either themselves prey so largely or exclusively upon the small invertebrata alluded to by Middendorff,—to say nothing of those very much smaller, upon which the phenomenon of phosphorescence so much more largely depends,—or be themselves so unrelentingly pursued by man for the sake of their oil. And secondly, without dwelling upon any such quantitative relations as the size of the microscopic 'animalcula' just alluded to may suggest, it is clear that the square area of the Mediterranean makes up a space for the extirpation from which even of so large an animal as a 'whale' a very considerable fleet would have been required. We know the numbers and the tonnage of the ships which, till the discovery of petroleum[2] in large and available quantities, formed the whaling fleets of quite recent times, 1849–1850, the American whalers in the Sea of Okhotsk alone numbering 250[3] three-masted vessels, with a minimum tonnage of 500 tons; but of any such whale-slaying machine having ever existed in the Mediterranean we have, within my knowledge, no record whatever. Now the capacity of the ancient writers for 'not marking withal' matters of interest to the modern naturalist, can scarcely be overrated; but it did not affect matters relating to war and the chase so much as such trifles as Stonehenge and the peaceful though colossal aqueduct near Nîmes[4]. And as a matter of fact, we find in those writers abundance of references made to the means employed for the capture of the tunny, a form of the chase which is in no way more exciting, more useful for illustration and metaphor, nor even more lucrative, than would that of the whale have been if it had been carried on to any appreciable extent in the large sea on the shores of which so much of the history of the world has been written and acted. The Greek word *κητεία* means a fishery, not of Cetacea, but of tunnies.

[1] The principal larger cetacea of the Mediterranean are piscivorous dolphins, such as *Delphinus tursio*, *Delphinus globiceps*, *Delphinus orca*; it is at least open to doubt whether such whales as the Balænopteræ and the sperm whale can be considered as anything more than occasional visitants of Mediterranean waters. See Wagner, 'Die Geographische Verbreitung der Säugethiere,' 'Abhandl. d. 2 Classe d. Ak. d. Wiss. München,' iv. Bd., Abth. i.; and Sundevall, 'Die Thierarten des Aristoteles,' 1863, p. 88; 'Aristoteles' Thierkunde,' Aubert und Wimmer, Bd. i. pp. 73–74, 1868.

[2] See Marsh, 'The Earth as modified by Man's Action,' 1874, p. 103.

[3] See Middendorff, *l. c.*, p. 849.

[4] Marsh, *l. c.*, pp. 426–427.

A story relating to the natural history of these true 'fishes' will show, in the way of a parallelism, the facility with which mistaken views may obtain currency, 'si modo imaginationem feriant aut intellectum vulgarium notionum nodis astringant,' quantitative measurements, statistics, relative proportions of masses to other things, and even literature itself, notwithstanding. In the Oxford University Museum we have a large skeleton of a tunny (*Scomber thynnus*), brought from Madeira, before my time, by my friend Dr. Acland. A foreign naturalist, whose name, under the circumstances, I think well to withhold, but whose reputation is commensurate with his very extensive performance, going over the Museum with me one day, remarked, after paying a not undeserved compliment to the skeleton, 'That fish never came from the Mediterranean.' I answered that, as a matter of fact, it had belonged to an ocean-going individual; but I also asked how it was possible to differentiate a Madeiran from a Mediterranean specimen. My friend answered, 'The Mediterranean is too closely fished by man to allow of any tunny attaining such dimensions.' I was silent, though very vivid recollections of long, however pleasant, days of coasting on those shores, without meeting any considerable number of vessels, or passing, as on the south coast of Asia Minor, any considerable towns except in ruins, might have conspired with my recollections of St. Paul being driven up and down for fourteen nights in Adria, to make me question this explanation. Some time after, I found that Cetti records tunnies of no less than from 1000 to 1800 lbs. as being caught now-a-days in the Sardinian fisheries [1]!

The results of investigation into the extent to which man's interference may have told injuriously upon the propagation of fish smaller in size, if not smaller in importance, such as the herring, may possibly show us that here too we have exaggerated our own powers for mischief. Not only is the sea a large field, but cyclical oscillations in the 'Frequenz' of its inhabitants are at least as possible, irrespectively of our interference, as are the similar variations observable in air-breathing animals; and many an animal, as for example the horse in South America, has become extinct even in recent, not to speak of earlier geological times, owing to quite other than human agencies. Man has no monopoly of

[1] See Lenz, 'Zoologie der alten Griechen und Römer,' 1856, p. 485.

destructive agencies, neither, if he had, would that, as it seems to me, prove that, 'though[1] living in physical nature he is not of her, that he is of more exalted parentage, and belongs to a higher order of existences.' He is not, in strictness of language, a 'cosmic,' a 'telluric,' a 'geological,' nor a 'supernatural agency.' He may ultimately obtain, as prophesied by Mr. Wallace[2], such a mastery of the dry land as to supersede on that portion of the world's surface the agency of natural selection; but he cannot even there effect cosmical changes in the climate, and as regards the sea, it is possible enough, as Mr. Moseley has suggested on the two concluding pages of his 'Notes by a Naturalist on the *Challenger*,' that when the present races of animals, plants, and men shall have perished, the deep-sea animals, at least, if not those of higher levels, 'will very possibly remain unchanged from their present condition[3].'

[1] Marsh, *l. c.*, p. 34.

[2] 'Natural Selection,' p. 326.

[3] Having been compelled to express dissent from Mr. Marsh's suggestion as to the phosphorescence of the Mediterranean having been a less striking phenomenon in ancient than it is in modern times, I cannot forbear to pay my poor meed of thanks to this writer for the pleasure and instruction which his works have afforded me. The 'Kulturpflanzen und Hausthiere' of Herr Victor Hehn resembles Mr. Marsh's work in dealing with the subject of man's action on organic nature in a way which attracts the attention and stimulates the thought at once of the politician, of the literary man, and of the man of science. I expressed my opinion upon the merits of the first edition of this work in the 'Academy' of August 15, 1872. A third edition of it appeared in 1877, considerably enlarged and improved. And it may be observed that, for dealing at all adequately with this subject, and indeed for avoiding very gross blundering in so dealing with it, a man must have some knowledge not only of purely scientific subjects, of the facts of history on the large scale, and of the results at least of philological inquiry, but also of the power which commercial legislation and commercial enterprise have for altering the distribution of the various vegetable and animal articles of trade; otherwise he may fall, as some have fallen, into the error of supposing commercial results to have been produced by changes in the laws, not of man, but of climate. I make this remark for, among other purposes, that of introducing another remark to the effect that it is much to be regretted a fresh edition of Dureau de la Malle's 'Économie Politique des Romains' should not be brought out in these days: it is a work of permanent value, though it bears the date of 1840. As works of a more exclusively scientific character, but still easily intelligible to persons possessed of a mastery of the rudiments of botany and zoology, and of cardinal importance in researches such as these, I will specify:—

De Candolle, 'Géographie Botanique raisonnée,' 1855.

Unger's 'Botanische Streifzüge,' in the 'Sitzungsberichte' of the Vienna Academy from 1857 to 1859 inclusively.

Isidore Geoffroy St. Hilaire, 'Histoire Naturelle Générale des Règnes Organiques,' tom. iii., 1862.

K. E. von Baer, 'Reden und Studien aus dem Gebiete der Naturwissenschaften,' four

Beginning at home, let us consider first of all what are the most prominent changes which man has effected in the landscape,

octavo volumes which appeared in the years 1864, 1873, and 1876, and contain much of geographical as well as of other interest. This illustrious scientist was for some years from 1839 onwards concerned, together with v. Helmersen, in bringing out, at the cost of the St. Petersburg Academy, a periodical, 'Beiträge zur Kenntniss des Russischen Reiches.' In one of the volumes (xviii. 1856, pp. 111–115) of this periodical, a short paper by v. Baer appears, the purport of which is shown by its title, 'Die Uralte Waldlosigkeit der Süd-russischen Steppe,' 'The Aboriginal want of Wood on the South Russian Steppe.' This paper was written in supplementation of a paper which had appeared in the fourth volume of the same periodical, 1841, pp. 163–198, with the same object of deprecating a useless and essentially nugatory attempt to make these steppes timber-bearing. From it I will give an extract, partly because it is so characteristic of the manner of the great biologist, and partly or mainly because it shows how pure natural history can be brought to bear upon political questions and may save a Government from engaging at great expense in chimerical undertakings. V. Baer says, *l. c.*:—'At that time (1841) I had forborne to bring up a piece of evidence (in favour of the South Steppe never having been wooded) which is much older than Herodotus; and the present communication has only just the purpose of putting out this evidence, for doing which I have had no earlier opportunity. This piece of evidence is furnished by the squirrels. They are found throughout the Russian empire, so far as trees are found to grow, even in the Caucasus, but with the exception of the Crimea and Kamtchatka, although both these peninsulas have the food which the squirrel wants, and the south coast of the Crimea has it in great abundance. Now from these facts the following conclusion can clearly be drawn, namely, that when these animals reached the southern borders of the forests in South Russia, and the eastern borders of the forests in Siberia, the wide expanse of the open South Russian steppes and also the bare levels northward in Kamtchatka were already in existence. When was it, it may be asked, that the squirrels came to these borders of the forests? I don't know, but that they did come to them before any historical period nobody will be inclined seriously to dispute.'

Oscar Peschel, in his 'Neue Probleme der vergleichenden Erdkunde,' 1876, p. 1881, adds in explanation of this curious and convincing argument, 'A climbing animal dependent for food upon seeds of the trees could not of course travel across the sunny plains of grass; and consequently the South Russian districts in question must have been treeless ever since there were squirrels on the south boundary of the Russian forests; and there can scarcely be any doubt that they were there thousands of years before the time of Herodotus.' Oscar Peschel gives no specific reference to v. Baer's works: and v. Baer himself, or his printer, curiously, a wrong one in his 'Autobiography,' p. 644. Nor have I found any reference to it in Professor Stieda's 'Karl Ernst von Baer, eine biographische Skizze,' 1878. I have therefore another justification for the giving of these details, and am glad if I have thus saved others trouble which I had to take for myself, not unhelped, however, herein, by the staff of the Bodleian Library.

If Oscar Peschel has made one trifling omission, he has *per contra* made some of the most important additions to geographical and anthropological knowledge, separately and combined, which have been made since the time of Ritter. I need scarcely specify his

'Völkerkunde,' 1874.

'Abhandlungen zur Erd- und Völkerkunde,' 2 vols., 1877–1878.

so far as the landscape is made up of organic elements, of our own country. I have not undertaken, and shall not attempt to speak

'Physische Erdkunde,' of which three fascicles have appeared in the present year.

The general principles to be found expounded in the works above specified have found a practical application in the particular question, Are the countries along the shores, and especially the eastern shores, of the Mediterranean to be looked upon as having been exhausted by man's interference with them in the way of agriculture, and so robbed of any chance of political rejuvenescence? And with this question is connected that which asks whether any perceptible change of climate has been effected in the same regions by the same agency. The literature of this controversy, which has been carried on obviously enough by partisans filled at least on one side with political bias, is, if we give only the most important memoirs, not very extensive, and may perhaps usefully find a place here.

C. Fraas, in his 'Klima und Pflanzenwelt,' 1847, takes the pessimistic view, which

J. P. Fallmerayer, in a review published in the same year apparently, and republished in his 'Gesammelte Werke,' 1861, ii. 462, endorses with a bitter readiness.

C. Fraas, in the 'Geschichte der Landbau und Forstwissenschaft,' München, 1865, had the opportunity of again expounding his opinions, p. 350 *et passim*, in his account of Liebig's views. Those views are to be found in

Liebig, 'Natural Laws of Husbandry,' Eng. Trans., 1863, and in his 'Chemische Briefe,' the ninth edition of which bears date 1878.

Oscar Fraas, possibly or presumably a relative of C. Fraas, from certain passages in his 'Aus dem Orient,' 1867–1868, would appear to hold similar views to those of his namesake; he speaks (vol. i. p. 213) in defiance of Arago's views, as expounded in 'Œuvres,' vol. viii., 'Notices Scientifiques,' vol. v. ed. 1859, p. 222, of a 'verändertes Clima der Nilländer,' and says (p. 215), what will be read with some surprise by Indian officials, 'Heutzutage erlahmt die Energie selbst eines kräftigen Europäer's unter der Sonne von Egypten . . . man erschlafft, wird träge und faul, man fängt an zu bummeln!' An excellent answer to all this is given by

Theobald Fischer, 'Beiträge zur Physischen Geographie des Mittelmeerländer, besonders Siciliens,' 1877, p. 154, *usque ad finem libri*, p. 167.

Fr. Unger, in his 'Wissenschaftliche Ergebnisse einer Reise in Griechenland,' 1862, has dealt similarly with this question at the conclusion of his small but excellent memoir, pp. 187–211.

The views of Victor Hehn, and those of the recently deceased botanist and author of an authoritative work, 'Die Vegetation auf die Erde,' 1872 (translated into French in 1877 by Tchihatcheff), viz. Grisebach, may be given in the words of the latter, when reviewing the former in the 'Göttinger gelehrte Anzeigen,' 1872, xlv. p. 1767. With these views we agree. They run thus:—'Mit Recht verwirft er die Meinung dass die klassischen Länder erschöpft seien und einer Erneuerung ihrer ehemaligen Blüthe keine natürliche Grundlagen mehr böten. Er trifft das Wesen der Sache, indem er sagt, dass ihr Klima, im Grossen aufgefasst, nicht vom Boden und seiner Vegetation, sondern von "weitgreifenden, meteorologischen Vorgängen" abhänge, die durch ihre geographische Lage bestimmt, "von Afrika und dem atlantischen Meere bis zum Aralen und Siberien reichen." Ebenso muss man sein eingehendes Verständniss dieser Frage anerkennen, wenn er im Bereich der Agrikultur-Chemie sich gegen die Ansicht ausspricht, dass der Boden Südeuropas durch seine alte Kultur an mineralischen Nahrungsstoffen erschöpft sei. Wie die lombardische Ebene durch die Alpenflüsse mit frischen Silicaten und Kalksalzen gespeist wird, so liefern die so

of such changes as those which the embankment of our rivers has effected, referring those of my hearers who may feel an interest in this particular change to Sir Christopher Wren's disquisition upon the subject, which may be found with very much else very well worth reading in the 'Parentalia,' p. 285. But I have to say that changes of proportionately equal magnitude have been effected in our landscape by the interposition of man in the way of introducing into it trees which, though now naturalised, are demonstrably not indigenous to our soil. The most striking of these changes are those which have been effected by the introduction of the common elm, *Ulmus campestris;* next, if indeed not equal in magnitude, those effected by the introduction of certain coniferæ; and then, at a long distance behind as regards numerical importance, those effected by the introduction of the horse-chestnut and the sycamore. I do not of course forget that such trees as the walnut, and a host of other trees which are now entering into the picturesque, if not into the economical aspect of Great Britain, are as foreign to our soil as their names remind us they are; but I am not delivering a treatise upon our forest trees, and I shall confine myself within the limits which the three or four trees or orders of trees specified in the preceding sentence mark out for me. Let me begin with the simpler cases, those of the horse-chestnut and the sycamore first. I should indeed be ungrateful, living as I do within such easy sight of the beautiful, if not unrivalled, horse-chestnuts of New College Gardens, if I did not express my sense of gratitude to the men who introduced that tree into England. There is, of course, as little question as to its non-indigenousness as there can be as to its beauty. Botanists, however, differ as widely as possible as to what its native land may have been. I have not been able to satisfy myself that Hehn, l. c., pp. 348 and 457, is right in saying that we owe the introduction of this tree into Europe to the Turks. All but certainly this was not the case if D. Hawkins, as cited by Fiedler in 'Reise durch alle Theile des Königreiches Griechenlands,' 1840, vol. i. p. 649, is right in saying that this tree grows wild on Pindus and Pelion. There are not wanting species on either

manigfaltig gegliederten Gebirgeketten, welche die Länder am Mittelmeere erfüllen, aus dem Innern ihrer Felsmassen unerschöpfliche und durch das fliessende Wasser stetig ausgebreitete Vorräthe, um die Erdkrumen der Thälen und Tiefebenen immer wieder auf Neue zu befruchten.'

side of the Greek Archipelago which no naturalist would divide or bifurcate, nor, I imagine, has the Greek Archipelago existed in its disconnecting discontinuity as long as the species *Æsculus hippocastanum.*

The sycamore is another undoubtedly non-indigenous tree, but it is thoroughly naturalised and abundant in certain parts of England; and notably in the Lake District it forms a very characteristic feature of the landscape, when it is massed round the equally distinctive old farm-houses. In the Lake District its leaves have assumed a somewhat darker colour than they ordinarily bear in the southern and midland counties; and its bark often exhibits what some naturalists would call a mimetic analogy to that of its fellow-countryman the Oriental plane. The sycamore has yet other claims upon our attention, as the readiness with which its seeds take root might have long ago destroyed, even to the eyes of the least observant, that 'idolon theatri molestissimum et ineptissimum' which taught that if a plant could be proved to be non-indigenous in a country it was useless to expect it to flourish there[1].

I will now turn to the Coniferae. In another place[2], I drew attention to the well-known and universally accepted fact, that till comparatively recent times the Scotch fir (*Pinus sylvestris*), the yew (*Taxus baccata*) and the juniper (*Juniperus communis*), had been the only representatives in these islands of the natural order *Coniferae.* I did not dwell then, and I will not dwell now, upon the greatness of the difference which has, in the last three hundred years, been effected in the general aspect of our country by our successive importations of the spruce, the larch, and the silver fir from other European countries, and the multitudinous trees belonging to the same order from North America, from North India, from California and Mexico, from Japan, from China, and from Chili, the names of which 'plants of the fir tribe suitable for the

[1] For an example of the operation of this notion, so opposed to the most obvious facts, see 'Viti (Fiji), by Berthold Seemann,' p. 426, where, apropos of the statement 'the cotton plant is not indigenous in Fiji,' we have the following note:—

'Most of the newspapers took this fact to be a serious drawback to the successful cultivation of cotton, quite forgetting that cotton is not indigenous to the United States and many other countries in which it flourishes. I made exactly the same statement (cotton is not indigenous in Fiji), but added that, notwithstanding, it had become almost wild in some parts, so well is the country adapted for its growth.'—B.S.

[2] 'British Barrows,' p. 724. Article XVII, pp. 326.

climate of the United Kingdom, cultivated by . . . . nurserymen and seed merchants,' fill up some sixty-six pages in a sale catalogue now before me. Any traveller, by rail or otherwise, can appreciate the greatness of the alteration which has been effected by man on nature, if he will but bear in mind the three trees just specified, and recollect as he sees the silver fir spreading out with its airy interspaces in the sky-line, and the larches and spruces clothing the hill-side in acres upon acres, that these trees were as little known to the untravelled Englishman of the times of the Tudors as were the 'Weymouth' pine, the Deodara, the Wellingtonia, or the Araucariae. The statesman, indeed, can read something of the political and commercial history of this kingdom in the trees which speak of the various countries, farther distant apart from each other than are 'China and Peru,' with which England has successively come into *rapport;* and the changes which he has suggested to him are scarcely, if at all, less complicated than those which the naturalist can show to have been similarly set up in the world of lower life represented by birds and insects. Since I wrote as above (l. c.) I became acquainted with an article on 'Coniferous Trees' in the October number of the 'Edinburgh Review' for 1864, to which I would beg to refer my hearers for a detailed and very interesting account of the successive successful acclimatisations of members of this natural order; and upon the ground thus sufficiently occupied I will not encroach. It is not uninteresting, and not entirely irrelevant either, to observe that Great Britain and Ireland were both richer in Coniferae in recent geological periods than they have been since those times down to those of the Stuarts. In the sunken forest at Cromer, in Norfolk, in a deposit[1] of a period immediately preceding the glacial, we find the spruce fir represented, together with nearly all the rest of the scanty list of really indigenous post-glacial English trees. In the Cromer forest we find the spruce represented, together with the Scotch fir, the yew, the oak, the elder, the birch, and the blackthorn. The ash has somehow failed to join itself on to this company; but we see it forming one of it, though the spruce in its turn is absent as well as all other trees, in many small copses or thickets in out-of-the-way parts of this country. Such, for example, are many mountain-lime-stone headlands in parts of the Principality,

[1] See 'Rudiments of Geology,' by Samuel Sharp, F.S.A., F.G.S. 2nd ed. 1876, p. 169.

where the Welshman—in spite of the traditional hatred for trees which his race, like some other ancient races, as, for example, the Spanish, is said to entertain—has allowed the ancient flora to remain, and left it unmixed with foreign importations. The intervention of the glacial period will easily account for the wiping out of the spruce from the list of post-glacially indigenous British trees; but it is not so easy to explain how it has been that the silver fir (*Abies pectinata*), which is found in the Scottish peat, was absent from at least historic Britain till the year 1603; and that the *Pinus mughus*, the *Taeda* of the Romans, should be found in the peat-bogs of Ireland, and should subsequently have become as thoroughly extinct there as the Irish elk, *Cervus megaceros*. On the other hand, it is not difficult to understand how it has been that the Scotch fir, with characteristic pertinacity and hardiness, followed up the retreating glacial forces more closely than even the 'Norway' spruce; for at this day it propagates itself, either by self-sown or by squirrel-sown seeds, much more surely and widely than does this equally or more than equally hardy tree.

I must not leave the subject of the Scotch fir without rectifying an error relating to it which various writers[1], from the time of Caesar's Greek translator down to those of Evelyn and of myself inclusively, have fallen into when writing about it. Julius Caesar, in an often-quoted and as often mistranslated passage[2], says of Britain, 'Materia cujusque generis, ut in Gallia, est *praeter fagum atque abietem;*' and these words are ordinarily taken to mean, 'There is wood of all kinds to be found in Britain, as in Gaul, *except* the beech and the fir.' Poor old Planudes of course blundered,

[1] Planudes fl. 1327 A.D. See p. 46 of Appendix to Cambridge edition of Cæsar's Works, 1706.

Evelyn, 'Silva, a Discourse of Forest Trees delivered in the Royal Society, Oct. 19, 1662.' Ed. Hunter, 1776, p. 139.

Hasted, 'Phil. Trans.,' vol. lxi., for year 1771, pt. 2, 1772, p. 166.

De Candolle, 'Géogr. Botanique,' pp. 154, 689. 1855.

Johns, 'Forest Trees of Great Britain,' p. 42.

Rolleston in 'British Barrows,' pp. 722–724. To do myself justice, I did not err so widely as my companions in this matter. I was as ignorant of Latin as they; but I accused Julius of only one blunder, while they accused him of two. If I had really believed that 'Caesar doth not wrong but with good cause' it would have been better for me. As it was I made a poorish 'explanation' for Julius as regarded the *abies*, but confessed that I felt some doubt as to the accuracy of his statement as to the beech. See Article XVII, pp. 325–6.

[2] De Bello Gallico, v. 12.

as a Constantinople monk of the fourteenth century was sure to blunder, 'reaping,' as Mr. Philip Smith has remarked apropos of his edition of the Anthology, 'the reward which often crowns the labours of bad editors who undertake great works;' and the words of Julius appear, l. c., in the following Greek dress: πᾶν εἶδος δὲ δένδρου παρ' αὐτοῖς, ὡς ἐν τῇ Γαλατίᾳ, πλὴν φηγοῦ τε καὶ πεύκης, φύεται. Evelyn, speaking of the fir (p. 139, l. c.), uses the following words: 'which with this so common tree (the beech) the great Caesar denies to be found in Britain; . . . . but certainly from a grand mistake, or rather, for that he had not travelled much up into the country.' Hasted (l. c.), in 1771, translates the words thus: 'This island has every kind of tree the same as Gaul *except* the fir and the beech.' Some scholars hold still that this is the right way of translating the words. But my friend Mr. J. P. Muirhead, the author of the Life of James Watt, pointed out to me that *praeter*, in the language of Julius, does by no means always mean *except*, but means sometimes simply *besides*. For example, when[1] Ariovistus stipulates that Caesar and he should meet and confer on horseback, each bringing ten assessors with him, Caesar's words run thus: 'Ariovistus, ut ex equis colloquerentur, et *praeter* se, denos ut ad colloquium adducerent, postulavit.' And we may learn from this single passage that it is as well to be quite sure of an author's meaning before we impute 'a grand mistake' to him, especially if he happen to be really a grand man. I may add that Cicero, in a single passage in the same connection as one which I shall have to refer to shortly for another purpose[2], uses the word *praeter* in both the senses, *except* and *besides*. His words, telling us how Verres bestowed himself, *somni, vini, stupri, plenus,* run thus: 'Vir accumberet nemo *praeter* (except) ipsum et praetextatum filium; tametsi recte dixerim *sine exceptione* virum quum isti essent neminem fuisse . . . . Mulieres autem nuptae nobiles *praeter* (besides) unam minorem Isidori filiam, &c., &c. Erat Pippa quaedam uxor . . . . Erat et Nice foemina.' My own natural history studies had familiarised me with the line of Plautus, Stich. iii. 460:—

'Mustela murem ut abstulit *praeter* pedes'—

and should have shown me that the *local* meaning of *praeter* is also its *general* meaning, and that it retains the idea of 'by the

[1] De Bello Gallico, i. 43.
[2] In Verrem, Act. ii. lib. v. 31.

side of,' even when by the aid of a negative, expressed or implied, it comes to be more conveniently translated by the word 'except.'

It would be perhaps showing as much over-anxiety to vindicate Caesar's accuracy, as has been shown of over-readiness to impute inaccuracy to him, if I were to point out, after Parlatore[1], that Caesar might have been familiar with the Scotch fir itself, *Pinus sylvestris*, even in Italy, to say nothing of the other European countries traversed by his victorious eagles. An historian who was or was not a professed botanist, might without any sensible man blaming him, speak now-a-days of all the common pines, 'Scotch,' 'umbrella,' 'cluster,' &c., as 'pines'; my present belief is that Julius would similarly have spoken of them all as *abietes*, and would probably have included the 'firs' proper under the same name as these 'pines.' But I wish hereby to retract the suggestion l. c. as to his having meant the silver fir, *Abies pectinata*, by the word *abies*, in the much vexed passage in question; though that suggestion was made in the best possible spirit, and is scientifically unanswerable as against those unhappy persons who feel malevolently towards Caesar, which I never did, and are at the same time, as I was when I made that suggestion, unable to translate him correctly[2].

I have yet a couple of points to mention regarding the use which man has made of the Coniferae, and the alteration which he has in comparatively recent times effected, firstly by, and secondly upon, the distribution of this order of trees.

The same number and the same article on Coniferous Trees in the 'Edinburgh Review,' October, 1864, gives the interesting history of the recovery by Brémontier and his followers, from the condition of blown sand, of the vast area (100,000 acres) in the Landes of Gascony, which should in justice to him bear his name.

[1] 'Études sur la Géographie Botanique de l'Italie,' p. 37, 1878.

[2] Mr. J. A. Froude cannot be accused of any want of loyalty to the subject of his 'Biography'; still we may say to him

Nec te tua plurima, Pentheu,
Labentem pietas, nec Apollinis infula texit—

for in that work, 'Caesar, a Sketch,' 1879, I read, at p. 271, of Britain when invaded B.C. 54, that 'the vegetation resembled that of France, save that he saw no beech and no spruce pine.' Caesar must have seen the beech, but not even Caesar could have looked either far enough forwards or far enough backwards to have seen the spruce fir growing in Britain.

His agency was the 'cluster' pine, elsewhere called the 'pouch' pine, the *Pinus pinaster* of the botanists [1]. The resinous and other

[1] The pine employed by Brémontier is the *Pinus maritima* of botanists; it is, however, as nearly allied to the *Pinus halepensis* as the two cedars *Deodara* and *atlantica* are to each other. And I used a picture enlarged from a drawing of Unger's ('Wissenschaftliche Ergebnisse,' p. 88) of a 'Pinus halepensis,' growing in Eubœa, to show the general habit of the tree which had proved so useful in the French Landes. It may be well, for my own credit at least, that I here explain that I had with me, and suspended in the lecture-room, a number of pictures in illustration of my subject. These I will herewith enumerate, stating the points they were intended to make intelligible to the eyes, thereby sparing the ears, of those who honoured me by coming to my lecture. I had with me—

Firstly, the picture just referred to, which was intended primarily to illustrate, as were some of the other pictures, the mischievous action of the goat, underwood being almost entirely absent; two goats being drawn browsing upon such shrubs as were left, and keeping them down to a line corresponding with what Ruskin calls in this country, where the old legal rule, *bidentibus exceptis*, still happily holds good in practical pasturage, the 'cattle line.' The great mass of the picture was occupied by the tall pines in question, and the bare, barren, and sunburnt native rocks, which irrigation and the prohibition of goats might cover with figs and olives.

Secondly, two pictures from Lepsius's Egyptian 'Denkmäler,' Abtheil. iii. 46, iv. 3, and iv. 126, represented goats and men allied in the unholy task of destroying the palm-trees of an enemy's country. In one of these pictures the goats had assumed the same arboreal habits which they are drawn as exhibiting in Hooker and Ball's 'Marocco,' p. 97, in the argan tree. This picture was also shown enlarged by permission of Sir Joseph Hooker. One of the pictures from the Egyptian monuments was of the time of the 12th dynasty, and therefore, Professor Rawlinson informs me, as early, according to Wilkinson, as from B.C. 2020 to B.C. 1860, or even, according to Brugsch, as from B.C. 2378 to B.C. 2200. It is of course important to know that the palm was so early as this a familiar object to Egyptian eyes, when, as I further learn from Professor Rawlinson, 'the earliest date-palms represented on Assyrian monuments belong' to no earlier a date than B.C. 833 to B.C. 858; and that even in Babylonia, where they now flourish far more than in the region corresponding to Assyria proper, the palm-trees have not monumental evidence for an earlier date than B.C. 1500. A cylinder from Babylonia, of uncertain but not earlier date than this, is figured in Professor Rawlinson's 'Ancient Monarchies,' vol. iii. p. 23, 2nd edition. These dates furnish something of an argument in favour of Unger's suggestion that the palm may have had its original home in Upper Egypt; and may make it seem more probable that the Assyrians learnt from the Egyptians, than the Egyptians from them, the art of cultivating this tree. Kämpfer ('Amœnitates Exoticæ,' p. 714) declares himself to be, as indeed the inhabitants of Egypt themselves were, of opinion that Arabia was the native home of the palm, and he dismisses the claims of a more westerly origin in the four plain words, *nam Africam non moramur*. We shall, however, go hereafter in detail into the claims of the 'Dark Continent.'

Thirdly, a picture of the gathering in of the date harvest in Persia, taken from Kämpfer's book just referred to, which was used to illustrate in connection with certain reports of the formation in Algeria of date plantations in regions previously barren (see Reclus, 'Earth,' i. p. 98, Eng. trans., 1871; Laurent, 'Mémoires sur le Sahara,' p. 85, 1859, *cit.* Marsh, *l. c.* p. 482) the power of man for producing happiness and enjoyment in localities previously but sandy, thirsty deserts.

Fourthly, a picture enlarged from one given in Martius' 'Historia Naturalis

products of this plantation form now an important article of commerce; their sale and the planting of more of the previously barren, shifting, sandy waste, received a great impulse, as did many alien interests, by the interruption to American imports caused by their great Civil War[1], and they occupy a large space in some of our various public exhibitions of economic products. Some little uncertainty appears to hang about the question as to the person to whom the chief credit of this work, which has been compared, and not unjustly, with that of the recovery of Holland from the empire of the sea, is really due. The 'Edinburgh' reviewer assigns it, apparently with good grounds for so doing, to M. Brémontier, and to a period beginning with the year 1789. Professor Koch[2],

Palmarum,' iii. 1823–1850; vol. iii. pl. 120, of the ruins of the ancient Agrigentum, with their modern surroundings. It is thus described by Martius himself, p. 249, note:—'Chamærops humilis, alia depressa, alia elata octodecimpedalis, in agro Agrigentino, antiquissimis ruinis celebri, depicta a Cl. Frid. Gaertner, architecta. Muros conspicis magnifici templi quod Jovi Olympio olim consecratum, nunc inopis palmæ, opuntiæ et agaves domicilium factum est. Junonis Lucinæ, Concordiæ et Herculis templa diruta remotiores tenent colles.' It would be difficult, except possibly by the introduction of the orange and olive into the picture, to give a more instructive view of a Mediterranean landscape as altered by man's interference. The ruins of what Pindar called the fairest city raised by earthly men, of what Virgil called 'maxima longè mœnia,' speak to man's power for destruction; the agave and the prickly pear tell of his discovery and utilisation of America; the fan-palm with its spreading, far-reaching roots and suckers stands as it did in the far-off times when the priscan inhabitants of Sicania fed upon its roots, as Cicero (In Verrem, Act. ii. lib. v. 38, 39) suggested they did before Ceres gave them in that very island the gift of Cerealia, and as it did in the much later days when Verres, by malversation and maladministration, reduced Roman sailors on the shores of what was called the granary of Rome, and was but a few days' sail from Rome, once again to pacify hunger by feeding on that characteristic Mediterranean plant. The importance which plants imported from the New World have assumed in the Old, forms a subject by itself; of the two just specified, besides their other applications, we learn from Admiral Smyth's still unsuperseded 'Memoirs of Sicily and its Islands,' 1834, p. 17, that they 'form impenetrable palisades for fortifications, and in the plains they present very serious obstructions to the operations of cavalry.'

[1] Lavergne's 'Économie Rurale de la France,' ed. iv. p. 296.

[2] Professor Koch, of Berlin, who seems to consider the planting of the vine to be the climax of attainment in the way of utilising a previously desolate region, writes thus of it, after visiting the spot: 'Weniger möchte es bekannt sein, dass unsere beliebten rothen Bordeaux-Weine ebenfalls in diesem Departement der Haiden wachsen, und dass der Boden vor nicht sehr langer Zeit hier erst für die Weinfelder urbar gemacht wurde. Die guten Weine wurden früher auf dem gegenüberliegenden Ufer der Gironde gewonnen.' p. 294.

See also Clavé, 'Études sur l'Économie Forestière,' 1862; cited by Marsh, *l. c.*, pp. 595–606; Reclus, 'Earth,' Eng. trans., i. 82; Edmond About, 'Le Progrès,' chap. vii; Lavergne, 'Économie Rurale de la France,' 1877, p. 297 *seqq.*

whilst mentioning (l. c. p. 293) Brémontier, couples with his name that of M. Desbiry, but adds that the greatest credit of all is due to M. Ivry, of Bordeaux, whom he visited himself in 1864, on his own plantations at Pian, and found to be still a vigorous man though eighty-six years of age. Professor Koch pays a meed of praise to the late Emperor Louis Napoleon for his exertions in the same direction and locality; and it is, I think, to another name connected with the Second Empire that the credit is, rightly or wrongly[1], assigned, of having enabled the wastes of Gascony to produce and to boast of the heterogeneous multitude of useful products displayed in our industrial exhibitions as being now manufactured out of the pine imported thither from Corsica.

It is in this same many-sided connection interesting to note, if we in these recent centuries have re-introduced several conifers which were indigenous, like the spruce, in the immediately pre-glacial, or like the silver fir, in the still later period of the deposition of the peat, but perished either before or during the pre-historic human period[2], and if we are still actively employed in adding to the number of species of this natural order in our landscape by importation from every quarter of the globe, from China to Chili, in proportions represented by a descriptive catalogue of more than 400 'plants of the fir tribe suitable for the climate of the United Kingdom,' we have, I think it may be shown, also considerably diminished the numbers of one of the few of our native representatives of this order. This is the yew (*Taxus baccata*). It is a tree which, though valuable to the turner, nevertheless grows too slowly to pay well in these days when the spirit which makes haste to be rich makes a 'vegetable manufactory' of the hill-sides of our Lake District (to use Wordsworth's prose), by covering them with the rapidly growing larch—to say nothing of the severe competition, even as a wood for the turner, to which the beautiful woods of New Zealand and other southern colonies now subject it. Formerly matters stood somewhat differently, when it could be said :—

> 'England were but a fling
> But for the eugh and the gray goose wing.'

[1] Wrongly very likely—in England we are content to ascribe the invention of the safety lamp to George Stephenson.

[2] See De Candolle, 'Géogr. Botan.' 807.

But when the invention of gunpowder and the application of it to the science and art of projectiles 'put me a caliver into Wart's hands,' the principal *raison d'être* of the yew-tree was destroyed. A man who drew a good bow, even if he drew it at a venture, had needs have 'the limbs, the thews, the stature, bulk, and big assemblage' of the men who won the battles of Agincourt, of 'Cressy red and fell Poitiers;' and if he were to put his arrows into the clout, he had needs have a steady and well co-ordinated eye, in addition to well co-ordinated and strong arms, to be effective. Such men were sent on many a campaign from England; and for the commencement of such campaigns, before the hardships of war had impaired the soldier's condition, no more efficient man-slaying machines could, when fair stock was taken of the relative deadliness of the available weapons, have been conceived of, till the rifled musket was discovered. But even English archers were liable to the influence of short rations, hard work, and weather; and as campaigns were not always settled in a few weeks, the firelock—a weapon which Feeble and Wart, even if they were not their 'craft's masters,' could, under the supervision of that admirably qualified musketry instructor 'Master Corporate' Bardolph, learn in a few weeks to use with as much effect as the most stalwart of tournament champions—displaced the bow and arrow, though not entirely till after the wars of the Roses. This displacement seems to have entailed the disappearance from many and many a locality of lines and avenues of yew-trees, of which here and there we still have a few representatives left us, and which, in such places as the combes in chalk districts, form in the way of contrast, and indeed also intrinsically, such a pleasant and interesting feature of the landscape [1].

Of the vastness of the change which the introduction of the common elm (*Ulmus campestris*) into Britain has produced in the landscape, any one who will count and compute the numbers of the trees visible in any one of our midland counties at one view

[1] Having above quoted Mr. Hasted to his disadvantage, I wish to make some compensation to his memory by here quoting a sentence of his with which I entirely agree, but which I had not read when I wrote as I have done in the text, relatively to the yew. It is the concluding sentence of the already quoted paper in the 'Philosophical Transactions' of 1771, and runs thus: 'Whoever has been much acquainted with the woods and tracts of ground lying on our chalky hills will surely never contend that the yew is not the indigenous growth of this country.'

will readily convince himself. It has, I think, been said already by some one, and may now be said again, that previously to the development of our railroad system all the experiences and sensations of the great majority of our rural fellow-countrymen were gained within an area limited by a horizon bounded by an uninteresting row of these hedgerow trees. Of the evidence for the belief that this tree was really imported by the Romans, and not known here previously by the Britons, however familiar it be to us Saxons, I have spoken elsewhere[1]. To the grounds for that belief, there stated, let me here add the authority, firstly, of the Cromer forest, in which no elm (not even the wych elm, of which I do not here speak) was found; and secondly, of Mr. Bentham[2], who says of it: 'In Britain it is the most frequent elm in central, southern, and eastern England, but in the north and the west only where planted. It is, indeed, doubtful, whether it be really indigenous anywhere in Britain.'

Man's increasing command over the inorganic world has, in yet another way and in another time, and that our own, very powerfully modified the botanical world around him; and as this particular instance of the efficiency for good and evil is a matter of some practical consequence, and one which is still a subject of discussion and comes into the sphere of legislative interference, I will mention some of the facts concerning it. I refer to the effects which the by-products of certain manufactories exercise upon the vegetation of the districts in which they are situated. One of the most interesting papers I have ever had the good fortune to listen to was one read by my friend Mr. Robert Garner, F.L.S., at the British Association Meeting held at Newcastle in the year 1863, and printed in the Report for that year at p. 114, as also in his 'North Staffordshire Tracts,' p. 10, reprinted from the 'Staffordshire Advertiser' of 1871. His words run thus[3]:—

'With respect to chemical impurities of the air, different plants have different susceptibilities for such influence, and the greater or less impurity of the atmosphere may indeed be shown from the effects on plants. Thus the rhododendron will flourish in an air fatal to the common laurel; wheat will luxuriate where a holly or oak will die. Some plants which appear naturally to luxuriate in the coal strata—as the oak, holly, or some ferns—die when the mines begin to be worked. Fortunately, annuals

[1] 'British Barrows,' pp. 721–722. Article XVII, p. 324.

[2] 'Handbook of the British Flora,' p. 746.

[3] British Association Report, *l. c.*

suffer least; for instance, corn and wheat do well where nothing else can, and perhaps the exhalations in question may even tend to ripen them. An increasing deterioration of the atmosphere in towns and mining districts may be estimated by means of plants as follows:—1. In the smallest degree of impurity, trees are destitute of the leafy lichens, and Ericæ, the Scotch fir, and the larch die. 2. Next, the common laurel, the Deodara cedar, the Irish arbutus, the laurustinus, and the yew die. 3. The araucaria, the thuia, the common cedar, the mezereon, and the Portugal laurel die. 4. The common holly, the rhododendron, the oak, and the elm die. 5. Annuals still live, and the almond, poplars, and many roses thrive, fruit-trees are barren, peas unproductive. 6. Hieracia, *Reseda lutea*, the elder, some saxifrages and sedums, with many syngenesious and cruciferous weeds, still luxuriate.'

The mountain and moorland plants are most, just as the nettle, the elder, the shepherd's purse, the sow-thistle are least susceptible of antihygienic influences; the former as well as the latter set of organisms showing the influence of habituation, both alike being unable to 'leave their place of birth; they cannot live in other earth,' or rather air. The presence of the former would be an infallible sign on the hygienometer; the presence of the latter encourages us not to despair[1].

[1] That man has sometimes the power of undoing the mischief he has done, even by the somewhat perilous, and often mischievous, action of legislation, a *précis* of the evidence taken and given before Royal Commissions on noxious vapours, and embodied in a Blue Book of last year's (1878) date, will abundantly show. This *précis* I take from a letter signed 'Edward Sullivan,' in the 'Times,' December 2, 1878. In this letter Mr. Sullivan says, in summing up for the defence of the alkali manufacturers:—

'As regards the injury done to the picturesque value of land by alkali manufacturers, I am afraid there is no doubt they must plead guilty. In some cases, especially in that of Sir Richard Brooke, the damage is most distressing; but there is a concurrence of evidence from Widnes, Weston, Runcorn, St. Helen's, Flint, and Hebburn, that during the last four years, since the passing of the Alkali Act of 1874, the damage has very much diminished, and that in districts where the number of works has not increased the present damage is inappreciable.

'At page 10 of the Report, Major Cross states he lives a mile and a half from the centre of Widnes. Since the passing of the Act of 1874, he had a fair crop of fruit, and roses and flowers grew luxuriantly.

'Page 11 (Runcorn). Mr. Wigg stated he had planted 1800 trees round his house, about a mile and a half from the nearest works, "which were all growing very well indeed."

'Page 11 (St. Helen's). Mr. Gamble produced two photographs of a plantation 1000 yards from the works, one taken in 1862 for the use of the Lords' Committee; the other, taken in 1876 at the same spot, showing a manifest improvement in growth and condition of trees.

'Page 11 (Flint). Mr. Muspratt stated that subsequently to the Act of 1874 vegetation was not affected at a greater distance than 200 yards. He instanced gardens containing elms and other trees flourishing within 500 yards, and old oaks growing luxuriantly within a mile of his works.

M. de Lavergne, in his work on the 'Économie rurale de la France depuis 1789,' does not mention the name of any individual

'As regards the depreciation in agricultural value caused by alkali works, a great deal is to be said.

'Pages 8 and 9 of the Report. Major Cross, "for seven years a member of the Widnes Local Board, and five years its chairman," states the average selling value of land in and about the present site of Widnes in 1854 not to have exceeded 50*l.* per acre. The greater part of the site of the town and works of Widnes was bought in 1860 at from 30*l.* to 40*l.* per acre. Since that time favourable sites within half a mile of Widnes have been sold at the rate of 1600*l.*, 2400*l.*, and 4800*l.* per acre.

'Land at Ditton, a mile and a half from Widnes, which in 1858 was not worth 60*l.* per acre, was sold for 300*l.*, and of late particular lots in Ditton and Cronton, the one being two miles and a half, the other three and a half, from Widnes, were sold at 600*l.* per acre. These purchases were made for building cottages, villas, &c. As regards letting land for agricultural purposes, Major Cross adduced several extracts from the poor-rate books, showing that the estimated rental of land situated near the works had steadily and often largely increased. For instance, at Cuerdley, on which the principal Widnes works are built, and which contains 1573 acres, mainly the property of Sir Richard Brooke, the estimated value of agricultural land per acre was, in 1861, 1*l.* 12*s.* 7*d.*; 1871, 1*l.* 16*s.* 3*d.*; 1875, 2*l.* 3*s.* At Ditton the value of land for agricultural purposes had risen during the same period from 1*l.* 13*s.* per acre to 3*l.* 5*s.* 7*d.* (page 9).

'Major Cross meets the allegation of the deteriorated value of farm produce, by stating that in the near neighbourhood of Widnes milk sells at from 3*d.* to 4*d.* a quart; hay at from 6*l.* to 8*l.* per ton. He states he has known hay and straw grown within a mile of Widnes fetch the highest price in the Liverpool market, and that in 1875 the Manchester and Liverpool Agricultural Society gave to the tenants of a farm of 80 acres within two miles of Widnes the prize for the best cultivated land.

'Page 10. "Mr. Wigg, while admitting the damage done in past times to Sir R. Brooke's estate, asserted that the value of his property, through the proximity of the alkali works, had enormously increased." That estate consists of 1200 acres on the Lancashire side and 5600 on the Cheshire side; and Mr. Wigg stated his reason for believing that the selling value of the Lancashire estate was at this moment greater than that of the two estates together in 1860.

'Mr. H. Beswick and Mr. H. Linaker, both agents to important estates near the works at Runcorn, Weston, and Widnes, and long and intimately connected with the district, bore witness to the same effect as Major Cross. Both, while admitting occasional visitations from gas, and consequent injury, declare that they have never had any difficulty in finding suitable tenants at invariably increased rates. "I can more readily," says Mr. Beswick, "let land at better rents within 5 or 6 miles of Runcorn than I can on other portions of Lord Cholmondeley's estates 20 miles away. . . Within the last few years I have refused 4*l.* a statute acre for land for agricultural purposes close to Widnes works. The rentals on the property in the neighbourhood of the works under my care have gradually increased during my time, but they have increased more rapidly during the last few years. The rental of two farms at Rock Savage, near to the Weston works, has increased from 1013*l.* in 1863 to 1503*l.* in 1876 and 1877. I regret that I cannot say the same for estates under my care at the distance of 20 miles." (Page 11.)

'I think, therefore, I may fairly assert that when the Report on Noxious Vapours, 1878, comes to be fairly examined and discussed, as most certainly it will be where so extensive an industry is at stake, it will prove that, great as may be the nuisance

as having been specially concerned in the great and successful undertaking of redeeming the Bordeaux Landes. But his remarks upon it[1] have so much of value in them, and touch upon so many of the multitudinous sides—historical, political, and economical—which this enterprise, and other State-supported enterprises, present to us when we study them in their entirety, that I think I may be allowed to quote them as they stand. After touching on the dangers which pines more than other woods are exposed to from the sparks which the railway train so readily and so fatally scatters in such dry and parched districts; but omitting the not inconsider-

complained of by the landowners of Lancashire, they have in the great majority of cases received a very substantial set-off in the increasing value of their land, both for rental and for sale.

'The alkali industry is a necessity in a manufacturing country. If it is an evil, it is a necessary one. Sulphuric acid, the base of all alkali products, may be called the heart of all manufacturing industries. The consumption of it is the surest gauge of their condition. There is scarcely a manufactured article in daily use that is not more or less dependent on it. To enhance the cost of its production by hasty or ill-judged legislation, would enhance the cost of half the industrial products of the country. It is not the greed of manufacturers that has increased the number of alkali works, but it is the increased trade of the country that has demanded an increased supply of an indispensable element of production.

'If new works had not sprung up at Widnes or St. Helen's, they would certainly have sprung up elsewhere. It is to be regretted that so many works have congregated at Widnes and St. Helen's. The consumption of coal alone, a million tons at the former and a million and a half at the latter annually, would of itself cause great nuisance to the neighbouring districts; but who, pray, is to blame for this evil? Not, certainly, the manufacturers who bought and leased the land offered them by the landowners, but the landowners who offered it.

'Complaints of injury done to trees, to the picturesque value of ornamental property, do not come with very good grace from the very proprietors who have sold and leased contiguous land at very high prices, for the expressed and avowed object of erecting and extending the works they now wish to destroy.

'Sir Richard Brooke, whose name most frequently occurs in the report, and who is undoubtedly the greatest sufferer in the picturesque value of his estate, has within the last few years leased land immediately opposite his house, at a very high rental, for the erection of alkali works and the deposit of alkali waste; and, I understand, has hundreds of acres more to be let for the same purpose: nor is he by any means the only landowner who has let and sold land expressly for the erection of alkali works.

'There is a general desire among alkali manufacturers to minimise the nuisance and injury caused by these works. Recent legislation has undeniably tended to that result, and any further legislation in the same direction that is reasonable and practicable will, I know, receive their hearty support; but it will be a fatal mistake if a somewhat onesided statement of local grievances should cause any hasty legislation that would destroy an industry that is absolutely indispensable to the manufacturing prosperity of the country.'

[1] Pages 297–300.

able, even if not complete, safeguard which the planting of lines of the *Robinia pseudacacia* on either side of the railroad would furnish; which he might very well have added, as this tree does such good service in this way in other parts of France: he dwells on the cost and the necessity of wells, and the State help in the way of subventions for this purpose; he alludes with some not unjustifiable bitterness, detectable again at pp. 453-461, to the 'lost opportunities' for good in the way of developing the resources of the Landes which the warlike folly of expenditure in Algeria has entailed; and finally, his allusions to the unhappy relations into which the Moors were successively brought with the Spaniards, with the French, and lastly with the Turks, are not without a singular interest and instructiveness. But M. Lavergne shall speak for himself and in his own language:—

'Un peu avant la révolution de 1789, au moment où tout s'éveillait à la fois, de grandes compagnies de défrichement se fondèrent, mais sans succès, pour avoir voulu aller trop vite; d'autres essais du même genre ont échoué plus récemment par la même cause. Il n'en a pas été de même des tentatives partielles faites en pleine connaissance de cause par les propriétaires du pays: plus d'une spéculation profitable s'est réalisée sans bruit sur des points isolés.

'Le chemin de fer de Bordeaux à Bayonne traverse maintenant les Landes dans toute leur longueur, et y apporte la puissance de l'industrie moderne. La valeur des terres a immédiatement doublé, triplé même, le long de la ligne, et tout le monde comprend que la solution du problème n'est plus qu'une question de temps. Rien n'était possible dans un pays sans chemins et sans eau: la compagnie du chemin de fer s'est engagée à ouvrir sur plusieurs points des routes munies de rails en bois, et si en même temps on parvient à créer de l'eau salubre, soit au moyen de puits ou de citernes[1], soit au moyen de canaux dérivés des étangs, le plus difficile sera fait; le reste viendra de soi. La plus grande partie des terres incultes sera sans doute semée en pins, chênes et chênes-liéges, et pour accélérer cette transformation, une loi récemment rendue permet à l'État de boiser les terrains communaux jusqu'à concurrence de six millions de francs. Les autres branches de la culture ne doivent cependant pas être négligées, et il faut leur faire aussi leur part, car le danger des incendies, si grand pour des bois résineux sous un soleil ardent, ne permet pas de couvrir le sol d'une forêt immense et continue: une simple étincelle du chemin de fer suffirait pour mettre le feu de Bordeaux à Bayonne.

'Les Landes peuvent être aussi productives que quelque contrée que ce soit, mais elles conserveront toujours un caractère spécial. La singularité de cette nature sera un de ses charmes. Les régions inhabitées ne se prêtent que lentement à l'habitation de l'homme, et le régime pastoral, qui multiplie les animaux, et par eux les engrais, y

---

[1] 'Il suffit, pour avoir de l'eau potable, de creuser des puits de cinq à six mètres de profondeur, avec des parois imperméables, et d'y introduire une couche de gravier. Chacun de ces puits coûte 600 francs. Les Landes en possèdent déjà une cinquantaine, et on calcule qu'il suffirait de 100,000 francs pour en doter toutes les communes qui en manquent.'

sera longtemps, avec le régime forestier, le principal instrument du progrès. Quand on mesure par la pensée cette vaste solitude, qui s'étend jusqu'aux portes d'une de nos plus grandes villes, on s'étonne que la France ait pu songer à coloniser des pays lointains, au lieu de porter ses efforts sur elle-même. Si le dixième de ce qu'a coûté l'Algérie avait été dépensé dans les Landes, on aurait obtenu de meilleurs résultats, et l'on aurait épargné bien des flots d'un sang généreux ; mais les stériles conquêtes de la guerre nous ont toujours beaucoup plus séduits que les créations fécondes de la paix. L'arrondissement de Mont de Marsan, bien qu'il renferme le chef-lieu du département, ne contient pas plus de 100,000 habitants sur 500,000 hectares, comme le Tel africain, et il s'y trouve plusieurs parties déjà très-peuplées et très-cultivées ; dans la Lande proprement dite, il n'y a pas plus de 10 habitants par 100 hectares, et quels habitants! Cette terre, qui sera un jour populeuse et florissante, n'offre à l'œil qu'un spectacle de désolation: c'est le désert tel qu'on va le chercher au delà des mers, avec son triste silence, sa végétation chétive et ses horizons infinis.

'La tradition raconte que, quand les Mores furent chassés d'Espagne, à la fin du seizième siècle, ils demandèrent à s'établir dans les Landes, avec l'espérance de les fertiliser. Les préjugés politiques et religieux ne le permirent pas. Non moins civilisés à cette époque que beaucoup de peuples chrétiens, les Mores connaissaient d'excellents procédés de culture qui marquent encore leur passage dans les plus riches provinces de la Péninsule. Les Landes seraient probablement devenues productives entre leurs mains, et ce qui leur restait de la barbarie musulmane aurait reculé devant les idées modernes de tolérance et d'égalité. S'ils ont tant dégénéré en Afrique où ils se sont réfugiés, c'est qu'ils y ont trouvé les Turcs, le plus destructeur de tous les peuples ; cette civilisation a péri tout entière faute d'un asile où elle pût se développer. Mais le royaume qui devait bientôt révoquer l'édit de Nantes et expulser de son sein des Chrétiens et des Français, ne pouvait s'ouvrir à des enfants de l'Islam étrangers et persécutés, et ce qui a puissamment contribué à ruiner l'Espagne ne pouvait contribuer à enrichir la France.'

Leaving now the subject of the introduction of foreign trees, and that of the unintentional destruction of our own, and taking up the subject of disforesting generally, I have to say that the literature of it has in these latter days become all but colossal; and that the moral of it all is just the reverse of that of the capitulary of Charlemagne[1], where it is ordained that wheresoever any good men and true are found to be available they may have forest land given them for clearing: 'ubicunque invenient utiles ullos homines iis detur silva ad extirpandum.' Two hundred and fifty pages of the second edition of Mr. Marsh's excellent work, 'The Earth as modified by Man's Action,' are devoted to this subject alone; the bibliography extending over *nine* pages, prefixed to his work, is very largely made up of the titles of works bearing upon it; and I hold in my hand a small, but closely printed, German octavo, which has some 280 pages devoted to the purpose of specifying the names and

[1] Cap. secund. Anni 813, sive Capitul. xxi. ed. Stephan. Baluzius, 1677, tom. i. p. 510, De Villicis regiis quod facere debent.

giving a few lines as to the scope of such works. Its own title is 'Die Bedeutung und Wichtigkeit des Waldes, Ursachen und Folgen der Entwaldung, die Wiederbewaldung, mit Rücksicht auf Pflanzenphysiologie, Klimatologie, Meteorologie, Forststatistik, Forstgeographie und die Förstlichen Verhältnisse aller Länder, für Forst- und Landwirthe, National-Oekonomen und alle Freunde des Waldes, aus der einschlagenden Literatur systematisch und kritisch nachgewiesen und bearbeitet von Freidrich Freiherrn v. Loffelholz-Colberg, königl. bayer. Oberförster.' Leipzig, 1872.

But in Herr v. Loffelholz-Colberg's list 'aller-Länder,' there is no mention of India nor of its forest or other departments, nor of their annual reports, nor of the names of (1) Balfour of Birdwood, (2) of Cleghorn, (3) of Dalzell, (4) of Danvers, (5) of Brandis, of J. L. Stewart, (6) of Colonel G. F. Pearson, or of Beddome, to each of whom, though unknown to me personally, I feel myself personally indebted. And extensive as is his bibliography, it admits of being supplemented by the specification not only of works which have appeared later, and in India, but of some of considerable importance which appeared earlier, and some of them in Europe of earlier date [1].

[1] For the Memoirs of the Indian authorities named above see:—(1) Revenue Department, No. 981, 1848; (2) Catalogue Bombay Products, 1862, and Journal, Society of Arts, Feb. 7, 1879; (3) Sind Forest Reports, 1858-1860; (4) Journal, Society of Arts, May 24, 1878; (5) Ocean Highways, Oct. 1872, and Systematic Works, p. 204; (6) Report on Forest Departments of India, 1872.

As regards other memoirs I find no mention of v. Baer's papers upon this very same question of the relation of woods to rainfall already referred to *supra*, in the 'Beiträge zur Kenntniss des Russischen Reiches,' iv. 1841, p. 190, xviii. p. 111, 1856. From the former of these two papers the following sentences may with some advantage be quoted, pp. 190–191:—'Noch viel weniger darf man glauben, dass nach dem Verhältnisse der Waldabnahme eines Landes auch die Wassermengen in seinen Flüssen abnehmen müsse. Es ist nicht unser Absicht den Einfluss ganz läugnen zu wollen; allein wir wollen nachdrücklich darauf aufmerksam machen, dass die Niederschläge aus der Luft nicht von den kleinern unter ihnen liegenden Localitäten abhangen, sondern von grossen ausgedehnten Verhältnissen, von vorherrschenden Luftzügen, von der Quantität Feuchtigkeit welche diese Luftzüge mitbringen, von der Differenz zweier einander berührender Luftmassen, dass diese Niederschläge es sind, die unsern Flüssen Nahrung geben, dass in unsern Breiten sie in Form des Schnees mehrere Monate hindurch aufgespeichert werden und endlich, dass in einem so flachen Lande wie Russland die Feuchtigkeit welche in Form von Regen und Schnee niederfällt, aus sehr weiter Ferne kommen kann. Dass unsere Flüsse und besonders das Gebiet der obern Wolga in trockenen Sommern wenig Wasser haben, hat seinen Grund vorzüglich darin, dass hier kein Gebirge ist, an welchem Niederschläge das ganze Jahr hindurch nothwendig erfolgen und eben deshalb hat es ohne Zweifel von

I show you yet another work, an English Parliamentary Report, of date 1875, Feb. 1, respecting the Production and Consumption

jeher einzelne Sommer gegeben, in denen das Wasser ungewöhnlich niedrig stand. Wir kennen Zeugnisse hierüber aus der Zeit Peters des Grossen, und ohne Zweifel wird man sie aus noch früherer Zeit finden wenn man darnach sucht.'

And to supplement a second time the bibliography of Herr Loffelholz-Colberg, I will say that the following quotation from the well-known and accomplished writer of the sixteenth century, Bernard Palissy, may fairly take its place with the foregoing more strictly scientific opinion of von Baer. Mr. Marsh shall introduce it for us (l. c., p. 303):—'In an imaginary dialogue in the "Recepte Véritable," the author, Palissy, having expressed his indignation at the folly of men in destroying the woods, his interlocutor defends the policy of felling them by citing the example of divers bishops, cardinals, priors, abbots, monkeries and chapters, who by cutting their woods have made three profits, the sale of the timber, the rent of the ground, and the "good portion" they received of the grain grown by the peasants upon it. To this argument Palissy replies: "I cannot enough detest this thing, and I call it not an error, but a curse and a calamity to all France; for when forests shall be cut, all arts shall cease, and they who practise them shall be driven out to eat grass with Nebuchadnezzar and the beasts of the field. I have divers times thought to set down in writing the arts which shall perish when there shall be no more wood; but when I had written down a great number, I did perceive that there could be no end of my writing, and having diligently considered, I found there was not any which could be followed without wood . . . . And truly I could well allege to thee a thousand reasons, but it is so cheap a philosophy, that the very chamber-wenches, if they do but think, may see that without wood it is not possible to exercise any manner of human art or cunning."'—Œuvres de Bernard Palissy, Paris, 1844, p. 82, first published in 1563.

I may do well to neglect chronological order and mention the work by Dr. J. C. Brown, a Fellow of the Royal Geographical Society, which appeared in 1876 under the title, 'Reboisement in France; or Records of the Replanting of the Alps, the Cevennes, and the Pyrenees with Trees, Herbage, and Bush, with a view to arresting and preventing the destructive consequences and effects of Torrents.' Dr. Brown has besides this and other works on kindred or on the same subjects, given us a work on 'The Hydrology of South Africa, or Details of the former Hydrographic Condition of the Cape of Good Hope, and of Causes of its present Aridity.'

Professor Ernst Ebermayer's work, 'Die Physikalischen Einwirkungen des Waldes,' being the 'Resultate der Förstlichen Versuchs-Stationen im Königreich Bayern,' Aschaffenburg, is of later date (1873) than the bibliographical *précis* of Loffelholz-Colberg, and would not therefore have been referred to by that writer as it ought to be by all subsequent writers on the same subject.

Professor Karl Koch's 'Vorlesungen über Dendrologie,' one third part of which is devoted to the subject of the 'Influence of Woods on the Health of Men, and on Climate,' is similarly of later date (1875) than the last edition of Mr. Marsh's 'The Earth as modified by Man's Action.'

Latest in order of time, but by no means last in order of merit, I must place Professor Wellington Gray's 'Notes on Tree-Planting and the Water Supply of the Deccan,' Aug. 1877, contained in the excellent 13th Annual Report of the Sanitary Commissioner for Bombay, Dr. T. G. Hewlett. The influence on climate of cosmical as compared with local agencies; of mountain and monsoon, that is, as compared with man's plantations; and on the other, the influence of the brute population of India,

of Timber in Foreign Countries, from which a very large amount of most useful information can be procured for the very moderate charge of elevenpence, one penny less than a shilling—a fact which would have rejoiced the heart of the late Mr. Joseph Hume. If in addition to this work we had rendered available to us the usufruct of the vast experience recorded in the Blue Books of the Indian Forestry and Sanitary Departments, in a volume of anything like the same size, I do not say of anything like the same price, the India Office would add considerably to the very large claims it has established upon the gratitude and acknowledgments both of men of science and men of action by the publication of those invaluable volumes.

I do not propose, indeed I do not dare, to attempt to give a summary of the results of very many volumes here alluded to, pleasant and even absorbing reading though many of them have proved themselves to be. I will not discuss the curious belief still prevalent in Spain, to the effect that trees breed birds, though somewhat similar articles of faith are not without adherents nearer home, merely observing, so that I may affront no one, that it would be truer to say that the destruction of trees leads to the banishment of birds, and thereby to the sexual, and in that case spontaneous, generation of insects. Nor will I speculate as to whether the hatred of a tree, of which you will be told in travelling in countries and districts at home and abroad (even in Sicily, see Fischer, l. c. p. 135), where the Celtic or other pristine ethnological element is still strong in the natives, is due to a hereditarily transmitted recollection of the days when, as the capitulary just quoted shows, man had to wage war against the forests, or a similarly transmitted recollection of the much more recent forest-laws and the feudal state of things contemporaneous with them. Neither, on the other hand,

the goats and the camels, as compared with the agency of the human inhabitants, who besides employing the two organic means for destruction just now mentioned, also 'hack, cut, and burn,' will be found instructively, though briefly, discussed in this essay. I take this opportunity of adding to this bibliography the names of three books with the contents of which I was not acquainted when I wrote as above. They are—

'Wald, Klima, und Wasser, von Dr. von Liburnau,' 1878. This little octavo is one of the Munich series of Science Primers, being Bd. xxix. of 'Die Naturkräfte, eine Naturwissenschaftliche Volksbibliothek.'

'Die förstlichen Verhältnisse Frankreichs, von Dr. A. v. Seckendorff,' 1879.

'Der Wald im Nationalen Wirthschaftsleben, von Ph. Geyer,' 1879.

will I content myself with simply repeating Mr. Marsh's summing up of the matter in the short way in which long words so often (literary critics notwithstanding) enable us to sum up the results of a long investigation, and saying with him that (p. 300 l. c.) the forest's 'general effect is to equilibrate caloric influences and moderate extremes of temperature.' But I will firstly, upon this occasion, repeat what I have often heard my late and much-lamented friend, Mr. Wm. Menzies, the author of the splendidly-illustrated book, 'Forest Trees and Woodland Scenery as described in Ancient and Modern Poets,' say, to the effect that England is after all as well-wooded a country as probably any other civilised one in the world, adding that Sir John Lubbock has, as I think, either in some volume which he has contributed to science, or in some return which he has extracted from Parliament, established the same fact. And, remarking that if we couple with this fact the consideration that this favourable numerical representation of trees is not due to the existence of large forests, we find therein an illustration of the working of certain peculiarities of our social and political condition as compared with those of other countries, which I leave to your consideration; I pass on, secondly, to say a few words as to the influence which trees exercise in the way of modifying climate locally by means of their leaves. Clearly this comes fairly under the title of my lecture. Man can cut down 'the goodly fir-trees' and other trees too, 'Laubholzer' as well as 'Nadelholzer,' of an entire country; he can burn them, and by his domesticated goats and cows and camels he can prevent their suckers and their seeds from replacing them by fresh plants. What consequences follow when the square area which a tree in full leaf represents is abolished? Firstly, whatever else may be disputed, there can be no doubt the loss of this square area means the loss of a very considerable area upon which dust and *particulate* matter can be caught and filtered out of the atmosphere. The more sticky the leaves, of course the more perfect the interception. And as modern investigations, such as those which Mr. John Simon, C.B., used to have carried on whilst in the Medical Department of the Privy Council Office, have taught all those who have ears to hear, even if not also eyes to see, that the germs of many or most infectious diseases are *particulate*[1], we can understand how it is that from so many

[1] We have such accounts from Ravenna and Beyrout; from the East and the

quarters of the world we have more or less well-established histories of belts or curtains of trees protecting towns from malarious and anti-sanitary influences.

Secondly, though doubt may be raised (e.g. by M. J. Bellucci, cit. 'Athenaeum,' March 14, 1874, p. 360) as to the giving off by trees of ozone into the air, there can be no doubt as to another mechanical effect besides the one already dwelt upon in the way of breaking the force and the fall of raindrops, and thereby preventing, *pro tanto*, the over-rapid flowing away of such rain and the over-violent washing away of the soil. Simple as this action is, it is, when coupled with the action of the roots and their spongioles, to which it gives a fairer chance of coming into play, one of the most important which a tree in leaf exercises. Finely divided rain sinks into the

West Indies, and from Guiana. Lord Mark Kerr (see 'Report on Measures adopted for Sanitary Improvements in India for June 1871 to June 1872,' p. 14) did much planting in Delhi in 1864, and, on coming eight years later to take stock of the effects of his hygienic work, was able to persuade himself that the almost entire disappearance of the Delhi boil was due to this particular cause. But the Indian Government had to report in the succeeding year's volume of the same series, p. 17, that they had not received from the authorities they had consulted 'reliable data to warrant any general conclusions being drawn as to the effect of trees and vegetation on these sores.' Still they proposed 'to institute a more particular inquiry into the matter, and to submit a Report on the investigations in due course.' Upon this subject something may be found in Mr. Menzies' 'Forest Trees and Woodland Scenery,' 1875, p. 101, q. v. *ibique ab ipso auctore necnon a me citata*. Since the appearance of Mr. Menzies' work the literature relating to the *Eucalyptus globulus* as an agency for 'purging the unwholesome air' has attained a great development. Especially to be recommended is a paper, 'The Eucalyptus near Rome,' by Dr. R. Angus Smith, F.R.S., published in the 'Proceedings' of the Literary and Philosophical Society of Manchester, vol. xv., No. 9, pp. 150–164, 1876, as also some papers in the 'Edinburgh Medical Journal,' February 1878, and May 1879, pp. 1052–1053, by Dr. Bell. And what is better even than good memoirs, good progress has been made in the way of actually planting this tree by no less conspicuous warriors than Garibaldi in the Roman marshes, and by Sir Garnet Wolseley in Cyprus. I have not, however, heard of any further development of the use of the *Helianthus annuus* as an anti-malarious agent, nor of the adoption of Mr. Menzies' recommendations of the employment of the horse-chestnut, the sycamore, or the balsam poplar and white poplar for the same purpose. To the references given l. c. may be added, as speaking in the same sense, Becquerel, 'Mém. Institut,' xxxv., 1866, p. 444, and Boudin, 'Géographie et Statistique Médicales,' vol. i. p. 229. Much has been written by the two last-named writers on the electrical action of trees; I will quote the following sentences from the latter of the two, l. c.: 'Enfin le déboisement doit être considéré comme équivalent à la destruction d'un nombre de paratonnerres égal au nombre d'arbres qu'on abat; c'est la modification de l'état électrique de tout un pays; c'est l'accumulation d'un des éléments indispensables à la formation de la grêle dans une localité où d'abord cet élément se dissipait inévitablement par l'action silencieuse et incessante des arbres. Les observations viennent à l'appui de ces déductions théoriques.'

soil, whilst rain which falls in larger masses runs off and forms torrents. The roots making up an interlacing fibrillar mass by their multitudinous divisions, entangle and detain the moisture which comes to them in capillary columns; and from the loaded sponge which they thus come to represent, they dole or issue out in rations the supplies necessary for keeping springs and streams in constant and perennial volume [1].

It is, I must say, a considerable marvel, that upon a third function of that part of a tree which man can affect, either by his own hands or through the intermediation of his domestic animals, with the greatest results in the way of mischief at the least cost of labour to himself, so much room for dispute and doubt should still be left open by the botanists. Upon this third function of the leaves, their power as evaporators, the most important perhaps of all their functions, both as regards the tree's own economy and as regards ours, it is little less than marvellous that a Professor of Botany should have to write thus in 1875. Professor Koch, however ('Vorlesungen über die Dendrologie,' 1875, p. 284), following Ebermayer, l. c., p. 183, says:—

'The question of the evaporation of water through the tissues of a plant is very like the question in medicine of the treatment of diseases. The more there is written about a disease, and the more we have so-called infallible remedies recommended for it one after the other, the less do we get of any real knowledge of it. There is scarcely a single point in the life of a plant on which so much, and indeed often so much that is intrinsically self-contradictory, can be specified as having been written, as this point of evaporation. Whilst Unger, and indeed certainly with right on his side, owns that a surface of (so much?) water gives off by evaporation three times as much as (an equal surface of?) a tree, Schleiden says that on the contrary the tree gives off three times as much as the open surface of water [2].'

It is true that Professor Koch goes on to say that nevertheless, as

[1] It is of course possible to exaggerate the preventive power of arboriculture, as of other beneficial agencies. If a mountain is sufficiently high, and can be blown upon by sea breezes as yet undeprived of the full proportion of moisture which a warm latitude can give them, you will have *from time to time* destructive torrents rushing down their sides, however well wooded they may be. But what is an occasional occurrence only in a well-wooded mountainous country, is a very common one in a district where the charcoal-burner, the wood-merchant, and the goat have been allowed to have their wasteful will unchecked. Homer's lines, Il. xi. 492–495, show that however striking the phenomenon he describes, it was nevertheless not so very common as the complaints with which so many of the Reports I have referred to prove it to be now in so many countries in Europe and elsewhere.

[2] The German words, which I have not attempted to translate quite literally, are as follows:—

'Mit der Verdunstung des Wassers durch die Pflanze geht es, wie in der Medizin

Sachs also has said, such observations and the results deduced from them have a scientific value. As it seems to me, they have not only a scientific value, as all observations which are reducible to weights and measures have, but that they have also a very distinctly appreciable practical value and applicability.

Anybody who will read the account given by my friend the Rev. Richard Abbay in 'Nature,' May 18, 1876, of the formation of a lake in a district in Australia, 150 miles from Sydney, and 2000′ above the level of the sea, subsequently to the destruction of the woodlands round about a particular area of depression, will be convinced that this occupation by water of what had been habitable land was not only posterior to, but caused by, the disforesting operations of the various agents specified, namely, squatters, grubs, cattle, sheep, and opossums, not unaided by disease of the trees themselves. The surplus of the water forming the lake corresponds to the enormous quantitative disproportion between the evaporating surface which it exposes when thus collected, and that which it would have exposed when dispersed through all the myriads of leaves which man and his allies had destroyed [1]. It is not, however, necessary to take such a long voyage as that to Sydney to get an unmistakeable illustration of the evaporating power of leaves. This power can be illustrated *e contrario* by observing the construction on the treeless Yorkshire or other English wolds of the perennial so-called 'dewponds.' It is not even necessary to travel as far as the nearest down or wold to make this observation, and fill in the necessary details as to extent of feeding ground to catch, and puddled ground to catch, the rainfall. A very simple experiment with plants no farther to fetch than cabbages, will show, as Professor Wellington Gray tells us (l. c. supra, p. 10), that 3000 square inches of their succulent leaves will give off as much as a pint of water per diem.

It may, however, be fairly objected that the rate of evaporation

mit den Krankheiten. Je mehr über eine Krankheit geschrieben ist und je mehr nach und nach sogenannte untrügliche Mittel empfohlen wurden, um so weniger ist sie erkannt. Kaum möchte über einen Gegenstand im lebenden Pflanze so viel, und zwar oft einander widersprechendes, geschrieben worden sein, als über die Verdunstung. Während Unger, und zwar wohl mit Recht, behauptet, dass eine Wasserfläche drei Mal so viel verdunstet, als der Baum, sagt Schleiden, dass umgekehrt dieser drei Mal so viel verdunste als die offene Wasserfläche.'

[1] See also Ebermayer, l. c. pp. 184, 185.

observable in an isolated mass of leaves, or in a single isolated tree, does not give us a measure of the rate at which the same process will go on in a wood when the exposed and evaporating surface is relatively so much smaller. And this difficulty, which lies in the geometrical nature of the case, may account for the great discrepancies in the estimates which various writers have given of the amount of watery vapour given off by masses of wood[1].

It must, however, be allowed that the cases in which the cutting down of trees, and the consequent putting into abeyance of the functions of their leaves, have been followed by the drying up of springs, are much more numerous, even if they are not better established, than those in which the reverse effect has been recorded, as by Mr. Abbay. The explanation of this apparently self-antagonising or capricious operation of the same primary cause is not far to seek. When a tree is cut down, the area once protected by its leaves is exposed to the uncounteracted action of the summer sun, and rainfall may run off it when thus hardened, just as it runs off an imperfectly thawed surface in the spring, or may sink away into chinks and fissures which that exposure may, and very often does, produce, and in either case such rainfall is lost to the summer-

[1] Professor Pfaff, for example (cit. Ebermayer, l. c. p. 186), gives us 120 kilogrammes as the entire amount evaporated by an oak with 700,000 leaves, each of a square surface of 2325 mill. during the period from May 18 to October 24.

Vaillant (cit. ibid.) gives the amount of watery vapour given off by an oak of 21 mètres height and 2·63 mètres girth at a height of 1 mètre above the ground, as 2000 kilogrammes on a fine day.

Hartig (cit. ibid.), the author of a 'Lehrbuch für Förster,' Stuttgart, 1861, calculates that a German morgen (= 2·3895 acres), carrying a thousand trees of nine different kinds of conifers and broad-leaved trees of twenty years' planting, exhales daily during the period of vegetation 3000 pounds weight of water.

Professor Prestwich, in his 'Water-bearing Strata,' 1851, p. 118, gives us as an estimate for the amount of watery vapour given off by the leaves of 'a tree of average size' two and a half gallons per diem.

Mr. Lawes (cit. in loc.), from 'Journal of Horticultural Society,' vol. v. pt. i., 1850, gives us as a foundation for an estimate of the relations between the amount of water taken in by vegetable organisms, with the matters it held in solution, and the solid residue thence extracted and retained by the plants for its uses or for ours, a statement to the effect that three plants of wheat or barley gave off 1½ gallon, 250 grains of water for every grain of solid residue in the adult plant.

Hellriegel, on the other hand (cit. Ebermayer, l. c. p. 187), gives us as his estimate that for the production of 1 lb. of dry barleycorns, 700 lbs. of water, inclusive of the water evaporated from the soil, are all that is necessary, and that other cerealia have their demands limited within somewhat similar proportions. *Intervalla vides humane commoda.*

dried fountain. If the water thus thrown upon the surface, thus modified, finds its way into a basin properly proportioned as to cubical, as to square area, and as to water-holding power, we may have a lake formed, as in the case related above by Mr. Abbay. It is, of course, more usual to find one or other, or two, or all, of these favourable conditions wanting, and in the more numerous class of cases we find that the diminution of wood and the diminution of water go hand in hand. I would go further than this, and aver that the diminution of wood and the diminution of water in the shape of ice may not only also go hand in hand, but may also be connected as cause and effect. M. Viollet-le-Duc, in his delightful work on 'Mont Blanc,' 1877 (translated by B. Bucknall, pp. 341 353), tells us that—

'Although the glaciers have been tending to diminish for the last forty years in a somewhat rapid ratio, which would seem to indicate an elevation of the mean temperature, the forests are quitting the heights where they still lingered, to take a lower position. Is there any connexion between these two results? We shall not endeavour to solve the problem.'

It is a little presumptuous to address one's-self to it after this deterrent warning. Still M. Viollet-le-Duc has (l. c. pp. 339, 377) shown us that the destruction of the forests is abundantly explained irrespectively of any inorganic agency by the mischievous action of man working as a goatherd and a woodcutter. His descriptions of these operations are couched in language of real pathos and eloquence, but scientifically it shows us that we need not look for any other cause for the disappearance or shrinking of the limits of the forests. The spruces and the larches, for such are the trees, being thus destroyed by the 'essentially destructive power' of man, how can their destruction be shown to entail the diminution of the glacier? I think the loss of these trees as evaporating agencies may be taken as a *vera ac sufficiens causa* for the diminution. A great deal of much interest has been written[1] upon the difference in the amount of watery vapour given off by various trees and by the cerealia, which last, and amongst which last, as might be expected from their deep roots and the amount of their *Stoffwechsel*, wheat-plants stand quantitatively pre-eminent. But for our present purpose it is sufficient to point out that the rays which strike on the mass of a glacier are, to say nothing of the other conditions of disadvantage which such a mass opposes to them, enormously *outnumbered* by the

[1] See Vogel, Pfaff, and Hartig, cit. Ebermayer, l. c., p. 185.

rays which strike on the needle-shaped leaves of an adjacent wood of ordinary acreage, made up of such trees as the spruce or the larch; and the vapour which is thus set free into the entire circumambient atmosphere alike of glacier and of wood, acts most potently in several ways in the direction of saving the glacier from wasting.

On the other hand, great as the influence of the evaporating power of trees and forests may be shown to be in some directions, it is possible enough to overrate it as regards such more than localised matters as the increase of the rainfall.

'It is,' says Dr. Brandis ('Ocean Highways,' Oct. 1872, p. 204), 'a widely spread notion, entertained by many writers who are competent to judge, that forests increase the rainfall, and that the denudation of a country in a warm climate diminishes its moisture. Much of what is known regarding the history and the present state of the countries round the Mediterranean seems to support this theory, but it has not yet been established by conclusive evidence.'

The important point seems to be that in *mountains* this influence may count for something considerable, whilst in the plains, howsoever well wooded, trees can act only as do other good radiators in the way of precipitating, not wind-borne moving vapour, but simply dew.

Mr. N. A. Dalzell, in the Report on the Sind Forest for 1859–1860, observes (par. 31):—

'Although it would be too hardy an assertion to say that the existence of forests in Sind causes any increase in the fall of rain, they certainly do so on the summits and tops of mountains;' and par. 35: 'In enumerating the benefits derived from forests, I make here no use of the fact that forests attract rain-clouds, because I do not think it applicable to plains, and because it is not yet clear that causes are not mistaken for effects, that is, whether it is the rain produces forests, or forests which produce rain; and certainly no inhabitant of Sind would consider it legitimate to decide that because a country is covered with wood, therefore it is wet.'

It is satisfactory to be able to add that the result of Professor Ebermayer's prolonged observations in Bavaria has brought him to the same conclusions as those of Dr. Dalzell, carried on in the very alien surroundings of Sind. Dr. Ebermayer's words on this subject, used in summing up the results of his researches, are (l. c., p. 202):—

'Auf Grund unserer Untersuchungen, glauben wir daher besuchtigt zu sein annehmen zu dürfen, dass in Ebenen von gleichern allgemeinen Charakter der Einfluss des Waldes auf die Regenmenge jedenfalls sehr gering ist, und dass er auch auf die procentische Regenvertheilung keine Einwirkung hat. Mit der Erhebung über die Meeresoberfläche nimmt die Bedeutung des Waldes bezüglich seines Einflusses auf die Regenmenge zu, er hat desshalb im Gebirge einen grösseren Werth als im Ebenen. Im Sommerhalbjahr ist die Einwirkung des Waldes auf die Regenmenge viel grösser als im Winterhalbjahr.'

Whatever the physical principles involved are, anybody may find beautiful illustrations of them, who will observe in a mountainous district how—

'The swimming vapour slopes athwart the glen,
Puts forth an arm and creeps from pine to pine,
And loiters slowly drawn[1],'

or how

'The light cloud smoulders on the summer crag[2],'

recollecting that the phrase 'Rauchen der Wälder' is used for the similar phenomenon when produced by trees, or who will finally in a lowland or other country stand and study the frost as it hangs itself on to such a tree as the birch often long before it has begun to whiten the ground around it.

[Since writing as above, the 'Observations Météorologiques faites de 1877–1878,' by M. Fautrat, published by the French 'Ministère de l'Agriculture et du Commerce: Administration des Forêts,' 1878, have come into my hands. This author, with the results of M. Mathieu's eleven years' observations at Nancy (for which see his 'Météorologie Comparée Agricole et Forestière,' published under the same auspices, February 1878) before him, as also the results of four years' observations in the Forest of Halatte, and of three years in the pinewoods of Ermenonville, has come to the following conclusions.

i. That when it rains more rain falls over a wooded than over a non-wooded area, and that whilst trees of all kinds possess the power of condensing vapour, broad-leaved trees produce less effect than is produced by the narrow-leaved Coniferae (pp. 14 and 16).

ii. That as regards the hydrometric condition of the air, the air over a wooded area contains more watery vapour (p. 18) than that over an unwooded area, but that the coniferae have more watery vapour in their circumambient atmosphere than the broad-leaved trees. M. Fautrat expresses, or rather expands, this fact in the following words:—

'If the vapour dissolved in the air was visible as are mists, we should see the forests surrounded with a vast screen of moisture, and around the Coniferae this envelope would be more marked than over the broad-leaved trees. What is the source of this vapour? Does it come from the soil; is it the result of evaporation

---

[1] Tennyson, 'Œnone.'

[2] Tennyson, 'Edwin Morris.'

from the leaves; or is it due in the Coniferae to the action of the thousands of points which the whorls of their leaves develop every year? *This is a complex question which the present data of physical science do not enable us to answer.* One thing one can say, and that is that the transpiration of the leaves cannot by itself produce this phenomenon. For, as a matter of fact, the transpiration in Coniferae is less active than it is in broad-leaved trees. This fact has been made clear by M. Grandeau in his "Essais historiques et critiques sur la Théorie de la Nutrition." (M. Fautrat might have added, 'as also by Hales, cit. Boussingault, "Ann. Chim. et Phys." sér. v. tom. xiii. 1878, p. 314, and Sachs, "Handbuch der Exp. Physiologie der Pflanzen," 1865, p. 225.') It then follows that if the vapour of water dissolved in such great abundance in the atmosphere enveloping the pines was the result of the evaporation from the trees, this phenomenon ought to be much more striking over the mass made up by the broad-leaved trees than in that made up by the Coniferae, whilst observation shows that exactly the contrary is the actual fact. We must therefore ascribe to the soil and *to other unknown causes* this remarkable property which pines have of attracting watery vapour.'

If it had appeared from M. Fautrat's tables that this excess of watery vapour was more marked in rainy than in dry times, it would have been easy to explain the fact by figuring to ourselves the all but infinite area which the fine films of water clothing every needle-shaped leaf of a coniferous tree would make up and offer for evaporation. For the leaves of our common Coniferae wet readily; and it is owing to this property I apprehend that they intercept as much as one-half the rain which falls upon them before it reaches the ground, whilst broad-leaved trees intercept but one-third. But, as it appears, the Coniferae possess the hygrometric advantage independently of the rainfall. And I have to say that the phenomenon in question, needing, as it thus confessedly does, some additional explanation besides and beyond that which our usually accepted views furnish, appears to me to become more intelligible by reference to the theory as to 'The Cause of Rain and its Allied Phenomena' which was put before the world in 1839, and subsequently published in a separate volume twenty years later, by Mr. G. A. Rowell. This theory may I think be stated as follows, the author of it having slightly modified it in 1872, and restated it in a 'Brief Essay on Meteorological Phenomena,' published in 1875. He supposes that the molecules of watery vapour are completely enveloped in a coating of electricity, to which they owe their buoyancy. This coating and this buoyancy he supposes to increase and decrease in ratio with the temperature of these molecules. Efficient conduction therefore of electricity will suffice on this theory to precipitate watery vapour either as rain, or as dew, or as mist. And I apprehend that Mr. Rowell would, in accordance with his own theory, look upon a fir-tree when shrouded, as M. Fautrat has described it, with a differentially thick envelope of vapour, as having thus clothed itself by virtue of the attractive effect of its myriad points. For electricity tending constantly to an equal distribution, so fast as the surcharge of electricity on the particles of vapour nearest the trees was carried away, so fast would the balance be redressed by supply from the particles more distantly placed. And thus in accordance with this theory, particles of watery vapour would be constantly setting in the direction of the conducting and attracting leaves and twigs. Becquerel's view, already quoted, according to which the plague of hail which has so often[1] been observed to follow upon the destruction of the woods of a country, is to be ascribed to the loss of the lightning-conductors which the cut-down trees represented while standing, and to the absence consequently of the incessant though insensible dissipating agency of the trees, appears to me to show that he would doubtless have allowed that Mr. Rowell's theory contains some, at least, of the elements of the true and complete theory of rain. It is not for me to meddle with memoirs in which neither living animal nor living vegetable organisms are concerned, otherwise I might have referred to Lord Rayleigh's paper in 'The Proceedings of the Royal Society,' March 13th, 1879, pp. 406, 409. But as regards the views they brought forward, and to a considerable extent as regards the whole question, I scarcely feel myself to be in a position to give any decided opinion.

That trees, like other beneficent agencies, do not fail to benefit themselves whilst thus benefiting the world at large, may be well gathered from the following passage from Professor Grandeau's work now in course of publication, 'Chimie et Physiologie appliquées à l'Agriculture et à la Sylviculture, 1879, Pt. I, la nutrition

[1] See a really pathetic account of this given as having been produced during his seven years' absence from Thüringen by Fischer at p. 164 of his charming 'Beiträge zur physischen Geographie der Mittelmeerländer,' 1877. Rain and hail-storms had become frequent, and the fishing brook had disappeared together with the wood of his boyhood. He adds:—

'Ich will gewiss damit nicht sagen, dass in jenen Gegend jetzt auch nur ein Tröffchen Regen weniger falle als früher, obwol auch das örtlich möglich, ja wahrscheinlich ist, aber der Vertheiler und Bewahrer der Feuchtigkeit fehlt und so können locale Ursachen zeitweilig Wirkung haben, die in Süd-Europa allgemeinen kosmischen, aber durch örtliche verstärkten zu zuschreiben est. Ich wurde recht lebhaft an Sicilien erinnert, aus dem ich eben heimkehrte.'

de la plante.' In summing up at p. 340 the results of his experiments, and after saying that the simplest and at the same time the best way of isolating a plant for purposes of experiment from the action of electricity, is to place it either under a metallic cage with large meshes, *or in the perimeter of a tree*, M. Grandeau proceeds as follows:—

'2° Les végétaux et en particulier les arbres, soutirent à leur profit l'électricité atmosphérique et isolent aussi complètement qu'une cage métallique la plante qu'ils dominent.

'3° L'isolation produite par un arbre élevé peut s'étendre notablement au delà du périmètre foliacé de l'arbre.

'4° Une plante soustraite à l'influence de l'électricité atmosphérique subit, dans son évolution et dans son développement, un retard et une diminution très-notables. Dans mes expériences, les quantités de substance vivante produites par les végétaux isolés ont été inférieures de 30 à 50 p. 100 à la production à l'air libre. La transformation du protoplasme chlorophyllien en glucose, en amidon, etc., paraît être tout particulièrement influencée par l'électricité atmosphérique.

'5° La floraison et la fructification subissent des modifications non moins grandes; sous cage isolante et sous les arbres, le nombre des fleurs, des fruits et le poids des graines ont été inférieurs de 40 à 50 p. 100. L'arrêt dans l'assimilation semble porter tout d'abord sur l'élaboration des principes hydrocarbonés.

'6° Le taux centésimal de substance sèche et le taux des cendres sont plus élevés en l'absence de l'électricité, les végétaux qui croissent hors cage s'étant constamment montrés plus riches en eau et plus pauvres en matières minérales que la plante de même espèce sous cage isolante.'

M. Celi's adaptation of one of Sir W. Thomson's apparatuses as an 'appareil pour expérimenter l'action de l'électricité sur les plantes vivantes,' cit. and figured by M. Grandeau in loco from 'Annales de Chimie et de Physique,' sér. v. tom. xv, October 1878, is well worthy of inspection in this connexion.]

The next part of my Lecture will be devoted to showing by the aid of three maps and one statistical table, how greatly man has modified the external aspect of the world he lives in by the introduction into the several parts of it of cultivated plants and domestic animals, previously, of course, unknown even in the wild state to such areas of its surface. The maps by their colours show the areas on which the parent stocks of the most valuable and now most widely spread of these acquisitions have, with more or less of approach to demonstration, been shown to be indigenous. The short table of statistics tells you in its second line that half of them came from one single 'quarter' of the globe, or in the language of modern zoogeographers from one single zoological 'region.' The table and the maps taken together show us how largely some quarters of the globe have been benefited by borrowing from others, or in the language of my subject, how largely they have been modified by man's interference.

The first of these maps is very closely similar to the one which shows on Mercator's projection the now more or less generally accepted zoogeographical regions of the earth's surface, the Palaearctic, to wit, the Ethiopian, the Oriental, the Australian, and the two regions of the New World, the Nearctic and the Neotropical; as given by Dr. Sclater, and in Wallace's great work on Geographical Distribution.

The second of these maps is an enlargement of that given by Professor Huxley in the 'Journal of the Ethnological Society of London,' June 7th, 1870, to illustrate and embody his views on the distribution of the principal modifications of mankind. This map serves, besides other useful purposes, that of limiting off, by a distinct colouration, a particular portion of the vast Palaearctic region which is specially important to the subject in hand, as it was either actually upon it, or upon regions closely adjacent to it within that region, that the parent stocks of the moiety of our cultivated plants and domesticated animals may either be found still living or may reasonably be supposed to have existed formerly. The particular subdivision of the Palaearctie Region has been coloured in a particular way by Professor Huxley, so as to indicate that upon it his 'Melanochroic' or dark-white variety of our species was living not in perfect purity of stock, but more or less peacefully intermingled with the Mongoloid and with his 'Xanthochroic' or fair-white varieties. The area thus peopled occupies itself on the map a district something of the shape of a tuning-fork, the two arms of which would form the northern and southern boundaries of the Mediterranean eastward from the longitudes of Albania and Tripoli; and would be carried by a broad base extending from the Caucasus over Syria and a part of north-west Arabia to the Red Sea, whilst its stem would cover Kurdistan, Khorassan, and North Persia, and end by bifurcating at a spot near Peshawur. The importance of this area is illustrated by the fact that a region very closely corresponding, if not quite coincident with it, is marked out upon quite different principles in the next map. A coincidence of much less intricacy, and therefore of much less cogency, though still not without a certain curious significance, is furnished to us by the fact that a certain island of blue colour, placed by Professor Huxley in the 'Dark Continent' of Africa to indicate the presence in Upper Egypt, Nubia, and Abyssinia of some traces of the Australioid type, corresponds with the area in that continent whence most or all of her few gifts of valuable cultivated plants and valuable domesticated animals have come to us, viz., the cotton plant, and, very probably, the date-palm; the ass, from the native stock *Asinus taeniopus;* and the cat, from the native stock *Felis maniculata.*

Of the two arms into which the eastward end of this area bifurcates, the upper or northward one would correspond with the Kuenlun range, and the southward with the Himalayas; Ladak, and part of the table-land of Thibet, lying between them. It is in the Kuenlun range that Jade mines are found.

The third map, being one of Johnston's Charts of the World on blank Mercator's Projection, has been coloured so as to illustrate the following facts in the distribution of certain plants and certain minerals connected with the ancient development and subsequent progress of human civilisation. One region is coloured as it is in the 'Planteogeographisk Atlas,' tav. ii, of Professor Schouw, Copenhagen, 1824, so as to show the distribution of the *Vitis vitifera* over the countries forming the northern and southern shores of the Mediterranean and Black Seas, over Asia Minor, Palestine, and Mesopotamia, over the lowlands both of Astrakhan and Turan, and along the southern slopes of the Himalayas, so as to end at the eastern extremity of that chain. In nearly the same latitude as that eastern extremity, and about in the same longitudinal line as the long axis of the Peninsula of Malacca, a spot of another colour marks the situation of the amber mines of Burmah[1], while four spots of yet a third colour in British Burmah, Banca[2], Celebes, and Khorassan[3], respectively indicate localities in

[1] For the Amber mines of Burmah see Balfour's 'Indian Cyclopaedia,' s. v., 1871; and Keith Johnston's 'Royal Atlas,' map. 28, in loco lat. 26° 20'.

[2] For the existence of tin together with copper in Burmah see Mortillet, 'Revue d'Anthropologie,' i. 1875, p. 653.

[3] For the similar collocation of the two metals which when combined make bronze in Khorassan and elsewhere in Central Asia south of the Caspian, see v. Baer, 'Archiv für Anthropologie,' ix. 4, p. 262, 1877. We know from the same irrefragable authority, Bulletin Acad. Sci. St. Pétersbourg. tom. xvii. p. 417–431, 1859, and tom. i., 1860, pp. 35–37, that the date-palm is still represented a little to the north of these deposits of tin and copper, at Sari, in the as yet Persian province of Mazanderan on the south

which copper and tin are still found in such proximity to each other and in such accessible abundance as to suggest that it is not improbable that in some one of those districts prehistoric man may have come upon the invention of bronze. A fourth colour marks the position of the Kuenlun Jade mines[1], whence, in still earlier than bronze times, stone weapons may with great probability be supposed to have been procured by man before he migrated into the jadeless regions westward. (See Article XXXIX, p. 686.)

The New World was coloured as it is in Schouw's tav. viii, l. c., to show the area of distribution of the Cactaceae, a region comprehending South America north of the Tropic of Capricorn, the Isthmus of Panama, the Peninsula of California up to 30° N. lat., the West Indian Archipelago, the northern shores of the Gulf of Mexico, and the strip of Gulfstream-washed North American coast between the Alleghanies and the Atlantic up to about 40° N. lat. From this area more than 25 per cent. of all our cultivated plants have been procured, as the annexed table shows; and, of course, since the time of Columbus.

This table (based, so far as it deals with the vegetable kingdom, mainly upon De Candolle's 'Géographie Botanique,' pp. 986–987) gives approximatively the proportions in which the several 'regions' of the globe established by that phytogeographer and by several zoogeographers have contributed to make up the lists of such cultivated plants and domesticated animals respectively as are of considerable, even if not always of cosmopolitan, importance.

Of (approximatively) 160 Cultivated Plants.	Per cent.	*Of (approximatively) 21 Domestic Mammals.* Per cent.	
The Palaearctic species are ..	50	50	are Palaearctic.
" Oriental "	25	14	" Oriental.
" African "	25	14	" African.
" Nearctic "	2·5	0	" Nearctic.
" Neotropical "	25	14	" Neotropical.
" Australian "	0	0	" Australian.

Of some of the great facts which these maps and this table put

shore of the Caspian. This tree is supposed to have been carried thither, as to so many other places, by the Arabs during their career of conquest, which contrasts to such advantage and in so many ways with that of other Mussulman conquerors.

[1] For an account of the Jade mines in the Kuenlun Range see Cayley, 'Macmillan's Magazine,' October, 1871; and for Jade generally, Rudler 'Popular Science Review,' October, 1879.

before you, half diagrammatically, the anthropologists, zoologists, and geographers[1] of the last quarter of the last century and the first third of this had possessed themselves; and following, at

[1] Pallas, Betrachtungen über die Beschaffenheit der Gebirge: an Address delivered Jan. 23, 1777. Zimmermann, 'Geographische Geschichte,' Bd. i. p. 114, 1778; Bd. iii. p. 250, 1783. Link, Die Urwelt und das Alterthum, i. p. 243 seqq., 1821.

There is perhaps no need for me to apologise for quoting the exact words of Pallas's Discourse, the less so as, though it appeared in two forms, one German, the other French, within a year of its being delivered, it is not, I think, a very common book. The issue which I quote from is that of 1778, the year in which his *Novae Species Glirium* appeared, six years later than the year in which the second volume of his *Spicilegia* with its wonderful Fasciculus XI. was published.

The difficulty in reading Pallas is to understand how his writings can bear the date they do. But he shall speak for himself:—'In den mittägigen Thälern dieses alten Landes muss man das erste Vaterland des menschlichen Geschlechts und des weissen Menschen suchen, die von dort in ganzen Nationen die glücklichen Gegenden von China, Persien und besonders Indien bevölkert haben, dessen Einwohner nach dem allgemeinen Geständniss unter allen Nationen die ersten gesitteten waren, und wo man vielleicht die Stammwurzeln der ersten Sprachen in Asia und Europa suchen muss. Selbst Tybet, eine der höchsten Gegenden Asiens dessen Einwohner, ihrem Vorgeben nach, von einer Ort Affen welche dieses Land zuerst bewohnten, abstammen (mit welche sie auch ohnedem einige Aehnlichkeit haben) Tybet hat die Verfeinerung seiner Sitten jenem Lehrern zu danken, die aus Indien dahin kamen.' Pallas adds as a note to this passage, 'Ich kann nicht umhin, hier zu bemerken, dass alle, so wohl in den nordischen, als in den mittägigen Ländern von dem Menschen zu Hausthieren gezämte Gattungen, in den gemassigten Erdstrichen des mittlern Asiens ursprünglich wild gefunden waren, das einige Kameel ausgenommen dessen beyden Abartungen nur in Africa gut fortkommen.' Pallas then proceeds to instance the wild ox, the buffalo, the wild sheep, the Bezoar goat and the Ibex, from a crossing of which he supposes our common domestic goat to have arisen; the wild boar and, as I believe, incorrectly, the wild cat (*Felis catus*), as being the parent stocks of their domesticated namesakes, and having their original homes in the mountains which occupy Central Asia and a part of Europe. He adds, 'Das zweybuckelige Kameel ist in den grossen Wüsten zwischen Tybet und China noch wild vorhanden.' Prejevalsky's 'From Kulja across the Tian-Shan to Lobnor' will be familiar in its English translation to most of us; his account of the wild camel is not more interesting as compared with this remark of Pallas' than is his account, p. 38, of the devouring of apples and apricots on the northern slopes of the Tian-Shan by wild boars, goats and deer, when compared with Tournefort's words ('Voyage du Levant,' Amsterdam, 1718, 4, t. 2, p. 129, cited by the Botanist Link, l. c., p. 234) describing a country which he visited and found to be 'Ein Land erfüllt mit natürlichen Weinbergen und Obstgärten wo Nussbäume, Aprikosenbäume, Pfirsichbäume, Birnbäume und Apfelbäume von selbst wachsen. Er setzt hinzu, man kann nicht zweifeln, dass hier einer von den Theilen Georgiens ist, wo, nach Strabo, alle Arten von Fruchten in Ueberfluss sind, welche die Erde ohne Cultur hervorbringt.' Georgia lies some distance away from Lobnor, but both alike lie well within the great mountain system with its outliers which is called 'Asiens Buckel' by the other writers, as also I apprehend within the modern 'Steppengebiet' of Grisebach.

whatever distance, the great Pallas, they insist upon the strength of the claims of that portion of Central Asia whence issue the great rivers Ganges and Indus, Tigris and Euphrates, and which they speak of as 'den grossen Buckel Asiens,' to be considered as the primitive home of man, mainly as it was, according to them, the original home of *all* our domestic animals and so many of our cultivated food-plants.

These writers and discoverers slightly overstated their case when they said that *all* our domestic animals could be referred to parent stocks indigenous to that region, though, as will be shortly shown hereafter, it would have been little beyond the truth if, instead of saying all the domestic animals absolutely, they had said all the domestic animals *which are absolutely indispensable to modern man's comfort and progress.* But their case for their particular thesis would have been greatly strengthened if they had known that jade in the form of stone implements had accompanied man together with the goat into Western Europe, and was found no nearer to the Swiss Lake Dwellings than are the Kuenlun mines pointed out on my map; if they had known that copper and tin could have been smelted together into bronze so readily either in Khorassan or in Burmah; if, to put however injudiciously my weakest point last, they had also known that amber—such a frequent accompaniment of prehistoric man—also lay within easy reach of his curious hands in this latter country. But prehistoric archæology has till lately made but little advance since the time of Lucretius. De Candolle ('Hist. des Sciences et des Savants,' p. 263, 1873), indeed, classes it as a discovery as new and as great as five others of the twenty or thirty years previous to 1873, viz., spectrum analysis, convertibility of force, the greater extent of glaciers in geological times, natural selection, and the alternation of (animal) generations; and the writers referred to knew not, and could not have known, the whole strength of their position. As regards my present purpose it is, in these but little later days, superfluous to point out how the discovery of mines whence prehistoric man must, or at least might, have furnished himself with his weapons, implements, and ornaments, actually upon or along the same mountain ranges, spurs, and valleys, in which he must, or at least might, have found in a wild state the animals which he has now around him as necessary and universal elements in his own social life, bears upon the extent, as measured by latitude and longitude as well as by other gauges, to which the world has been modified by his migrations and importations.

Let us now enumerate the twenty domesticated mammals which we possess, and which for practical purposes may be taken as making up a tale of about twenty or twenty-one; let me specify which amongst them belong, as regards their origin, to the Palaearctic region, and to the restricted portion of it already dwelt upon and defined, as the maps show you; and thirdly, leaving considerations of locality and of number, let me contrast the value of nine, ten, or eleven mammals which man domesticated in that district with that of the others acquired from[1] or contributed by all the other regions of the globe taken together.

Our twenty-one chief domesticated mammals may be enumerated in something like order of merit and necessity to us as follows: the dog, the cow, the sheep, the pig, the horse, the cat, the goat, the ass, the camel, the dromedary, the buffalo, the alpaca, the vicugna, the reindeer, the zebu, the banteng, the yak, the ferret, the rabbit, the mongoose, and the guinea-pig, omitting some few species the importance of which as being locally limited to very small areas, and as consisting of individuals numerically few, is too small to make it necessary to notice them. Representatives of more than one-half of this list can be fairly claimed by the Palaearctic centre of creation as owing their parentage to stocks native to its soil; this half consisting of the dog, the cow, the sheep, the pig, the horse, the goat, the camel, the dromedary, the reindeer, the ferret, and the rabbit. I have said 'representatives' of one-half of this list because it is more than probable that some of our breeds of domestic dogs and of pigs may have been reclaimed from wild parent-stocks in other regions of the world. There can, however, be no reasonable doubt that the great majority of the domestic breeds known till comparatively recent times in Europe, of each of these two animals, the dog and the pig, were drawn from parent-stocks living in

[1] It is a curious point in mythology that, so far as my memory serves me, no god nor demigod should have the credit assigned him of having domesticated any animal except the horse. Of course this fact, if fact it be, shows two things with more or less probability; firstly, namely, that these acquisitions were made in very far-off times, not merely in 'the ages before morality,' but in those much earlier ones, 'the ages before history;' and secondly, that the acquisition of the horse was made in later days than the domestication of the other animals in question.

the Palaeartic Region, and this is all that is necessary for my present argument.

As regards the ox, the sheep, the horse, and the goat, I cannot think that with our present knowledge of zoogeography there can be any question that their parent-stocks were Palaearctic animals; and I am further prepared to express my belief that further investigation will render it highly probable that it was in that particular though very extensive part of the Palaearctic Region spoken of vaguely as 'Asiens Buckel,' or 'Hochasien,' and comprehending portions of all the great mountain ranges from the Caucasus proper to the northern side of the Hindoo Koosh, and from the Taurus to the Altai mountains, that these several parent-stocks were brought under the influence of domestication. Wild animals are still to be found in some one or other or in several spots within that area from which we have no *à priori* reason for doubting that man might in the course of ages have educed the three last-named of the four domestic animals, the ox, the sheep, the horse, and the goat; and that a wild ox existed in the regions in which the Old Testament writers lived, not only their writings, but the Assyrian sculptures, and not only the Assyrian sculptures, but geological remains testify. The case, however, for the ox having been first domesticated in Central Asia, is the weakest of the four, and it may be well to take it first. The Rev. William Houghton has in his memoir on the domestic mammalia of the Assyrian sculptures ('Trans. Soc. Bibl. Archaeology,' v. i. st. i. p. 2, 1876, and ibid. 1877, p. 54) given us a very spirited drawing from one of the Assyrian sculptures representing the hunting and the killing of the wild ox. What is of special value in this sculpture is for our purpose the presence between the shoulder-blades of a hump, which is present in so many other of the larger Ruminantia, but which, as Mr. Houghton remarks, reminds us of the Indian zebu, and of the fact that there are no specific differences between these two oxen underlying their soft parts. There can be no doubt that the figure is intended to represent a wild animal. The Accadians, who were in the habit of giving names to animals which referred to the countries whence they obtained them, gave names to the ox, which Professor Sayce thinks must refer to the country between the Euphrates and Syria and to Phœnicia. The bulls of Bashan, and possibly of the Taurus range, may be rightly recalled to our memories by these names. The European names for the ox, on the other hand, are said by M. Joly (cit. Isidore St. Hilaire, 'Hist. Nat. Gén.' iii. p. 89) to have an Asiatic origin, and M. A. Pictet ('Des Origines Indo-Européennes,' pp. 330–343) has declared his views to the same effect. This, however, is only what would have been expected in the European languages of the Aryan division. What is of importance as regards the domestication of the ox is to note that, though such languages as the Finnic may use loan words taken from Aryan tongues to express the general idea of Ox (=Bovine animal), they frequently have true Turanian vocables to denote such particularities as we have in view when we speak of heifers, calves, cows, bulls, and the 'ox,' *sensu strictori*, confirming in the last matter the statement of Strabo (vii. 4, 8) that castration was learnt from the eastern Europeans and Sarmatians. There is in fact a good deal of evidence for a view which should hold either that the Turanian races domesticated the wild ox, or rather the wild calf, independently; or that the human species did this great work before the differentiation into Aryan-speaking and Turanian-speaking men was carried out. That the Scythian breed of cattle should have been hornless in the time of Herodotus (iv. 29) appears to me to be explicable, not on the hypothesis taken up by later observers that it is an effect of cold, but as being a result of long-sustained domestication; and if what Hehn, p. 413, l. c., suggests as to the South Russian breed of small red steppe cattle being descendants of those Scythian oxen is true, we should have a further confirmation of this view furnished in their persistency. There is, at any rate, another breed of cattle in the South Russian steppes, which goes by the name of the 'Kalmuc' cow, and is supposed to have accompanied the Mongolian or Tartar hordes in their invasion of Europe.

Some writers, in defiance of the arguments that have just been glanced at, and of many others, have advocated the claims of Africa to be considered the parent country of the domestic ox. The main fact, as it seems to me, which has induced or seduced them rather into this conclusion, is the great extent to which boviculture has developed itself through the length and breadth of the 'Dark Continent.' But without wasting words in pointing out the curious conclusions to which this reasoning would lead us in other cases, I would refer such persons to Middendorff's account of the development which this same boviculture has attained in Siberia, and to his statement that not only have the nomads of the southern steppes, the Buráts, the Mongols, and the Kirghiz, herds numbering thousands and tens of thousands wintering out in the open, but that even the Jakuts (by, it is true, taking more care of their cattle) have, from being simply nomads, become a pastoral people of distinction, and even 'improved cattle-breeders!' ('Sibirische Reise,' iv. 2, 2, p. 1323.)

Coming, in the second place, to the consideration of the sheep, I must allow that considerable hesitation has been expressed by many writers as to the question of its parent-stock; and that doubt may be not altogether unreasonably felt as to whether that stock may not have become extinct, as the parent-stock of the cow has all but entirely done. But what I know of the deerlike agility and watchfulness of some of our European mountain breeds of sheep, and in the second place what I see of the smaller size of the animal as giving it a less severe battle to fight for its survival, makes me slow to think that their parent-stock need be thought likely to have perished as has that of the larger ruminant. And setting this view aside, we may say that either the Mouflon (*Ovis musimon* and *cyprius*), with a range from Majorca to Cyprus, and not without footings, occupied by such varieties as *Ovis orientalis* and *Ovis vignei*, on the mainland on various points of the mountain-ranges of the Taurus and of Armenia to those of Tibet; or the Argali, *Ovis fera Sibirica* s. *Ovis argali*, with an all but equally extensive range from the Pamir plateau above Samarcand and Bokhara to the Sea of Okhotsk as *Ovis nivicola*, or *Ovis polii*, must be credited with having given to the world this inestimable gift. If it shall really turn out to be correct that a true Argali, that is to say a variety of wild sheep, in which both sexes carry horns, had been found in the Taurus, as Ainsworth (cit. A. Wagner, 'Die Geographische Verbreitung der Säugethiere,' Abhandl. d. ii. kl. d. Ak. d. Wiss. München, Bd. iv. Abth. i. p. 139) and Ritter ('Erdkunde,' xi. 506), have averred is the case, the claims of the Argali would to some persons, I apprehend, appear to be stronger than they might if its range should, as I incline to think it will, be shown to be confined to the more easterly limits just given. But under any and all circumstances, the fact that the female Mouflons have no horns, whilst the female Argalis have them, though smaller in size no doubt than those of the male, when coupled with the fact that in the older breeds of domestic sheep both sexes carry horns, appears to me to be conclusive in favour of the Central Asiatic Wild Sheep. As regards the Natural History arguments I shall content myself, and I daresay others also, by referring[1] to the already quoted eleventh fascicle of Pallas's 'Spicilegia,' and to Isidore Geoffroy St. Hilaire's 'Histoire Naturelle,' iii. pp. 86–87, ibique citata, but I would add a couple of facts from the linguistic side of the mass of arguments available for deciding the question. The first of these is as follows:—The early Accadian inhabitants of the plains of Babylonia, when they gave an epithet to an animal, very frequently chose it from the locality whence they supposed the animal to have been derived. And the epithet which they bestowed upon the sheep was 'num,' or 'numma[2],' which means 'the highlands,' and which, as applied by people living in those wide plains, and as being applied by them to the wolf also, has a very obvious significance. It is true, as anybody may convince himself by consulting Bochart's 'Hierozoicon,' ii. 2, p. 516, that poets and other writers, Aryans and Semites, Greeks, Romans, and Arabians indifferently, have connected the sheep, as they saw its habits, with mountainous scenery and surroundings; what is of special importance in the epithet as used in the Accadian column of the bilingual Assyrian inscriptions is, that it was used in such a country and in such early, not to say such unpoetical, times.

My second linguistic fact tells, as it seems to me, strongly in favour of not merely the Asiatic but of the Mongolian origin of the domestic sheep; it appears, I mean, to point to a more or less limited area in the wide field of Asia as having been the particular spot, or at any rate one of the particular spots, where a wild sheep

[1] I may add a few words from the already quoted memoir by Andreas Wagner, l. c., p. 137: 'Hochasien ist recht eigentlich das Vaterland der *Wildschafe* und *Wildziege*, die hier in zahlreicher Menge und in sehr verschiedenen Formen vorhanden sind. Ob diese alle gesonderte Arten oder nicht vielmehr viele von ihnen nur Rassen von Hauptarten ausmachen, ist eine Frage die noch lange nicht beantwortet est.' Mr. Wallace's suggestion ('Geographical Distribution,' vol. i. p. 232) that the vast plateau of Central Asia may, in comparatively recent geological times, have been much less elevated, and may then have been much more fertile than it is now, deserves more than this simple mention.

[2] For these facts see the Rev. W. Houghton 'On the Mammalia of the Assyrian Sculptures,' Trans. Soc. Biblical Archaeology, v. 1, 1876, pp. 3–7, ibid. 2, 1877, p. 42. 'Gleanings from the Natural History of the Ancients,' 1879, pp. 12–89.

was brought under domestication. This fact as given by Ahlquist in his interesting work, 'Die Kulturwörter der Westfinnischen Sprachen,' 1875, p. 14, is to the effect that the Tatars, by which word he means presumably Turkic and Tungusic tribes in the neighbourhood of the Lake Baikal, have words of their own for ram and ewe, *täkä*, to wit, and *sarik*, which the Tscheremissians, who live now as far away from that lake as is the river Volga, use as loan words. It is, I submit, not easy to imagine that a word would have maintained its life thus intact and vigorous if the thing which it represents had not been part of the national life of the tribe using and retaining it. And this suggestion gains in force when we learn from the same authority, l. c., that the Hungarian language has adopted Slavonic words for the ewe, the ram, and the lamb, and find him deducing from this the conclusion that the Hungarians, albeit a steppe tribe, had not been shepherds before they came into relation with the Slavs. It may have been due to this, but it may also have been owing to a prepotency either in the Aryan language or in the pastoral craft of the Slav race. For except upon one or other of these latter hypotheses, it is difficult to see why the Tscheremissians on the Volga should have retained their Mongolian names for the ewe and ram, whilst not only the Hungarians but the Ostjaks, the Vogals, the Mordvins, the Syrians, and the Wotjaks, from the Volga to the Irtisch, should be using more or less modified Slavonian words for the same things. Anyhow, that a lowly, organised language, such as the Tataric, should have words of its own for the domestic ewe and ram, is a point of great significance, especially when we consider that these Tatars lived around the spurs of the Altai range on the lower and middle zones of which the Argali was then, as now, available for the purposes of domestication.

Thirdly, of the horse. The fossil or semi-fossil bones of the horse, *Equus caballus*, are found in the lower Thames valley gravels under our feet, and from this area of the world's surface all the way to the regions round the Lake Baikal; and in this latter district the horse is found, as I think may be safely said, in a wild state at the present day. It is true that a very large number of naturalists of the first rank, such as Mr. Darwin and Mr. Wallace, have acquiesced in the view which teaches that the so-called 'Tarpan' is but a 'feral' animal, the offspring of runaway stallions and mares from the steppe droves. But it is also true that the small number of naturalists of the first rank *who have travelled over the Russian steppes,* viz. the younger Gmelin, Pallas, and Middendorff, are of the contrary opinion; and that whilst acknowledging that the steppe horse, like, perhaps, all other domestic animals except the sheep, may lapse into feral habits, they hold to the view that the true 'Tarpan' is a descendant of the pristine wild stock, whilst the 'Musin' is but a steppe horse run wild [1].

The main argument for the descent of the wild horses of the steppes from the domestic or semi-domesticated stocks of the Turanian nomads, rests on the fact that a great variety of colour is observed to exist in the free droves. This, however, appears to me to prove nothing more than that the tame and wild varieties breed freely together [2]. I myself, long ago, succeeded in obtaining numbers of feral rabbits, parti-coloured with white, on an area already occupied by the ordinary English wild rabbit. The feral rabbits never attained an equality in numbers with the gray stock, but being spared in shooting, whilst the wild stock was not, they maintained themselves for a considerable number of years in what was for themselves as against predatory attacks of various kinds an only too conspicuous prominence. But nobody would have argued

[1] See Middendorff, 'Sibirische Reise,' iv. 2, 2, pp. 1308–1321; Gmelin, 'Reise durch Russland,' i. 45, 1770, and for drawing Tab. ix.

It may be well, for several reasons, to give the exact facts as to the opinions which Pallas held at various times respecting the feral or the truly and aboriginally wild character of the so-called wild horse of the Steppes. In 1769 (see 'Voyages de Pallas,' French translation, 1788, vol. i. p. 324) Pallas inclined to the view of the Tarpan being simply a feral race; and he repeated this opinion in 1773 (see l. c., vol. v. p. 90). But in 1776, in the eleventh fascicle of his 'Spicilegia Zoologica,' p. 5, he expresses himself to the following effect: 'Equi feri in campis Bessarabicis circaque Tanaïn et per omnem Tatariam magnam in desertis vagantur gregatim, *magnam quidem partem* fugitivis Nomadum equis permixti atque multiplicati; ideoque versicolores; *aliqui tamen habitu toto a cicuratis adeo discrepantes ut primitiva de stirpe feros esse dubitari vix posset.* Conf. de iis qui ad Tanaïn atque in eremo inter Volgam et Jaikum habentur S. G. Gmelin [the younger Gmelin], Reisen durch Russland, vol i. p. 44 seq., et Itinerarii nostri, vol. i. p. 211; et vol. iii. part ii. p. 513.' See also the posthumously (1831) published 'Zoographia Rosso-Asiatica,' vol. i. p. 260.

To these references I would add the 'Geographische Geschichte,' i. p. 181, 1778, of the zoologist Zimmermann. Writing only two years after the appearance of Pallas's Memoir just cited, Zimmermann not only entirely accepted the view given above in italics, but l. c., p. 204, speaks in not exaggerated terms of Pallas as 'der erste aller von mir gekannten Reisenden.'

[2] The Mongols and Kalmucks, from superstitious motives, take great pains to secure various colours for their domestic horses, sheep, and goats. Hence some of the variety in the feral horses. See Pallas, 'Mongol. Volk,' i. pp. 117, 178, 179.

from this that no wild stock could be held to exist on that area. Still though we may follow the highly trustworthy naturalists and travellers just mentioned as to the persistence of the aboriginal horse in a wild state on the Turanian steppes, we have yet to show that it is probable that it was on those steppes rather than in any other part of the wide area over which the true wild horse once ranged that it became reduced to domestication. And here again the Accadian inscriptions come to our assistance; the horse being called there (see the Rev. William Houghton, l. c., 1876, p. 3) 'imiru Kur-ra,' 'the animal from the East.' We see from this that these ancient Turanians claimed, and had their claim acknowledged, that the taming of the horse was an achievement wrought out in the cradle of their race. I have sometimes thought that the ascription by the Greeks of this feat to Poseidon may be similarly taken to indicate that they had some sort of dim conviction that the horse had come to them from the countries beyond the Egean. This, however, may be an overstraining of the value of such hints. But the history of the horse, whether dug out of Pile-dwellings and Neolithic interments, or out of records such as those in Genesis and Exodus, show that it came comparatively late into use, as a domestic animal at least, in the regions to the west of the Central Asiatic plains [1].

The fourth of the domesticated animals, which I have spoken of as having in great probability had a Central Asiatic origin, the goat, namely, has its claims, supported by the vast majority of naturalists without any hesitation. The wild *Capra aegagrus* of the Taurus, of the Caucasus, of the Persian mountains, and of Kirghiz and Tatar districts, 'possibly mingled,' says Mr. Darwin, 'Domesticated Animals and Cultivated Plants,' i. p. 105, 'with the allied Indian species, *Capra Falconeri*,' may be safely taken as the parent-stock of this animal. The Tibetan and Angoran varieties of the goat, by their well-deserved reputation, may seem, even in these days and under the light thrown on the subject by the book just quoted, to lend some support to Col. Hamilton Smith's principle [2], that where the largest and most energetic breeds of a race exist, there we may look for their original habitation.

[1] See further, Lenormant, 'Premières Civilisations,' tom. i. p. 322; Ahlquist, 'Die Kulturwörter der Westfinnischen Sprachen,' 1878, p. 9; 'Spectator,' April 27, 1878, *ibique a me citata*.

[2] These are Col. Hamilton Smith's views (Nat. Library, 'Dogs,' vol. ii. p. 163, cit.

It is thus seen that four out of the twenty-one domesticated Mammalia may, with very considerable probability, be supposed to have been first domesticated in Central Asia, and though the non-cosmopolitanism of the two camels, *Camelus bactrianus* and *Camelus dromedarius*, renders them less available for my present purpose, that, viz., of pointing out the great changes which man has effected in transporting into all parts of the world what he found only in some more or less circumscribed portions of it, the facts of the Central Asiatic origin of the two-humped variety or species, and of the South-western Asiatic, or at least Arabic, origin of the one-humped dromedary, bear not a little on the whole question.

I do not omit the dog and the pig[1] from the list of the animals

Rev. Wm. Houghton, l. c.). Speaking of the possible derivation of the greyhound from an Asiatic home 'somewhere to the westward of the great Asiatic mountain chains where the easternmost Bactrian and Persian plains commence, and where the steppes of the Scythic nations spread towards the north,' Col. Hamilton Smith says, 'when we look to the present proofs of this conclusion and assume that where the largest and most energetic breeds of the race exist, there may we look for their original habitations, we then find, to the east of the Indus, the very large greyhounds of the Deccan, to the west of it the powerful Persian breed, and to the north of the Caspian the great rough greyhound of Tartary and Russia, and thence we may infer that they were carried by the migrating colonies westward across the Hellespont, and by earlier Celtic and later Teutonic tribes along the levels of Northern Germany as far as Britain.' It is curious that Colonel H. Smith should not in this connexion have mentioned the Thibetan dog, figured by himself, l. c., with the tan-coloured supra-orbital stripe, common so significatively to this variety *and* to the Mexican Alco. For the Thibetan mastiff has long been known to be one of the largest varieties of the species, and quite recently (see 'Times,' Dec. 26, 1879) Mr. Baber, the consular resident at Szechuen, is reported as writing of them as the largest dogs he had ever seen.

[1] That the Central Asiatic wild boar lends itself readily to domestication is thus expressed by Pallas, 'Zoographia Rosso-Asiatica,' p. 269: 'Porcelli cicurari assuescunt facile et cum domesticis generant.' And Radde's words ('Reisen im Süden von Ost-Sibirien,' 1862, i. 236) are as much or more to the point, as they apply to adult animals: 'So muss ich gestehen, dass sie sehr friedlicher Natur sind und es mir mehrmals passirte mittelalte Wildschweine sich mir bis auf vier Faden weite nahen zu sehen.' If the so-called 'wild' boar is so tame as to allow this so many centuries after the invention of gunpowder, it is easy to understand that it may have been much more amenable to man's influence thousands of years before that discovery. As regards the dog, it seems probable that even within the limits of the Central Asiatic region we are dealing with, two very distinct wild stocks may have furnished corresponding tame ones. The large Indian dog, or Hyrcanian dog of the ancients, may very reasonably be supposed (as suggested by Fitzinger) to have been the parent-stock of the modern Thibetan mastiff, whilst Pallas says that the Kalmuck domestic dog is so like the jackal of the same region that it is impossible not to consider them identical. See 'Spicilegia Zoologica,' Fasc. xi.

which there is good reason, to my judgment, for thinking were domesticated in Central Asia, because I do not think they were domesticated within that area, but because, I cannot deny, that it is probable they were also domesticated elsewhere. But it may fairly be suggested that the art, skill, and craft of domesticating these and the other six animals having been first learnt in Central Asia, spread thence; and that thus all or nearly all the acquisitions which man has made in the way of domestication, may thus owe their origin—if not in the way of actual blood-lineage, yet in that of being the fruits of man's experience acquired there—to the district in question.

I pass by a natural transition to point out very shortly, not the cardinal necessity of the possession of the sheep, goat, ox, horse, camel, pig, and dog, for food and clothing, for locomotion, and for carrying on the processes of the hunting, of the pastoral and of the agricultural life, but how that necessity has been unconsciously recognised by man in certain of his earliest institutions.

Of these seven mammals, six are now distributed over the face of the whole habitable world; but long before this had become the case with any one of them, except possibly the dog, man had expressed unconsciously, if not quite inarticulately, his recognition of their value by using them in one way or another for one or another of his most sacred rights and ceremonies. The single Latin word 'Suovetaurilia,' denoting a particular kind of sacrifice of the swine, the sheep, and the ox, which is figured on many a tablet found in this as in other countries, and was performed at great crises of Rome's fate, may suffice as regards the three animals which speak so plainly to our eyes in those sculptures. To Eastern and to Western people it was indifferent (see Exod. xii. 5, Ps. l. 9, and classical writers *passim*) whether sheep or goats were taken out of the fold for this purpose. As regards the dog, Livy (xl. 6) tells us that in the Purification of a Macedonian army the two halves of a dog's body were placed, one on one side, one on the other, of the road along which the soldiers were passed. Similarly, we are told by the Arab Ahmed Ibn-Fozlan, who must have witnessed the proceeding with a good deal of repulsion, that a dog was cut in half and put into the ship in which a Norse chief was burnt in the tenth century on the banks of the Volga (see Anderson, 'Proc. Scot. Soc. Antiq.,' May 13, 1872, p. 522); and I have myself taken up, not without some effort in overcoming a certain reluctance, the bones of a dog who was keeping his mistress faithful company in a grave undoubtedly of the earliest Neolithic period in England [1].

As regards the horse, Achilles, fresh from his conversation with Xanthus and Balius, tells the Trojans (Il. xxi. 132) that even their wonted sacrifices of horses will not profit them; the Mongols (see Howorth's 'History of the Mongols,' i. 262, 289; and Yule's 'Marco Polo,' i. 265, cit. in loco), the Lusitanians (Livy, Epit. 49), and the Norsemen (see Ibn Fozlan, l. c.), all alike sacrificed horses on great occasions.

I have not found, nor did I expect to find, any account of the sacrificing of the camel, either in Semitic or classical literature; if, however, it be a sound principle that races as yet uncivilised would be likely to sacrifice or otherwise deprive themselves upon great occasions of the services of their oldest and most valued domesticated animals [2], we ought to be able to show that the Central Asiatic nomads did so by the 'ships of their deserts.' And I find in Mr. Howorth's valuable 'History of the Mongols,' i. p. 426, the following passage:—

[1] See 'British Barrows,' p. 518, 1877; 'Journal Anthropological Institute,' October, 1875, p. 157; Article XVIII, p. 394.

[2] As I am speaking of animals domesticated in Central Asia, I have not mentioned the ass which, as Dr. Sclater has shown ('Proc. Zool. Soc.' 1862, p. 164), owns as its parent-stock the *Asinus taeniopus* of Abyssinia. Its history gives, however, an illustration of the principles enunciated above at least as striking as those of any of the eight Asiatic mammals just specified. From the references made to this animal in the Pentateuch, it would appear to have been domesticated in the region there treated of before either horse or camel, though subsequently to the ox. Pindar's reference to it as used for sacrifice by the Hyperboreans (Od. Pyth. x. l. 52) will be to persons who will bear in mind its African origin almost as convincing evidence of the great antiquity of the date of its domestication as its appearance on the oldest Egyptian monuments of the Fourth Dynasty. Hecatombs, such as Pindar speaks of, are, numerically, figured on one tomb, reproduced for us by Lepsius. That the ass should so early have been introduced into Hyperborean regions even by a poet is a little surprising, considering that the horse, which is so much better suited for such climates, was already available there; but besides being surprising it is also significant. For the sacrificial and ceremonial use of this animal, see Orelli's 'Excursus ad Tacit. Hist.' v. 3, vol. ii, 1848, of his edition of the great historian, ibique citata; Dean Stanley's 'Jewish Church,' i. 96, ibique citata; 'Pindar, ed. Dissen and Schneidewin,' sect. ii. 1847, p. 353, ibique citata. For the linguistic Palaeontology of the name, see Lenormant, 'Origines de Civilisation,' i. 319. For the use of the animal by the modern Hyperboreans see Middendorff, 'Sibirische Reise,' iv. 2, 2, p. 1322, where, however, that naturalist, albeit reckoning 'Pferdekenntniss und Pferdezucht als seiner Specialität,' or one of them, leaves the difficulty above hinted at unexplained.

'Ssanang Setzen now goes on to tell a story which crystallises for us a very curious phase of old Mongol manners. Altan Khakan had a son called Pubet Paidshi. The young man died, and his mother determined to kill 100 boys and 100 foals of camels, which were to be buried with him, and to accompany him as an escort to the other world. She had killed over forty boys when a tumult arose among the people.'

Here I think I may leave this part of my subject, the significance of this series of facts being sufficiently self-evident. For as against these seven domesticated mammals which Central Asia may with so much probability claim as being her gifts to mankind, inasmuch as she either herself furnished their parent-stocks, or at any rate furnished the necessary opportunities for gaining the knowledge subsequently used in domesticating similar stocks elsewhere, what can all the rest of the habitable globe set either as regards cosmopolitanism or as regards importance? As regards importance the other thirteen are all but insignificant; as regards cosmopolitanism, universal importation, that is, either for purposes of practical utility or 'animi voluptatisque caussa,' as Caesar put it, we can mention but the African cat and the African ass.

I come now to the consideration of the facts and views with which botanists have supplied us as to the original homes of our cultivated plants. Our own inspection and recollection of the landscapes of the various countries in which we have travelled will enable us to estimate the greatness of the change, which man's migrations and transportations have effected in the sphere of all his labour under the sun. And I will begin what I have to say under this head by the apparent paradox that the argument which our cultivated plants furnish us with for determining the locality whence man issued to occupy the world and subdue it, and alter its external appearance, would, like some other arguments, have appealed with greater force to one of the civilised races of antiquity than it does at first sight to us. It is, herein also like some other arguments, cogent for all that. Let us state it. Fifty per cent. of our cultivated plants have been shown by De Candolle, 'Géographie Botanique,' pp. 986, 987, and by Élisée Reclus, 'The Ocean' (English Trans. ii. chap. 27, 292), following him, to belong to 'Europe' and 'Asie septentrionale et occidentale,' that is to say, to the Palaeartic Region of Zoogeography. So far the figures are equal for cultivated plants and for domestic animals, and I do not feel it necessary to dwell upon the differences which the other proportional numbers show as regards Africa proper and South

America. What is of importance, however, to point out, is that to anybody living, not merely before the time of Columbus—whose discovery has been said to have acted upon the Old World much as the approximation of a new heavenly body, planet, or other, might act upon the whole earth—but before the time, say, of Tacitus and Agricola, what Africa and India had given him in the way of cultivated plants, would have seemed just as insignificant as what, putting the ass and the gallinacean birds out of sight, they had given him in the way of domestic animals. He might, if living in Italy, have said, as did Columella (iii. 9. 5, cit. Hehn, p. 423, l. c.), 'Curae mortalium obsequentissima est Italia, quae paene totius orbis fruges adhibito studio colonorum ferre didicit,' and pointed out beforehand the airy inaccuracy of Goldsmith's apostrophe to that country in his 'Traveller.' He might, I am inclined to think, with the evidence available to him, have pointed out, and correctly, that the middle zone of deciduous trees which girdled then, as now, so many of the Italian hills with a belt of chestnuts, and much, therefore, of its distinctive character, was due to the intercourse of Rome with Pontus and Galatia in pre-Christian times. And he might have drawn thence the same conclusions which we may, I think, also draw as to the area on the world's surface whence man set forth westward on his career of occupation, having, as he had, available for his wants, vegetables, plants, and trees of no less value, and of no less prominence in the landscape, than are these of Palaeartic, though not of Italian, origin, viz. wheat, barley, rye, oats, spelt, buckwheat, millet (*Panicum*), peas, beans, hemp, flax, cabbage, turnip, plum, walnut, vine, cherry, olive. Of tea, coffee, sugar, even of rice, of oranges, and of several other of the gifts of the Indian region; or of coffee, or any one of the three, or four if we include *Musa ensete*, now flourishingly growing in Sicily, gifts of Africa proper, a man living at that time had as little knowledge as he could have had of the gifts to come from the still undiscovered New World, of the potato, of maize, of the pineapple, to which his all alien stone pine was to lend its name, of the equally incorrectly named artichoke, of the tomato, now somewhat variously obtrusive or intrusive in Mediterranean regions, or of tobacco, or of the prickly pear, or of the agave, though of the two latter in reference to what was then, and is still, such a large part of human activities, it can be said, as by

Admiral Smyth (p. 17 of his 'Memoir of Sicily and its Islands,' 1824), that they 'form impenetrable palisades for fortifications, and in the plains present very serious obstructions to the operations of cavalry.'

My third map, with the distribution of the vine after Schouw, should be compared with my picture from Kaempfer's 'Amoenitates Exoticae,' Fasc. iv. p. 711, 1712, of what he calls, p. 714, the *Messis dactylifera*, the date-harvest of Persia, and speaks of as being 'lusus magis quam labores.' The distributional limits of the 'fruitful' vine and the 'fruiting' date-palm now, as of yore, overlap each other, as was pointed out by Arago in his 'Mémoire sur l'État Thermométrique du Globe terrestre' ('Œuvres,' v. 216, ed. 1858) in Palestine, when from this fact he, with much ingenuity, argued that 3300 years have not appreciably altered the climate of Palestine. For

'la limite thermométrique en moins de la datte diffère très peu de la limite thermométrique en plus de la vigne;'

and, what makes the argument, especially to those who have Kaempfer's picture of the luxuriant date-harvest before their eyes, entirely and beautifully perfect, he further (p. 217, l. c.) tells us,

'à Abusheer (Bushire) en Perse, dont la température moyenne ne surpasse certainement pas 23°, on ne peut, suivant Niebuhr, cultiver la vigne que dans les fossés ou à l'abri de l'action directe des rayons du soleil.'

A more simple, but also a more conclusive proof that the Syrian climate has not materially changed within the historic period cannot be imagined [1].

I began this Lecture with details as to the distribution of pines

[1] It is strange to find that Arago could, when dealing with France, have swerved so far from the line of evidence he employed as to Palestine, as to have told the Chamber of Deputies (February 27, 1836), 'Vous serez peut-être étonnés d'entendre que dans les environs de Paris, il y a quelques siècles, il faisait beaucoup plus chaud qu'aujourd'hui.' vol. xii. 'Œuvres, Mélanges,' p. 434. But for the context one might have been tempted to take the last of the words just quoted as applying to the month of February only; and in all gravity the title of chapitre xix. in the memoir already quoted, vol. viii. 'Œuvres,' vol. v. 'Nat. Scient.' p. 239, 'Observations prouvant que l'ancien climat se maintient dans une partie des Gaules,' might seem to justify such an interpretation of words spoken under some provocation in debate. And the more so as a few pages previously (p. 214) we find Arago recognising the essential deceptiveness which must attach to 'une foule de documents historiques' in the following words: 'On remarquera que je devrai résoudre le problème que je me suis posé sans avoir recours à des chiffres certains, à des observations numériques. L'invention des thermomètres ne remonte guère qu'à l'année 1590; on doit même ajouter qu'avant 1700 ces instruments n'étaient ni exactes ni comparables.'

and firs by man's agency; I may fitly close those details by attempting something as regards that of one of the palm tribe. For, though Leopold von Buch was wrong in holding that the two natural orders were altogether mutually exclusive as regards natural geographical distribution, as a voyage in the Mediterranean, or the sight of Martius' picture of *Brahea dulcis* (vol. iii. taf. 162) side by side with a true pine in Mexico, teaches us, there can be no doubt that Caesar and his countrymen were, speaking generally, right in holding the fir and the beech to be as characteristic of Gaul and Britain as their repeated allusions and their coins show them to have thought the palm was of Palestine and the adjacent countries, at least eastward and southward.

What then do we know, firstly, as to the original home or botanical region to which the date-palm, *Phoenix dactylifera*, belongs? and secondly, what can we surmise as to the particular spot in that area in which that tree was first made available as a cultivated plant, and subjected to those human influences which three of my pictures are intended to illustrate?

As to the first of these questions there is no doubt, and no occasion for any very lengthy answer. The region which Grisebach names, after its principal constituent element, simply 'Sahara,' and which stretches over more than ninety degrees of longitude from Macaronesia to Multania, from the Canaries, that is, to the Great Desert of Rajputana, and which comprehends not only the Sahara strictly so-called, but cis-Saharan Africa also, from the longitude (E. 10°) of Tunis eastward, and not only Old Egypt and Arabia, but young 'Egypt,' or Sinde also, is the botanical region of the date-palm. Sir Joseph Hooker ('Morocco and the Great Alas,' 1879, p. 409) has pointed out that there are many Canarian plants which form an exceedingly interesting group, the members of which, though chiefly Egypto-Arabian, are found to extend in some instances even into Western India, and he suggests that 'it is not unreasonable to suppose that such have covered Africa in a sub-tropical latitude, and thus reached the Canaries under conditions now operating.' Other plants, therefore, if not other trees, may have spread over the same area, whether by man's aid or without it, and may be taken as equally characteristic of it, even though they may not need so much 'water to their feet and fire to their heads.' It is, *per contra*, I may remark, by a surplusage of water to the head and a noxious quantity of heat to the feet that the latitudinal limits, south and north parallels, of the date-palm are given. If, as Dr. Daubeny suggested ('Lectures on Climate,' 1863, p. 86), we have, as in certain truly tropical (and continental) countries, heavy falls of rain during that particular time of the year when the pollen should be carried to the pistilliferous flower, this latter will not be fertilised (unless by man's interference), the dioecious character of its flowers putting it thus, as it does also *Borassus flabelli*, at a serious disadvantage as compared with the cocoanut-palm[1], *Cocos nucifera*, whose company they, in consequence perhaps of a sense of their inferiority, appear to avoid.

On the other hand, the requirement of a mean temperature of from 70° to 81·5° F. excludes the date-palm from bearing dates, except under specially favourable, and therefore only locally prevalent conditions, eked out by human protection, on the north shores of the Mediterranean[2]; all the way from Alexandretta,

[1] It is not only the 'tempest's wrath,' but also the 'battle's rage,' which the dioecious character of the date-palm helps in the work of destruction. The pictures from Lepsius's Egyptian Denkmäler which I have copied for this Lecture show that this was known in the time of those 'great old houses and fights fought long ago.' History tells us that Norman and Saracen (see Admiral Smyth's 'Sicily,' p. 19, Martius, iii. p. 262), Anjou and Arabian generals have, each alike, in defiance either of the letter or of the spirit of their professed religion, or of both, cut down the male palms, and so prevented *pro tanto* the reproduction of the tree with 360 uses to mankind. The modern Arabs, according to Rohlfs, 'Afrikanische Reisen,' Aufl. 2, 1869, p. 70, cit. Hehn, l. c. p. 513, appear sometimes even in very severe military operations or devastations to spare the palm even when cutting down other fruit trees. But Abd-el-Kader appears to have had some transgressions even as to palm-trees on his conscience to repent of. The solitary palm, the existence of which von Baer reports to us on a certain peninsula on the south shore of the Caspian, called in our maps the Peninsula of Mejankal, but in his apparently, and curiously, the Peninsula of Potemkin, is, I should think, a solitary survivor of some such proceedings as those figured in my Egyptian pictures. Von Baer himself looks upon it as a survivor of companions not destroyed by the art and malice of man, but by local refrigeration, due to the extinction of certain volcanoes which were active even in comparatively recent times. Verecunde dissentio.

[2] Martius writes on this subject, l. c. iii. p. 263, as follows: 'Haec igitur habuimus quae de incremento, quod arbor illa capit in imperio florae per Europam meridionalem patenti, diceremus. Ex quibus intelligi potest *omnino ut nascatur* arte effici, cogitandumque nobis esse eam plures culturae gradus intra fines quos occupaverit percurrere. Quae si ad summum ascenderit flores emittit, fructusque dulcis et boni saporis edit, et si *manu et arte accedente fecundetur*, etiam semina ad propagandum idonea gignit; quod fit in Hispaniae parte ad meridiem versus remotissima, in Sicilia, in Graeciae promontoriis maxime ad meridiem vergentibus, et in insula Cypro (nimirum sub lat. bor. 35° et medio calore annuo 18° C. ad 20° C.). In altera zona flores quidem

where it still grows, to Gibraltar. The solitary, and for this as for other reasons unfertile, palms which we still see here and there in the Aegean and along the region of the west and north shores of Asia Minor, short of the Black Sea eastward, and which still strike us as being something as alien to that landscape as was the seedling-palm at Apollo's Delian temple to the eyes of the much-travelled Ulysses (Odyss. vi. 162), have been planted there not as 'food-plants,' but 'animi voluptatisque caussa.'

As regards the particular and single spot in the vast botanical region, if particular and single spot there really was, upon the longitudinally vast area of which the date-palm was brought under that human influence which has since caused it to effloresce into so many varieties, very various opinions have been advanced, and I propose to add a fresh one to their number. It may appear at first sight that such a discussion and such an attempt have in themselves an intrinsic futility. We do not need to refer to King Juba's report of his exploratory voyage to the Canaries to learn that the date-palm will bear dates even in an oceanic and uninhabited island, and some persons may think that we need only, like the wits of Charles's time, to study ourselves and our sensations to see how the forefathers of the Guanches, when they in some post-Juban or post-Augustan period occupied the island, would, under the stimulus of hunger alone, come to learn the art of date-culture, even if they had not brought the knowledge of it with them. Still, I think, on the doctrine of chances, or, what comes to the same thing, the principle 'Frustra fit per plura quod fieri potest per pauciora,' as well as upon certain concrete arguments furnished by the Egyptian monuments on the one hand, and by certain curious but still life-like and truth-like stories on the other,

et fructus fert, sed fructuum caro non plane excolitur, quum acerbi sit saporis, fructificatio nulla, semina cassa : huc pertinet tractus littorum maris Mediterranei in Gallia meridionali, in Italia, in Sardinia, item regionis Dalmatiae, Insulae Ionicae, Graeciaeque septentrionalis. Cujus zonae terminum septentrionalem posueris fortasse 41° 20′–45° lat. bor. In tertia linea palam durat quidem sub divo, sed flores aut raros aut nullos emittit : immo frondescit tantum ; cujus zonae terminus septentrionalis tendit ut commemoravi, per insulas lacus Verbeni sub lat. bor. 46° media anni temperie a 12° usque ad 13° C. Arbor hic provivere potest, etiam si hiemis temperies interdum sub frigoris gradum deprimatur dummodo ne nimis (forsan ad—3° vel 4° C.) accedat, quo frigoris etiam mali medicae, citri, aurantii, et myrti extingui atque opprimi solent. Superior altitudinis terminus in monte Aetnae usque ad pedum 1400 vel 1680, teste viro cl. Philippio, adscendit.'

which I find in Herodotus, though other writers have not quoted him *ad hoc*, that it is not unreasonable to suggest yet another site for the one where man first intermeddled with the self-preservation and the species-preservation of the date-palm [1].

Kaempfer, from whose opinion I dissent with the greatest reluctance when I consider the thoroughness with which that model traveller availed himself of his opportunities, and the abundance of those opportunities themselves, gives us his views as to the place in which the palm in question was first cultivated by man, in the following words (p. 714) of his 'Amoenitates Exoticae,' Fascic. iv. 3, published in 1714: 'Ejus patria in Asia quidem, nam Africam non moramur.'

Ritter ('Erdkunde,' Theil xiii. p. 771 seqq.) considerably narrows this area by selecting the Babylonian Nabataeans in the valley of the Tigris and Euphrates as having been the people who discovered and first practised the art of improving the date-palm. But Professor Rawlinson, in a letter to me, gives 'B. C. 1500, or even earlier' as the possible date of a probably early Babylonian cylinder figured with palms in his 'Ancient Monarchies,' iii. p. 23, 2nd ed., and 'B. C. 883' as the earliest date for Assyrian figures representing palms; whilst the Egyptian Twelfth Dynasty, which possessed the tree, carries us back to from 1860 B. C. to 2200 B. C., according to Wilkinson and Brugsch respectively.

[1] It is a little amusing to find twenty-two pages, 289–311, of Seemann's 'Popular History of Palms' devoted to discussing the questions whether the date-palm was an 'endemic (genuine) member of the Canarian Flora,' and 'whether it was indigenous to the Canary Islands.' This book was, however, published in 1856, and though something, and perhaps too much, was even then ascribed to 'occasional causes' in the explaining of anomalies in geographical distribution, a good deal has been learnt since that time which would have rendered that dozen of pages impossible. It is remarkable that the author did not use the arguments supplied him by Dr. Carl Bolle in support of the Atlantic hypothesis, which since those days has been buried as deeply as the Atlantic itself was supposed to have been. Of course another question, not raised indeed by Dr. Seemann, as to whether the art of artificially cultivating the date could have originated in what we now know to be oceanic islands and spread thence eastward is, by the knowledge we have since 1859 gained as to 'Man and Nature' in their independent as well as in their mutually interacting operations, rendered all but an impertinence. We (see Darwin, 'Animals and Plants under Domestication,' i. p. 328, 2nd ed.) 'do not believe that any edible or valuable plant except the Canary grass has been derived from an oceanic or uninhabited island.' It is only just *not* an impossibility that the date-palm should have been so derived; if it had been, this would indeed have been something more surprising than all the usefulness of the tree, than all its beauty, and even than all the blunders which have been made about it.

Unger, 'Sitzungsberichte k. Akad. Wiss. Wien,' Bd. xxiii. Hft. i. p. 204, 1857, suggested the countries on the eastern side of the Persian Gulf as the centre whence in the very earliest times of commerce and international intercourse this plant was carried over Arabia, Persia, Hindustan, and North Africa. But he, in a later Memoir, published after travel in Egypt, ib. xxxviii. pp. 75, 104–106, 1859, quotes Delile as averring that, *valeat quantum valeat*, the Egyptians themselves considered that Arabia Felix was the original country of the date-palm; and by twice (ll.cc.) mentioning the fact that Egypt itself is called not only the land of the sycamore, but also the land of the palm-tree, he would appear to assign the same weight to that tradition which I have felt justified in assigning to those embodied in the Accadian Inscriptions. Unger himself suggests, though very guardedly, that the date may have been imported into Lower from Upper Egypt. He is, as such a botanist would be sure to be, careful to disclaim any acceptance of the cogency which others have assigned to an argument based on the luxuriance of growth which the tree does attain in the locality in question. 'There is nothing in all this, however, to hinder us from supposing that the palm does so flourish there, because in its migration from the north southwards it came in the latter place for the first time upon the soil best suited to it.'

Martius, on the other hand (l. c. iii. 263), uses this very argument for assigning the original site of the date-palm to the southern part of Tunis, 'Blad el-Dscherid,' as he writes the name of the locality, *h. e. arida terra*, 'falso nuncupata Biledulgerid,' as he adds, 'Beled el-Jerid,' I may add as named in Johnston's Royal Atlas in lat. N. 34°, long. S. 10°. 'Quo loco,' says Martius[1], 'solidae conspici-

[1] In the same African connection in Martius's grand book I find the two following passages, which are in themselves a lecture on the extent to which man has modified the landscape of Southern and Northern Africa, both by acclimatising there plants, some useful merely, some beautiful, some both, from 'regions Caesar never knew,' China, namely, and America. The maize might have been added to the importations specified in those quotations. Speaking of the date-palm Martius says (p. 264): 'In Promontorium Bonae Spei introducta, nunc per calidiorum regionum hortos sparsa et *una cum Solano tuberoso, Tritico rep.* colitur.' Speaking of the North Coast and the *planities Tadschurae*, he writes: 'Palma illic est splendidissimum decus sylvarum *Citri aurantiorum quae Opuntiis cinguntur*.' The potato, the orange, and the hedges of opuntias set round them were as little known to 'all the world' of the Mediterranean as the gas, the coal, the glaze of *our* pottery, and the tea, coffee, and tobacco, which, though sold by the *épicier* in every English hamlet, and making up, as some persons will say, but a Philistine tale, are yet become absolute necessaries of life even

entur palmarum sylvae *tanquam in prima patria gnatae.* Earum fructus sunt frequentissimi et sapidissimi.'

Professor Robert Hartmann ('Die Nigritier,' pp. 116, 117, 1876) gives the most recent account with which I am acquainted of the date-palm as cultivated in Africa. His remarks as to the existence in Africa of really wild forms of *Phoenix*, *e.g. Phoenix spinosa s. humilis*, the 'Kjom-kom' of Senegal, with small well-flavoured fruits, and the *Phoenix reclinata*, a very variable form, to set off against the *Phoenix sylvestris indica* which has so often, though not correctly, been said to be botanically indistinguishable from the cultivated *Phoenix dactylifera*, are specially valuable. He insists, as I had also done previously to becoming acquainted with his views, upon the priority of date, which the Egyptian monuments, with date-palms figured upon them, can show us compared with the Assyrian or Babylonian similarly adorned. The only argument which I can imagine—I have not seen or read of its being suggested by any one else—to be likely to be set against this one based upon the monuments, is one, partly indeed based upon ancient Egyptian records, but partly also upon stories recorded for us, with every indication of their being true, by Herodotus. It might run thus. Brugsch (cit. Unger, l. c. 1839, p. 106; 'Geographie der alter Aegypten,' p. 74) tells us that palm-wine is enumerated in the Egyptian Tribute-lists as having been one of the articles received from Babylonia. Herodotus, i. 193, informs us that wine was made from dates in Babylonia; and in a couple of passages, iii. 20, 22, he relates what has become, since his time, the very commonplace

to the most cultured of mankind. [Since writing as above I have met with an address delivered September 24, 1879, by the traveller Nachtigal before the German Association for the Advancement of Science at Baden-Baden. In this Address, delivered in deprecation of certain schemes for the utilisation of certain parts of the Sahara, Herr Nachtigal insists that whatever other results might accrue from the letting in of the waters of the Mediterranean upon the salt marshes of the district referred to by Martius, as cited in the text above, the ruin of the date-culture, the most valuable treasure of that region, would probably be one also. For 'the date-palm,' says Herr Nachtigal, 'wants fresh water for its roots, solar rays for its crown, and fears rain and atmospheric moisture. It is well known that date-plantations in the neighbourhood of the sea produce only second-rate fruit; and there is some ground for doubting whether the regions exposed to the doubtful benefits of the Mediterranean are really the regions which produce the best dates in the world and thereby have earned the name Beled el-Dscherid, that is, literally, the Land of the Date-palm. Would it not be rash to endanger a cultivation, the produce of which is counted by millions of money, for very uncertain results?']

occurrence of a superiorly civilised assailing an inferiorly civilised race by means of strong drink. He tells us how Cambyses sent a cask of palm-wine, presumably brought with him from his own country, as a present to the Aethiopians, previously called 'blameless' by Homer. The Egyptians, also, according to Herodotus, ii. 86, employed palm-wine (probably, when we compare this passage with the others already cited, from Babylonia) in the process of embalming. I have set up this argument, but I think I may knock it down, and thereby save some of my friends some trouble, by observing that in England we ought not to think that because a country shows pre-eminent skill in manufacturing raw material, that therefore that raw material must even have been grown, not to say, originally found growing wild, in that country. Fusel oil, for example, a product analogous in its operation to palm-wine, is manufactured in this land out of potatoes; but potatoes are not thereby shown to have been first cultivated either in Great Britain or Ireland.

I gather from Martius that 'Celsius in Hierobotanico operam dat ut Palaestinam tanquam veram hujus arboris patriam esse ostendat.'

I, in my turn, venture to advocate the claims of the Nasamones who dwelt around the south-eastern extremity of the *Syrtis major*, now known as the Gulf of Sidra (long. E. 20°), to be considered as the race which first cultivated the palm; and with them I should couple those of the Garamantes of Fezzan. What I have to say about them is based mainly upon the apparently truthful and certainly life-like account which Herodotus gives of them in three or four passages, i. 32 and iv. 172, 182, 183, none of which Martius refers to in his enumeration of profane writers in contradistinction to the sacred writers who mention date-palms referred to by Celsius; but partly also upon a single passage of Diodorus Siculus, iii. 4. We find thus that the Nasamones were a numerous and powerful, but certainly a very far from civilised people. They combined polygamy with polyandry, much as the Massagetae did at the same time. Some of their other practices combine several of the notes of a priscan people, such as the veneration of ancestors, and the regard for justice which has made the words *Trollorum fides* proverbial; and finally those social feelings which are indicated by the words, i. 32, *ἀνδρῶν δυναστέων παῖδας ὑβριστάς*, and which

Nillson[1] has averred to be eminently characteristic of savages. I subjoin the entire passage, iv. 172[2], for several reasons, in the original Greek; and I submit that a people who embodied so much of wild life in their social condition, could have learnt little from any of the nations to the east of them, whether Egyptian, Arabian, Assyrian, or Persian. But as regards their dealing with the date-palms, we have this remarkable statement made by Herodotus, iv. 172 and 186: 'In the summer they leave their flocks by the sea-side, and go up to the district Augila, to get in the harvest of the date-palms which grow there in great abundance, and are of great size, all being fruit-bearing.' Now Herodotus, and, as he tells us, i. 193, the Greeks of his time generally, were acquainted with the bisexual dioecious character of the palm and the fig; that the Babylonians used artificial means for securing the impregnation of the pistilliferous trees he tells us in loco; and we know that those latter were, as they are (see Kaempfer, l. c. p. 672) still, to be found in Persia as they are (see Martius, l. c. p. 264) still in Egypt. Some considerable weight, therefore, may fairly be assigned to his statement, iv. 172, to the effect that at Augila (as also probably, see iv. 183, in the country of the Garamantes) there were none but

[1] 'Early Inhabitants of Scandinavia,' Eng. Trans. ed. Lubbock, p. 167: 'Aristocracy is strongly developed amongst all savage nations.'

[2] As regards the size of the Fezzan dates, the dates of the Garamantes referred to by Herodotus, iv. 183, as living ten days' journey from Augila westwards, and as having φοίνικες καρποφόροι πολλοὶ κατάπερ καὶ ἐν τοῖσι ἑτέροισι, we have the following information from Dr. Ed. Vogel, cit. Seemann, l. c. pp. 285, 286: 'The largest date of Fezzan (which is also the best) is 21½ Parisian lines and 10 in diameter, the smallest 7½ by 5.' Lyon, in his 'Narrative of Travels in North Africa,' 1821, p. 72, tells us, 'the dates of Sockan in Fezzan are of a quality far superior to any produced in the north of Africa.' Herodotus, iv. 172: Αὐσχισέων δὲ τούτων τὸ πρὸς ἑσπέρης ἔχονται Νασαμῶνες, ἔθνος ἐὸν πολλόν· οἳ τὸ θέρος καταλείποντες ἐπὶ τῇ θαλάσσῃ τὰ πρόβατα, ἀναβαίνουσι ἐς Αὔγιλα χῶρον, ὀπωριεῦντες τοὺς φοίνικας· οἱ δὲ πολλοὶ καὶ ἀμφιλαφέες πεφύκασι, πάντες ἐόντες καρποφόροι· τοὺς δὲ ἀττελέβους ἐπεὰν θηρεύσωσι, αὐήναντες πρὸς τὸν ἥλιον καταλέουσι, καὶ ἔπειτα ἐπὶ γάλα ἐπιπάσσοντες πίνουσι. γυναῖκας δὲ νομίζοντες πολλὰς ἔχειν ἕκαστος ἐπίκοινον αὐτέων ποιεῦνται τὴν μίξιν· τρόπῳ παραπλησίῳ τῷ καὶ Μασσαγέται, ἐπεὰν σκίπωνα προστήσωνται μίσγονται. πρῶτον δὴ γαμέοντος Νασαμῶνος ἀνδρὸς νόμος ἐστὶ τὴν νύμφην νυκτὶ τῇ πρώτῃ διὰ πάντων διεξελθεῖν τῶν δαιτυμόνων μισγομένην· τῶν δ' ὡς ἕκαστός οἱ μιχθῇ, διδοῖ δῶρον τὸ ἂν ἔχῃ φερόμενος ἐξ οἴκου. Ὁρκίοισι δὲ καὶ μαντικῇ χρέωνται τοιῇδε. Ὀμνύουσι μὲν τοὺς παρὰ σφίσιν ἄνδρας δικαιοτάτους καὶ ἀρίστους λεγομένους γενέσθαι, τούτους, τῶν τύμβων ἁπτόμενοι. Μαντεύονται δὲ ἐπὶ τῶν προγόνων φοιτέοντες τὰ σήματα, καὶ κατευξάμενοι ἐπικατακοιμέωνται· τὸ δ' ἂν ἴδῃ ἐν τῇ ὄψει ἐνύπνιον, τούτῳ χρᾶται. Πίστισι δὲ τοιῇσίδε χρέωνται· ἐκ τῆς χειρὸς διδοῖ πιεῖν, καὶ αὐτὸς ἐκ τῆς τοῦ ἑτέρου πίνει· ἢν δὲ μὴ ἔχωσι ὑγρὸν μηδὲν, οἱ δὲ τῆς χαμάθεν σποδοῦ λαβόντες λείχουσι.

these latter pistilliferous trees. Of course this statement would need supplementation by one which he may very well have supposed his readers would take for granted, to the effect that the Nasamones (and probably the Garamantes) brought the male flowers from a distance, carefully selecting those *liberaliori quodam vigore ac pleniori habitu*, just as Kaempfer, p. 672, tells us the Persian date-farmers did; this being, in fact, pretty nearly the whole of what is required in the way of cultivating the date-palm. The palms resorted to, at least by the Nasamones, were large; they could not, therefore, have been wild date-palms; and being thus proved to be more or less under the care of man, they are, secondly, proved to have been even more under that care and more dependent upon it than cultivated palms elsewhere, inasmuch as the pollen necessary for fertilising their flowers had to be brought to them from a distance, the bridging over of which could only be effected by man's intervention at fixed intervals. My argument, in other words, lies in the fact that a tribe, which, being of very priscan habits and customs, cannot be supposed to have borrowed much from its more civilised neighbours, was, nevertheless, credited in the time of Herodotus with possessing groves of cultivated and exclusively female date-palms, which bore large and, we may perhaps infer, excellent dates, as they still continue to do.

We have furnished to us in modern times a verifiable history very closely parallel with that which I here suggest; the *Elaeis guineensis* is undoubtedly, as a cultivated plant, an acquisition of negro minds; and as Hartmann says, l. c. p. 118, this acquisition has been made for us by a race which still carries on the practice of human sacrifices; and that in sight of European factories and European steamboats, much as the Nasamones, whom I suppose to have discovered the cultivation of another palm, carried on their polyandry almost within sight of the Egyptian pyramids. 'The thing that hath been is the thing that shall be.'

The picture before you from Kaempfer's 'Amoenitates Exoticae,' p. 711, Tab. iii. Fasc. iv. 1711, coupled with his comment[1] upon the scene of enjoyment which it represents, and in which the palm-

[1] 'Hi sunt palmicolarum in messe, ut sic loquar, dactylifera lusus magis quam labores, neutiquam cum nostratium agricolarum infinitis occupationibus comparandi. Heu ilias hic laborum! dum agros effringimus, subaramus et resulcamus; dum occamus et liramus, runcamus et refarrimus. . . Secus sentias de ambrosiis dapibus Persarum et Arabum; hae gratis omnino et solo almae naturae munere conferuntur.'

trees play so essential a part, may remind us of Linnaeus's often-quoted saying, 'Man *dwells naturally* within the tropics, and lives on the fruit of the palm-tree; he *exists* in other parts of the world, and makes shift to feed on corn and flesh.' But it may suggest a little more than this. It may cause us to think seriously on the question what will be, not the effect on external nature which man's action will produce, but what will be the effect which external nature will produce upon man, if by some recrudescence of a glacial period, either in a geological sense, or in the economic sense, which an exhaustion of our supply of Nearctic as well as Palaearctic coal would, in the absence of any substitute, bring about, we should be driven southwards, and become tropico- instead of cosmopolitan. What will be the effect of the easy terms upon which life can be maintained in the tropics upon the species which has hitherto never developed a lasting civilisation except under the stimulation 'curis acuens mortalia corda' of northern latitudes or mountain elevation[1]? How will it fare with intellectual culture when and where, not to speak any further of our date-palm, the cocoanut-palm, the banana, the breadfruit, will make exertion so all but superfluous for the *dura a stirpe genera* who now govern the world? If we are to guide ourselves as we peer into the twilight of the future by what we can see going on in the broad Mediterranean noonday of the present, the example of the idle Corsican is not altogether encouraging. A Corsican family, we are told by their French fellow-citizens[2], with a couple of dozen of chestnut-trees, and with a herd of goats which 'find themselves,' to the great disgust of all botanists, have no aspirations left to satisfy beyond that of being able to buy a gun, to the great disgust of all sportsmen. In a matter of prophesying, Sir, the argument from authority and authorities has its legitimate place, and upon the present occasion it happens to have a very legitimate time[3]. I have in a work on 'Hereditary Genius,' published in the year 1869, found it stated that 'No Englishman of the nineteenth century is purely nomadic:' and that even the most so among them have also inherited many civilised cravings which are necessarily starved, and thus entail

[1] Wallace, 'Natural Selection,' p. 318; and Bonstetten, 'L'homme du Midi et l'homme du Nord,' 1826, passim.

[2] Hehn, l. c. p. 346.

[3] [On the evening when this lecture was delivered the chair was occupied by Francis Galton, Esq., F.R.S., author of a work on Hereditary Genius.—EDITOR.]

personal discomfort and create the required stimulus for their gratification, when they are tempted to let themselves lapse into savage Corsican sloth. In the thousands of years which may yet intervene between us and the necessity for a southward exodus, these cravings and uneasinesses will have become more inseparably a part of our nature than even the most optimistically-minded member of the London School Board can as yet assert they have become. I have not far to look for another authority who will assure us that the desire and appetite for intellectual enjoyment may become as really a 'constitutional demand' as those lower stimuli which in 'old, unhappy, far-off times' enabled man to subdue other gregarious animals to his own uses, and, so aided, to overrun victoriously the whole globe. Your secretary, Mr. Bates, after eleven years of absence from England, to which the world owes his charming work the 'Naturalist on the River Amazon,' and after seeing many tribes living in the happy position in which a moderate amount of light work will produce for the simple, peaceful, and friendly people all the necessaries of their simple life (l. c. vol. ii. p. 137 of the 'Mundurucus'), found yet (p. 416)—

'after three years of renewed experience of England, how incomparably superior is civilised life, where feelings, tastes, and intellect find abundant nourishment, to the spiritual sterility of half-savage existence, even if it were passed in the garden of Eden. What has struck me,' says Mr. Bates, 'powerfully is the immeasurably greater diversity and interest of human character and social conditions in a single civilised nation, than in equatorial South America, where three distinct races of man live together. The superiority of the bleak north to tropical regions, however, is only in their social aspect, for I hold to the opinion that although humanity can reach an advanced state of culture only by battling with the inclemencies of nature in high latitudes, it is under the equator alone that the perfect race of the future will attain to complete fruition of man's beautiful heritage, the earth [1].'

[1] V. Baer, who after making himself in his earliest years a prince among biologists, became in his later years a not inconsiderable geographer, expressed himself in Russian so long ago as 1848 in one of the geographical manuals of the Geographical Society of Russia to much the same effect as the two writers above quoted. His words were translated into German no earlier than 1873, and stand as follows in his 'Studien aus dem Gebiete der Naturwissenschaft,' Theil ii. Hälfte i. pp. 45–46 :—

'Mit recht propheziet daher aus dieser Productions-Kraft der Tropenwelt ein geistreicher Botaniker, Herr Meier in Königsberg, dass der Mensch, in der civilisirten Welt rasch sich mehrend, in die heisse Zone zurückwandern werde. Jamaica allein, so gross ungefähr als das Königreich Sachsen, werde vielleicht 25, ganz gewiss aber 12½ Mal so viel Menschen ernähren können als Sachsen. Und wie viele, setzen wir hinzu, die Waldfläche Brasiliens! Verkehrt genug nennt man diesen Boden einen jungfräulichen. Er trug nur für den Menschen bisher wenig Frucht. Dagegen hat der Haushalt der Natur Jahrtausende hindurch in ihm organischen Stoff aufge-

It is something like an anti-climax to suggest that even when man is in the tropics and is surrounded there with all the luscious temptations which the cultivation of those latitudes will give him on such easy terms, he will still be beset with certain urgent needs in the way of supplying his bodily wants as well as his cravings for intellectual excitement and employments. For it is a mistake to think that the craving for flesh and even for fatty foods becomes at all obsolete in tropical countries, or that man is at all less of a flesh-feeder in the regions which are now at least the selected localities of the most typical flesh-feeders, from Carnivora in his own class—through the vertebrate snakes down to Arachnida in the Invertebrata—than he is in the picturesque wilds where the flesh-

speichert für die Menschen, die noch kommen sollen, sowie in andern Gegenden früher, als die Erdrinde sich bildete, in ihr Steinkohlen vergraben wurden als ungeheure Magazine von Brennstoff für eine Zeit, in welcher das vermehrte Menschengeschlecht den Waldwuchs sehr beschränkt haben wird. Aber der Mensch, der aus Europa zurückwandert in die Heimath, aus der er ursprünglich ausgewandert ist, bringt einen Gewinn mit, den er unter den Tropen nirgends erlangt hat, *die Liebe zur Arbeit, die Schätze der Wissenschaft, die Künste der Industrie und die Einsicht in die Bedürfnisse eines geordneten Staatslebens.* Damit könnte er freilich die arbeitscheuen Naturzustände der früher dort ansässigen Völker erdrücken. Aber man darf hoffen, dass unter dessen auch die humane Gesinnung immer mehr sich fest gesetzt haben wird, dass der weiter vorgeschrittene Mensch erkennt, dass er kein Recht hat, den unentwickelten jüngern Bruder zu unterdrücken, sondern die Verpflichtung, ihn schonend weiter zu bilden; dass die Erde ein grosses Waisenhaus ist in welchem die sogenannten Wilden die zahlreichen Waisen sind. Man darf erwarten, dass unter den Tropen, wo weniger Zeit für die Production der Nahrungsmittel erfordert wird, wo die Natur sie an Bäumen reifen lässt, die geistige Bildung viel allgemeinen werden muss als im Norden. In der That hat doch in Mittel-Europa, ich spreche nicht einmal von unserem Norden, nur der kleinste Theil der Bewohner Musse genug, um die geistigen Anlagen, die in ihm schlummern, auszubilden, während die bei weitem grössere Anzahl das ganze Jahr hindurch beschäftigt ist, den Nahrungsstoff zu bereiten. Wie viel mehr Musse hat schon die arbeitende Klasse in Italien! Auch hat sie nicht aufgehört, an Kunst und Wissenschaft sich zu ergötzen, und wird dafür von uns Nordländern mit Unrecht, wie ich glaube, träge genannt. *Europa scheint mir also für die Geschichte der Menschheit, wenn wir sie in grossen Umrissen überblicken, die hohe Schule, wo sie zur Arbeit gezwungen wurde und geistige Beschäftigung lieben lernte.* Möchten unsre Nachkommen der 30sten und 300sten Generation, wenn sie im üppigen Ceylon oder in der ewig gleichmässigen Temperatur der Südsee-Inseln im Schatten der Palmen über die Schicksale der Menschheit nachdenken, anerkennen, *dass wir die Schulzeit im Norden nicht schlecht verwendeten, sondern geistige Gaben auf sie vererbt haben, die unter den Tropen nicht gedeihen konnten, denn noch jezt lebt der Naturmensch dort in sorgenloser Kindheit.* Möchten sie, wenn sie wissenschaftliche Reisen in den Norden unternehmen um den Schnee mit eigenen Augen zu erblicken, mit dankbarer Achtung auf die Ruinen unserer Schul- und Arbeitshäusen sehen.'

Mr. Herbert Spencer speaks to the same effect in his 'Principles of Biology,' vol. ii. pp. 502–3.

furnishing Cheviot sheep are so abundantly forthcoming as to enter even into the landscape. It were a still greater and more serious mistake if any one were to compare, for succulence or sapidity, the flesh-food as yet procurable in the tropics with that which we have furnished us in every well-ordered house, and even hostel, in the United Kingdom of the chilly and rainy isles.

The subject is not altogether romantic, as I have already acknowledged; there is the more reason therefore for putting its practical side prominently forward, and thereby, as we may hope, doing something, however humble, for the bettering of man's estate. That it is not altogether visionary to hope for some improvement in this direction, or to strive to make acquisitions in the way of domestication under a tropical of the same kind as those which our forefathers made under a Central Asiatic sun, the following utterance of the late Dr. J. E. Gray, of the British Museum, an authority untainted with enthusiasm, may be taken as showing. Speaking at the 1864 (Bath) Meeting of the British Association (see Report of Address, p. 83, in Transactions of Sections) of our at present available domestic animals, Dr. Gray said:—

'An attentive study of the list, and of the peculiarities of the animals composing it, induces me to believe that, in attempting to introduce new domestic animals into some of our colonies, it would be desirable not to confine ourselves to the European breeds, but to ascertain whether some of the domestic races of Asia or Africa might not be better adapted to the climate and other conditions of the colony, although for reasons, to which I have before adverted, it would neither be worth the trouble, nor consistent with good policy, to attempt their introduction here.

'There is evidently ample room for such experiments, which might be advantageously made, for instance, in the colonies of the coast of Africa, where our horse, ass, oxen, sheep, and goats, and even dogs, have greatly degenerated, where the horse and the ass live only for a brief period, where the flesh of the ox and sheep is described as bad and rare, and the flesh of the goat, which is more common, is said to be tasteless and stringy. The pig alone, of all our domestic animals, seems to bear the change with equanimity: and the produce of the 'milch pig' is often sold to passengers of the mail packets, and the ships on the stations, as the milk of the cow, or even the goat, is rarely to be obtained. Unfortunately both the white and the black inhabitants are merely sojourners in the land, and do not seem to possess sufficient energy or inclination to make the experiment themselves.'

There is a more serious aspect or rather prospect of our future relation to the animal world. In this realm of activity, as in some others, we have of late been very rapidly extending our responsibilities. A man needs not to have spent years in the Malay Archipelago as Mr. Wallace has done, nor in the very different surroundings of Siberia as Middendorff has done, nor, Sir, in those of South Africa, to be convinced that the numbers of domesticated animals, I do not say of species of domesticated animals, will assuredly, and at no such very distant period, gain a relative magnitude of which our forefathers, who so patiently won them for us from savagery, could have had no conception. And that earlier than the attainment of this relative preponderance, the domestic animals on this world's surface will be nearly the only large land animals left upon it, and that the wild ones will be but pigmy vermin, '*winzige Ungeziefer*' in Middendorff's words, or, at least, less noble animals, is equally evident. For example, we can see as regards the lion, the king of beasts, that the breach-loading rifle is now rapidly completing what the smooth-bore, with flint and steel, began; for whereas he loses his life by his boldness in coming out into the open, we have in one part of the Old World the tiger, and in another the hyaena, substituted for him, a change in neither case much or at all for the better.

I have no reason for doubting that in these days we all consciously strive to act up to what has been spoken of, though not wholly correctly, as 'the new commandment of the nineteenth century,' 'Thou shalt not be cruel;' and I sincerely trust, that as regards all animals, domestic and wild, whether in the fields or in the streets and shambles, whether in the woods or within walls, this commandment may, like some others, attain greater extension in practice, as its many-sided applicability becomes more and more manifest. But I think that, even without our intending it, the extension of domestication has increased the sum total of lower animal happiness. A South African traveller, Sir, whose authority you will not repudiate, and we shall not even question, has told us (Galton, 'Domestication of Animals,' Trans. Ethnog. Soc. iii. N. S. 1865, p. 122), from his own observation of the still very really wild life of those regions, that it is not after all such unmixed happiness as persons might think, who have never crouched by night by the side of pools in that thirsty land, and watched how nightly drinking, even of water, may lead to much misery. 'The life of all beasts,' says that writer, 'in their wild state, is an exceedingly anxious one. From my own recollection, I believe that every antelope in South Africa has to run for its life every one or two days upon an average, and that he starts or gallops under the influence of a false alarm many times in the day.' Surely whatever the biped, who can foresee and ponderate, may think of the lot and the future of the domestic Ruminants, their lot, to themselves, as they are not troubled with anticipations, totals up an aggregate of comfort and even of enjoyment far exceeding that which the majority of wild graminivorous creatures of similar bulk ever obtain. A flock of well-fed Cheviots, on a snowy moor, in all their hornlessness and helplessness as against violence, shows the traveller that he is in a country whence wolves have entirely disappeared; would their lot be happier if they were exposed not merely to the winds and sleets of Northumbria, but also to the attacks of wolves to which even in France and Germany they would be liable?

We need not, however, travel in South Africa, as you have done, to prove the point that dog-fights and bull-fights, cockpits and shambles notwithstanding, domestication has, on the whole, increased the sum of the happiness of the lower animals. Let us by an easy effort of imagination figure to ourselves what would become of the flocks and herds of sheep and oxen, 'even very much cattle,' which are now living with as large a share of enjoyments as, and a very much larger share of leisure at least than many of their masters, if those masters were one and all to be swept away by some epidemic. Suppose, as Dr. Roberts in his memoir on 'Spontaneous Generation' (p. 39) has suggested, that the ferment which produces some one or other of our worse forms of infectious disease should 'sport,' as it is playfully styled, or vary, as a peach may sport or vary into a nectarine; and then suppose that the increased malignity and infectiousness with which it might thus become endowed, should as entirely destroy our own species within these Islands, as of late years disease has been known to entirely depopulate certain Polynesian islets, or as some analogously-developed disease may be supposed to have exterminated the horse in South America within recent geological periods. There can be very little doubt in the mind of anybody who has much experience of the power of combination for mischief which dogs can, independently of men, develope, even in a civilised and thickly populated country, that in a few days after our disappearance they would be masters of the country. The mere desire for blood which is so eminently characteristic of the musteline carnivores would very shortly and certainly show itself again in our old servants in their

Saturnalia; and in a very short time the entire race of sheep, except in a few mountain districts, would have been as wastefully slaughtered for their blood and fat as flocks and herds have been and still are slaughtered by us in Australia or South America. Oxen would hold out a little longer than sheep, and pigs, I incline to think, longer than either. But that a great diminution of the sum total of brute enjoyment, and, if such a thing there be, of brute happiness also, would take place after we had disappeared, I think needs no demonstration, especially to anybody who, without any experience of any canine mutiny, has ever studied the phenomena of a dog-show or listened at night to the opera which its denizens perform. The various races which, without exactly being domesticated, stand yet on the borderland separating wild from domesticated life, would also very shortly and very sharply have brought home to them the fact of their being more dependent on man than perhaps either they or we have entirely recognised. Rabbits and hares, pheasants and partridges, if they had reason, would reasonably regret the times when they viewed, with something perhaps of disgust, the slouching form of the gamekeeper with his double-barrelled shot-gun perambulating the ridings in the woods and skirting their sunny boundaries. Cats and weasels would with little less delay than the dogs make the life of quadrupeds just specified as miserable as that of the sheep and ox had already been made; and would, after the lapse of a year or two, with the aid of hawks and corvidae of several kinds, greatly thin their numbers. The river embankments on the lower Thames, lastly, which excited the admiration of Sir Christopher Wren, and were referred by him to the time of the Romans, and also those on many other rivers, having no one to repair any of the breaches which floods would make in them, would before very long allow a very large acreage of land to become swamp, marsh, and lagune; not only thus, on the one hand, depriving many species of animals of their means of subsistence, but also on the other introducing predatory birds, such as gulls, and accelerating the disappearance of many others which really hold their own in such neighbourhoods even now only by man's protection and thanks to his presence.

The purview of this prophecy extends no further than the precincts of the British Islands; in continental countries organic nature would more completely resort to the condition it was in before it began to be modified by man's interference; the *Regnum Hominis* would not be succeeded by the *Regnum Canum familiarium*, but by that of *Canum luporum*; and generally the larger *ferae naturae*, both those which eat others and those which are eaten by others, would resume an importance even in the landscape which their extirpation within our four seas has rendered an impossibility for all future time short of the time when the Channel will once again become dry land.

In concluding a Lecture the title of which might serve for the often-to-be-repeated title of many successive and closely printed volumes, let me take as a text the following words from Victor Hehn's book ('Kulturpflanzen und Hausthiere,' 3rd edition, 1877; Berlin; p. 435), to which I owe more even than I have expressed: 'Was die moderne Welt von der alten unterscheidet ist Naturwissenschaft, Technik und Naturalökonomie'—what makes the modern world to differ from the old is natural science, command of apparatus, and political economy. As regards this last differential peculiarity, I have to remark that Herr Victor Hehn's last edition bears the date of 1877, and that, consequently, he cannot have had colonial tariffs either of Melbourne or of Canada before his eyes; nor, though living in Berlin, could he have heard the words uttered there only ten days ago, though they were in an authoritative voice (see 'Times,' May 2nd); nor, finally, could he have been present at a meeting attended in Paris by the representatives of no less than fifty-eight Chambers of Commerce on the very day before, the first, that is, not of April, but of May in this very year of grace 1879. Otherwise I cannot but think that Herr Hehn would not have said the political economy of the present, either as put out in words, or as carried out in practice, was so very different from that of ancient times. To any one at all thick of sight or hard of hearing the proportions of any such difference are wholly inappreciable. I turned to what was one of the favourite studies of my youth, my Aristophanes, and I find Dicaeopolis[1], to adapting whose

[1] Acharn. 33-36:—

τὸν δ' ἐμὸν δῆμον ποθῶν,
ὃς οὐδεπώποτ' εἶπεν, ἄνθρακας πρίω,
ἀλλ' αὐτὸς ἔφερε πάντα χὡ πρίων ἀπῆν.

Cato and Varro appear, according to the passage given in Hehn, p. 425, to have been similarly in the dark, the first of these averring, 2. 5, in words very nearly reproducing that of Dicaeopolis, 'Patrem familias vendacem non emacem esse oportet,' whilst the

name Prince Bismarck would, I apprehend, as little object as it would seem he does to adopting his principles, sighing for the time when he would get back to his farm, the articles consumed in which at least were 'reserved for native industry.' The amount of difference between these views and those of the statesman just mentioned, or those of M. Pouyer-Quertier, or of another countryman of MM. Quesnay, Turgot, and Chevalier who is reported in the same 'Times' of Friday, May 2nd, no time having been lost in giving these valuable views to the world, to have averred that an increase in the imports denoted the impoverishment of a country, I must, as did Captain Lemuel Gulliver under somewhat similar circumstances in Laputa, profess myself to be 'not skilful enough to comprehend.' What is shown seems to me to be that in modern not less than in ancient times men will run their heads against the multiplication table, and that for the passing moment, at least, it is not always the heads which come off second best in the encounter.

Of the second difference between the old world and the new which our command of methods and means, our recognition of the futility of attempting enterprises with a *manus nuda* and an *intellectus sibi permissus*, has created, the gas, glass, and coal around us in this room speak, and I need not.

As regards the third great point of contrast upon which Herr Hehn insists, that of natural science, we are all probably at one with him. Our agreement may be illustrated by contrasting the different factors which two poets, each an artist capable of taking a wide view with due perspective and proportion of the sum of man's activities, have in ancient and modern times respectively enumerated as making up that sum. When Juvenal specifies what he means by 'Quidquid agunt homines,' the comprehensive title of his satires, he enumerates nothing—because, I suppose, he considered all else beneath the dignity of a poet—but

> 'Votum, timor, ira, voluptas,
> Gaudia, discursus'—

large enough matters, but imponderables all of them. Contrast these items,—I purposely speak in Philistine phraseology—with

latter enjoined, 1. 22. 1, in words which the Chambers of Commerce aforesaid re-echoed in their modified Roman tongue, 'Quae nasci in fundo ac fieri a domesticis poterunt, eorum ne quid ematur.'

those which our present Poet-Laureate enumerates in epexegesis of the 'march of mind;' there we have the line,

'In the steamship, in the railway, in the thoughts that shake mankind,'

—ponderables and imponderables severally holding their due mutual proportion. And from this line I can pass in this place by a natural and locally suggested transition to what I believe to be as large a difference between the ancient and modern world as either of the two last touched upon. The whole of the old world, of the *orbis veteribus notus*, of πᾶσα ἡ οἰκουμένη, was but a small fragment as measured by the geographer when compared with the world dealt with by our emigration agents and Custom-house officers. The discovery of America has been said to have exercised much the sort of influence upon the old world, socially and politically, that the approximation to our globe of some new planet would exercise astronomically; and since those 'spacious times of great Elizabeth' China, Japan, Australia, and Polynesia have each entered into the circle of influences acting upon and acted on by the world as known to the classical writers. In speaking of any district beyond those in relation with the valleys of the Euphrates, the Danube, the Rhine, the Rhone, and the shores of the Mediterranean and Black Sea, the ancients would but say in really pathetic antithesis:

'Longa procul longis via dividit invia terris.'

The Brindisi mail brings every manager of a museum, as well as every secretary for the colonies, into weekly relation with 'regions Caesar never knew,' by agencies of which he never dreamt and of which in our own times the greatest perhaps of his successors, fortunately for us, as he is reported to have remarked in Plymouth Sound, never learnt to avail himself. And it is in reference to the all-pervading intercommunication which the application of steam to navigation has rendered possible that I wish to utter two concluding sentences, not respecting the vast contrast which it has set up between the present and all preceding centuries, but respecting the contrast which it will shortly have created between the present and all future times. Before this application had established highways on the ocean and invented machinery which,

'Spurning sails and scorning oars,
Keeps faith with time on distant shores,'

it was possible on many an oceanic island to recover links which had fallen out of the chain of evidence as to the origin of species which the older and larger continents of dry land had furnished; it was possible also to elucidate the origin, humble and lowly enough, of our own civilisation by what we could see, and not less by what we might fail to discover, in the inchoate civilisations, in similar localities, of semi-savage men. The lines of intercommunication between the most distant parts of our globe, which the navigator with, in his own language, 'a steam-engine under his foot,' is daily weaving into a more and more nearly all-encompassing web, will very shortly have introduced so much of the most recent results of our modern civilisation into what were but lately the most secluded of localities as to rob them of that value and interest for the pursuers of the knowledge specified, which they up to a few years ago so eminently possessed.

These few years—for they will be but few—to come, have a great responsibility put upon them in the way of preserving those perishable and destructible links in the history of the past, which may be made incandescent and luminous for the advancement of knowledge, and to some not inconsiderable extent for the benefiting of man's estate.

In this work the Society, which has honoured me by inviting me to address them this evening, has borne a distinguished part in the past, and I cannot doubt, but, on the contrary, have many reasons for believing, that it will bear an increasingly important one in the future.

XLIII.

BIOLOGICAL TRAINING AND STUDIES,

AN ADDRESS DELIVERED TO THE BIOLOGICAL SECTION OF THE BRITISH ASSOCIATION FOR THE ADVANCEMENT OF SCIENCE.

Amongst the duties of a President of a Section the delivery of an Address has in these latter days somehow come to be reckoned; and that I may interpose myself for but as short a time as possible between your attention and the papers announced to you for reading upon your list, I will begin what I have to say without any further preface.

I wish first to make a few observations as to the kind of preparation which is indispensable, as it seems to me, as a preliminary to an adequate and intelligent comprehension of the problems of biology; or, in other words, to an adequate and intelligent comprehension of the discussions which will take place in this room and in the two other rooms which will be assigned to, and occupied by, the department of Ethnology and Anthropology, and that of Physiology pure and proper, with Anatomy.

Having made these observations, I propose, in the second place, to enumerate the subjects which appear likely to occupy prominent places in our forthcoming discussions; and thirdly, I will, if your patience allows me, conclude with some remarks as to certain of the benefits which may be expected, as having been constantly observed to flow from a due and full devotion to biological study.

In the first place, then, I wish to say that though the problems of biology have much of what is called general interest (that is to say, of interest for all persons) attaching to them,—as, indeed, how could they fail to have, including as they do the natural history of our own and of all other species of living organisms, whether animal or vegetable?—some special preparation must be gone through if that general interest is to be thoroughly and intelligently gratified. I would compare the realm of biology to a vast landscape in a cultivated country of which extensive views may be obtained from an eminence, but for the full and thorough appreciation of which it is necessary that the gazer should himself have cultivated some portion, however small, of the expanse at his feet. It is, of course, a matter of regret to think that persons can be found who look upon an actual landscape without any thought or knowledge as to how the various factors which make up its complex beauty have come to co-operate, how the hand of man is recognisable here, how the dip of the strata is visible there, and how their alternation is detectable in another place as the potent agency in giving its distinctive features; but I take it that real and permanent, however imperfect, pleasure may be drawn from the contemplation of scenery by persons who are ignorant of all these things. I do not think this is the case when we here deal with *coup d'œil* views of biology. The amount of the special knowledge, the extent of the special training need not necessarily be great; but some such special knowledge and training there must be if the problems and argumentations familiar to the professed biologist are to be understood and grasped by persons whose whole lives are not devoted to the subject, so as to form for them acquisitions of real and vital knowledge.

The microscope has done very much (indeed I may say it has done almost all that is necessary) for enabling all persons to obtain the necessary minimum of practical and personal acquaintance with the arrangements of the natural world of which I am speaking. The glass trough used in Edinburgh, the invention of John Goodsir, whose genius showed itself, as genius often does show itself, in simple inventions, can be made into a miniature aquarium (I purposely use a word which calls up the idea of an indoors apparatus, wishing thereby to show how the means I recommend are within the reach of all persons); and in it, lying as it does horizontally and underlain as it is by a condenser, animal and vegetable organisms can be observed at any and at all hours, and continuously, and with tolerably high magnifying-powers even whilst undisturbed. Thus is gained an admirable field for the self-discipline in question. The microscope which should be used by preference for exploring and watching such an aquarium should be such a one as is figured in Quekett's work on the Microscope (p. 58, fig. 36), as consisting of a stem with a stout steadying base, and of a horizontal arm some 9 inches long, which can carry indifferently simple lenses or a compound body. I think of the two it is better that the aquarium should be horizontal rather than the microscope; and those who think with me in this matter can nevertheless combine for themselves the advantages of the horizontal position of the instrument with those of the horizontal position of the objects observed by modifying the eyepiece in the way figured by Quekett (p. 381, fig. 266). It would be a long task to enumerate fully all the scientific lessons which may be gathered, first, and all the educational agencies, secondly, which may be set and kept in movement by a person who possesses himself of this simple apparatus. The mutual interdependence of the animal and vegetable kingdoms, their *solidarité* as the French have called it, and as the Germans have called it too, copying herein the French, is one of the first lessons the observer has forced upon him; the influence of physical and chemical agencies upon the growth and development of living beings he soon finds strikingly illustrated; the mysterious process of development itself is readily observable in the eggs of the common water-snails and in those of freshwater fish, so that the way in which the various organs and systems of organs are chiselled out, built up, and finally packed together and stratified can be taken note of in these yet transparent representatives of these great sub-kingdoms which all the while are undisturbed and at peace: and all these points of large interest are but a few of many which these small means will enable any one to master for himself in the concrete actuality, and thoroughly. The necessity for carefulness and truthfulness in recording what is seen, the necessity for keeping in such records what one observes quite distinct from what one infers, the necessity for patience and punctuality, are lessons which, from having a moral factor as well as a scientific one in their composition, I may specify as belonging to the educational lessons which may be gathered from such a course of study.

I have been speaking of the microscope as an instrument of education, and I wish before leaving the subject to utter one caution as to its use when this particular object of education is in view. If a subject is to act educationally, it must be understood thoroughly; and if a subject is to be understood thoroughly, it must form one segment or stretch in a continuous chain of known facts. Ἀρκτέον ἀπὸ τῶν γνωρίμων, said one of the greatest of educators; you must start from some previously existing basis of knowledge, and keep your communications with it uninterrupted if your knowledge is not to be unreal. And my concrete application of these generalities is contained in the advice that no sudden jump be made from observations carried on with the naked eye to observations carried on with the highest powers of the microscope. I am speaking of the course to be pursued by beginners, and beginners we all were once; and if our places are to be filled (and filled they will be) by better men, as we hope, than ourselves, they will have to be filled, we also hope, by men who have yet to become beginners. It is in their interest I have been speaking; and I say that a beginner does not ordinarily get an intelligent conception of the revelations of the microscope except in Bacon's words, *Ascendendo continenter et gradatim,* by progressing gradually from observations with the naked eye through observations dependent upon dissecting-lenses, doublets by preference, and the lower powers of the compound microscope, up to observations to be made with the higher and highest magnifying-powers. Unless he ascends by gradations from organs and systems to structures and tissues and cells, his wonder and admiration at the results of the ultimate microscope analysis, of what he had but a moment before knowledge of only in the concrete and by the naked eye, is likely to be but unintelligent.

There are three other agencies which can be set into activity with nearly as little trouble and difficulty as the simple apparatus of which I have just been speaking, and which will, like it, secure for us the necessary preliminary discipline or '*Propädeutik*' for the rational comprehension of Biology. These are Local Museums, Local Field Clubs, and Local Natural Histories. Local authorities, persons of local influence, should engage and interest themselves in the starting into life of the two former of these agencies; and if some such person as Gilbert White could be found in each county to write the Natural History of its Selborne, I know not at what cost it would not be well to retain his services. As the world is governed, upon each particular area of its surface there is to be found a certain percentage of the population occupying it who have special calls for particular lines of study. It is the interest of each country to have such means and such institutions in being as will render it possible to detect the existence of persons gifted with such special vocations, to give the talent thus entrusted to them fair scope for development, and to render smaller the risk of their dying mute and inglorious. A young man who is possessed of a talent for Natural Science and Physical Inquiry generally, may have the knowledge of this predisposition made known to himself and to others, for the first time, by his introduction to a well-arranged Local Museum. In such an institution, either all at once, or gradually, the conviction may spring up within him that the investigation of physical problems is the line of investigation to which he should be content to devote himself, relinquishing the pursuit of other things; and then, if the museum in question is really a well-arranged one, a recruit may be thereby won for the growing army of physical investigators, and one more man saved from the misery of finding, when he has been taken into some other career, that he has, somehow or other, mistaken his profession, and made of his career one life-long mistake.

Here comes the question, What is a well-arranged museum? The answer is, a well-arranged museum, for the particular purpose of which we are speaking, is one in which the natural objects which belong to the locality, and which have already struck upon the eye of such a person as the one contemplated, are clearly explained in a well-arranged catalogue. The curiosity which is the mother of science is not awakened for the first time in the museum, but out of doors, in the wood, by the side of the brook, on the hillside, by scarped cliff and quarried stone; it is the function of the museum, by rendering possible the intellectual pleasure, which grows out of the surprise with which a novice first notes the working of his faculty of inspiration, to prevent this curiosity from degenerating into the mere woodman's craft of the gamekeeper, or the rough empiricism of the farmer. The first step to be taken in a course of natural instruction is the providing of means whereby the faculties of observation and of verification may be called into activity; and the first exercise the student should be set down to is that of recognising, in the actual thing itself, the various properties and peculiarities which some good book or some good catalogue tells him are observable in it. This is the first step, and, as in some other matters, *ce n'est que le premier pas qui coûte.* And it need not cost much. There is a name familiar to Section D, and, indeed, not likely for a long while to be forgotten by members of the British

Association generally, extrinsic means as well as the intrinsic merits of the well-loved man conspiring to keep his memory fresh among us, and the bearer of that name, Edward Forbes, has left it as his opinion that 'It is to the development of the provincial museums that, I believe, we must in future look for the extension of intellectual pursuits throughout the land.' (Lecture 'On the Educational Uses of Museums,' delivered at the Museum of Practical Geology and published in 1863. Cited by Toynbee, 'Hints on the Formation of Local Museums,' 1863, p. 46.) With the words of Edward Forbes I might do well to end what I have to say, but I should like to say a word as to the policy of confining the contents of a local museum to the natural-history specimens of the particular locality. No doubt the first thing to be done is the collection of the local specimens, and this alike in the interest of the potential Cuviers and Hugh Millers who may be born in the district, and in the interest of the man of science who may visit the place when on his travels. But so long as a specimen from the antipodes or from whatever corner of our world be really valuable, and be duly catalogued before it is admitted into the museum, so that the lesson it has to teach may be learnable, I do not see my way towards advising that foreign specimens be excluded. It is to my mind more important that all specimens should be catalogued as soon as received, than that any should be rejected when offered.

I must not occupy your time further with this portion of my address. Let me first say that a person who wishes to know what a Field Club can do for its members, and not for them only, but for the world at large, will do well to purchase one, or any number more than one, of the 'Transactions of the Tyneside Naturalist's Field Club;' and that if there be any person who thinks that White's 'Selborne' relates to a time and place so far off that there can be no truth in the book, and who yet would like to try upon himself the working of the fourth of the disciplinary agencies of which I have spoken, that, namely, of reading some local Natural History on the spot of which it treats, and comparing it with the things themselves *in situ*, let him repair to Weymouth, and work and walk up and down its cliffs and valleys with Mr. Damon's book in his hands.

I shall not be suspected in this place and upon this occasion, nor, as I hope, upon any other, of a wish to depreciate the value of scientific instruction as an engine for training the mind; but neither, on the other hand, should I wish to depreciate the value of literary culture, my view of the relations of these two gymnastics of the mind being the very simple, obvious, and natural one that they should be harmoniously combined—

> Alterius sic
> Altera poscit opem res, et conjurat amice.

I know it may be said that there are difficulties in the way, and especially practical difficulties; but I have always observed that people who are good at finding out difficulties, and especially practical difficulties, are like people who are good at finding out excuses,—good at finding out very little else. The various ways of getting over these difficulties are obvious enough, and have been hinted at or fully expressed by several writers of greater or less authority on many occasions. It is, however, of some consequence that I should here say what I believe has not been said before, namely, that a purely and exclusively literary education, imperfect and one-sided as it is, is still a better thing than a system of scientific instruction (to abuse the use of the word for a moment) in which there should be no courses of practical familiarising with natural objects, verification, and experimentation. A purely literary training, say, in dialectics, or what we are pleased to call logic, to take a flagrant and glaring instance first, does confer certain lower advantages upon the person who goes through it without any discipline in the practical investigation of actual problems. By going through such a training attentively, a man with a good memory and a little freedom from over-scrupulousness, can convert his mind into an arsenal of quips, quirks, retorts, and epigrams, out of which he can, at his own pleasure, discharge a *mitraille* of chopped straw and chaff-like arguments, against which no man of ordinary fairness of mind can, for the moment, make head. It is true that such sophists gain this dexterity at the cost of losing, in every case, the power of fairly and fully appreciating or investigating truth, of losing in many cases the faculty of sustaining and maintaining serious attention to any subject, and of losing in some cases even the power of writing. A well-known character in an age happily, though only recently, gone by, who may be taken as a Caesar worthy of such Antonies, used to speak of a pen as his torpedo. Still they have their reward, they succeed now and then in convincing juries, and they are formidable at dinner-tables. It would not be fair, however, not to say that a purely literary training can do much better things than this. By a purely classical education a man, from being forced into seeing and feeling that other men could look upon the world, moral, social, and physical, with other (even if not with larger) eyes than ours, attains a certain flexibility of mind which enables him to enter into the thoughts of other and living men; and this is a very desirable attainment. And, finally, though I should be sorry to hold with a French writer that the style makes the man, the benefit of being early familiarised with writings which the peculiar social condition of the classical times, so well pointed out by De Tocqueville ('De la Démocratie en Amérique,' i. 15), conspired and contributed not a little to make models of style, is not to be despised. Such a familiarity may not confer the power of imitating or rivalling such compositions, but it may confer the power of appreciating their excellences, the one power appearing to us to be analogous to the power of the experimenter, and the other to that of the pure observer in Natural Science; and we should undervalue neither.

Masters of Science, it must be confessed, are not always masters of style; let not the single instance of last night tempt you to generalise, it is but a single instance, the writings of the man whom we in this section are most of us likely to look upon as our master in Science have been spoken of by our President in his recently published volume as 'intellectual pemmican[1];' and if scientific reading and teaching is to be divorced from scientific observation of natural objects and processes, it is better that a man, young or old, should have in his memory something which is perfect of its kind, entire and unmutilated, such as the opening sentences of the 'Brutus' of Cicero, which Tacitus, I think, must have had in his memory when he wrote his obituary of Agricola, or as the opening sentences of the 'Republic' of Plato, or the conclusion of the 'Ajax' of Sophocles, than that he should have his memory laden with a consignment of scientific phrases which, *ex hypothesi*, have for him no vital reality. I have already said that I am strongly of opinion that literary should always be combined with scientific instruction in a perfect educational course; these

[1] [Professor Huxley was President of the British Association at the meeting at which this Address was delivered.—EDITOR.]

somewhat lengthy remarks refer therefore only to systems in which it is proposed that we should have not only a bifurcation but a radical separation of studies and students; and the moral of this may be summed up by saying that a purely scientific education must be a thoroughly practical one, familiarising the student with actual things as well as with words and symbols. It was upon the solid ground that Antaeus learnt the art of wrestling; it was only when he allowed himself to be lifted from it that he was strangled by Hercules.

Coming now to the second part of my address, I beg to say that the word Biology is at present used in two senses, one wider, the other more restricted. In this latter sense the word becomes equivalent to the older, and till recently more currently used word 'Physiology.' It is in the wider sense that the word is used when we speak of this as being the Section of Biology: and this wider sense is a very wide one, for it comprehends, first, Animal and Vegetable Physiology and Anatomy; secondly, Ethnology and Anthropology; and thirdly, Scientific Zoology and Classificatory Botany, inclusively of the Distribution of Species. It may have been possible in former times for a single individual of great powers of assimilation to keep himself abreast of, and on a level with, the advance of knowledge along all these various lines of investigation; but in those times knowledge was not, and could not, owing to difficulties of intercommunication, the dearness of books, the costliness or non-existence of instruments, have been increased at the rate at which it is now being, year by year, increased; and the entire mass of actually existing and acquired knowledge was of course much smaller, though man's power of mastering it was no smaller than at present. It would now be an indication of very great ignorance if anybody should pretend that his own stock of information could furnish him with something in each one of the several departments of knowledge I have just mentioned, which should be worthy of being laid before such an assembly as this. As will have been expected, I shall not presume to do more than glance at the vegetable kingdom, large as is the space in the landscape of life which it makes. What I propose to do is merely to draw your attention to a very few of the topics of leading interest, which are at the present moment being, or rather will shortly begin to be, discussed by experts in the Department of Physiology and

Anatomy, in the Department of Ethnology and Anthropology, and, thirdly, in the Department of Scientific Zoology.

Under the head and in the Department of Physiology proper and Anatomy, our list of papers and, I am happy to add, the circle of faces around us suggest to us the following subjects as being the topics of main interest for the present year:—the questions of Spontaneous Generation; that of the influence of organised particles in the production of disease; that of the influence of particular nervous and chemical agencies upon functions; that of the localisation of cerebral functions; that of the production and, indeed, of the entire *rôle* in the economy of creation of such substances as fat and albumen; and, finally, that of the cost at which the work of the animal machine is carried on.

The question of Spontaneous Generation touches upon certain susceptibilities which lie outside the realm of science. In this place, however, we have to do only with scientific arguments, and I trust that the Section will support the Committee in their wish to exclude from our discussions all extraneous considerations. Truth is one; all roads which really lead to it will assuredly converge sooner or later: our business is to see that the one we are ourselves concerned with is properly laid out and metalled.

Upon this matter I am glad to be able to fortify myself by two authorities; and first of these I will place an utterance of Archbishop Whately, which may be found in the second volume of his Life, pp. 56–68, and appears to have been uttered by him, aet. 57, an. 1844: 'A person possessing real faith will be fully convinced that whatever suppressed physical fact appears to militate against his religion will be proved by physical investigation either to be unreal or else reconcilable with his religion. If I were to found a church, one of my articles would be that it is not allowable to bring forward Scripture or any religious considerations at all to prove or disprove any physical theory or any but religious and moral considerations.' My second quotation shall be taken from the great work of one of the first, as I apprehend, of living theologians, John Macleod Campbell, 'The Nature of the Atonement,' pp. xxxii, xxxiii, Introd., and it runs thus: There are 'other minds whose habits of pure scientific investigation are to them a temptation to approach the claim of the kingdom of God on our faith by a wrong path, causing them to ask for a kind of evidence not proper to the subject, and so hindering their weighing fairly what belongs to it. No scientific study of the phenomena which imply a reign of law could ever have issued in the discovery of the kingdom of God. But neither can it issue in any discovery which contradicts the existence of that kingdom; nor can any mind in the light of the kingdom of God hesitate to conclude that if such seeming contradictions arise there is implied the presence of error either as to facts or as to conclusions from the facts.' These are valuable words and weighty testimonies. But in a matter of this importance one must not forbear to point out what may seem to be wanting even in the dicta of such men as the two I have quoted. Neither of them has allowed the possibility of error attaching itself to the utterances of more than one of the two parties in such issues as those contemplated. Neither appears to have thought of the cases in which religious men, if not theologians, have brought woe on the world because of the offences they have with ill-considered enunciations created. And whilst fully sympathising with all that the Archbishop and Mr. Campbell have said, I must say that they appear to me to have left something unsaid; and this something may be wrapped up in the caution that there may be faults on both sides. But at any rate this Section cannot be considered a fit place for the correction of errors save of the physical kind; and all other considerations are for this week and in this place extraneous. In some other week or in some other place it will be, if it has not already been, our duty to give them our best attention.

To come now to the kind of considerations which are the proper business of Section D: let me say that for the discussion of Spontaneous Generation very refined means of observation, and, besides these, very refined means of experimentation are necessary. And I shall act in the spirit of the advice I have already alluded to as given to the world by one of her greatest teachers, if I put before you a simple but a yet undecided question for the solution of which analogous means of a far less delicate character would appear to be, but as yet have not proved themselves to be, sufficient. Thus shall we come to see very plainly some of the bearings, and a few of the difficulties, of the more difficult of the two questions. What an uneducated person might acquiesce in hearing spoken of as spontaneous generation, takes place very constantly under our very eyes, when a plot of ground which has for many years, or even generations, been devoted to carrying some particular vegetable growth, whether grass or trees, has that particular growth removed from it. When such a clearing is effected, we often see a rich or even a rank vegetation of a kind previously not growing on the spot spring up upon it. The like phenomenon is often to be noted on other surfaces newly exposed, as in railway-cuttings and other escarpments, and along the beds of canals or streams, which are laid bare by the turning of the water out of its channel. Fumitory, rocket, knot-grass or cowgrass (*Polygonum aviculare*), and other such weeds, must often have been noted by every one of us here in England as coming into and occupying such recently disturbed territories in force; whilst in America the destruction of a forest of one kind of wood, such as the oak or the chestnut, has often been observed to be followed by an upgrowth of young forest trees of quite another kind, such as the white pine—albeit no such tree had been seen for generations growing near enough to the spot to make the transport of its seeds to the spot seem a likely thing. In one case referred to by Mr. Marsh, 'Man and Nature,' p. 289, the hickory, *Carya porcina*, a kind of walnut, was remarked as succeeding a displaced and destroyed plantation of the white pine. Now the advocates of spontaneous generation must not suspect me of hinting that there is any question, except in the minds of the grossly ignorant, of the operation of any such agency as spontaneous generation here; no one would suggest that the seeds of the *Polygonum aviculare*, to say nothing of those of the hickory, were produced spontaneously; but what I do say is, that the question of how these seeds came there is just the very analogue of the one which they and their opponents have to deal with. And it is not definitely settled at this very moment. Let us glance at the instructive historical parallel it offers. For the very gross and palpable facts of which I have just spoken there are two explanations offered in works of considerable authority. The one which has perhaps the greatest currency and commands the largest amount of acceptance is that which, in the words of De Candolle, regards *la couche de terre végétale d'un pays comme un magasin de graines*, and supposes that in hot summers and autumns, such as the present, the fissures in the ground, which have proved so fatal this year to the young partridges, swallow up a multitude of seeds, which are restored again to life when the deep strata into which they are thus introduced, and in which they are sealed up as the chasms close up, come in any way to be laid open to the unimpeded action of the sun and moisture. Squirrels, again, and some birds resembling herein the rodent mammalia, bury seeds and forget to dig them up again; and it is supposed that they may bury them so deep as to be protected from the two physical agencies just mentioned. Now germination cannot take place in the absence of oxygen; and I would add that well-sinkers know to their cost how often the superficial strata of the earth are surcharged with carbonic acid. The rival explanation and the less popular (I do not say the less scientific) looks to the agency of transportation as occurring constantly, and sufficing to explain the facts. By accepting this explanation, we save ourselves from running counter to certain experiments, some of which were carried out, if I mistake not, under the auspices of this Section (see 'Brit. Assoc. Reports'), and which appear to curtail considerably the time during which seeds retain their vitality, and to multiply considerably the number of conditions which must be in force to allow of such retention for periods far shorter than those which have to be accounted for. A better instance of the expediency of checking the interpretations based merely upon observations, however accurately made, by putting into action experiments, cannot be furnished than by recording the fact put on record by Mr. Bentham, when discussing this question in his last year's Address to the Linnean Society:—

'Hitherto direct observation has, as far as I am aware, only produced negative results, of which a strong instance has been communicated to me by Dr. Hooker. In deepening the lake in Kew Gardens they uncovered the bed of an old piece of water, upon which there came up a plentiful crop of Typha, a plant not observed in the immediate vicinity; and it was therefore concluded that the seed must have been in the soil. To try the question, Dr. Hooker had six Ward's cases filled with some of the soil remaining uncovered close to that which had produced the Typha, and carefully watched; but not a single Typha came up in any one of them.' (Note in President's Address, May 24, 1869, p. lxxii of 'Linnean Society's Proceedings.')

To this I would add that experiments with a positive result, and that positive result in favour of the second hypothesis, if hypothesis it can be called, are being constantly tried in our colonies for us, and on a large scale. I had taken and written here of the *Polygonum aviculare*, the 'knot' or 'cowgrass'—having learnt, on the authority of Dr. Hooker and Mr. Travers (see 'Natural History Review,' January, 1864, p. 124, Oct. 1864, p. 619), that it abounds in New Zealand, along the roadside, just as it does in England—as a glaring instance, and one which would illustrate the real value of the second explanation even to an unscientific man and to an unassisted eye. But on Saturday last I received by post one of those evidences which make an Englishman proud in thinking that whithersoever ships can float thither shall the English language, English manners, and English science be carried, in the shape of the second volume of the 'Transactions' of the New Zealand Institute, full, like the first, from the beginning to its last page of thoroughly good matter. In that volume, having looked at the table of its contents, I turned to a paper by Mr. T. Kirk on the 'Naturalised Plants of New Zealand,' and in this, at p. 142, I find that Mr. T. Kirk prefers to regard the *Polygonum aviculare* of New Zealand as indigenous in New Zealand. Hence that illustration which would have been a good one falls from my hands. And I must in fairness add, that because one agency is proved to be a *vera causa*, it is not thereby proved that no other can by any possibility be competent simultaneously to produce the same effect, whatever the Schoolmen with the law of Parsimony ringing in their ears may have said to the contrary. I have dwelt upon this subject at this length with the purpose of showing how much difficulty may beset the settlement of even a comparatively simple question which involves only the use of the unassisted eye, or at most of a simple lens. The *a fortiori* argument I leave you to draw for yourselves, with the simple remark that the question of spontaneous generation is now at least one to be decided by the microscope, and by the employment of its highest powers in alliance with other apparatus of all but equal complexity.

We come, in the second place, to say a word as to the extent of the influence which organic and living particles, of microscopic minuteness but solid for all that, have been supposed, and in some instances at least have been proved, to exercise upon the genesis and genesiology of disease, and so upon the fortunes of our race, and our means for bettering our condition, and that of our fellows.

I need not refer to Dr. Sanderson's valuable Report (just published in the Privy Council's Medical Officer's 'Blue Book,' Twelfth Report, 1870, p. 229) upon those contagion particles which he proposes to call by the convenient name, slightly modified from one invented by Professor Béchamp, of 'Microzymes'; for Dr. Sanderson is here to refer to the matter for himself and for us; and when this meeting is over we shall all do well to lay to heart what he may tell us here and now, and, besides this, to study his already printed views upon the matter. It may be perhaps my business to remind you that these views, so far as they are identical with Professor Hallier's as to the importance of those most minute of living organisms, the micrococcus of his nomenclature, the microzymes of Mr. Simon's 'Blue Book,' were passed in review as to their botanical correctness by a predecessor of mine in this honourable office—namely, by the Rev. J. M. Berkeley, at the Meeting held two years ago at Norwich; and that some of the bearings of the theory and of the facts, howsoever interpreted, upon the Theory of Evolution, were touched upon by Dr. Child in his interesting volume of 'Physiological Essays,' p. 148, published last year. It would not perhaps be exactly my business to express my dissent from any of these results or views put forward by any of these investigators I have mentioned; but I wish to point out to the general public that none of these inquirers would affirm that the agencies shown by them to be potent in the causation of *certain* diseases were types and models of the agencies which are, did we but know it, could we but detect them, potent in the causation of *all* diseases. Many diseases, though, possibly enough, not the majority of the strictly infectious diseases, are due to material agents quite distinct in nature from any self-multiplying bodies, cytoid or colloid. To say nothing of the effects of certain elements (and elements, it will be recollected, in their singleness and simple atomicity, have, as the world happens to be constituted and governed, never been honoured with the office of harbouring life) which when volatized, as mercury, arsenic, and phosphorus may be, or indeed which, when simply dissolved, may be most ruinous to life, there are, I make no doubt, animal poisons produced in and by animals, and acting upon animal bodies, which are neither organised nor living, neither cytoid nor colloid. Dr. Charlton Bastian is not likely to underrate the importance of such agents, howsoever produced, in the economy, or rather in the waste, of Nature; yet from his very careful record of his own very closely observed and personal experience we can gather that he would not demur to conceding that non-vitalised, however much animalised, exhalations may be only too powerful in producing attacks, and those sudden and violent and fever-like attacks, of disease. Dr. Bastian tells us ('Phil. Trans.' for 1866, vol. 196, pt. ii. pp. 583, 584) that whensoever he employed himself in the dissection of a particular nematoid worm, the *Ascaris megalocephala*, he found occasion to observe, and that in himself, and very closely, the genesiology of a spasmodic and catarrhal affection, not unlike hay-fever as it seems to me, but under circumstances which appear to preclude the possibility of any living organisms being the cause—as they have been supposed, and by no less an authority than Helmholtz, to be—of the malady just mentioned. For in Dr. Bastian's case this affection was produced, not only when the *Ascaris megalocephala* was dissected when fresh, but 'after it had been preserved *in methylated spirit for two years, and even then macerated in a solution of chloride of lime for several hours before it was submitted to examination.*' Could any microzyme or megalozyme have survived such an amount of antizymotic treatment—such a pickling as this? This is not exactly a medical association, and I have entered upon this discussion not altogether without a wish to show how subjects of apparently the most purely scientific and special interest, as Mycology and Helminthology (the natural history, that is to say, and the morphology of the lowest plants and of the lowest Vermes), may, when we least expect it, come or be brought to bear upon matters of the most immediate and pressing practical importance. And in this spirit I must say a word upon the way in which the pathology of snake-bites bears upon the matters I have been speaking of, and the extent of the debt which practical men owe to such societies as our Ray Society, and to such publications as their colossal volume on the snakes of India, in which Dr. Günther's views as to the real history of the striking and terrible yet instructive phenomena alluded to are combined ('Reptiles of British India,' Ray Society, 1864, p. 167). That the snake-poison is an animal poison is plain enough; that it is fatal to men and animals everybody knows; but I rather think that these two facts relative to it are not equally notorious, rich in light though they be, viz. that the potency of this particular animal poison varies in direct ratio to the quantity imbibed or infused, just as though it were so much alcohol, or so much alcoholic tincture of musk or cantharides; or secondly, that its potency varies in direct ratio to another varying standard, viz. the size of the animal producing it. Now the vaccine matter from the arm of a child is as potent as the vaccine matter from the arm of any giant would be; and whether a grain or a gramme of it be used will make no difference, so long as it be used rightly. There is a contrast, indeed, between the *modus operandi* of these two animal poisons. I would add that in the 'Edinburgh Monthly Medical Journal' for the present month there is a very valuable paper, one of a series of papers, indeed, of the like character, by Dr. Fayrer, where at page 247, among much of anatomical and other interest, I find the following important statement: 'This poison may be diluted with water, or even ammonia or alcohol, without destroying its deadly properties. It may be kept for months or years, dried between slips of glass, and still retain its virulence. It is capable of absorption through delicate membranes, and therefore it cannot be applied to any mucous surfaces, though no doubt its virulence is much diminished by endosmosis[1]. It appears to act by a catalytic form; that is, it kills by some occult influence on the nerve centres.' There is such a thing as an ignorance which is wiser than knowledge, *for the time*, of course, only; such an ignorance is wisely confessed to in these words of Dr. Fayrer's. An explanation may be true for some, yet not thereby necessarily for all, the facts within even a single sphere of study; even a true explanation may have but a very limited application, as a tangent cannot touch a circle at more than a single point. The memoirs published in our own reports by Dr. B. W. Richardson, on the action of the nitrites, and those published by Dr. A. Crum Brown and Dr. T. R. Fraser, there and elsewhere, on the connexion between chemical constitution and physiological action, deserve especial study as bearing on the other side of this discussion; whilst Professor Lister's papers show how the reference of certain diseases to vitalistic agencies may become of most vital importance in practice. There exists, as is well known, a tendency to resolve all physiological into physico-chemical phenomena: undoubtedly many have been, and some more

[1] Diapedesis may account for what virulence remains, and the poison may therefore possibly be a cytoid.

may still remain to be, so resolved; but the public may rest assured that in the kingdom of Biology no desire for a rectification of frontiers will ever be called out by any such attempts at, or successes in the way of, encroachment; and that where physics and chemistry can show that physico-chemical agencies are sufficient to account for the phenomena, there their claim upon the territory will be acceded to, as in the cases we have been glancing at; and where such claims cannot be established and fail to come up to the quantitative requirements of strict science, as in the cases of continuous and of discontinuous development or self-multiplication of a contagious germ, and in some others, they will be disallowed.

Pathology has of late made a return to Physiology for much service she has received, and this in the following directions. Dr. W. Ogle has thrown much light on the physiology of the cervical sympathetic nervous system by his record of a pathological history to be found in the recently issued volume (vol. lii.) of the 'Medico-Chirurgical Transactions.' The rough and cruel experimentation of war has had its vivisections utilised for the elucidation of the physiology of nerves, and especially of their trophic function, in the valuable volume issued by the American Sanitary Commission, under the editorship of Dr. Austin Flint. Dr. Broadbent has done something towards elucidating the question of the localisation of functions in particular parts of the cerebral convolutions, which was so extensively and so very exhaustively discussed at Norwich, by his paper in our most useful and comprehensive Journal of Anatomy and Physiology,' May 1870, 'On the Cerebral Convolutions of a Deaf and Dumb Woman.'

I take this opportunity of mentioning two valuable papers on the very practical question of the influence of the vagus upon the heart's action. One of these is a German paper by a gentleman who is a zoologist and comparative anatomist as well as a physiologist, Dr. A. B. Meyer; 'Das Hemmungsnerven-System des Herzens' is the title of his memoir, a separate publication as I think: the other is an abstract of a paper [I have not seen the paper published *in extenso* as yet] by Dr. Rutherford, 'On the Influence of the Vagus upon the Vascular System,' published in the journal just referred to. Especially do I think Dr. Rutherford's view as to the vagus acting centrifugally as regards the stomach, and carrying stimulus, not thither but thence, to the medulla oblongata, which stimulus is then radiated downwards by a route formed distally by the splanchnic nerve, so as to produce inhibitory vascular dilatation in the neighbourhood of the peptic cells, as worthy of attention[1].

A considerable number of the papers which will be read before this Section, indeed a considerable part of the Section itself, will be devoted to the Natural History of man. Nothing, I apprehend, is more distinctive of the present phase of that 'proper study of mankind' than the now accomplished formation of a close alliance between the students of archaeology strict and proper and the biologist with the express purpose of jointly occupying and cultivating that vast territory. Literature and art and the products of the arts furnish each their data to the ethnologist and anthropologist in addition to those which it is the business of the anatomist, the physiologist, the palaeontologist, and the physical geographer to be acquainted with; nor can any conclusion attained to by following up any single one of those lines of investigation be considered as definitely absolved from the condition of the provisional until it has been shown that it can never be put into opposition with any conclusion legitimately arrived at along any other of the routes specified. In political alliances the shortcomings of one party necessarily hamper and check the advance of the other; a failure in the means or in the perseverance of one party may bring the joint enterprise to a premature close; mutual forbearance, not to dwell longer upon extreme cases, may finally be as effectual in slackening progress as even mutual jealousies. No such disadvantages attach themselves to the alliance of literature with science, as the German 'Archiv für Anthropologie,' issued to the world under the joint management of Ecker the biologist and Lindenschmit the antiquary, will show any one who consults its pages, replete with many-sided but not superficial, multifarious but never inaccurate, information.

The antiquary is a little prone, if he will allow me to say so, when left alone, to make himself but a connoisseur; the historian, whilst striving to avoid the Scylla of judicial dulness, slides into the Charybdis of political partisanship; and the biologist not rarely

[1] Since writing as above I have seen, but have not read, a paper by Dr. Coats in Ludwig's 'Arbeiten aus der physiologischen Anstalt zu Leipzig' for 1870 which would seem to treat of this subject. The Würzburg Physiological Laboratory Reports for 1867-68 contain, as is well known, a series of papers upon it.

shows himself a little cold to matters of moral and social interest whilst absorbed in the enthusiasm of speciality. The combination of minds varying in bent is found efficacious in correcting these aberrations; and by this combination we obtain that white and dry light which is so comforting to the eye of the truth-loving student, to say nothing as to its being so much stronger than the coloured rays which the work of one isolated student may sometimes have cast upon it from the work of another. It would be invidious to speculate, and I have forborne from suggesting, whether the literary contingent in the conquering though composite army has learnt more from observation of the methods and evolutions of the scientific contingent, or the scientific more from the observation of the literary; it is, however, neither invidious nor superfluous to congratulate the general public upon the necessity which these, like other allies, have been reduced to, of adopting one common code of signals, and discarding the exclusive use of their several and distinctive technicalities. Subjects of a universal interest have thus come to be treated, and that by persons now amongst us, in a language universally 'understanded of the people.' I have been careful to include the palaeontologist amongst the scientific specialists whose peculiar researches have cast a helpful and indeed an indispensable light upon the history of the fates and fortunes of our species. But it is not organic science only which anthropology impresses into its service; and it would be the sheerest ingratitude to forget the help which the mineralogist gives us in assigning the source whence the jade celt has come or could come, or to omit an acknowledgment of the toil of the analytical chemist, who has given the percentage of the tin in the bronze celt, or in the so-called 'leaden' and therefore Roman coffin.

I am very well aware that many persons who have honoured me by listening to the last few sentences have been thinking that it is at least premature to attempt to harmonise the two classes of evidence in question; and that the best advice that can be given to the two sets of workers severally is, that they should work independently of each other. Craniography is said, and by irrefragable authority, to be a most deceptive guide; works and articles on ethnology tell us stories of skulls being labelled, even in museums of the first order of merit, with such Janus-like tickets as

'Etruscan Tyrol *or* Inca Peruvian;' and one of the most celebrated anthropotomists of the day has been so impressed with the fact that Peruvian as well as Javanese and Ethiopian skulls may be found on living shoulders within the precincts of a single German university town, that he has busied himself with forming a pseudo-typical ethnological series from the source and area just indicated. Great has been the scandal thence accruing to craniography, and the collector of skulls has thence come to be looked upon as a dilettante with singular ghoul-like propensities, which are pardonable only because they relate to savage races of modern days, or to cemeteries several hundred years old, but which are not to be regarded as being seriously scientific. Now to me the existence of such a way of estimating such a work appears to argue a sad amount of ignorance of the laws of the logic of practical life, or, indeed, of the chapters on 'approximate generalisations,' which any man, however unpractical, can read in a treatise on logic. A man's features and physiognomy are instinctively and intuitively, or, if you prefer so to put it, as a result of the accumulated social experiences of generations of men, taken as a more or less valuable and trustworthy indication of his character; were this not so, photographers would not, as I apprehend, and hope they do, make fortunes; yet the face is at least as often fallacious as an index of the mind as the skull is fallacious as an index of race. The story of the misconception by a physiognomist of the character of Socrates is familiar to us, as I think, from Lemprière's Dictionary; and it may serve to parallel the story which Blumenbach and Tilesius tell us of the exact correspondence of the proportions of a skull from Nukahiva with those of the Apollo Belvidere. The living faces in a gaol, again, to put the same argument upon other grounds, are as dangerous to judge from as are the skulls in a museum; yet every detective is something like a professor of physiognomy, and most of them could write a good commentary on Lavater. The true state of the case may, perhaps, be represented thus:—A person who has had a large series of crania through his hands, of the authenticity of which, as to place and data, he has himself had evidence, might express himself, perhaps, somewhat to the following effect if he were asked whether he had gathered from his examination of such a series any confidence as to his power of referring to, or excluding from, any such series any skull which he had not seen before. He might say,

'The human, like other highly organised types of life, admits of great variety; aberrant forms arise, even in our own species, under conditions of the greatest uniformity possible to humanity: amongst savages great variety exists (see Bates, "Naturalist on the Amazons," ii. p. 129), even though they all of them may live the same "dull grey life" and die the same "apathetic end;" and consequently it may never, except in the case of Australian or Esquimaux, and perhaps a few other crania, be quite safe to pledge one's-self as to the nationality of a single skull. Still there is such a thing as craniographical type; and if half a dozen sets, consisting of ten crania apiece, each assortment having been taken from the cemeteries of some well-marked nationality, were set before me, I would venture to say, after consultation and comparison, that it might be possible to show that unassisted cranioscopy, if not invariably right, even under such favourable circumstances, was nevertheless not wrong in a very large number of cases.' If it is true on the one hand that *in generalibus latet error*, it is true on the other that security is given us by the examination of large numbers for the accuracy and reliability of our averages, a principle which Gratiolet informs us is thoroughly recognised in Chinese metaphysics, and which he has formulated in the following words:—'L'invariabilité dans le milieu s'applique à tout. La vérité n'est point dans un seul fait mais dans tous les faits; elle est dans les moyennes, c'est-à-dire dans une suite d'abstractions formulées après le plus grand nombre d'observations possibles.' ('Mémoire sur les Plis cérébraux,' p. 93.) The natural history sciences do not usually admit of the strictness which says that an exception, so far from proving a rule, proves it to be a bad one; rather are we wise in saying that in them at least the universality of assertion is in an inverse ratio to that of knowledge, and that the sweeping statements dear, as Aristotle long ago remarked ('Rhetoric,' ii. 21. 9 and 10; ii. 22. 1), to a class which he contrasts with the educated, are abhorrent to the mind of organic nature. It is true enough, as is sometimes said, that when opinions and assertions are always hedged in by qualifications, the style becomes embarrassed, and the meaning occasionally hard to be understood; but this difficulty is one which lies in the very nature of the case, and the real excellence of style does not consist in its lulling the attention and relieving the memory by throwing an alliterative ring on the ear, but in the furnishing a closely fitting dress to thought, and an accurate representation of actual fact.

If we are told that the attempt to harmonise the results, not merely of cranioscopy, but of any and all natural science investigation, with the results of literary and linguistic research, is needless and even futile, this is simply equivalent to saying that one or other of these methods is worthless. For as Truth is one, if two routes purporting both alike to lead to it do not sooner or later converge and harmonise, this can only be because one or other of them fails to impinge upon the goal. It is true that by certain lines of investigation light is thrown upon a problem only at a single point, and that all further prosecution of investigation along that line will but lead us off at a tangent. Still the throwing of even a single ray upon a dark surface is an achievement with a value of its own; and it is a cardinal rule in our sciences never to ignore the existence of seemingly contradictory data, in whatsoever quarter they may show themselves. For what would be said of an investigator of a subject such as physical geography, who should declare that he would pay no attention except to a single set of data, when he was discussing whether a particular archipelago had been formed by upheaval, or should be held to be the fragments and remnants of a disrupted continent; and that if geological evidence was in crying discord with his interpretation of the facts of the distribution of species, it was not his business to reconcile them? He would be held to have neglected his business, as you may see by a reference to Mr. Bentham's 'Address to the Linnean Society,' May 24, 1869 ('Linn. Soc. Proc.' for 1869, p. xcii[1]).

The argument from identity of customs and practices to identity of race is liable to much the same objections, and to much the same fallacies, as is the argument from identity of cranial conformation. The case may be found admirably stated in Mr. Tylor's work on the 'Early History of Mankind,' p. 276, ed. 2; and I may say that the means of bringing the problem home to one's-self may be found by

[1] The following references to passages of the kind referred to above as to the untrustworthiness of craniographical evidence may be useful:—Geographisches Jahrbuch, 1866, p. 481. Hyrtl, 'Topograph. Anatomie,' i. p. 13. Henle, 'System. Anat.' i. 198. Krause, i. 2, p. 251. 'Archiv für Anthropologie,' Hölder, *ibid.* ii. 1, p. 60. See also His and Rütimeyer, and Ecker in their systematic works severally, the 'Crania Helvetica' and the 'Crania Germaniae meridionalis.'

a visit to any collection of flint implements. In such a collection, as Mr. Tylor has pointed out, p. 205, we are very soon impressed with the marked uniformity which characterises these implements, whether modern or thousands of years old, whether found on this side of the world or the other. For example, a flint arrow-head which came into my hands a short time back, through the kindness of Lord Antrim, after having done duty in these iron times as a charm at the bottom of a water-tub for cattle in Ireland, was pointed out or at to me by a very distinguished Canadian naturalist, who was visiting Oxford the other day, as being closely similar to the weapons manufactured by the Canadian Indians. Now after such an experience one may do well to ask in Mr. Tylor's words ('Early History,' p. 206),—

'How, then, is this remarkable uniformity to be explained? The principle that man does the same thing under the same circumstances will account for much, but it is very doubtful whether it can be stretched far enough to account for even the greater proportion of the facts in question. The other side of the argument is, of course, that resemblance is due to connexion, and the truth is made up of the two, though in what proportions we do not know. It may be that, though the problem is too obscure to be worked out alone, the uniformity of development in different regions of the Stone age may some day be successfully brought in with other lines of argument, based on deep-lying agreements in culture which tend to centralise the early history of races of very unlike appearances, and living in widely distant ages and countries.'

If the psychological identity of our species may explain the identity of certain customs, its physiological identity may explain certain others. Some of this latter class are of a curious kind, and relate not to matters of social or family, but to matters of purely personal and individual interest, concerning as they do the sensibility, and with it all the other functions of the living body. Such customs are the wearing of labrets or lip-rings, nose-rings, and, if I may add it without offence, of certain other rings inserted in the wide region supplied by the fifth or trifacial nerve[1]. A physiological explanation may lie at the base of these practices, which appear to put at the disposal of the persons who adopt them a perennial means for setting up an irritation, whence reflex con-

[1] See 'Medicine in Modern Times,' p. 57, Article XL, p. 698.

sequences in the course of reflex nutrition and reflex secretion, as of gastric juice, may flow. A curious book was written, or at least published, on the subject of these practices, and others akin to them, in 1653, by Dr. John Bulwer, a benevolent doctor, who paid attention to the care of the deaf and dumb previously, I think it is stated, to Dr. Wallis, and who consequently, with proper pride, if this precedence really belongs to him, signs himself 'J. B. cognomento Chirosophus.' The title of the book is 'Anthropometamorphosis; Man Transformed, or the Artificial Changeling.' I was made acquainted with its existence by my friend Mr. Tomlinson, of Worcester College, from the library of which society I procured a copy for consultation: the book is not rare I think, but I believe it is little known; it contains much that is curious, and it is, inasmuch as it was written more than 200 years ago, ὅτ' ἀκήρατος ἦν ἔτι λειμών, from some, though not from all points of view, the more valuable. It is, I apprehend, to some of these customs, as well as to others, that Zimmermann (not the author of the work on Solitude, but Zimmermann the zoologist) alludes in a rather amusing passage, which may be found in the third volume of his larger work on the 'Distribution of Species and on Zoology' (see p. 257). I speak of the passage as amusing; it is more than that, or I would not quote it; indeed you will not see that it is particularly amusing unless I tell you that volumes ii and iii are of date 1783, and are dedicated to his own father, whilst volume i, of date 1778, is dedicated to 'His Most Serene Highness and Lord, Ferdinand Duke of Brunswick, my Most Gracious Lord.' Its quality of amusingness depends upon these dates, and the speculations they set us to make as to how the Serene Duke, his 'Most Gracious Lord,' had offended the man of science in the interval between 1778 and 1783. It runs thus:—'If you argue from similarity of customs and ceremonies to identity of origin of two tribes under comparison, you must first show that these customs are not such as would naturally tend to the amelioration of the conditions of the inhabitants in the two countries under consideration, and would probably therefore, or can naturally, suggest themselves to each of the races in question. Or there may be customs founded on innate folly and stupidity, and thus, for your argument to be valid, you must show that, of two peoples widely separated, each cannot by any chance come into its own country to adopt the like foolish and stupid customs. For whilst two wise heads are to make out, each independently of the other and contemporaneously, a wise discovery or invention, it is much more likely, on the calculation of chances, and considering the much greater number of fools and blockheads ("Thoren und Dummköpfen"), that in two countries widely apart closely similar follies should be simultaneously invented. And then, if the inventing fool happens to be a man of influence and consideration, *which is, by the way, an exceedingly frequent coincidence*, both the nations are likely to adopt the same foolish practice, and the historian and antiquary, after the lapse of some centuries, is likely to draw from this coincidence the conclusion that the two nations both sprang from the same stock.' Judge and speculate for yourselves how the spirit which breathes in this passage was excited, but note its scientific value too. We must not forget that it is possible, in thought at least, to dissociate the psychological unity of man from his specific identity even; and that, as regards identity of race, it is only reasonable to expect that when similar needs are pressing, similar means for meeting them are not unlikely to be devised independently by members of two tribes who have for ages been separated from their original stocks. The question to be asked is, does the contrivance about which we are speculating combine, or does it not combine in itself so large a number of converging adaptations as to render it upon the calculation of chances unlikely that it should have been independently invented? Yet this very obvious principle has been neglected, or Lindenschmit would not have found it necessary to say that, by laying too much stress upon certain points of national identity in the stones used for the formation of cromlechs or dolmens, the Hünenvolk might be made out to have chosen to settle only in those parts of Germany where erratic blocks of granite or other such large stones could be found! ('Archiv für Anthropologie,' iii. p. 115, 1868.)

Sir John Lubbock's recently published work on 'The Origin of Civilisation' may, I anticipate, cause the history and genealogy of manners and customs to enter largely into the composition of our lists of papers. There is no need for me, as the author of the book is here himself, to occupy your time in recommending his work; but I may be allowed to say that the utility of such pursuits as those which Sir John Lubbock's book treats of receives some little illustration from the fact that, as we learn from him and from Mr. Tylor, the human mind blunders and errs and deceives itself in these subjects in just the same way as it does in the kindred, though more immediately arising, pressing, and important matters of social and political life. In these latter spheres of observation we are apt occasionally to mistake one of those intermittent reactions of opinion, produced as eddies are produced in a river by the deposit of sand and mud at angles in its onward course, for a deliberate giving up of the principles upon which all previous progress has been dependent. The straws which float upon the surface of a backwater may be taken as proofs that the river is about to flow upwards, and a feeble oarsman in a light boat may be deceived for some moments by the backward drifting of his small craft. Now an analogous blunder is often made in matters of purely historical interest; and we may do well to learn from the experience thus cheaply earned. 'The history of the human race has,' says Sir John Lubbock, p. 322, *l. c.*, 'I feel satisfied, on the whole been one of progress: I do not of course mean to say that every race is necessarily advancing; on the contrary, most of the lower are almost stationary:' but Sir John regards these as exceptional instances, and points out that if the past history of man had been one of deterioration, we have but a groundless expectation of future improvement; whilst on the other, if the past has been one of progress, we may fairly hope that the future will be so also.

Mr. Tylor's words are equally to the purpose, though, as forming the end of a chapter merely and not the end of a book, they are less enthusiastic in tone (p. 193, Tylor, 'Early History of Mankind'). They run thus:—

'To judge from experience, it would seem that the world, when it has once got a firmer grasp of new knowledge or a new art, is very loath to lose it altogether, especially when it relates to matters important to man in general, for the conduct of his daily life, and the satisfaction of his daily wants, things that come home to men's "business and bosoms." An inspection of the geographical distribution of art and knowledge among mankind seems to give some grounds for the belief that the history of the lower races, as of the higher, is not the history of a course of degeneration or even of equal oscillations to and fro, but of a movement which, in spite of frequent stops and relapses, has on the whole been forward; that there has been from age to age a growth in man's power over nature, which no degrading influences have been able permanently to check.'

I must not trespass into the province of the botanist, but I should be glad to say that no easier method of learning how the natural history sciences can be made to bear upon the history of man, as a whole, can be devised than that furnished by the perusal of such memoirs as those of Unger's upon the plants used for food by man. The very heading and title of the paper I am specially referring to appears to me to have an ambiguity about it which, in itself, is not a little instructive. In that title, 'Botanische Streifzüge auf dem Gebiete der Cultur-Geschichte,' the latter word may be taken, I imagine, etymologically at least, to refer either to culture proper, or to floriculture, or to agriculture. At any rate, the paper itself may be read in the Sitzungsberichte of the Vienna Academy for 1859; it has, I suppose, superseded the interesting chapters in Link's 'Urwelt und Alterthum,' of date 1821; and it is not unlikely, I apprehend, to be itself, in its turn, superseded also.

Coming, in the third place, to Zoology, I suppose I shall be justified in saying that the largest issue which has been raised in the current year, an issue for the examination of the data for deciding which the two months of July and August which are just past may have furnished persons now present with opportunities, is the question of the kinship of the Ascidians to the Vertebrata. There is or was nothing better established till the appearance of Kowalewsky's paper, now about four years ago, than the existence of a wide gulf between the two great divisions of the animal kingdom, the Vertebrata and the Invertebrata: nothing could be more revolutionary than the views which would obviously rise out of his facts; and within the present year these facts have been abundantly confirmed by Prof. Kupfer, whose very clearly written and beautifully illustrated paper has just appeared in the current number of Schultze's 'Archiv für microscopische Anatomie.' Kupfer's researches have been carried on upon *Ascidia canina;* but they more than confirm the accuracy of what Kowalewsky had stated to take place in *Ascidia mammillata*, and which may be summed up briefly thus:—In the larval Ascidian we have in its caudal appendages an axis skeleton clearly analogous, if not essentially homologous, to the chorda dorsalis of the vertebrate embryo, as consisting, like it, of rows of internally placed cells, and giving insertion by its sheath to muscles. We have further the nervous system and the digestive taking up in such embryos much the same positions relatively to each other, and to this molluscan chorda dorsalis, that are taken up by the confessedly homologous system in the Vertebrata; we have the nervous system originating in the same fashion, and closing up like the vertebrate myelencephalon out of the early form of a lamellar furrow into that of a closed tube; we have, finally, the respiratory and digestive inlets holding the vertebrate relationship of continuity with, instead of the invertebrate of dislocation and separation from, each other. Such are the facts; but I am not convinced that they will bear the interpretation that has been put upon them; though I must say the possession of this chorda dorsalis by the active locomotor larva of the Ascidian which one day settles down into such immobility lends not a little probability to Mr. Herbert Spencer's view of the genesis of the segmented vertebral column in animals undoubtedly vertebrate. But on this view I should not be inconsistent with myself, inasmuch as, to waive other considerations, the chorda dorsalis in each case would be considered as an adaptive or teleological modification, not a sign of morphological kinship[1]. Much perplexity may or must arise here; and whilst entertaining these views, I felt myself bound to examine myself strictly to find whether in not taking them up, I might not be giving way to that reactionary reluctance to accept new ideas which advancing years so frequently bring with them; but a recent paper, by Lacaze-Duthiers, published in the 'Comptes Rendus' for May 30, 1870, and translated in the 'Annals and Magazine of Natural History' for July 1870, would justify me, I think, in calling that reluctance by another name. For in that paper the renowned malacologist just mentioned has brought to light the fact that there is another sessile and solitary Ascidian, the *Molgula tubulosa*, which goes through no such tadpole-like stage as had been supposed to be gone through by all Ascidians except the Salpae, which is never active and never puts out the activity which is so remarkable in the other Ascidians, but settles down and remains sedentary immediately after it is set free from the egg-capsule,

[1] See, however, Mr. Herbert Spencer's Appendix D to his principles of Biology, pt. iv. chap. xvi. This appendix was printed in 1865, but not published till December 1869. I had not seen it when I wrote as above.

neither enjoying a Wanderjahr *nor possessing a chorda dorsalis.* We are not surprised after this that M. Lacaze-Duthiers observes that 'although embryology may and must furnish valuable information by itself, it may also, in some cases, lead us into the gravest errors.' Mr. Hancock, of Newcastle-upon-Tyne, has sent us a paper upon this subject, which will be read duly and duly noted by us.

Leaving Malacology, which has not in the United Kingdom obtained the same hold as yet upon the public mind that it has on the Continent, where, like Entomology, there and here, it has a periodical or two devoted to the recording of the discoveries of its votaries, I have much pleasure in directing attention to two short papers by Siebold in the 'Zeitschrift für wissenschaftliche Zoologie' (xx. 2, 1870), on parthenogenesis in *Polistes gallica* v. *diadema*, and on paedogenesis in the *Strepsiptera*. In each of these short papers Siebold informs us that adequate room and time could not be given them in the Innsbruck meeting held just a year ago, or in the report of the meeting. It is to me a matter of difficulty to think what there could have been of greater value than those papers in a section of Wissenschaftliche Zoologie; it will be to all present a matter of congratulation to learn, from the venerable professor's papers, that he will shortly favour us with a new work on the subject of parthenogenesis. A fresh instance of parthenogenesis in Diptera, viz. in *Chironomus*, has just been put upon record in the St. Petersburg Imperial Academy's Memoirs (xv. 8, January 13, 1870).

The subject of the geographical distribution of the various forms of vegetable and animal life over the surface of the globe, and in the various media, air, earth, water—fresh and salt, whether deep or shallow—has always been, and will always remain, one of the most interesting subsections of biology. It was the contemplation of a simple case of geographical distribution in the Galapagos archipelago which brought the author of the 'Origin of Species' face to face with the problem which the title of his work embodies; and it is impossible that sets of analogous and of more complicated facts (many of which, be it recollected, such as the combination now being effected between our own fauna and flora and those of Australia and New Zealand, are patent to the observations of the least observing) should not, since the appearance of that book, force the serious consideration of the explanation it offers upon the thoughts of all who think at all. The wonders of the deep-sea fauna will, I apprehend, form one, the commensalism of Professor Van Beneden another, subject of discussion, and furnish an opportunity for receiving instruction to all of us. The one set of observations is a striking exemplification of the way in which organisms have become suited to inorganic environments; the other is an all but equally striking exemplification of the way in which organisms can fit and adapt themselves to each other. The current journals have[1], as was their duty, made us acquainted with what has been done in both of these directions; and I am happy to say that in the case of the deep-sea explorations, as in that of parthenogenesis and spontaneous generation, a new work, giving a connected and general view of the entire subject, is announced for publication.

One instance of the large proportions of the questions which the facts of geographical distribution bear upon, is furnished to us in the address recently delivered before the Geological Society by its president, who is also our president, and who may have forgotten to refer to his own work (see 'Nature,' No. 24, 1870). Another may be found in the demonstration which Dr. Günther, contrary to our ordinarily taught doctrines, has given us ('Zool. Soc. Trans.' vol. vi. pt. 7, 1868, p. 307) of the partial identity of the fish-faunas of the Atlantic and Pacific coasts of Central America; many, thirdly, are furnished to us by Mr. Wallace's works *passim.*

It would be superfluous, after introducing even thus hurriedly to your notice so large a series of interesting and important subjects as being subjects with which we shall forthwith begin to deal in this Section, to say anything at length as to the advantages which may reasonably be expected to accrue from the study of Biology. I may put its claims before you in a rough way by saying that I should be rejoiced indeed if, when money comes to be granted by the Association for the following up the various lines of biological research upon which certain of its members are engaged, we could hope to obtain a hundredth, or I might say a thousandth part of the amount of money which has in the past year been lost to the State and to individuals through ignorance or disregard of biological laws now well established. I need say nothing of the

[1] See 'Nature,' No. 39, July 28, 1870, and 'Royal Society's Proceedings,' August 1870, for deep-sea explorations, and 'Academy,' September 10, 1870, for commensalism.

suffering or death which anti-sanitary conditions entail, as surely as, though less palpably and rapidly than, a fire or a battle; and I might, if there were time for it, take my stand simply upon what is measurable by money. This I will not do, as it is less pleasant to speak of what has been lost than of that which has been or may be gained. And of this latter let me speak in a few words, and under two heads—the intellectual and the moral gains accruing from a study of the Natural History Sciences. As to the intellectual gains, the real psychologist and the true logician know very well that the discourse on method which comes from a man who is an actual investigator is worth, even though it be but short and packed away in an Introduction or an Appendix, or though it cover but a couple of pages in the middle of a book, like the 'Regulae Philosophandi' of Newton, more than whole columns of the 'Sophistical Dialectic' of the ancient Schoolman and his modern followers. 'If you wish your son to become a logician,' said Johnson, 'let him study Chillingworth'—meaning thereby that real vital knowledge of the art and science can arise only out of the practice of reasoning; and as to the value of actual experimentation as a qualification for writing about method, Claude Bernard and Berthelot are, and I trust will long remain, living examples of what Descartes and Pascal, their fellow-countrymen, are illustrious departed examples. (See Janet, 'Revue des Deux-Mondes,' tome lxii. p. 910, 1866.)

I pass on now to say a word on the working of natural science studies upon the faculty of attention, the faculty which has very often and very truly been spoken of as forming the connecting-link between the intellectual and the moral elements of our immaterial nature. I am able to illustrate their beneficial working in producing carefulness and in enforcing perseverance, by a story turning upon the use of, or rather upon the need for, a word. Von Baer, now the Nestor of biologists, after a long argumentation ('Mém. Acad. Imp. Sci. St. Pétersbourg,' 1859, p. 340), of the value which characterises his argumentations generally, as to the affinities of certain oceanic races, proceeds to consider how it is that certain of his predecessors in that sphere, or, rather, in that hemisphere, as Mr. Wallace has taught us Oceania is very nearly, had so lamentably failed in attaining or coming anywhere near to the truth. This failure is ascribed to something which he calls 'Ungenirtheit,' a word which you will not find in a German dictionary, the thing itself not being, Von Baer says, German either. I am happy not to be able to find an exact equivalent for this word in any single English vocable; the opposite quality shows itself in facing conscientiously 'the drudgery of details, without which drudgery,' Dr. Temple tells us ('Nine Schools Commission Report,' vol. ii. p. 311), 'nothing worth doing was ever yet done.' Mr. Mill, I would add, speaks to the same effect, and even more appositely, as far as our purpose and our vocations are concerned, in his wise 'Inaugural Address at St. Andrews,' p. 50. For the utter incompatibility of an ἀταλαίπωρος ζήτησις (these two words give a Thucydidean rendering of '*Ungenirtheit*') with the successful investigation of natural problems, I would refer any man of thought, even though he be not a biologist, to a consideration of the way in which problems as simple at first sight as the question of the feeding or non-feeding of the salmon in fresh water (see Dr. McIntosh, 'Linn. Soc. Proc.' vii. p. 148), or that of the agencies whereby certain molluscs and annelids bore their way into wood, clay, or rocks, must be investigated. It is easy to gather from such a consideration how severe are the requirements made by natural science investigations upon the liveliness and continuousness with which we must keep our faculty of attention at work.

I shall speak of but one of the many purely moral benefits which may be reasonably regarded either as the fruit of a devotion to or as a preliminary to success in natural science. Of this I will speak in the words of Helmholtz, taking those words from a report of them as spoken at the meeting of the German Association for the Advancement of Science, which was held last year at Innsbruck. There Professor Helmholtz, in speaking of the distinctive characteristics of German scientific men, and of their truthfulness in particular, is reported to have used the following words :—

> 'Es hat diesen Vorzug auch wesentlich zu verdanken der *Sittenstrenge* und *der uneigennutzige Begeisterung* welche die Männer der Wissenschaft beherrscht und beseelt hat, und welche sie nicht gekehrt hat an äussere Vortheile und gesellschaftliche Meinungen.'

These words are, I think, to the effect that the characteristics in question are in reality to be ascribed to the *severe simplicity of manners* and *to the absence of a spirit of self-seeking* which form the guiding and inspiring principles of their men of science, and prevent them from giving themselves up to the pursuit of mere worldly advantages, and from paying undue homage to the prejudices of society. I think *Sittenstrenge* may be considered as more or less adequately rendered by the words *severe simplicity of manners;* at any rate, as things are known by their opposites, let me say that it is the exact contradictory of that '*profound idleness and luxuriousness*' which, we are told by an excellent authority (the Rev. Mark Pattison, 'Suggestions on Academic Organization,' p. 241),—for whose accuracy I would vouch in this matter were there any need so to do,—'*have corrupted the nature*' of a large class of young men amongst ourselves; whilst *the absence of a spirit of self-seeking* is, in its turn, the contradictory of a certain character which Mr. Mill (l. c. p. 90) has said to be one of the commonest amongst us adults, and to which Mr. Matthew Arnold has assigned the very convenient epithet of 'Philistine.' Investigation as to whether these undesirable tendencies are really becoming more rife amongst us, might be carried on with advantage in a place such as this[1], in the way of inquiries addressed to colonists returning home after a successful sojourn abroad. Such persons are able to note differences without prejudice, and, *ex hypothesi,* with unjaundiced eyes, which we are apt to overlook, as they may have grown up gradually and slowly. But, perhaps, researches of this kind are not quite precisely the particular kind of investigation with which we should busy ourselves; neither would the leaders of fashion, the persons with whom all the responsibility for this illimitable mischief rests, be very likely to listen to any statistics of ours, their ears being filled with very different sounds from any that, as I hope, will ever come from Section D. Whether men of science in England are more or less amenable to blame in this matter than the rest of their countrymen, it does not become us to say; but it does become and concern us to recollect that we have particular and special reasons, and those not far to seek, nor dependent on authority alone, for believing and acting upon the belief that real success in our course of life is incompatible with a spirit of self-seeking and with habits of even refined self-indulgence.

[1] [Liverpool, where this Address was delivered.—EDITOR.]

XLIV.

ADDRESS ON ANTHROPOLOGY,

DELIVERED TO THE ANTHROPOLOGICAL DEPARTMENT OF THE BRITISH ASSOCIATION FOR THE ADVANCEMENT OF SCIENCE.

Some few weeks ago Mr. James Parker, of Oxford, invited me to visit your Somersetshire caves, in the company of the Warwickshire Naturalists' and Archaeologists' Field Club. It struck me that I should do well, as I was to preside over the Anthropological Department at this British Association Meeting, if I tried to learn as much as I could of the relics and of the surroundings of the Prehistoric inhabitants of your neighbourhood; and for this, as well as for other reasons, I gladly accepted the invitation. During that pleasant midsummer excursion, I was more than once impressed with the similarity which its incidents bore to those of the undertaking in which we are now engaged, and, indeed, to those of the study of Anthropology generally. First, the organisation of the expedition had entailed some considerable amount of labour upon those who had charged themselves with that duty; and, secondly, a thorough exploration of the recesses and sinuosities of the several caves which we explored devolved upon us not only a good deal of exertion, but even some slight amount of risk; for the passages and galleries along which we worked our way were sometimes low and narrow, often steep, and nearly always slippery. Thirdly, the outline of the regions explored bore quite different aspects accordingly as we lighted them up or had them lit up for us in one or in another of several different ways.

If in any segment of these caves the outside daylight could anyhow find a zigzag way down some shaft into the interior, that segment wore a general aspect more comfortable to the eye, and so to the mind, than others not so illuminated. These latter regions, again, varied greatly *inter se*, according to the various artificial means employed for lighting them up. The means ordinarily used for this end made their outlines look a little colder and harder than the reality itself, cold and hard though this was; whilst under certain other modes of illumination employed (it is true, only occasionally, and for purposes of effect, not *ex necessitate*) the self-same outlines looked somewhat lurid. But, howsoever produced and howsoever affecting us, the light was light nevertheless, and, on the whole, we preferred it a good deal to the darkness. It is never well to press a metaphor too far nor too closely; so I will now lay aside my parable, though it admits of some further extension, and take up the actual business of the Department.

It may be well to lay before the Department, first of all, the titles of a few of the principal subjects upon which we have papers prepared for us; and after, or indeed during the enumeration of these specimens of what will prove, I can assure you, a very valuable series of memoirs, we can proceed, as will be naturally suggested, to those general considerations with which it is customary to open the transactions of such assemblages as ours.

First among our contributors I must mention the President of the London Anthropological Institute, in which Institute the Ethnological Society of 1844 and the Anthropological Society of 1863 are united. Colonel Lane Fox has told us ('Archaeologia,' xlii. p. 45, 1869) that it was whilst serving on the Sub-committee of Small Arms in 1851 that he had his attention drawn to the principle of continuity by observing the very slow gradations of progress that were taking place at that time in the military weapons of our own country. Out of those labours of his on that Sub-committee other benefits have arisen to the country at large, of which it is not my province to speak. What I have to speak of is his suggestion, put out with greater definiteness in his invaluable Lecture on Primitive Warfare, delivered before the United Service Institution, June 5, 1868 (p. 15), to the effect that his find at Cissbury furnishes the links which were wanting to connect the Palaeolithic with the Neolithic Celt types. Sir John Lubbock [1] and Mr. Evans [2] have told us that they do not see their way to accepting this view; and Mr. James Geikie, who holds that the palaeolithic deposits are of pre-glacial and inter-glacial age, is almost necessitated, *ex hypothesi*, to repudiate any such transition.

[1] Nilsson's 'Primitive Scandinavia,' Editor's Introd. p. 24.

[2] 'Flint Implements,' p. 72.

He does so (pp. 436–438 of his work on the Great Ice Age) in language which shows us that Colonel Lane Fox's lecture just referred to, with its diagram No. 1 (printed, it is true, for private circulation), could not have met his eye. Colonel Lane Fox's paper will relate to further explorations carried on at Cissbury during the present year by a Committee of the Anthropological Institute with the kind permission of Major Wisden, the owner of the soil. It will raise more than one large question for us to address ourselves to. I shall, when Colonel Lane Fox's paper comes before the Department, contribute towards its discussion by showing a number of flints from Cissbury, given me by my friend Mr. Ballard, of Broadwater.

Mr. Pengelly will, on Monday, give us an account of the 'Anthropological Discoveries in Kent's Cavern.' A more interesting subject will not often have been treated in a more interesting manner.

Polynesia and Australasia generally have always been an interesting field for the anthropologist. Our recent acquisition of Fiji makes it doubly interesting to us just now; and a flood of literature has burst forth upon us to meet that interest.

Professor Dr. Carl E. Meinicke is to be heartily congratulated on having, in the present year, brought out a work on the islands of the Pacific ('Die Inseln der Stillen Oceans, eine geographische Monographie,' Erster Theil, Melanesien und Neuseeland, Leipzig, 1875), in which he can, with not unbecoming pride, say that he is still working upon the same principles which guided him nearly fifty years ago in the composition of his works on the continent of Australia and the South-Sea races. Though I possess Professor Meinicke's works, I am not as yet entirely in possession of all his views; but so far as I can see, they are well worthy of attention. I do not hesitate, however, at all in saying that the most important contribution to the ethnology of Polynesia which has been made recently is the article on that subject in the 'Contemporary Review' for February 1873, by the Rev. S. Whitmee, of Samoa. And I may say that I am not without hopes that we shall be favoured with some papers upon the ethnology, anthropology, and future prospects of the Polynesian race by other persons eminently qualified to speak upon the subject, as having spent many years usefully among them, and on the spot. I observe that writers who have little respect for most things else, and by no means too much for themselves, speak still with something like appreciation of the work done in those regions by the London Missionary Society; and we here shall value highly any papers which we may be favoured with from men who have had such long and such favourable opportunities for forming opinions on matters which touch at once our national and our scientific responsibilities.

What question can be of closer concernment than that of the possibility of rescuing the inhabitants of Polynesia from that gradual sliding into extinction which some writers appear to acquiesce in as the natural fate of such races? As a text for our discussions upon this subject, I will here quote to the Department a passage from the continuation of Waitz's 'Anthropologie' by Dr. Gerland—the author, be it remembered, of a special Monograph upon the Causes of the Decrease and Dying-out of Native Races, which appeared in 1868 ('Ueber das Aussterben der Naturvölker,' Leipzig), and has been often referred to by writers on anthropology since that year, and is referred to by himself in the passage I now lay before you. It runs thus ('Anthropologie der Naturvölker,' von Dr. Theodor Waitz, fortgesetzt von Dr. Georg Gerland, 1872, vol. ii. pp. 512, 513):—

'The decrease of the Polynesian populations is not now going on as fast as it was in the first half of the century; it has in some localities entirely ceased, whilst in others the indigeneous population is actually on the increase[1]. From this it is clear that the causes for that disappearance of the native races which we discussed at length in the little book above referred to, are now less or no longer operative. For, on the one hand, the natives have adapted themselves more to the influences of civilisation; they are not so amenable as they were at first to the action of diseases, although we still from time to time have instances to the contrary at the

[1] See 'Times' of Saturday, August 21, 1875, p. 6, where the Natal correspondent, writing of the Caffres, tells us, 'we shall have to begin civilising the natives some day. We had better have begun with them ten years ago at 200,000 strong, than now at 350,000; but we had better begin with them now at 350,000 than ten years hence when they may number half-a-million.' Since writing as above I have received through my friend the Rev. W. Wyatt Gill a long extract from a paper written in 1861, by the Rev. A. W. Murray. This paper fully confirms Gerland's more recent views as to the prospects of the native races. Mr. Murray, having spent forty years in Polynesia, has the best possible right to be heard upon it.

present moment (see, for example, "Ev. Miss. Mag." 1867, p. 300, Cheever, 295) [or, I may add, our own recent information as to the destructive outbreak of measles in Fiji]; they have become more able to respond to the efforts to raise their mental and moral status than they were; and, with the advance of civilisation, they have begun to avail themselves more of the remedial agencies which it brings with it. On the other hand, we cannot ignore the fact that the Europeans themselves, in spite of many important exceptions, have nevertheless done a very great deal for the natives, and are always doing more and more for them. Whilst in this matter the English Government deserves great praise, and whilst Sir George Grey has done more for the Polynesians than almost any other man, the missionaries nevertheless stand in the very first rank amongst the benefactors of these races, with their unwearied self-sacrificing activity; and Russel ("Polynesia," Edinb., 1840) is entirely right in saying that all the progress which the Polynesians have made was really set on foot by the missionaries. They have had the greatest influence upon the civilisation of the natives; they have taken their part and protected them when they could; they have further given them the fast foothold, the new fresh object, motive, and meaning for their whole existence, of which they stood so much in need. The Polynesians have often declared to the missionaries, "If you had not come, we should have perished;" and they would have perished if their country had not been so discovered. The resources of their physical life were exhausted; and they had none of the moral nor ideal support for the needs of their spiritual nature which they stood so urgently in need of, as they had already attained a grade of culture too high to allow of their living without some support of that kind. It is true that extraneous circumstances have often, especially in the outset, brought about their conversion—as, for example, the authority of their chiefs, the force of example, as also, on the other hand, the occurrence of misfortune, great mortality, the loss of a battle, after which they wished to make the experiment of worshipping a new god (Russel, pp. 886, 390). And it is also true that the missionaries have introduced them to an exceedingly bigoted and often little-elevated form of Christianity; but even this has been a fortunate circumstance; for just the comprehensibility, the plain appeal to the senses, of this new religion took hold of the imagination of these races, and they could take hold of it with their understanding; and, howsoever it may have been put before them, it was immeasurably above the level of heathenism, and considerably above that of Mahommedanism. Whatever the dogmas taught were, the ethics of Christianity were taught with them; and in most cases the missionaries gave, at the same time, in their lives striking examples of the value of those ethics; and the fact of their maintenance and exemplification was the main thing.'

Mr. Bagehot has been quoted by Mr. Darwin, in his 'Descent of Man,' ed. 1, vol. i. p. 239, ed. 2, p. 182, as saying that 'it is a curious fact that savages did not formerly waste away before the classical nations, as they do now before the modern civilised nations; had they done so the old moralists would have mused over the event; but there is no lament in any writer of that period over the perishing barbarians.' On reading this for the first, and indeed for a second time, I was much impressed with its beauty and originality; but beauty and originality do not impress men permanently unless they be coupled with certain other qualities. And I wish to remark upon this statement, first, that it is exceedingly unsafe to argue from the silence of any writer, ancient or modern, to the non-existence of the non-mentioned thing. I do not recollect any mention in the ancient writers of Stonehenge, nor can I call to mind at this moment any catalogue of the vocabularies of the Cimbri and Teutones, of the Ligures and Iberians, with whom the ancients were brought into prolonged contact. These little omissions are much to be regretted, as, if they had been filled up, a great many very interesting problems would thus have been settled for us which we have not as yet settled for ourselves. But these omissions do not justify us in thinking that Stonehenge is an erection of post-Roman times, nor in holding that any of the strange races mentioned were devoid of a language. And, secondly, what we know of the classical nations dates from a time when the 'merciless bronze' had begun to give way to the 'dark gleaming' steel. But long before the displacement of bronze weapons by iron ones, the bronze had had abundant time to displace both stone weapons and the people who used them. And it is plain enough to suggest that one reason why the old moralists did not muse over the disappearance of the aboriginal races lies in the fact that these races had neither a contemporary Homer to sing their history, nor an Evans to interpret their weapons after their extinction. The actual Homeric poems deal with a region thickly peopled and long subdued by a Greek-speaking metal-using race. Rhodes and Crete were as different then from what Fiji and New Guinea are now, as Merion and Idomeneus are from Thakombau and Rauparahu. But, thirdly, let us ask, as the philosophers did with regard to the fish and its weight in and out of the bucket of water, Are the facts about which we are to inquire really facts? Now I am not going to plunge into the excursuses appended to editions of Herodotus, nor to discuss the history of the Minyae, or of any other race of which we know as little. But I will just quote a few verses from a beautiful passage in Job which appear to me to give as exact a description of a barbarous race perishing and outcast, as could be given now by a poetical observer in Australia or California. Speaking of such a race the poet says:—

'For want and famine they were solitary, fleeing into the wilderness in former time desolate and waste. Who cut up mallows by the bushes, and juniper roots for their meat. They were driven forth from among men, (they cried after them as after a thief;) to dwell in the cliffs of the valleys, in caves of the earth, and in the rocks. Among the bushes they brayed; under the nettles they were gathered together. They were children of fools, yea, children of base men: they were viler than the earth' (Job, chap. xxx. ver. 3–8).

I opine that these unhappy savages must have 'wasted away' under these conditions, and that there is no need, with such actual *verae causae* at hand, to postulate the working of any 'mysterious' agency, any inscrutable poisonous action 'of the breath of' civilisation. What is mysterious to me is not civilisation, but the fact that people who are in relation with it do not act up to its behests. And what is the mystery to me is not how an epidemic can, when introduced amongst helpless Polynesians, work havoc, but how it is that epidemics should be allowed to do so here in England from time to time. We are but some four years away from the last small-pox epidemic, of the management, or rather mismanagement, of which I had myself some little opportunity of taking stock; and what we saw then in England renders it a little superfluous to search for recondite causes to account for depopulation in countries without Local Boards. You owe much in Bristol to your able, energetic, and eminently successful officer of health, Dr. David Davies. I hope he may favour us with his views upon this very interesting subject, and may, knowing, as he well does, how much energy and knowledge is required for the reduction of a rate of mortality, tell us how much wickedness, perversity, and ignorance is necessary for increasing such a rate, whether in Great or in Greater Britain. I think that he will tell us that what is mysterious is not the power of the principles of action I have just mentioned, but the toleration of them. Such, at least, are my views[1].

We have several philological papers promised us. Amongst them will be one by the Rev. John Earle, who is known to you in this neighbourhood as living near Bath, and who is known to people not so pleasantly situated on the earth's surface as you are, as the author of a Handbook of the English tongue. I shall, as he will be present hereafter to speak on philology, spare myself and you the trouble of any remarks on that truly natural science, observing merely that Dr. Farrar[2] and Professor Häckel[3] are both agreed upon one point, namely that the adoption of natural-history methods by the students of languages has opened up for them a fresh career of importance and interest and usefulness.

Somersetshire is not without its historian; and the possibility of his coming renders it unadvisable for me to say anything now as to the relation of history to our subject upon the present occasion.

[1] Since I wrote as above, we have received the news of the murder of Commodore Goodenough at Santa Cruz. Commodore Goodenough was one of those persons to have met whom makes a man feel himself distinctly the better for his interviews and intercourse. He was not only a typical representative of what is called 'Armed Science,' he not only possessed the eye to watch and the arm to strike, happily so common in our two services, but he added to all this a cultivation and refinement duly set forth and typified by manners which were

'not idle but the fruit
Of loyal nature and of noble mind.'

It is indeed a 'puzzling world,' as it has been forcibly phrased, in which such a man loses his life, and we lose his power for good, through the act of what Wordsworth calls

'A savage, loathsome, vengeful, and impure.'

Still Corfe Castle is near enough to Bristol to prevent us from forgetting that we ourselves were once as treacherous and murderous as the modern Papuans, and that less than 900 years ago. If we have improved, there is hope for them.

[2] Farrar on the 'Growth of Language,' pp. 17, 18, 'Journal of Philology,' 1868.

[3] Häckel, 'Anthropogenie,' 1874, p. 361.

If, however, the Department can find time to listen to me a second time, I shall be glad to read a short paper myself upon this very subject, mainly in the hope of getting Mr. Freeman to speak upon it also.

I come now (perhaps I should have come before) to the consideration of the subject of craniology and craniography. Of the value of the entirety of the physical history of a race there is no question; but two very widely opposed views exist as to the value of skull-measuring to the ethnographer. According to the views of one school, craniography and ethnography are all but convertible terms; another set of teachers insist upon the great width of the limits within which normal human crania from one and the same race may oscillate, and upon the small value which, under such circumstances, we can attach to differences expressed in tenths of inches or even of centimetres. As usual, the truth will not be found to lie in either extreme view. For the proper performance of a craniographic estimation, two very different processes are necessary: one is the carrying out and recording a number of measurements; the other is the artistic appreciation of the general impressions as to contour and type which the survey of a series of skulls produces upon one. I have often thought that the work of conducting an examination for a scholarship or fellowship is very similarly dependent, when it is properly carried out, upon the employment of two methods—one being the system of marking, the other that of getting a general impression as to the power of the several candidates; and I would wish to be understood to mean by this illustration not only that the two lines of inquiry are both dependent upon the combination and counterchecking of two different methods, but also that their results, like the results of some other human investigations, must not be always, even though they may be sometimes, considered to be free from all and any need for qualification. Persons like M. Broca and Professor Aeby, who have carried out the most extensive series of measurements, are not the persons who express themselves in the strongest language as to craniography being the universal solvent in ethnography or anthropology. Aeby, for example, in his 'Schädelformen der Menschen und der Affen,' 1867, p. 61, says:—'Aus dem gesagten geht hervor, dass die Stellung der Anthropologie gegenüber den Schädelformen eine ausserordentlich schwierige ist;' and the perpetual contradiction of the results of the skull-measurements carried out by others, which his paper (published in last year's 'Archiv für Anthropologie,' pp. 12, 14, 20) abounds in, furnishes a practical commentary upon the just quoted words. And Broca's words are especially worth quoting, from the 'Bulletin de la Société d'Anthropologie de Paris,' Nov. 6, 1873, p. 824:—'Dans l'état actuel de nos connaissances la craniologie ne peut avoir la prétention de voler de ses propres ailes, et de substituer ses diagnostics aux notions fournies par l'ethnologie et par l'archéologie.'

I would venture to say that the way in which a person with the command of a considerable number of skulls procured from some one district in modern times, or from some one kind of tumulus or sepulchre in prehistoric times, would naturally address himself to the work of arranging them in a museum, furnishes us with a concrete illustration of the true limits of craniography. I say 'a person with the command of a considerable number of skulls;' for, valuable as a single skull may be, and often is, as furnishing the missing link in a gradational series, one or two skulls by themselves do not justify us (except in rare instances, which I will hereinafter specify) in predicating anything as to their nationality. Greater rashness has never been shown, even in a realm of science in which rashness has only recently been proceeded against under an Alien Act, than in certain speculations as to the immigration of races into various corners of the world, based upon the casual discovery in such places of single skulls, which skulls were identified, on the ground of their individual characters, as having belonged to races shown on no other evidence to have ever set foot there.

It is, of course, possible enough for a skilled craniographer to be right in referring even a single skull to some particular nationality; an Australian or an Eskimo, or an Andamanese might be so referred with some confidence; but all such successes should be recorded with the reservation suggested by the words, *ubi eorum qui perierunt?* and by the English line, 'the many fail, the one succeeds.' They are the shots which have hit, and have been recorded. But if it is unsafe to base any ethnographic conclusions upon the examination of one or two skulls, it is not so when we can examine about ten times as many—ten, that is to say, or twenty, the locality and the dates of which are known as certain quantities. A craniographer thus fortunate casts his eye over the entire series, and selects from it one or more which correspond to one of the great types based by Retzius not merely upon consideration of proportionate lengths and breadths, but also upon the artistic considerations of type, curve, and contour. He measures the skulls thus selected, and so furnishes himself with a check which even the most practised eye cannot safely dispense with. He then proceeds to satisfy himself as to whether the entire series is referable to one alone of the two great typical forms of Brachycephaly or Dolichocephaly, or whether both types are represented in it, and if so, in what proportions and with what admixture of intermediate forms. With a number of Peruvian, or, indeed, of Western American skulls generally, of Australian, of Tasmanian, of Eskimo, of Veddah, of Andamanese crania before him, the craniographer would nearly always, setting aside a few abnormally aberrant (which are frequently morbid) specimens, refer them all to one single type[1].

Matters would be very different when the craniographer came to deal with a mixed race like our own, or like the population of Switzerland, the investigation into the craniology of which has resulted in the production of the invaluable 'Crania Helvetica' of His and Rütimeyer. At once, upon the first inspection of a series of crania, or, indeed, of heads, from such a race, it is evident that some are referable to one, some to another, of one, two, or three typical forms, and that a residue remains whose existence and character is perhaps explained and expressed by calling them 'Mischformen.' Then arises a most interesting question—Has the result of intercrossing been such as to give a preponderance to these 'Mischformen?' or has it not rather been such as in the ultimate resort, whilst still testified to by the presence of intermediating and interconnecting links, to have left the originally distinct forms still in

[1] It is not by any means entirely correct to say that there is no variety observable among races living in isolated savage purity. The good people of Baden who, when they first saw them, said all the Bashkirs in a regiment brought up to the Rhine in 1813 by the Russians were as like to each other as twins, found, in the course of a few weeks, that they could distinguish them readily and sharply enough (see Ecker, 'Crania Germaniae Occid.' p. 2; 'Archiv für Anthrop.' v. p. 485, 1872). And real naturalists, such as Mr. Bates, practised in the discrimination of zoological differences, express themselves as struck rather with the amount of unlikeness than with that of likeness which prevails amongst savage tribes of the greatest simplicity of life and the most entire freedom from crossing with other races. But these observations relate to the *living heads*, not to the skulls.

something like their original independence, and in the possession of an unoverwhelmed numerical representation? The latter of these two alternative possibilities is certainly often to be seen realised within the limits of a modern so-called 'English' or so-called 'British' family; and His has laid this down as being the result of the investigations above mentioned into the Ethnology of Switzerland. At the same time it is of cardinal importance to note that His has recorded, though only in a footnote, that the skulls which combine the characters of his two best-defined types, the 'Sion-Typus' to wit, and the 'Disentis-Typus,' in the 'Mischform,' which he calls 'Sion-Disentis Mischlinge,' are the most capacious of the entire series of the 'Crania Helvetica,' exceeding, not by their maximum only, but by their average capacity also, the corresponding capacities of every one of the pure Swiss types[1]. Intercrossing, therefore, is an agency which in one set of cases may operate in the way of enhancing individual evolution, whilst in another it so divides its influence as to allow of the maintenance of two types in their distinctness. Both these results are of equal biological, the latter is of pre-eminent archaeological interest. Retzius[2] was of opinion, and, with a few qualifications, I think, more recent Swedish Ethnologists would agree, that the modern dolichocephalic Swedish cranium was very closely affined to, if not an exact reproduction of the Swedish cranium of the Stone Period; and Virchow[3] holds that the modern brachycephalic Danish skull is similarly related to the Danish skull of the same period. There can be no doubt that the Swedish cranium is very closely similar indeed to the Anglo-Saxon; and the skulls which still conform to that type amongst us will be by most men supposed to be the legitimate representatives of the followers of Hengest and Horsa, just as the modern Swedes, whose country has been less subjected to disturbing agencies, must be held to be the lineal descendants of the original occupiers of their soil. I am inclined to think that the permanence of the brachycephalic stock and type in Denmark has also its bearing upon the Ethnography of this country. In the Round-Barrow or Bronze Period in this country, sub-spheroidal crania (that is to say, crania of a totally different

[1] See Dr. Beddoe, 'Mem. Soc. Anth. Lond.' iii. p. 552; Huth, p. 308, 1875; D. Wilson, cit. Brace, 'Races of the Old World,' p. 380; and His, 'Crania Helvetica.'

[2] 'Ethnologische Schriften,' p. 7.

[3] 'Archiv für Anthropologie,' iv. pp. 71 and 80.

shape and type from those which are found in exclusive possession of the older and longer barrows) are found in great abundance, sometimes, as in the South, in exclusive possession of the sepulchre, sometimes in company, as in the North, with skulls of the older type. The skulls are often strikingly like those of the same type from the Danish tumuli. On this coincidence I should not stake much, were it not confirmed by other indications. And foremost amongst these indications I should place the fact of the 'Tree-interments,' as they have been called (interments, that is, in coffins made out of the trunk of a tree), of this country, and of Denmark, being so closely alike. The well-known monoxylic coffin from Gristhorpe contained, together with other relics closely similar to the relics found at Treenhoi, in South Jutland, in a similar coffin, a skull which, as I can testify from a cast given me by my friend Mr. H. S. Harland, might very well pass for that of a brachycephalic Dane of the Neolithic period. Canon Greenwell discovered a similar monoxylic coffin at Skipton, in Yorkshire; and two others have been recorded from the same county—one from the neighbourhood of Driffield, the other from that of Thornborough. Evidence, again, is drawn from Col. Lane Fox's opinion that the earthworks which form such striking objects for inquiry here and there on the East-Riding Wolds must, considering that the art of war has been the same in its broad features in all ages, have been thrown up by an invading force advancing from the east coast. Now we do know that England was not only made England by immigration from that corner or angle where the Cimbric Peninsula joins the mainland, but that long after that change of her name this country was successfully invaded from that Peninsula itself. And what Swegen and Cnut did some four hundred and fifty years after the time of Hengist and Horsa, it is not unreasonable to suppose other warriors and other tribes from the same locality may have done perhaps twice or thrice as many centuries further back in time than the Saxon Conquest. The huge proportions of the Cimbri, Teutones, and Ambrones are just what the skeletons of the British Round-Barrow folk enable us now to reproduce for ourselves. It is much to be regretted that from the vast slaughters of Aquae Sextiae and Vercellae, no relics have been preserved which might have enabled us to say whether Boiorix and his companions had the cephalic proportions of Neolithic Danes, or those very different contours which we are familiar with from Saxon graves throughout England, and from the so-called 'Danes' graves' of Yorkshire. Whatever might be the result of such a discovery and such a comparison, I think it would in neither event justify the application of the term 'Kymric' to the particular form of skull to which Retzius and Broca have assigned it.

Some years ago I noticed the absence of the brachycephalic British type of skull from an extensive series of Romano-British skulls which had come into my hands; and subsequently to my doing this, Canon Greenwell pointed out to me that such skulls as we had from late Keltic cemeteries, belonging to the comparatively short period which elapsed between the end of the Bronze Period and the establishment of Roman rule in Great Britain, seemed to have reverted mostly to the prae-Bronze dolichocephalic type. This latter type, the 'kumbecephalic type' of Professor Daniel Wilson, manifests a singular vitality, as the late and much lamented Professor Phillips pointed out long ago at a Meeting of this Association held at Swansea—the dark-haired variety, which is very ordinarily the longer-headed and the shorter-statured variety of our countrymen, being represented in very great abundance in those regions of England which can be shown, by irrefragable and multifold evidence, to have been most thoroughly permeated, imbibed, and metamorphosed by the infusion of Saxons and Danes, in the districts, to wit, of Derby, Leicester, Stamford, and Loughborough. How, and in what way, this type of man—one to which some of the most valuable men now bearing the name of Englishmen, which they once abhorred, belong—has contrived to reassert itself, we may, if I am rightly informed, hear some discussion in this department. Before leaving this part of my subject I would say that the Danish type of head still survives amongst us; but it is to my thinking not by any means so common, at least in the Midland counties, as the dark-haired type of which we have just been speaking. And I would add that I hope I may find that the views which I have here hinted at will be found to be in accord with the extensive researches of Dr. Beddoe, a gentleman who worthily represents and upholds the interests of Anthropology in this city, the city of Prichard, and who is considered to be more or less disqualified for occupying the post which I now hold, mainly from the fact that he has occupied it before, and that the rules of the British Association, like the laws of England, have more or less of an abhorrence of perpetuities.

The largest result which craniometry and cubage of skulls have attained is, to my thinking, the demonstration of the following facts, viz., first, that the cubical contents of many skulls from the earliest sepultures from which we have any skulls at all, are larger considerably than the average cubical contents of modern European skulls; and secondly, that the female skulls of those times did not contrast to that disadvantage with the skulls of their male contemporaries which the average female skulls of modern days do, when subjected to a similar comparison[1]. Dr. Thurnam demonstrated the former of these facts, as regards the skulls from the Long and the Round Barrows of Wiltshire, in the Memoirs of the London Anthropological Society for 1865; and the names of Lez Eyzies and Cro-Magnon, and of the Caverne de l'Homme Mort, to which we may add that of Solutré, remind us that the first of these facts has been confirmed, and the second both indicated and abundantly commented upon by M. Broca.

The impression which these facts make upon one, when one first comes to realise them, is closely similar to that which is made by the first realisation to the mind of the existence of a subtropical Flora in Greenland in Miocene times. All our anticipations are precisely reversed, and in each case by a weight of demonstration equivalent to such a work; there is no possibility in either case of any mistake; and we acknowledge that all that we had expected is absent, and that where we had looked for poverty and pinching there we come upon luxuriant and exuberant growth. The comparisons we draw in either case between the past and the present are not wholly to the advantage of the latter: still such are the facts. Philologists will thank me for reminding them of Mr. Chauncy Wright's brilliant suggestions that the large relative size of brain to body which distinguishes, and always, so far as we know, has distinguished the human species as compared with the species most nearly related to it, may be explained by the psychological tenet that the smallest proficiency in the faculty of language

[1] The subequality of the male and female skulls in the less civilised of modern races was pointed out as long ago as 1845, by Retzius in Müller's 'Archiv,' p. 89, and was commented upon by Huschke, of Jena, in his 'Schädel, Hirn und Seele,' pp. 48–51, in 1854.

may 'require more brain power than the greatest in any other direction,' and that 'we do not know and have no means of knowing what is the quantity of intellectual power as measured by brains which even the simplest use of language requires[1].'

And for the explanation of the pre-eminently large size of the brains of these particular representatives of our species, the tenants of prehistoric sepulchres, we have to bear in mind, first, that they were, as the smallness of their numbers and the largeness of the tumuli lodging them may be taken to prove, the chiefs of their tribes; and, secondly, that modern savages have been known, and prehistoric savages may therefore be supposed, to have occasionally elected their chiefs to their chieftainships upon grounds furnished by their superior fitness for such posts—that is to say, for their superior energy and ability. Some persons may find it difficult to believe this, though such facts are deposed to by most thoroughly trustworthy travellers, such as Baron Osten Sacken (referred to by Von Baer, in the 'Report' of the famous Anthropological Congress at Göttingen in 1861, p. 22). And they may object to accepting it, for, among other reasons, this reason—to wit, that Mr. Galton had shown us in his 'Men of Science, their Nature and Nurture,' p. 98, that men of great energy and activity (that is to say, just the very men fitted to act as leaders of and to commend themselves to savages[2]) have ordinarily smaller-sized heads than men possessed of intellectual power dissociated from those qualities.

The objection I specify, as well as those which I allude to, may have too much weight assigned to them; but we can waive this discussion and put our feet on firm ground when we say that in all savage communities the chiefs have a larger share of food and other comforts, such as there are in savage life, and have consequently better and larger frames—or, as the Rev. S. Whitmee puts it (*l. c.*), when observing on the fact as noticed by him in Polynesia, a more 'portly bearing.' This (which, as the size of the brain increases within certain proportions with the increase of the size of

[1] The bibliographer will thank me also for pointing out to him that the important paper in the 'North-American Review,' for October, 1870, p. 295, from which I have just quoted, has actually escaped the wonderfully exhaustive research of Dr. Seidlitz (see his 'Darwin'sche Theorie,' 1875).

[2] An interesting and instructive story in illustration of the kind of qualities which do recommend a man to savages, is told us by Sir Bartle Frere in his pamphlet, 'Christianity suited to all forms of Civilization,' pp. 12–14.

the body, is a material fact in every sense) has been testified to by a multitude of other observers, and is, to my mind, one of the most distinctive marks of savagery as opposed to civilisation. It is only in times of civilisation that men of the puny stature of Tydeus or Agesilaus are allowed their proper place in the management of affairs. And men of such physical size, coupled with such mental calibre, may take comfort, if they need it, from the purely quantitative consideration, that large as are the individual skulls from prehistoric graves, and high, too, as is the average obtained from a number of them, it has nevertheless not been shown that the largest individual skulls of those days were larger than, or, indeed, as large as the best skulls of our own days; whilst the high average capacity which the former series shows is readily explicable by the very obvious consideration that the poorer specimens of humanity, if allowed to live at all in those days, were, at any rate, when dead not allowed sepulture in the 'tombs of the kings,' from which nearly exclusively we obtain our prehistoric crania. M. Broca[1] has given us yet further ground for retaining our self-complacency by showing, from his extensive series of measurements of the crania from successive epochs in Parisian burial-places, that the average capacity has gone on steadily increasing.

It may be suggested that a large brain, as calculated by the cubage of the skull, may nevertheless have been a comparatively lowly organised one, from having its molecular constitution qualitatively inferior from the neuroglia being developed to the disadvantage of the neurine, or from having its convolutions few and simple, and being thus poorer in the aggregate mass of its grey vesicular matter. It is, perhaps, impossible to dispose absolutely of either of these suggestions. But, as regards the first, it seems to me to be exceedingly improbable that such could have been the case. For in cases where an overgrowth of neuroglia has given the brain increase of bulk without giving it increase of its true nervous elements, the Scotch proverb, 'Muckle brain, little wit,' applies; and the relatively inferior intelligence of the owners of such brains as seen nowadays may, on the principle of continuity, be supposed to have attached to the owners of such brains in

[1] See his paper, 'Bull. Soc. Anthrop. de Paris,' t. iii. ser. i. 1862, p. 102; or his collected 'Mémoires,' vol. i p. 348, 1871.

former times. But those times were times of a severer struggle for existence than even the present; and inferior intelligences, and specially the inferior quickness and readiness observable in such cases, it may well be supposed, would have fared worse then than now. There is, however, no need for this supposition; for, as a matter of fact, the brain-case of brains so hypertrophied[1] has a very readily recognisable shape of its own, and this shape is not the shape of the Cro-Magnon skull, nor indeed of any of the prehistoric skulls with which I am acquainted.

As regards the second suggestion, to the effect that a large brain-case may have contained a brain the convolutions of which were simple, broad, and coarse, and which made up by consequence a sheet of grey matter of less square area than that made up in a brain of similar size but of more complex and slenderer convolutions, I have to say that it is possible this may have been the case, but that it seems to me by no means likely. Very large skulls are sometimes found amongst collections purporting to have come from very savage or degraded races; such a skull may be seen in the London College of Surgeons with a label, '5357 D. Bushman, G. Williams. Presented by Sir John Lubbock[2];' and, from what Professor Marshall and Gratiolet have taught us as to other Bushman brains, smaller, it is true, in size, we may be inclined to think that the brain which this large skull once contained may nevertheless have been much simpler in its convolutions than a European brain of similar size would be. This skull, however, is an isolated instance of such proportions amongst Bushman skulls, so far, at least, as I have been able to discover; whilst the skulls of prehistoric times, though not invariably, are yet most ordinarily large skulls. A large brain with coarse convolutions puts its possessor at a disadvantage in the struggle for existence, as its greater size is not compensated by greater dynamical activity; and hence I should be slow to explain the large size of ancient skulls by suggesting that they contained brains of this negative character.

[1] I may, perhaps, be allowed to express here my surprise at the statement made by Messrs. Wilks and Moxon, in their very valuable 'Pathological Anatomy,' pp. 217, 218, to the effect that they have not met with such cases of Cerebral Hypertrophy. They were common enough at the Children's Hospital in Great Ormond Street when I was attached to it.

[2] [This skull is evidently 1299 of the new Catalogue of Crania prepared by Professor Flower, where it is named 'The Cranium of a Koranna.' —EDITOR.]

And I am glad to see that M. Broca is emphatically of this opinion, and that, after a judicious statement of the whole case, he expresses himself thus ('Revue d'Anthropologie,' ii. 1, 38): 'Rien ne permet donc de supposer que les rapports de la masse encéphalique avec l'intelligence fussent autres chez eux que chez nous.'

It is by a reference to the greater severity of the struggle for existence, and to the lesser degree to which the principle of division of labour was carried out in olden days, that M. Broca, in his paper on the Caverne de l'Homme Mort just quoted from, explains the fact of the subequality of the skulls in the two sexes. This is an adequate explanation of the facts; but to the facts as already stated, I can add from my own experience the fact that though the female skulls of prehistoric times are often, they are not always equal, or nearly, to those of the male sex of those times; and, secondly, that whatever the relative size of the head, the limbs and trunk of the female portion of those tribes were, as is still the case with modern savages, very usually disproportionately smaller than those of the male. This is readily enough explicable by a reference to the operations of causes exemplifications of the working of which are unhappily not far to seek now, and may be found in any detail you please in those anthropologically interesting (however otherwise unpleasant) documents, the Police Reports.

Having before my mind the liability we are all under fallaciously to content ourselves with recording the shots which hit, I must not omit to say that one at least of the more recently propounded doctrines in Craniology does not seem to me to be firmly established. This is the doctrine of 'occipital dolichocephaly' being a characteristic of the lower races of modern days and of prehistoric races as compared with modern civilised races. I have not been able to convince myself by my own measurements of the tenability of this position; and I observe that Ihering has expressed himself to the same effect, appending his measurements in proof of his statements in his paper, 'Zur Reform der Craniometrie,' published in the 'Zeitschrift für Ethnologie' for 1873. The careful and extensive measurements of Aeby[1] and Weisbach[2] have shown that the occipital region enjoys wider limits of oscillation than either of the other divisions of the cranial vault. I have some regret in saying

[1] Aeby, 'Schädelform des Menschen und der Affen,' pp. 11, 12, and 128.

[2] Weisbach, 'Die Schädelform der Roumanen,' p. 32, 1869.

this, partly because writers on such subjects as 'Literature and Dogma' have already made use of the phrase 'occipitally dolicho-cephalic,' as if it represented one of the permanent acquisitions of science; and I say it with even more regret, as it concerns the deservedly honoured names of Gratiolet and of Broca, to whom Anthropology owes so much. What is true in the doctrine relates, among other things, to what is matter of common observation as to the fore part of the head rather than to anything which is really constant in the back part of the skull. This matter of common observation is to the effect that when the ear is 'well forward' in the head, we do ill to augur well of the intelligence of its owner. Now the fore part of the brain is irrigated by the carotid arteries, which, though smaller in calibre during the first years of life, during which the brain so nearly attains its full size, than they are in the adult, are nevertheless relatively large even in those early days, and are both absolutely and relatively to the brain which they have to nourish, much larger than the vertebral arteries, which feed its posterior lobes. It is easy therefore to see that a brain in which the fore part supplied by the carotids has been stinted of due supplies of food, or however stunted in growth, is a brain the entire length and breadth of which is likely to be ill-nourished. As I have never seen reason to believe in any cerebral localisation which was not explicable by a reference to vascular irrigation, it was with much pleasure that I read the remarks of Messrs. Wilks and Moxon in their recently published 'Pathological Anatomy,' pp. 207, 208, as to the indications furnished by the distribution of the Pacchio-nian bodies as to differences existing in the blood-currents on the back and those on the fore part of the brain. These remarks are the more valuable, as mere hydraulics, Professor Clifton assures me, would not have so clearly pointed out what the physiological up-growths seem to indicate. Any increase, again, in the length of the posterior cerebral arteries is *pro tanto* a disadvantage to the parts they feed. If the blood-current, as these facts seem to show, is slower in the posterior lobes of the brain, it is, upon purely physical principles of endosmosis and exosmosis, plain that these segments of the brain are less efficient organs for the mind to work with; and here again 'occipital dolichocephaly' would have a justification, though one founded on the facts of the nutrition of the brain-cells, not on the proportions of the brain-case. In many

3 M 2

(but not in all) parts of Continental Europe, again, the epithet 'long-headed' would not have the laudatory connotation which, thanks to our Saxon blood, and in spite of the existence amongst us of other varieties of dolichocephaly, it still retains here. And the brachycephalic head which, abroad[1] at least, is ordinarily a more capacious one, and carried on more vigorous shoulders and by more vigorous owners altogether than the dolichocephalic, strikes a man who has been used to live amongst dolichocephali by nothing more forcibly, when he first comes to take notice of it, than by the nearness of the external ear to the back of the head; and this may be said to constitute an artistic occipital brachycephalism. But this does not imply that the converse condition is to be found conversely correlated, nor does it justify the use of the phrase 'occipital dolichocephaly' in any etymological, nor even in any ethnographical sense.

I shall now content myself, as far as craniology is concerned, by an enumeration of some at least of the various recent memoirs upon the subject which appear to me to be of pre-eminent value. And foremost amongst these I will mention Professor Cleland's long and elaborate scientific and artistic paper on the Variations of the Human Skull, which appeared in the Philosophical Transactions for 1869. Next I will name Ecker's admirable, though shorter, memoir on Cranial Curvature, which appeared in the 'Archiv für Anthropologie,' a journal already owing much to his labours, in the year 1871. Aeby's writings I have already referred to, and Ihering's, to be found in recent numbers of the 'Archiv für Anthropologie' and the 'Zeitschrift für Ethnologie,' deserve your notice. Professor Bischoff's paper on the Mutual Relations of the horizontal circumference of the Skull and of its contents to each other and to the weight of the Brain, has not, as I think, obtained the notice which it deserves. It is to be found in the Proceedings of the Royal Society of Munich for 1864, the same year which witnessed the publication of the now constantly quoted 'Crania Helvetica' of Professors His and Rütimeyer. Some of the most important results contained in this work, and much important matter besides, were made available to the exclusively

[1] See upon this point:—Broca, 'Bull. Soc. Anth.' Paris, ii. p. 648, 1861; *ibid.* Dec. 5, 1872; Virchow, 'Archiv für Anth.' v. p. 535; 'Zeitschrift für Ethnol.' iv. 2, p. 36; 'Sammlungen,' ix. 193, p. 45, 1874; Beddoe, 'Mem. Anth. Soc. Lond.' ii. p. 350.

English reader by Professor Huxley, two years later, in the 'Prehistoric Remains of Caithness.' I have made a list, perhaps not an exhaustive one, but containing some dozen memoirs by Dr. Beddoe, and having read them or nearly all of them, I can with a very safe conscience recommend you all to do the like. I can say nearly the same as regards Broca and Virchow, adding that the former of these two savants has set the other two with whom I have coupled him an excellent example, by collecting and publishing his papers in consecutive volumes.

But I should forget not only what is due to the place in which I am speaking, but what is due to the subject I am here concerned with, if in speaking of its literature, I omitted the name of your own townsman, Prichard. He has been called, and, I think, justly, the 'father of modern Anthropology.' I am but putting the same thing in other words, and adding something more specific to it, when I compare his works to those of Gibbon and Thirlwall, and say that they have attained, and seem likely to maintain permanently, a position and importance commensurate with that of the 'stately and undecaying' productions of those great English historians. Subsequently to the first appearance of those histories other works have appeared by other authors, who have dealt in them with the same periods of time. I have no wish to depreciate those works; their authors have not rarely rectified a slip and corrected an error into which their great predecessors had fallen. Nay, more, the later comers have by no means neglected to avail themselves of the advantages which the increase of knowledge and the vast political experience of the last thirty years have put at their disposal, and they have thus occasionally had opportunities of showing more of the true proportions and relations of even great events and catastrophes; still the older works retain a lasting value, and will remain as solid testimonies to English intellect and English capacity for large undertakings as long as our now rapidly extending language and literature live. The same may be most truthfully said of Prichard's 'Researches into the Physical History of Mankind.' An increase of knowledge may supply us with fresh and with stronger arguments than he could command for some of the great conclusions for which he contended; such, notably, has been the case in the question (though 'question' it can no longer be called) of the Unity of the human species; and by the employment of the philosophy of continuity and the doctrine of evolution, with which the world was not made acquainted till more than ten years after Prichard's death, many a weaker man than he has been enabled to bind into more readily manageable burdens the vast collections of facts with which he had to deal. Still his works remain, massive, impressive, enduring—much as the headlands along our southern coast stand out in the distance in their own grand outlines, whilst a close and minute inspection is necessary for the discernment of the forts and fosses added to them, indeed dug out of their substance, in recent times. If we consider what the condition of the subject was when Prichard addressed himself to it, we shall be the better qualified to take and make an estimate of his merits. This Prichard has himself described to us, in a passage to be found in the preface to the third volume of the third edition of the 'Physical History,' published in the year 1841, and reminding one forcibly of a similar utterance of Aristotle's, at the end of one of his logical treatises ('Soph. Elench.' cap. xxxiv. 6). These are his words:—

'No other writer has surveyed the same field, or any great part of it, from a similar point of view. . . . The lucubrations of Herder and other diffuse writers of the same description, while some of them possess a merit of their own, are not concerned in the same design, or directed towards the same scope. Their object is to portray national character as resulting from combined influences—physical, moral, and political. They abound in generalisations, often in the speculative flights of a discursive fancy, and afford little or no aid for the close induction from facts which is the aim of the present work. Nor have these inquiries often come within the view of writers on Geography, though the history of the globe is very incomplete without that of its human inhabitants.'

A generation has scarcely passed away since these words were published in 1841; we are living in 1875; yet what a change has been effected in the condition of Anthropological literature! The existence of such a dignified quarterly as the 'Archiv für Anthropologie,' bearing on its titlepage in alphabetical order the honoured names of V. Baer, of Desor, of Ecker, of Hellwald, of His, of Lindenschmit, of Lucae, of Rütimeyer, of Schaaffhausen, of Semper, of Virchow, of Vogt, and of Welcker, is in itself perhaps the most striking evidence of the advance made in this time, as being the most distinctly ponderable, and in every sense the largest, Anthropological publication of the day.

Archaeology, which but a short time back was studied in a way which admirably qualified its devotees for being called 'connoisseurs,' but which scarcely qualified them for being called men of Science, has by its alliance with Natural History and its adoption of Natural History methods, and its availing itself of the light afforded by the great Natural History principles just alluded to, entered on a new career. There is, as regards Natural History, Anatomy, and Pathology, nothing left to be desired for the conjoint scheme represented by the periodical just mentioned, where we have V. Baer for the first and Virchow for the last, and the other names specified for the rest of these subjects; whilst Archaeology, the other party in the alliance, is very adequately represented by Lindenschmit alone. But when I recollect that Prichard published a work 'On the Eastern Origin of the Celtic Nations' ten years before the volume of 'Researches,' from which I have just quoted, and that this work has been spoken of as the work 'which has made the greatest advance in Comparative Philology during the present century,' I cannot but feel that the 'Redaction' of the 'Archiv für Anthropologie' have not as yet learnt all that may be learnt from the Bristol Ethnologist; and they would do well to add to the very strong staff represented on their titlepage the name of some one, or the names of more than one, comparative philologist. This the Berlin 'Zeitschrift für Ethnologie' has done.

Of the possible curative application of some of the leading principles of modern Anthropology to some of the prevalent errors of the day, I should be glad to be allowed to say a few words. The most important lesson as regards the future (I do not say the *immediate* future) which the modern study of Human Progress (for such all men who think, except the Duke of Argyll, are now agreed is the study of Anthropology) teaches is the folly and impossibility of attempting to break abruptly with the past. This principle is now enforced with persistent iteration from many Anthropological platforms; and I cannot but think it might advantageously be substituted in certain portfolios for the older maxim, 'Whatever is certainly new is certainly false,' a maxim which seems at first sight somewhat like it, but which, as being based on pure ignorance of the past and teaching only distrust of the future, is really quite different from it. I am not sure that Prichard ever put forward the former of these two doctrines, though it is just the doctrine which would have commended itself to his large, philosophical, many-sided, well-balanced judgment. He died in 1848—the very year which perhaps, of all save one in history, and that one the year 1793 (a year in which he was yet a child), showed in the most palpable way the absurdity of attempting to make civilisation by pattern, and of hoping to produce a wholesome future in any other way than that of evolution from the past.

What have been called the senile, what could equally well have been called the cynical, Ethics of Pessimism, had not in Prichard's time found any advocates in this country; indeed, so far as I have observed, they are of a more recent importation than most other modern heresies. I do not deny that at times it is possible to give way to certain pressing temptations to think that we are living in a certainly deteriorated and a surely deteriorating age, and that it is hopeless and useless to set up, or look up to, aspirations or ideals. When, for example, we take stock of the avidity with which we have, all of us, within the last twelve months read the memoirs of a man whom one of his reviewers has called a 'high-toned aristocrat,' but whom I should call by quite another set of epithets, we may think that we are not, after all, so much the better for the 3000 years which separate us from the time when it was considered foul play for a man to enact the part of a familiar friend, to eat of another man's bread, and then to lay great wait for him. Or can we, in these days, bear the contrast to this miserable spectacle of mean treachery and paltry disloyalty, which is forced upon us in the same history by the conduct of the chivalrous son of Zeruiah, who, when he had fought against Rabbah and taken the city of waters, sent for his king who had tarried in Jerusalem, lest that city should thenceforward bear the name, not of David, but of Joab? Or again, as I have been asked, have we got very far above the level of sentiment and sympathy which Helen, an unimpeachable witness, tells us the Trojan Hector had attained to and manifested in his treatment of her,

'With tender feeling and with gentle words'?

Would the utterances of any modern epic poet have so surely brought tears into the eyes of the noble-hearted boy depicted by

Mr. Hughes, as the passage of Homer just alluded to, and characterised by him 'as the most touching thing in Homer, perhaps in all profane poetry put together'? What answer can be made to all this by those who maintain that the old times were not better than these, who maintain the doctrine of Progress, and hold that man has been gradually improving from the earliest times, and may be expected to go on thus advancing in the future? An answer based upon the employment of simple scientific method, and upon the observance of a very simple scientific rule—upon, to wit, the simple method of taking averages, and the simple rule of enumerating all the circumstances of the case. Noble actions, when we come to count them up, were not, after all, so very common in the olden times; and side by side with them there existed, and indeed flourished, intertwined with them, practices which the moral sense of all civilised nations has now definitely repudiated. It is a disagreeable task, that of learning the whole truth; but it is unfair to draw dark conclusions as to the future, based on evidence drawn from an exclusive contemplation of the bright side of the past. A French work, published only last year, was recommended to me recently by an eminent scholar as containing a good account of the intellectual and moral condition of the Romans under the Empire. I have the book, but have not been able to find in it any mention of the gladiatorial shows, though one might have thought the words *Panem et Circenses* might have suggested that those exhibitions entered as factors of some importance into the formation of the Roman character. It is impossible to go beyond that in the way of looking only at the bright side of things. Still we ourselves have less difficulty in recollecting that there were 300 Spartans sacrificed to the law-abiding instincts of their race at Thermopylae, than in producing, when asked for them, the numbers of Helots whom Spartan policy massacred in cold blood not so many years after, or those of the Melians and Mitylenaeans whom the polished and cultivated Athenians butchered in the same way, and about the same time, with as little or far less justification for doing so. Homer, whom I have quoted above, lived, it is true, some centuries earlier, but living even then he might have spared more than the five words contained in a single line (176 of Iliad xxiii.) to express reprobation for the slaughter of the twelve Trojan youths at the pyre of Patroclus. The Romans could applaud Terence's line, 'Homo sum, humani nihil a me alienum puto;' but it did not strike them till the time of Seneca that these noble words were incompatible with the existence of gladiatorial shows, nor till the time of Honorius did they legally abolish those abominations. Mutinies and rebellions are not altogether free from unpleasant incidents even in our days; but the execution of 6000 captives from a Servile War, in the way that Crassus executed his prisoners after the final defeat of Spartacus, viz. by the slow torture of crucifixion, is, owing to the advance of civilisation, no longer a possibility. If the road from Capua to Rome witnessed this colossal atrocity, there are still preserved for us in its near neighbourhood the remains of Herculaneum and Pompeii to show us what foul broad-daylight exuberance could be allowed by the public conscience of the time of Titus and Agricola to that other form which sits 'hard by hate.' The man who in those days contributed his factor to the formation of a better public opinion, did so at much greater risk than any of us can incur now by the like line of action. Much of what was most cruel, much of what was most foul in the daily life of the time, had, M. Gaston Boissier notwithstanding, the sanction of their state religion and the indorsement of their statesmen and emperors to support it. There was no public press in other lands to appeal to from the falsified verdicts of a sophisticated or a terrorised community. Though then as now,

'Mankind were one in spirit,'

freedom of intercommunication was non-existent; no one could have added to the words just quoted from Lowell their complemental words,

'And an instinct bears along,
Round the earth's electric circle the swift flash of right or wrong.'

The solidarity of nations had not, perhaps could not have been dreamt of—the physical pre-requisites for that, as for many another non-physical good, being wanting.

Under all these disadvantages men were still found who were capable of aspiration, of hope for, and of love of better things; and by constant striving after their own ideal, they helped in securing for us the very really improved material, mental, and moral positions which we enjoy. What they did before, we have to do for those who will come after us.

XLV.

THE EXAMINATION SYSTEM AND THE PRE-REQUISITES OF CANDIDATES:

AN ADDRESS DELIVERED AT THE ST. MARY'S HOSPITAL MEDICAL SCHOOL.

I WAS some time back honoured with an invitation similar in character to that of which I am now availing myself, and which, I take it, I owe to the suggestion of my much valued friend and former pupil, your Dean, Dr. Shepherd. What I had to do then was a good deal easier of performance that what I have to do now. Then, as now, I was told that something in the shape, and if possible, of the nature of an address would be expected from me; but upon that occasion the address was to come first, and the distribution of prizes second in order, and I felt that there was little need to be over-anxious as to the former of my duties, as the minds of all present—of candidates successful and unsuccessful, of their admirers and sympathising friends—would be intent upon coming with all speed to the second part of the business. And whilst I was glad, on the one hand, to be free from any very heavy sense of responsibility, I should have been sorry, on the other, to have interposed myself, or my remarks, at any length between any of my audience and the very pure and yet, I apprehend, very intense enjoyment which witnessing a public recognition of merit in a young friend or relative confers. I have comforted myself, however, whilst considering what I have to do this day, by thinking that it was quite within my competence to secure for myself the merit of being brief, and that if I were to write down what I wished to say, and to confine myself strictly to my manuscript, I could count upon giving satisfaction in that way at least.

I take it that a person who is put into the honourable position which I this day hold, has generally something which he is glad to have the opportunity of saying from so good a standing-point or vantage-ground. Accordingly, I wish to say a few words upon one subject, amongst others, to which I have of late devoted a very considerable amount of my very fully occupied time—to wit, the Examination System and the pre-requisites of candidature. The system of requiring certificates of attendance at lectures and upon other courses of instruction, and the examination system, are two very distinct means intended to co-operate towards one end—that, namely, of securing to the public that its future physicians and surgeons shall, firstly, have had certain opportunities, and, secondly, have made certain use of them. I must not fail to add, what will disabuse any, either old or young, of the idea that I have anything very revolutionary to propose—namely, that these two systems are necessary as much in the interests of the future doctor as in those of his future patients, and that without some such apparatus and machinery as that which they represent, many a man would lose chances of forming in himself habits of attention, indispensable for any success, or indeed any virtue, which chances may never recur. Indeed, it is a matter of common observation, at least of possible common observation, and I would it were matter of more common remark, that it is necessary not only for the acquirement of habits which belong only in part to our intellectual nature, but even for the acquirement of purely intellectual accomplishments, that certain kinds of work should be done at certain times and fixed periods in a boy's, in a young, or in an older man's life. Languages can be gained with greater facility before the age of twenty-five than they can ever afterwards; and the same may, I am inclined to believe, be said of the power of recognising and recollecting specific differences in zoology and botany. It is well that the same natural restriction does not invariably exist as to the power of mastering that of which, however, it is of more importance that early mastery should be gained than of almost any other subject—to wit, mathematics. These are the words of a man whose experience and success as an educator has been great, not intended, it is true, for the particular case we deal with, and applying only in the way of analogy, but forcible in that way and appropriate :—

'There is but a certain time allotted for each thing to be done that we have to do, whether it belongs to this world or the other, and if we pass the allotted time it is too late for that work to be done. If you are idle here at school, it may for a short time make little difference, perhaps no perceptible difference at all. But you know perfectly well that that is not so always. After a time it becomes too late to recover what you have not chosen to take when it was within your reach. There are things which can be learnt when you are not twelve which can never be learnt so well afterwards; there are still more which must be learnt before you are seventeen or eighteen, or you can never really learn them at all. You may afterwards wish very much that you had not missed the chance; but your wishes will not give you back the power that is gone; you are too late. And the same holds good long after. Each time of life, as it comes, marks off the foundations of certain studies as done with; if you have not laid them by that time you never can. And precisely the same thing is true of other things besides studies[1].'

So much for the good to be expected and attained by the observance of a regular curriculum. It admits of no question. What does, I think, admit of question, and what is, I think, capable of improvement, is the method for securing such observance. The method at present in vogue for this end is known as the 'signing up' of certificates of attendance at lectures. Now such a certificate can only really depose to the fact that a student was present at the delivery of particular sets of discourses or demonstrations; it cannot depose to his having profited by them. What the public wish to be assured of is the latter matter, and its results in the shape of the possession by him of a certain amount of attainments and dexterity. But this can only be done by an examination held by one set of authorities or another; and the very first and the very last principle of any and every examination which deserves the name is the principle of English law—*De non existentibus et de non apparentibus eadem est ratio.* Nothing, in the words of the examination statutes of my own University, should influence the result of the examination except what forms part of or directly results from the examination itself. I do not question the good to be had from attending lectures; I am well assured that good lectures, not over numerous, bless both him who gives and him who takes. The giver is benefited by having to put his knowledge into a compact form,

[1] Temple, 'Sermons,' Ser. II., Serm. xl. p. 308. 1871.

so as to be readily transmissible and communicable in public; the hearer is benefited by obtaining orally, or rather auditorally, what it would have cost him more time to obtain, if indeed he could have obtained it at all, by reading or otherwise. Besides, hearing is what is known as a 'natural process,' and no improvement in the way of printing can ever entirely supersede it, or make it what it is sometimes said to be—namely, 'a barbarous anachronism.' I say nothing of the advantage to be drawn from contact with a living personal source of knowledge, though it is clear enough that a few striking expressions delivered by an earnest man *viva voce* may awaken more thought and create more lively interest than a whole volume of print, however well illustrated. For as iron sharpeneth iron, so man the face of his friend. I see, know, and gratefully recollect the benefits of lectures; but the more excellent an institution is the more is it likely to be injured by compulsory enactments intended to govern or protect it. Attendance at morning or evening chapel, or both, is an excellent practice; but the making it compulsory has a very sure, I do not say an invariable, tendency to rob it of its beneficial effect. It is a more edifying sight to see a single individual going to such a service spontaneously,—as I am told a very distinguished statesman, the junior member for a constituency not a hundred miles hence, may be seen doing in all weathers,—than to see a whole college of young men hurrying to Divine service to have their attendance upon it entered as upon a roll-call. I am glad to think that the answering a roll-call pure and simple is allowed in some of our colleges to stand as an alternative for attendance in chapel. Now compulsory attendance at lectures, like compulsory attendance at chapels, aims at attaining something which I believe to be distinctly good, but which I also know to be as distinctly not securable by it. I cannot see the wisdom of aiming at the unattainable, and as testimony to bodily attendance is the only result really attained by the process of 'signing up,' I should distinctly limit the bearing of the documents I refer to to the scope really attainable—to wit, the scope of a roll-call. I am informed that in one of the largest of continental countries, the system of signing for students is entirely given up, it having been so much abused, and that if a student only passes his examinations well he need not have attended a single lecture. Hereby, however, I submit that the public are robbed of some of the security which they have a right to demand. No examination, however large a factor of it the practical part may be, can give entirely satisfactory proof that a man knows his subject thoroughly and practically: the elasticity of words, the power of verbal memory, the possibility of 'preparing' and 'grafting' a candidate for examination, as, in America and elsewhere, a mine is 'prepared' and 'grafted' for unwary speculators, are not all the heads under which sources of fallacy in examinations might be enumerated. Hence I should wish to secure for the public what a system of roll-calls can secure—namely, the attendance of a student in a particular spot where particular opportunities for learning particular things should be available for him in a particular order and succession. This system of requirements should be made to tally with the system of examinations, and thereby teachers and pupils, examiners and candidates, would all alike be relieved from much that is onerous, unreal, and a snare. The examination system would dignify the system of the roll-call, which, indeed, as aiming at something definitely attainable and attaining it, even if nothing more, would at any rate possess the dignity which truth possesses, that of '*incorrupta fides nudaque veritas.*' Common sense would consider the advantage, sense of duty would enforce the necessity, of using opportunities whilst they were available, and the two systems, that of examinations and that of the pre-requisites for them, would be brought into a more harmonious and less burdensome solidarity than they at present enjoy.

I have been speaking of the duties of young men and of learners, but do not suppose that I think that older men and teachers, like myself, have not their duties too. I know that I have mine, and that I often perform them very much otherwise than I should and wish to do. One hears talk sometimes which makes one think that the talker supposes that morality belongs to one sphere and science to another, and that the two may impinge upon or collide with one another, but cannot otherwise influence each other. This is an entire mistake. 'Faith' and 'duty' are words which may, when we see them on the outside of a tract, prepare us for finding ethical and other disquisitions *in pari materie* within its covers; but faith and duty, faithfulness and thoroughness, are also things which can no more be left out of the world of scientific work than they can or ought to be left out of 'that other world' to which I have just alluded. Examining is scientific work; indeed, in these days, it is a work which occupies a very large portion of the time of many a scientific man, whether to his benefit or that of science I do not stop to discuss. But I submit, and without any fear of contradiction, that there is no work which calls for more exercise of conscience; no work—not even that of the judge on the bench—which, when well done, illustrates more completely the truth of the old doctrine, 'In justice, all moral virtue lies involved.' An examiner has many temptations to strive against: the temptation to idleness; to give way to weariness; to meet the sameness of his subject-matter with perfunctoriness in dealing with it; to give way to feelings of pique when he finds that his own pet views or papers are entirely unknown to the examinee. Of course a strong and upright man resists all these temptations; but strength and uprightness are largely or entirely moral qualities. I need not labour, however, at what is self-evident. Let me say a few words about the way in which a man's moral nature is, or ought to be, called into activity, not now when he is engaged in testing, but when he is engaged in communicating or acquiring, knowledge. As regards the duties of a teacher when teaching, he is bound to beware of leaving any one side of a question, any one set of facts, in neglect and inadequately expounded. Imperfection of exposition in a teacher, not only produces, in the second generation, so to say, imperfection of investigation in the hearer, but—as words terribly shoot back, like the Tartar's bow, mightily entangling and perverting the judgment—such imperfection and want of fulness in the communication reacts by producing imperfection and want of fulness in investigation in the teacher himself. It is (trite remark) difficult to estimate the consequences of any one action; but it is easy to see that an example set by a person, himself set in authority, of slovenliness and inadequacy in his methods of work may hurt the consciences of his younger brethren, and have widely and lengthily ramifying consequences in neutralising chances for neutralising evil and suffering.

I have said thus much about the responsibilities of students and teachers *in praesenti*. I will, with your permission, say a little more upon the responsibilities which will gather round the former *in futuro*.

Of the students of any hospital at any one time we may safely say that, making a small deduction for accidental relinquishments of the career and for other disturbing causes, we may look upon all as likely, in a few months, to come into positions in which the lives of fellow-creatures of their own species will be largely dependent upon their decisions—i. e., upon the knowledge they have stored up, the power of applying it which they have gained, and the resolution they may have for duly using both. Some of those now present may be fortunate enough to come into their responsibilities in or upon areas not destitute of professional colleagues, of whose counsel and advice they may be glad to avail themselves. Some, however, may have to be the sole and unsupported representatives of medical and surgical science in some isolated country locality. These are large powers and large responsibilities: it is but commonplace to say that a consideration of their magnitude, as it looms out in the future, should make the opportunities and the *ex hypothesi* irreplaceable advantages of the present seem doubly valuable. This is, I say, a commonplace remark; but it is as well to repeat it, for all that. It is not difficult to imagine, indeed it is easy to bring proof, that the deepest regret, and, more, the most lasting remorse, may be produced by the thought that a little more attention to some particular line of practice, to some particular set of cases, to some new or some old modes of a curative kind, might have enabled a man to save a life which has slipped away for the want and in the absence of the knowledge and the power which might have been obtained thus, but has not been. Such considerations will readily suggest themselves to every private conscience in greater detail than it is well for me to attempt. Public opinion, in English-speaking countries at least, on both sides of the Atlantic and in that newer Southern world, attaches what certain eminent though anonymous publicists are wont to write of as an exaggerated, but what I should speak of as a due and proper, value to human life and human suffering. And upon private conscience more or less enlightened, and public opinion when properly awakened, morality and its sanctions rest securely. I say, 'public opinion when properly awakened;' for though systematic writers, in these latter days, lay abundant weight upon the indispensability of the existence of activity in public opinion to the sustentation of morality, I am not clear that the teaching of philosophers has as yet begun to exercise all the influence in this direction which there is no doubt, and which it is much to be desired, it will do shortly. I have often occasion to note that men who are individually 'upright' are yet sluggish and negligent, and even cowardly in the work of contributing their factor to the formation of a healthy atmosphere or medium of public opinion. Yet, if it was true in former times that *contemptu famae, contemnuntur virtutes,* it is undoubtedly the fact that the progress of thought in more modern times has made the moral to be drawn from those words of more pressing urgency than many persons as yet feel it to be. A man, for example, is guilty of some dereliction of duty; he sells himself, let us suppose, or whatever rights of property he may still retain in the commodity he calls himself, for the vote or votes of one or more beer-sellers, and for the seat on one or other side of one or other House of Parliament which that vote or votes may directly or indirectly gain or keep for him. A man's own conscience is supposed to punish him enough for an action or actions such as this, but it is most wrongly supposed so to do. It is the duty, and a duty too often pretermitted, of everybody who recognises a bad action as being a bad action, to speak of it as being such, and as meriting general reprobation. Without such speaking out, morality grows faint, and may be asphyxiated for want of what is the 'vital air'—to use the language of the older physiological chemists—of the atmosphere in which it lives. Those who, like medical men, see and know much of the natural history and habits of their fellows have many opportunities of helping towards creating a healthy tone of social feeling; and it is possible enough to do one's duty in this way without entering upon a course of extravagant aggression or crusading.

A few words to point out what I have come to think should be the main guiding principles necessary for him who would secure real success in the practice of a really noble profession. I say *real* success; and I will say that what the world calls success is, perhaps, not so often dissociated from this *real* success as a few glaring instances might make one think previous to counting them up. And I believe that the rule, 'Put yourself in his place,' based on what modern philosophers call the principle of 'altruism,' but what is found expressed plainly enough in much older language than theirs, is the rule which, if I were confined to the choice of one single guiding maxim to be given to a young doctor just entering upon the responsibilities of practice, I should choose for that maxim. Sympathy is truly called a divine gift, and it does assuredly give a superhuman power. I pre-suppose, of course, patience in investigation and carefulness in ratiocination ; but such is often the obscurity, intricacy, and complexity of a medical problem, that in the ultimate resort, it is upon intuition rather than upon syllogism that its true solution depends. It is to the man who has the touch of genius, that strength of imagination, which enables him to put himself in his patient's place, and thus do full justice to him, that there 'ariseth up light in the darkness.' It is for the want of all this that 'great men are not always wise.' There is an example suited to all men, suited eminently to medical men, as it is contained in the history of One who, though now known to us by other names, was known to our Saxon forefathers as the 'Healer.' He spent a life to the neglect of Himself in combating the wickedness and in alleviating the misery of others. As regards these two lines of labour, it is, I think, possible to maintain that men's instincts or inclinations have led them, in looking at the history of this life, to give too little prominence to the severity of the outspoken unsparing denunciations of an evil generation which it records. It is not possible to maintain that men have gone astray in the importance they have assigned to the manifestations they have recognised in it of vast pity and boundless sympathy.

XLVI.

THE RELATIVE VALUE OF CLASSICAL AND SCIENTIFIC TRAINING:

BEING A REVIEW OF DR. MAX V. PETTENKOFER'S WORK 'WODURCH DIE HUMANISTISCHEN GYMNASIEN FÜR DIE UNIVERSITÄT VERBEREITEN.'

THE German-reading public can possess itself at a very trifling cost of a very weighty opinion as to the relative value of classical and of scientific training, by the purchase of an address delivered last December, in Munich, by Professor Max von Pettenkofer, in his capacity of Rector or Chancellor of the University for the time being. There is in existence an English document (we fear we cannot speak of it as a *publication*) in the shape of a report, laid before the authorities of the Owens College, Manchester, which has appended to it a name nearly, or quite, as familiar to the student and readers of 'Nature' as Pettenkofer's—viz. that of Professor Roscoe, and in which the same process of 'ponderation' is applied to the classical 'Gymnasia' and the modern 'Real-Gymnasien' severally. Von Pettenkofer, who is not referred to in that report, shall here speak for himself, and we may say at once, that after stating more or less fully the objections which are ordinarily urged against the classical system, he declares himself an adherent of the party which stands *super antiquas vias*. The two delegates of the Owens College appear to incline in the same direction somewhat, but are more eclectic and more careful in balancing their utterances as to the possibility of combining the two systems than either Von Pettenkofer, whom we shall forthwith cite on the one, or than Helmholtz, whom they cite on the other side.

The argument from authority has a legitimate place in questions concerning such matters as the genesis of culture and as the existence of capacity and capabilities; for in such questions neither the facts themselves nor the mode of their origination can be always looked upon as beyond the region of probability. But as we are writing in a scientific periodical, we will begin at least with something which admits of being quantitatively estimated; and we will do this by giving the time-tables of the classical (*Humanistischen*) and of the modern (*Real-Gymnasien*) schools in Bavaria, as we find them in Von Pettenkofer's address (pp. 5 and 18).

In classical schools, out of 99 hours per week:—

8 hours per week are given to			German.
26	"	"	Latin.
22	"	"	Greek.
8	"	"	French.

(I. e. 64 hours, or 65 per cent., are given to languages, three-fourths being Latin and Greek, and one-fourth German and French.)

17 hours per week are given to			Mathematics.
10	"	"	History.
8	"	"	Religious Instruction.

In 'Real-Gymnasien,' out of 112 hours per week:—

9 hours per week are given to			German.
14	"	"	Latin.
13	"	"	French.
4	"	"	English.

(I. e. 40 hours, or 33 per cent., are given to languages, of which time only one-third is given to one ancient language, one-third to French, the other two-thirds to German.)

27 hours per week are given to Mathematics.

(I. e. algebra, elementary geometry, trigonometry, descriptive and analytical geometry and higher analysis, taking 22 per cent. of the whole number.)

4 hours per week are given to			History.
19	"	"	Natural Science and Geography.
24	"	"	Drawing and Modelling.
8	"	"	Religious Instruction.

The 'Real-Gymnasien' are thus seen to exact 25 per cent. more hours than the classical schools; and it is by this increase on the one hand, coupled with a curtailment of the quota assigned to languages on the other, that time is found for mathematics and for natural science, with the drawing and modelling so indispensable to it. Von Pettenkofer deprecates the making of any material increase in the number of hours to be spent in the Gymnasien, on the undeniable ground that the day is no longer and man no stronger now than were the days and the men of 2000 years ago; and the space for such additamenta as must be made to the curriculum must be found by bettering the methods and means for communicating instruction, and effecting thus an economy of time. The Bavarian chemist and hygienist does not himself suggest any ways and means whereby this economy may be effected, and presumptuous though it be, we will attempt to supplement this deficiency by saying that such an economy might be effected in England and English schools by applying one or other, or all three, of the following lines of treatment to the classical curriculum, even without cutting its Greek adrift. Latin and Greek, to put the boldest suggestion first, might be studied in certain, and those not a few, cases, as literatures and not as philologies; or, as a second alternative, when some training in philology is to be retained at whatever cost, such training might be made more intelligible, and so less distasteful and wasteful of time, by making the study of it comparative, as recommended by Professor Max Müller in his evidence before the Commission just referred to; or thirdly, synthetical scholarship, in the way of verse-making, should be considered as a luxury and refinement to be reserved for the delectation and cultivation of those few who, in any age, show any aptitude for it, and synthetical scholarship in the way even of writing Latin prose might, due precautions having been taken, be dispensed with in the cases of youths who, whilst wholly incapable in that, had shown some capacity in some other line. Our 'due precautions' should consist in the multiplying the practice of synthetical scholarship in the way of translation from Latin into English. We know the horror which these suggestions will excite in the breasts of schoolmasters of the type represented by the gentleman who told the Commissioners already referred to, that if he were set to teach history in set lessons, he 'should not know how to do it.' But we believe that by the adoption of any one of the three lines of action just glanced at, space and time might be found for the introduction of the natural sciences into the curriculum of any public school, and that at once without injury to the dignity of either the one or the other of the two sets of studies, and without injury to the physical or mental health of the learners.

But it is time, perhaps, that we should let Von Pettenkofer speak for himself; and this he does (at p. 12, l.c.) to the following effect:—'I am convinced that philology and mathematics furnish precisely the material for teaching and intellectual discipline which is essential for our Gymnasia, and I look upon the material furnished by other sciences as mere accessories. I know that in putting forward this view, which I do not do now for the first time, I put myself into opposition with the tide of opinion which is prevalent just at present, and which anticipates great advantages from the introduction of additional subjects of instruction, and especially from the introduction of instruction in natural science into "Latin schools (Lateinschulen) and Gymnasia."' Further on (p. 16) he proceeds as follows:—'The results of actual experience appear to me to favour my views. In other parts of Germany experiments have now, for a long while, been made with Gymnasia and similar institutions, in which much natural science is taught. But I cannot as yet discover that any remarkable number of persons who have subsequently distinguished themselves in natural science have come from these schools. In this matter reliable statistics of the pupils leaving (*der Abiturienten*) a Berlin Gymnasium, the so-called "Old Cologne Gymnasium," in which natural science has for a long while formed part of the curriculum, would be very instructive. Distinguished men come, from time to time, from this Gymnasium, but certainly not in greater numbers than from any other classical (*Humanistischen*) Gymnasium where no natural science at all is taught. It would long ago have been a notorious fact if a disproportionate number of the younger professors of natural science in the Prussian Universities could have been shown to have been formerly students in the Cologne Gymnasium.' We imagine that this 'Old Cologne Gymnasium' thus referred to by Von Pettenkofer is none other than the 'mixed' (*simultan*) school described by Mr. Matthew Arnold under the name of the *Friedrich Wilhelm's Gymnasium* at Cologne, in his 'Schools and Universities of the Continent,' pp. 218–221; and but that more antagonism and less familiarity subsisted between North and South Germany six months ago than, we are happy to think, subsists now, we apprehend that more would have been made of the history of this institution by the Munich Professor. For Dr. Jaeger, the director of this mixed school, who, as he had been refused a nomination to another school, the Bielefeld Gymnasium, by the Education Minister, on account of his politics, cannot be suspected of reactionary leanings, spoke to Mr. Arnold in the following sense (see p. 221, l. c.):—'It was the universal conviction with those competent to form an opinion, that the *Realschulen* were not at present successful institutions. He declared that the boys in the corresponding forms of the classical school beat the Realschule boys in matters which both do alike, such as history, geography, the mother tongue, and even French, though to French the Realschule boys devote far more time than their comrades of the classical school. The reason for this, Dr. Jaeger affirms, is that the classical training strengthens a boy's mind so much more. This is what, as I have already said, the chief school authorities everywhere in France and Germany testify. In Switzerland you do not hear the same story.'

With regard to Switzerland, we learn from the Owens College Report, above mentioned, that Professor Zellner, of the Polytechnic School at Zurich, holds that the establishment of 'Real Gymnasia,' or High Schools of Science, to take equal rank with the old classical Gymnasia, and to put pure and applied science on the same footing for educational purposes as that which the classics enjoy in these schools, is a desirable thing, but that he allows that by the introduction of a 'bifurcation' system into the older schools, they might be made equal to meeting all modern requirements. Helmholtz, on the other hand, may in the same report be found pleading strongly for 'the foundation on equal terms of complete academic institutions for science' as a 'counteraction of the tendency of classical men to lean on authority alone.'

'Philological culture,' says the eminent physiologist of Heidelberg, 'has an ill effect on those who are to devote themselves to science; the philologist is too much dependent on authority and books, he cannot observe for himself, or rely upon his own conclusions, and having only been accustomed to consider the laws of grammar, all of which have their exceptions, he cannot understand the invariable character of physical laws.' Granting with all respect the premises laid down by Professor Helmholtz, we should demur to the conclusion which he would base upon them, and profess ourselves unable to see that, because particular institutions have a tendency to dwarf and stunt particular faculties, they should therefore be left undisturbed to do this evil work uncounteracted. And still leaving the premises unimpugned, we should set up a cross-indictment to the effect that if classical studies left the student of them unacquainted with the invariability of natural laws, physical studies leave the student unacquainted with the variability of men's minds. But, so far as the business of life consists in having to do business and hold intercourse with our fellow-men, this acquaintance with the variability of men's minds is simply the particular kind of knowledge which is not only the most practically useful and marketable of all kinds of knowledge, but is precisely the kind which, by common consent, is allowed to characterise if not to institute 'culture.'

Lord Lyttelton, however, and the Endowed Schools Commissioners would appear to be in favour of the establishment of locally distinct schools for the two sets of studies and of students, and herein to be at one with Helmholtz. The Owens College Delegates, on the other hand, are, like ourselves, in favour of a system of bifurcation, which would not necessarily keep apart persons of different mental conformation who might be much benefited by mutual contact. They have come to this conclusion mainly for reasons based on observations and testimony given in Germany. Our peculiar social organisation makes the question more complex for us; but we, too, have our experience as well as the Germans; and time has shown that an Englishman, whose reputation as an educationalist is equal to that of Helmholtz as a physicist, may, in this very matter, be as far wrong as we believe that great physicist to be. In 1864 Dr. Temple told the Public Commissioners (see Report, vol. ii. p. 312) that he should 'not consider it wise to follow the Cheltenham and Marlborough examples by attaching to the public schools modern departments. The classical work would lose, the other work would not gain!' In 1867 we find a distinguished Rugby master, the Rev. J. M. Wilson, speaking to the following effect of the results produced by the changes set on foot in accordance with the proposals of the Public Schools Commissioners, and earnestly and honestly carried out. 'Lastly, what are the general results of the introduction of scientific teaching in the opinion of the body of the masters? In brief it is this: that the School as a whole is better for it, and that the scholarship is not worse. . . . . This is the testimony of classical masters, by no means specially favourable to science, who are in a position which enables them to judge. . . . . It is believed that no master in Rugby School would wish to give up natural science and recur to the old curriculum.'

XLVII.

THE EARTH-CLOSET SYSTEM.

I VENTURE to think that Von Pettenkofer's opinion of the earth-closet system may be worthy the attention of the readers of 'The Lancet' at the present moment. It may be found to the following effect in the 'Zeitschrift für Biologie,' 1867, bd. iii. hft. ii. and iii. p. 298 :—

'As to salubrity, I not only do not look forward to any gain as being likely to arise from disinfection with earth and peat, but, on the other hand, I fear the greatest danger from it, especially as regards cholera. If it is, as it is now pretty generally believed to be, actually the fact that the porosity of the soil and its impregnation with excreta do, at all events at particular seasons, bring about a local disposition for cholera, and that the immunity from cholera which a rocky soil enjoys depends upon the circumstance that such soils cannot be so impregnated, I cannot see my way towards recommending the disinfection of privies with earth and peat.'

Now, I apprehend that it will be allowed that the same line of reasoning will apply to the localisation and diffusion of typhoid fever. And I would add that certain observations which I made recently in a fever-stricken village, with the aid of the light which Dr. Budd's and Dr. T. K. Chambers's writings had given me, have induced me to think that of the two recognised foci for infection—the bespattered privy and the contaminated well—the former may be the one which is more commonly at work. For though it is said that the larger proportion of women- and children-sufferers points to the water, of which they are said to drink more, being the cause at work, the facts are, not that the women and children drink more water in tea, &c. than the men, but that they get less beer; whilst many of the men in our semi-savage villages never use a privy at all, or at least not habitually. This last is the true differentiating condition. What applies, however, to the wood-work and contents of a privy applies to the like elements in the constitution of the earth-closet, so far as disinfection or the want of disinfection is concerned. And if I am told that the earth-closet is inoffensive, and that the privy is fetid, I answer that a rattlesnake is none the less dangerous because its rattle is removed; and that, for anything shown or known to the contrary, odour is to infection, deodorisation to disinfection, what the noise of the serpent is to its bite.

I believe now, as I said some years ago in an article in the 'Quarterly Journal of Science' for April, 1866, page 189, that some modification of the latrine system, securing all the advantages and avoiding all the dangers of the water-closet system proper, may be contrived for, and safely entrusted to, even the poorest and most careless of our populations. Upon this point Liebig's opinion will be of interest, and the more so, perhaps, inasmuch as, with a curious neglect of accuracy, the illustrious chemist is often alleged to be an opponent of systems for the removal of sewage by the cheapest mode of carriage—namely, that by suspension in water. His real opinion may be found in a letter addressed to Dr. Varrentrap, May 1, 1866, and published by that writer in his very valuable work, 'Ueber Werth oder Unwerth der Wasserclosette,' p. 178, Berlin, 1868. Baron Liebig, after stating that he agrees with Mr. Lindley's plans for the drainage of Frankfort in all essential points, says:—

'I am of opinion that of the present means for the removal of sewage, the one which is based upon a water-supply, distributed at the rate of six cubic feet (about thirty-six gallons) per head, is the safest and cheapest method for the removal of all impurity both of house and of street water . . . For the purposes of agriculture it is of particular importance that the contents of the sewers should not be conducted into the sewers, but should be used for manuring.'

Since writing the above, I have performed each of the following five experiments several times:—

1. Having added to five drops of liquor ammoniæ (London Pharm.) 100 cubic centimetres of distilled water, I connected the glass jar containing this solution with an aspirator, between which and the jar some of Nessler's reagent for the detection of ammonia was placed in a tube with several bulbs. A Woulff's bottle, containing sulphuric acid, was adapted to the distal side of the jar, so as to secure the passing of the aspirated air through the acid. Very shortly after the commencement of aspiration the test fluid

became yellow and turbid, and finally threw down a very abundant red precipitate.

2. The same quantity of solution of ammonia having been poured into a jar of the same size as the one employed in experiment No. 1, dried earth was poured into the jar up to about the level which the 100 cubic centimetres of water occupied in the other jar. The jar having been similarly connected on either side, and aspirated, Nessler's reagent became turbid and yellowish, but gave no very distinct precipitate, and none at all of a red colour.

3. Ashes from a coal fire having been substituted for the dried earth of experiment No. 2, this experiment was again tried, with the result of the formation of scarcely any precipitate in the bulbs containing Nessler's reagent.

4 and 5. These experiments consisted in repeating experiments Nos. 2 and 3 severally, with the addition to the earth and ashes respectively of as much water as the jars would receive into the space already partially filled with the solid substances specified. In each case a large quantity of yellowish-red precipitate was formed in the test fluid upon aspiration. The precipitate was much less dense and abundant than that produced in experiment No. 1, and took a much longer aspiration before it was formed. It was formed much more rapidly by the air aspirated from the wetted earth than by that from the wetted ashes.

It will be asked, Do not these experiments show that ashes and earth are, each of them in their respective order, superior to water for use in closets? I think not; for, firstly, it is not certain that ammonia is the cause, or a necessary co-efficient of the cause, of miasma, any more than it was proved formerly by Daniell that hydro-sulphuric acid was the cause of malaria. Secondly, there is much reason to believe that it is precisely when the earth receives choleraic and typhoid evacuations, and should, *ex hypothesi*, disinfect them, that they become most deadly. (See Pettenkofer, 'Zeitschrift für Biologie,' 1865, p. 357, *et passim*; Liebermeister, 'Deutsche Klinik,' Feb. 17, 1866; Varrentrap, loc. cit., p. 101; Parkes's 'Hygiene,' pp. 254, 593, ed. i.) The healthiness of Alexander's armies has been ascribed to his practice of frequently changing his camping-ground, and army surgeons nowadays recommend the like practice, or disinfect the ground itself, as the French did in the

Crimea. Thirdly, these experiments show very plainly how seriously wetting impairs the power which earth and ashes have of retaining gases in their pores; and this, which the *rationale* of their operation, as ordinarily taught, would have led us to anticipate, has been practically confirmed by Dr. Mouat's actual experience of the working of earth-closets in India. (See 'Report on Gaols of Lower Provinces,' p. 144, 1868.) Now 'slops,' or fluid refuse of all kinds, must be got rid of somehow and some way; and either they are thrown upon the earth or ash 'conservancy,' where it exists, and, producing the physico-chemical solecism to which I have alluded, they reproduce the horrid Manchester 'middens' (q.v., or, by preference, Mr. George Greaves's account of them in the 'Quarterly Journal of Science'), or they have a system of sewers to receive them, along which the solid matters, little if indeed at all less noxious than they, might just as well pass too. Fourthly, the double journey, in and out of town, which earth must go through, is a great drawback upon the merits which it may possess; whilst the fact that coals do come into our towns as it is may tempt us to advocate the employment of the ashes they are converted into for the purposes in question. In the house of the rich man, where the bodies producing refuse are not very greatly out of proportion, or, possibly enough, only equal in number to the fires producing ashes, and where a separate system for liquid refuse is but a trifling item in the expense of house-building, an ash 'conservancy' is a possibility. But the quantitative considerations, to which I have just alluded, show *a priori* that it is as inapplicable in theory to the needs of a poor population as, I believe, actual trial of it on the large scale has shown, and always will show, it to be in fact.

XLVIII.

ON TYPHOID OR ENTERIC FEVER IN INDIAN GAOLS, AND ON THE RELATIONS OF THAT DISEASE AND OF CHOLERA TO THE DRY-EARTH SYSTEM OF CONSERVANCY.

Dr. Buchanan, in the Appendix (No. 4, pp. 96, 97, 106) to the Twelfth Report of the Medical Officer of the Privy Council (for 1869, published in 1870), has thrown doubt upon the validity of certain reasonings of mine (in 'The Lancet,' March 20th, 1869, pp. 411, 412) as to the possibility of a dry-earth system of conservancy favouring the spread of typhoid (or enteric or pythogenic) fever. One of my arguments having been based upon the fact that fever spread in the gaols of India in spite of the introduction into them of that system for dealing with excreta, Dr. Buchanan objects that this argument is of no cogency unless it be shown that the Indian fever in question is enteric and not typhus fever. When I wrote I was of opinion that this Indian gaol fever was enteric and not typhus, and further investigation has confirmed me in this belief.

In 1869 I did not think it necessary to lay any emphasis upon this point, for I thought it was a settled, received, and established belief. Dr. Murchison, in his chapter on the 'Geographical Distribution of Typhus Fever' (p. 58 of 'A Treatise on the Continued Fevers of Great Britain,' 1862), had said that there were 'no authentic records of typhus, such as we see it in this country, having been met with in Asia, Africa, or the tropical parts of America.' If, in the same connexion, Dr. Murchison allows that the 'Pali disease' and the true bubonic plague may be analogous to, if not identical with, typhus, I have to say that between typhus as thus locally represented and the gaol fevers as described in the Indian gaols there is scarcely a single point in common; and if

Dr. Morehead's suggestion, which Dr. Murchison thinks 'not unreasonable,' as to the superaddition of an infectious to a malarious factor in the formation in such cases as those of the Agra gaol, be accepted, I have to say, in the second place, that it makes my case stronger instead of weaker. For what we *know* of such combinations of an 'earth-born' or 'earth-sown'—to use Dr. Bryden's expressive synonym for *malarious* poison—with a personally infectious element is based upon experience of two diseases only—typhoid, to wit, and cholera,—and of these two diseases precisely it is known that they spread 'by the operation of decomposing excreta.' We *suppose* the same to be the case with dysentery. But I must say that the supposition of remittent fever becoming infectious appears to me to be purely hypothetical, and, as yet at least, far too unsubstantial to carry even the weight of an objection without tottering.

Boudin, Guyon, and La Roche are quoted by Keith Johnston, in the second edition of the 'Physical Atlas of Natural Phenomena,' 1856, pl. 35, p. 121, to the effect that 'typhus' was limited to the northern temperate zone. The date of this work reminds me that in those days it was thought necessary (as by myself in a Report on Smyrna, p. 59) to insist on the reality of the distinction, then a comparatively novel one, between typhus and typhoid fevers; and I have referred to Keith Johnston's article, not so much for his own or his authorities' opinions as to what they called typhus, as for the sake of quoting his valuable and suggestive, even though not wholly accurate, remark—'The geographical and climatal limits of typhus in Europe and America will be found to correspond nearly with those of the glutinous Cerealia and potato.' The organic world, whether pathological or physiological, whether animal or vegetable, must be looked at as a whole. Facility in colligating facts can always be obtained; the power of detecting their *rationale* can sometimes be obtained by the employment of this method. I should not, however, accept Keith Johnston's suggestion of this particular geographical phytographical correspondence; a nearer, though still only partially correct, boundary would be obtained by taking the northern limit of the Palmaceae as the southern limit of true typhus; a botanist who would supply us with a botanical expression for the words 'annual isotherm of 68° Fahr.' would very nearly meet the case.

Leaving now the subject of typhus and its existence or non-existence in India, and coming to the question of typhoid fever, I must express my surprise at Dr. Buchanan's having said (p. 106) that 'nothing about enteric fever can be brought into evidence from India'; for Dr. Murchison quotes a cloud of authorities (p. 407 *op. cit.*) in support of his statement that 'in India it is now known to be far from uncommon'; and I suppose that a disease not uncommon outside may be taken as likely to be not uncommon inside the walls of gaols. I will quote one authority whom Dr. Murchison has not quoted—namely, Dr. James Annesley, the author of a work on the 'Causes, Nature, and Treatment of the more prevalent Diseases of India and of Warm Climates generally,' a first edition of which appeared with plates, in quarto, in 1828, and a second without plates, in octavo, in 1841. At page 457 of the first edition, and page 547 of the second, under the head 'On Organic Changes in Fevers,' I find the following words, for the latter part of which plate xxii. of the first edition furnishes a good illustration:—

'Marks of disease of the large and small intestines are generally confined to their internal tunics. The duodenum, jejunum, and ileum, especially the duodenum and termination of the ileum, very frequently are diseased in their mucous surface, which is inflamed in patches, sometimes covered with a muco-purulent secretion and studded with small ulcerations, particularly the termination of the ileum. Occasionally the mucous surface is of a brick-red or purplish shade of colour, apparently ecchymosed, and covered with a bloody sanies, and readily detached from the subjacent texture. In several cases the ulcerations—which sometimes are large and far apart, at other times small and agglomerated, especially the former—have nearly penetrated the tunics of the intestines, and, in a very few cases, I have observed this occurrence actually to have supervened, the contents of the bowels being partly effused into the peritoneal cavity, and having produced peritonitis.'

I come now to evidence which, as being of more recent publication than my communications in 'The Lancet' of March, 1869, was not available when I wrote, but which shows beyond all doubt that enteric or typhoid fever is one at least of the gaol-fevers of India. In Dr. De Renzy's last Report on the Sanitary Administration of the Punjab[1] (for a recently received copy of which I am indebted, as I imagine, to his kindness), I find, at page 127, the following statement relative to one of the gaols brought forward in evidence by me, in 'The Lancet,' in March, 1869:—

[1] For 1869. Published at the Lawrence Press, Lahore, 1870.

'In an extract from Dr. Fairweather's valuable report on the fevers which prevailed in the spring of the year in the Rawulpindi gaol, which is printed, will be found conclusive proof that typhus fever must henceforth be regarded as one of the endemic diseases of that district.'

On the next page we find that Dr. De Renzy uses the most modern nomenclature, and contrasts typhus with typhoid and relapsing fevers. Upon reading this, I determined to come forward, though a most 'unwilling witness[1],' yet without delay—as, from not having seen any notice of this Report, I believed, and believe, I must have been one of the first persons in England to receive a copy of it,—and state that I had received information which led me to believe, in accordance with Dr. Buchanan's suggestion, that typhus—a disease as little, though also as much, affected in its contagiousness and spread by the use of earth-closets as small-pox—did really exist, though alongside of typhoid, in one of the Indian gaols of which I had written. But on turning to Dr. Fairweather's Report, at page 80 (Appendix) of Dr. De Renzy's volume, I could not thereby convince myself that it did furnish conclusive proof of the existence of typhus in that gaol. In Dr. Fairweather's Report notes are given of eight cases, of which five only ended fatally, and with a post-mortem examination. Of these five fatally ending cases, we find it recorded of one (Case 6) that 'his symptoms were more those of enteric than of typhus fever'; and as the intestines, in the account of the post-mortem examination, are described as 'having the lower part of the ileum one mass of ulceration, enlargement and thickening of Peyer's patches,' we need only remark that the diagnosis formed during life was very abundantly confirmed by the autopsy, but that neither seems to

[1] I see from one of Dr. Buchanan's notes, p. 96, that some objection can be raised to my employment of the words 'an unwilling witness' when speaking of a person who, like Dr. Mouat, comes forward and publishes, with the truthfulness which we expect and find in such officials, facts which are, or seem to be, scarcely reconcilable with theories he has advocated. I am not aware that ordinary usage attaches any offensive insinuation to the words alluded to and employed above. Still I may express regret for having employed a phrase which could be misinterpreted. After a 'very critical reading,' however, of my letter in 'The Lancet' of March 20th, 1869, I think that another of my expressions used there does require amendment. Though John Hunter said that nothing was so difficult as to know when a fact was a fact, I still am sorry that I said that Dr. Mouat's *facts* had been called in question, as that expression does, I must acknowledge, admit of a harsh interpretation. I do not suppose Dr. Mouat puts this upon it; I should regret it if any future commentator should do so.

prove that the case was anything but what it was supposed to be—namely, enteric fever. A second of the five died with general peritonitis, consequent upon perforation of the intestine of one of 'three small corroding ulcers in ileum, one on ileo-colic valve, one a few inches further back, and a third about a foot from ileo-colic valve.' The mesenteric glands are stated to have shown no signs of disease; but the man had been ill six weeks; and this fact, which explains the former one so far as it needs explaining, is at the same time incompatible with the hypothesis of typhus, and explicable upon that of typhoid. Of the third of the five cases (Case 7) it is recorded that he was admitted into the hospital on April 7th 'with continued fever, and all that can be recollected concerning him is that he was much troubled with cough and with pain in the right side.' The man died on June 5th, and the autopsy revealed tubercles crowding the left lung, vomicae in the right, and ulcers with elevated hardened edges in the ileum. This appears to have been a case of tuberculosis. Two of the five fatal cases (Cases 3 and 4) remain to be dealt with. In neither of them was there any intestinal or mesenteric mischief; and neither of them, therefore, could have been 'enteric' fever. The persistence of typhus spots is about the only post-mortem appearance pathognomonic of typhus; and this is said to have been absent in one of the two cases (Case 3), whilst of the other case (Case 4) we are told it began as 'intermittent fever.' With regard to the three remaining cases of the eight in which there was a favourable termination, it is more difficult to speak positively either way as to typhus or typhoid. But with reference to the value of the rash as a diagnostic mark, in two of these three cases, I must say, considering what we know of the protean variations which typhoid fever may, even in our chilly climate, put on as to the character of the cutaneous eruption accompanying it[1]—variations consisting in the development of a 'scarlet rash,' of 'petechiae,' of 'vibices,' of 'sudamina,' in addition to the typical lenticularly raised rose-spots—that it is open to me or to any one else to demur to any one of these cases being positively set down as certainly not enteric. With reference, lastly, to the third of the three cases which survived (Case 8), it is well to mark that the case lasted more than an

[1] See Murchison, 'Treatise on Continued Fevers,' pp. 474, 519; and in 'The Lancet,' Dec. 10, 1870.

entire month; and that Dr. Fairweather says of it that it looked 'like a complication of a dysentery, typhus, and typhoid.' Of the eight cases, therefore, which were taken as proving conclusively the presence of typhus in India, two appear to have been certainly typhoid, the post-mortem examination proving the disease to have been this; a third is the case last mentioned; a fourth appears to have been a case of tuberculosis; a fifth, though covered with eruption like that of typhus during life, did not retain it after; a sixth began as a case of intermittent fever; and the two others who survived may possibly have been cases of typhoid or of remittent.

It would appear from the 'Report on Measures adopted for Sanitary Improvements in India from June, 1869, to June, 1870,' recently published, p. 85, that there are two reports extant upon the fever in the Rawulpindee Gaol, by Dr. Lyons, the medical officer in charge. It would be interesting to see these.

Dr. Fairweather has prefixed to his detailed accounts of the eight cases, previously referred to, a short account of the general characteristics of the disease. Amongst these is noted the absence of yellowness of the conjunctivae, which may show that the disease known in Great Britain as 'relapsing fever' was not a factor in the epidemic. Dr. Murchison says of the disorder, indeed (*op. cit.*, p. 301), that it is unknown in India, and, indeed, in all tropical countries.

Certain remarks as to the general characters of the epidemic which Dr. De Renzy makes in various parts of his report appear to me to be more easily reconcilable with the hypothesis of its having been typhoid than with one which should suppose it to have been either relapsing or typhus fever. These remarks are as follows:—

1. It prevailed in the latter half of the year. See Dr. De Renzy, *loc. cit.*, p. 121. See also 'Report on Sanitary Improvements in India,' 1870, p. 30; and Dr. Bryden's 'Report on Cholera Epidemic,' pp. 20, 74.

2. It appeared first in villages (see Dr. De Renzy, p. 121, and note to p. 87 A of the Appendix, and 'Report of Sanitary Improvements,' p. 45), and Dr. Anstie tells us ('Notes on Epidemics,' p. 70) that typhoid is 'the special epidemic of the slumbrous conservative rural districts.'

3. Its severity did not vary concomitantly with that of the famine. (Dr. De Renzy, *loc. cit.*, p. 132.)

4. It affected more women than men (Dr. De Renzy, *loc. cit.*, p. 151; 'Report on Sanitary Improvements,' *loc. cit.*, p. 83), as I should expect typhoid would usually in India, and have noted it to do in England, explaining it in both cases by the greater time spent by the women at home, and the larger dose of poison which they thus have the certainty of imbibing if the air or the water or both were soiled in the immediate neighbourhood of their dwelling.

As Dr. Buchanan says (p. 106), 'It has already been shown that nothing about enteric fever can be brought into evidence from India'; alongside of this sentence (which without some qualification limiting its scope was never justified, and should now be simply retracted) ought to be set a few equally short as well as a few longer sentences from the latest and most authoritative publication on the diseases of India and their sanitation—I mean the 'Report on Measures adopted for Sanitary Improvements in India from June, 1869, to June, 1870.' At p. 30 I read: 'The disease [typhoid] is manifestly, Dr. Ranking thinks, on the increase in India, and is daily acquiring greater importance.'

At p. 61 I find notices of it amongst our soldiers at Muttra, at Mean Meer, at Jullundur, and at Kurrachee; and I suppose that, if it is found in barracks in so many places, there is some little probability of its finding its way into gaols.

At p. 60:—

'During 1868 the number of cases of typhoid was considerable. Out of the total of 91 deaths from fever, 35, more than one-third, are ascribed to the enteric or typhoid form of disease. *But exactitude in the statistics of fever is difficult from similarity of different types. The liability of the British soldier in India to typhoid fever has not hitherto attracted the attention it deserves.*'

On the same page a death from typhoid fever is recorded as occurring at Rawulpindee in the 1—6th Regiment. This is to be noted, firstly, as Rawulpindee is the place in the gaol of which there were eighty-eight fever deaths in 1868; and, secondly, as in this Report the contagious fever inside the gaols is not distinctly spoken of as typhoid, but only the fever affecting the soldiers (see pp. 19, 78, 82). Against this abstinence from giving the fever a name must be set the disclosures of Dr. Fairweather's postmortems. Much the same applies to the utterances of this Report (p. 15) as to Peshawur, another of the places the gaols of which were referred to by me and Dr. Buchanan after Dr. De Renzy (see Dr. Buchanan's 'Report,' p. 96, note), thirty-nine cases of death among the soldiers being ascribed to a 'continued fever, which bears the strongest resemblance to, if it be not identical with, the enteric fever of Europe.' Though the Report never names the gaol fever, it would appear that it was regarded as identical with that raging outside, for we have great stress laid on the befoulment of the water-supply of the whole population; and at p. 281 we are told, 'all classes appear to suffer from this bad water. *Prisoners in gaol have nothing else to drink.*'

The gaol fever of the third gaol, that of Umballah, in which a fever was found to spread in spite of an earth conservancy, Dr. De Renzy appears to hold (p. 128), upon the Report of Dr. Bateson, to have been relapsing fever. But the precautions he lays stress upon in the case of the town of Ambalah (Umballah), pp. 133–135, are such as would be efficient against enteric rather than against typhus or its congener, relapsing fever. Two cases of typhoid, one verified by a post-mortem, are reported by Dr. De Renzy (p. 125) from Simla.

I will now proceed to comment upon another statement of Dr. Buchanan's. At p. 91 of his Report, after quoting the opinions of seven gentlemen—the Rev. Henry Moule, Mr. Oswald Foster, Mr. Garnett of Lancaster, the Governor of Dorchester Gaol, the late Dr. Meyer of Broadmoor, Surgeon-Major Wyatt, and Captain Mervin Drake,—he says: 'From this favourable expression of opinion there are some very few dissentients;' and he proceeds to enumerate three—namely, Dr. Geo. Johnson, Prof. Pettenkofer, and myself. I have to say that to our three names Dr. Buchanan should have added: Firstly, eight names of the Army Sanitary Commissioners—namely, those of General J. H. Grant, Captain Douglas Galton, Dr. John Sutherland, H. H. Massy, Esq., T. A. L. Murray, Esq., W. E. Baker, Esq., Sir J. R. Martin, and Robert Rawlinson, Esq. For the Commission composed of these eight gentlemen issued, on March 8, 1869, a memorandum on a 'Report and Order of the Madras Government upon the Dry-Earth System of Sewage in the Madras Presidency,' which was circulated more or less freely in England, which was republished in

Madras in June 1869, and which, as the 'Report of Sanitary Improvements in India' (so often quoted) very mildly puts it (p. 132), 'threw the weight of their experience into the scale of sewerage by water-carriage.' This memorandum should, I think, be made more accessible in England than it is at present. We have as much need of its plainly given common-sense instructions as the people in Madras; and we have, I apprehend, in our capacities of ratepayers and taxpayers, an equal claim to have them made available for the use of Local Boards and other Sewer authorities. The Army Sanitary Commissioners, I am glad to observe, have nothing to retract of their opinions as to the dry-earth system; and in their answers to Dr. Cunningham's request for instructions in his capacity of Sanitary Commissioner with the Government of India, given in the Report just quoted, at p. 206, and of date July 8, 1870, they refer the Commissioner to the memorandum of March 8, 1869, and add to it the following words:—

'It follows that to trust to dry-earth conservancy for improving the health of towns, while ordinary station or town drainage is permitted to soak away in cesspits or on the surface, is simply to poison the subsoil with sewer water, which, if collected and conveyed in drain-pipes, would become a valuable manure. The question may now be considered as settled by scientific investigation, that the sewage of inhabited buildings should be treated as a single element, whether as regards health or agriculture; and also that to divide this sewage into two parts, and to remove the parts separately, is, as we have stated elsewhere, to pay double where one payment would answer every purpose.'

The same Sanitary Commission is reported, in the same volume (p. 38), as having decided that the dry-earth system could not be generally introduced into a large city like Bombay; and (at p. 13) as declaring it unnecessary to discuss a certain scheme for the removal of the sewage of Calcutta, which the Justices of the Peace for that city—with whose names and numbers I am unacquainted—had previously rejected, for this reason amongst others, namely, that it was 'simply a system of dry-earth conservancy.'

Secondly: The Rivers Pollution Commissioners, appointed in 1868—namely, Sir William Denison, Dr. Frankland, and John Chalmers Morton,—ought to have their opinion at least referred to. In their report, published this year, they say (p. 50): 'We can have no hesitation in pronouncing the dry-earth system, however suitable for institutions, villages, and camps, where personal or official regulations can be enforced, entirely unfitted to the circumstances of large towns.'

Thirdly: I think Dr. Parkes's name must be added to those of the 'very few dissentients,' who will thus, without counting the Calcutta magistracy, come to number seventeen, as against the seven gentlemen named by Dr. Buchanan. For in the third edition of Dr. Parkes's 'Practical Hygiene,' which bears date April, 1869, I find the following words, which do not appear in the edition of 1864:—

> 'There is one evident objection to all these dry plans—namely, that the excreta are retained about our houses for some time. No doubt when mixed with earth they are inodorous, and it is presumed harmless, but of this no evidence has been given. What would be the result of cholera or typhoid discharges received in earth and allowed to remain in the house?'

I have upon other occasions pointed out times and places in which a dry-earth system of conservancy may have claims upon our favourable consideration: those times, I have always held, are not times of epidemics; those places do not lie within the *enceinte* of large towns. In several of the Reports which I have consulted I have found more or less favourable mention made of the system; these passages I have not thought it my business to quote, for reasons which may be readily imagined. I propose at a future opportunity to offer a further communication on this subject, in which, I trust, the readers of 'The Lancet' will find something of greater interest than they can, I fear, have found in this interminable array of authorities. The immoderate length to which my paper has reached may be justified, I hope, upon the principle that a man who goes a little out of his own proper line, as I suppose I do by meddling with hygiene, ought to be doubly sure of his ground. I trust I have shown that I have not been wanting in this *abundans cautela*.

APPENDIX.

NOTES ON THE SITE OF ROMAN POTTERY WORKS AT THE MYNCHERY UPON THE SEWAGE FARM, NEAR OXFORD.

In the year 1879, in the course of trenching and levelling operations at the Mynchery or site of the nunnery near Littlemore, Oxford, undertaken by the Local Board in connection with the formation of a Sewage Farm, the engineers employed by the Local Board exposed in the farm-yard a number of human skeletons with some mediæval stone coffins. The bodies found undisturbed were all lying with the head to the west and the feet to the east, and in the graves numerous pavement tiles were found similar to those of Plantagenet times exposed in making excavations in All Souls Chapel. The skeletons were of both sexes and of different ages, and may have been the bodies of persons who had died in a poor-house attached to the priory of the Benedictine nuns, dissolved by Henry the Eighth, which formerly stood on this spot.

As in many of the fields a considerable quantity of broken Roman pottery was also found, both on the surface and buried to the depth of one and two feet, it was considered desirable that the ground should be carefully examined in order to see if evidence could be obtained that the pottery had been manufactured in the locality. Permission having been given to Dr. Rolleston by the Local Board to continue the excavations, Mr. Edgar S. Cobbold, who took great interest in the discovery, gave very material assistance, and in the course of time four kilns (one a double one in a good state of preservation) in which the pottery had been made, together with a well and pieces of the burnt clay on which the pots were placed for the process of baking, were exposed. In the rubbish heaps adjoining the kilns some moulds were found, and burnt clay retaining the marks of the workmen's fingers. As a rule the pottery was much broken, but a few perfect pieces and others almost perfect were found, and of these some were contained in the well. The pots or shards consisted of jars, bottles, vases, flat round dishes or mortaria, and a saucer or small cup. Two uninjured jars were procured 2.4″ high and 8″ in maximum width, and a furrow ran round each jar at the plane of maximum width. The pottery has been preserved in the University Museum, Oxford. The mortaria were frosted internally with fine quartz-sand. Coarse flattish pieces of burnt clay marked as if with impressions of straw, like a cream-cheese, were found in considerable abundance; they were believed to be the wrappers used in covering the pots over during the baking. A few fine shards with an encrusting filagree pattern were also obtained. In colour the pots and shards were either red, grey, or black and yellowish, usually unglazed. Some pieces had a hard bright glaze not unlike Etruscan, but there was no attempt at the red figures one sees on real Etruscan ware. As a rule they were like the Roman pottery made at Castor and Upchurch. Pieces of Samian-like ware were also obtained ornamented with figures of wild animals, which were interesting as showing a not very successful attempt at copying the continental Samian. The shards were found not only in immediate proximity to the kilns, but scattered over various parts of the farm. Along with them were a good many masses of half-burnt or less than half-burnt clay. A sort of mask representing a semi-comic face was found with some of the shards, and along with it some teeth of bos and cervus. Potters' marks, not names, but consisting of circles, crosses, horizontal and vertical lines, were seen on some of the shards.

The well was situated to the north of No. 1 kiln. Nos. 1 and 2 kilns were to the north of a quarry made in the year 1878, and traces of two other kilns were to its south. Two of the kilns, one of which was double, were made of rough stones set in clay, and a third was made of clay burnt *in situ*. On a plan of the kilns constructed by Mr. Cobbold it is stated that the double kiln had a floor of burnt clay with holes in it, and contained coarser rubbish than the outside, also two whole pots and many large pieces with a layer of white ash at the bottom. The two divisions of this kiln were separated by a partition of rough stones, and they both opened into a common passage. This kiln including its passage was about 8 feet long, and the two divisions, including the intermediate partition wall, were about 7 feet wide. The kilns had apparently been dug out of the natural sand, and were covered over with black earth containing shards and rubbish; and in the earth adjoining the double kiln two human skeletons and a few ox-bones were found. The human bodies had apparently been flung into a rubbish-pit close to the double kiln, a potter's field wherein strangers or slaves were buried.

NOTES ON ARCHÆOLOGICAL DISCOVERIES MADE AT WYTHAM, BERKSHIRE.

In the year 1869, in the course of digging gravel on Lord Abingdon's property a little to the north of the village of Wytham, graves containing human remains were found, and between that date and March 1878 many others were exposed. The graves were in the gravel, 5 feet or even 11 feet deep, and in some instances so far down as even to reach the subjacent clay. One grave is noted as about 3 feet long; another 7 feet 7 inches long and 4 feet 10 inches in greatest width; a third 10 feet long and 5 feet 6 inches wide. Some of the graves were filled in with humus, others with gravel; probably the gravel dug out in making the grave had been thrown in again.

In some instances it is noted that the 'head was lying at the east rather than towards the east;' in one the head was at the south with the jaw pointing east; in one the head was at the south-west, the feet at the north-east; in another the skeleton was lying W.N.W. by E.S.E. The bodies were in the contracted or semi-contracted position, with the knees approximated to the face; and it is noted that the skeleton or head was lying sometimes on its right side, sometimes on its left, and with the hands sometimes up to the face. The skeletons were of both sexes and of different ages, from old age to childhood. One grave on the south side of the gravel-pit, exhumed in January 1870, contained the skeleton of a child about six years of age. As many as three skeletons were in one instance exposed in a single grave, but usually only a single body was in each grave. The skulls are described as 'large, very like my globose Romano-British type, and like the Sion type of His and Rütimeyer.'

The following objects are noted as having been found in graves along with the bodies:—the bones of a pig, the split tusk of a boar, another split tusk (1. 1. 19. 70) with a hole worked through it in company with a worked flint, the horn of a red-deer, pottery both coarse and fine in texture, amber-beads which had probably formed a necklace, and charcoal about the head.

Additional details are given in the following extracts from Dr. Rolleston's notes:—

'January 28, 1870. Found a large stone like a segment of a mill-stone, with many traces of burning, above the gravel, with another smaller stone 4′ 2″ below the surface in mould. Also many animal bones, some burnt through and through. A large quantity of bones of cow, also sheep's bones and horn cores, also pigs' and stags' bones; also a bone double-pointed needle. Subjacent to the earth containing these objects was a grave 8′ deep, the width across level of head 56″; the length of the grave was 7′ 7″. The length of skeleton from astragalus to top of head 33″. Head at S.W., feet at N.E.; black carbonaceous matter a little above vertex. Left hand close to face, proximal end of metacarpals in relation with teeth, but digits pointing downwards. Right ulna in line and under leg bones, extended and in articulation with humerus. Head of right humerus lying on right upper maxilla. Skull and whole body on left side, the right scapula lying close to coracoid of left scapula, which coracoid was looking upwards. Body curled round. Patella present, knees 8″ from upper jaw. Distal end of left radius 1″ from upper jaw. Right fibula projected beyond distal end of right femur proximally to head, outer trochanter in relation to os calcis. Two highest cervical vertebræ between condyles of lower jaw. Spines of dorsal vertebræ pointing upwards close to left coracoid. Centres of lumbar vertebræ much eaten away. Ribs pressed away from head. Sacral spines looking upwards. The left tibia the most superficial of the long bones.

'June 6, 1870. Went over to Wytham with result of bringing back one human skull with long bones from a grave 8 feet deep, 10 feet long, 5 ft. 6 in. broad, in the gravel: with lower end of worked celt, and two more or less doubtful worked flints, and pig's jaw; also a great number of various styles of pottery, (1) Saxon with thumb marks, (2) very coarse British, (3) much Roman, (4) some angular pottery with vandyck figuring, very like Saxon, but said not to be Saxon by Prof. Phillips. Also bones of cow, sheep, deer, horse, pig; also bones of hippopotamus from the gravel. The skeleton was that of an adult man, head of humerus detached, teeth much worn, exostosis on right upper jaw and on right mandible, and tartar on teeth. Both patellæ *in situ*. Lying W.N.W. by E.S.E., doubled up, from head to knee 21 inches. Both arms with elbow flexed, humerus parallel with upper vertebræ, lumbar vertebræ with spines looking upwards. Between chin and knee, flint, pig's jaw. The left os innominatum is much blackened internally, also the head of the corresponding femur; to a less extent also the right innominate bone and femur. Probably a fire was lighted in the grave superficial to the body and the fine ashes subsequently worked their way into the bones more or less cracked by the heat. Femur 18·2″ inches long; tibia 14 inches. Stature calculated by length of femur 5′ 6″, by length of femur and tibia together 5′ 5·4″. The stature is lower than is usual in the brachycephalic Britons. Extreme length of skull 7·1; extreme breadth 5·8; cephalic index 82. The conceptacula cerebelli touch the same surface with the teeth. Teeth all present, but much worn. Lower jaw very powerful. Sutures extensively obliterated externally. Frontal sinuses and supraciliary ridges largely developed. In contour the skull is typically brachycephalic; the posterior half of the parietals dip precipitately, and though the upper occipital squama does lie in a plane a little posterior to that of the posterior half of the parietals, it nevertheless does not project behind the inion, which forms a coarse transverse ridge without any differentiation of the lineæ supremæ.'

In March 1871, R. Johnson and W. B. Clarke, Esqrs. found the skeletons of three dogs in a hole, about 4 feet below the surface of the ground and about 2 feet in the gravel. At the bottom of the hole the earth was black and mixed with charcoal, and several stones were found which showed traces of fire: the dogs' skeletons were immediately on the top of this blackened earth. The skeletons were close together as if they had been thrown into the hole in a heap.

Messrs. Johnson and Clarke say that close to the dogs in the same hole were—

'remains of ox, pig, roe-deer, pike and some other sort of fish, teal, crane, wild swan, and horse, and various sorts of pottery. One mass apparently consisted of woody fibres and oxidized iron. We also found half a spindle-whorl, and Johnson picked up a bone needle on a rubbish-heap close by, which may have come from the alluvium or gravel. The long bones of the ox were split and broken, as if for the purpose of getting the marrow out. Some few pieces of bone show signs of having passed through the fire.'

The following note on a skeleton exposed in a grave opened in 1875 was made by W. Hatchett Jackson, Esq.:—

'Wytham, Sept. 17, 1875. The body lay with the head pointing S. E. by S. It was stretched at full length, perfectly straight. The head was about 1 ft. 6 in. below the surface, and lay with the upper part of the body in mould, while the legs and feet lay in the gravel. The feet were about 2 ft. 3 in. below the surface, as nearly as could be judged.

'The head was fallen forwards on the breast. The thyroid cartilage was slightly ossified, but tumbled to pieces in the carriage, the bony matter being barely of paper thickness. The left arm lay along the side of the body, palm uppermost. The right arm had been folded across the body, and its fingers lay partly in, partly across the pelvic cavity. When I saw the body, the workmen had broken one leg, and the feet were gone from both legs. The position of the bones however showed that the left leg had lain across the right, in the position of the recumbent figures commonly called Knight Crusaders. The workmen had saved some of the bones of the feet. The skull and pelvis had been cracked.

'No pottery or any other remains lay upon the body, or very near it. A small piece of broken pottery lay in the gravel.'

The skeleton in relation to which the amber-beads were found was exposed on March 18, 1878, in a grave about one foot in depth, the superficial gravel being slightly excavated. The head was at the south, the jaw pointing east; body on right side, hips and knees bent. On this find Dr. Rolleston has noted:—

'This is a semi-contracted position such as is usually supposed to characterise the British skeleton. But some Saxon bodies, or Teutonic at least, are found in the East Riding thus contracted, and there can be no doubt that the amber-beads with worked flattened approximal surfaces are Saxon.'

In addition to the graves two larger openings were found in the gravel. In a plan of the Wytham gravel-pits, prepared in January 1870 by Sir Henry Dryden, the one of these pits which is said not to be excavated is marked on the east side of the pit: another and larger pit was on the south side. On it Sir H. Dryden has made the following note:—

'Opening excavated Jan. 26, and appeared to be the remainder of a circular pit, probably covered in and used as a habitation. The floor was hardened gravel, partly a naturally hard bed in the gravel, partly hardened artificially and by exposure. In digging out the humus, which had collected in it, teeth of ox and deer and a good deal of pottery were found, but throughout the field the humus is charged with similar pottery. On the floor a portion of a bivalve shell was found.'

Most of the skeletons and other objects collected at Wytham are preserved in the Oxford University Museum.

NOTES ON ARCHÆOLOGICAL DISCOVERIES AT YARNTON, OXFORDSHIRE.

During the years 1875 to 1877, both inclusive, several human skeletons were exposed in an ancient burial-place to the south-west of the church of Yarnton, Oxfordshire, on the north side of the railway station. The excavations leading to the discovery of the skeletons were made for the purpose of getting gravel as ballast in connection with railway works then going on in that district.

Whilst removing the soil two ditches of a semicircular form were opened into. These ditches were situated in close proximity to each other, and, as appears from a plan constructed by Sir H. Dryden, had the concavity of the semi-lune turned in opposite directions. The ditches were filled with black earth, and in and near them the graves containing the skeletons were found. One ditch is noted (May 13–15, 1876) as 'about 7′ 10″ wide, and about 1′ 6″ below the surface, with sloping sides, and may consequently have been a little less wide when first dug out. Its depth is from 5′ to 5′ 10″.' This ditch was exposed for a space of about 60′, running nearly north and south, but trending a little eastwards. On May 13, 1876, Dr. Rolleston noted :—

'We found a skeleton in a grave sunk through the side of the above ditch in such a way as to show that the ditch had been filled up, and the grave sunk into partly the filling-up of the ditch and partly into the side or escarpment of it. The skeleton was lying so near the slope of the ditch that it would have been broken into pieces in the digging out of the ditch. There can be no reasonable doubt that the ditch was the earlier of the two. This is confirmed by the fact that in another of these ditches two bodies were found, one on the top of the other, in 1875. The buriers struck upon a ditch as a place to dig a grave in, and sank this grave as it happened within the limits of the ditch on either side. I think the ditches may have been simply places for living in. As a general rule we have found very little in them, and that very little, animal bones—ox, dog, goat.'

The graves were between two and three feet from the surface of the ground. They measured from two to three feet in length and were about two feet in breadth. The direction of the grave varied in different instances, and the bodies had been laid in them with the head in varying positions. Thus it is noted in one that the head was at the east; in another that it was at the north-east; in a third that the grave was north-by-south, and the face was at the south; in a fourth that the grave was east and west, with the face to the north; in a fifth the feet were north-east-by-north, the head at the opposite end. In those graves where the skeleton had not been disturbed by the workmen, it was seen that the body had been buried in the contracted position, the knees being approximated to the face.

The skeletons were of both sexes and at various periods of life, some being children. One child's skeleton is noted as lying with the head at the south by south-west, the feet were north by north-east, the face was looking southwards. The body was lying on the right side, the right knee was more bent up than the left, one hand was up at the face.

Various articles were found in the graves along with the skeletons. In one a broken urn was close to the skull; in another pieces of pottery and cows' bones; in a third a bone-pin, 8 inches long, between the arms; in a fourth a drinking-cup or food-vessel and a flint were in front of the face; in a fifth an iron spear (1. 29. 10. 76) was immediately in front of the face, and it projected about ten inches beyond the skull; and in the same grave an iron knife or dagger (1. 29. 10. 76) nearly a foot long and an inch and a half wide extended from the elbow by the side of the body downwards to the side of the left os innominatum: its handle end was towards the feet, its point towards the head. It is also stated that a bronze torque or collar was found by the workmen with one of the skeletons. This collar was acquired by John Evans, Esq., F.R.S.

Details of some of the graves are given in the following extracts from Dr. Rolleston's notes :—

'March 23, 1876. The grave was about 3 feet deep and 3 feet long by 2 ft. 3 in. broad, the long axis lying north by south. The body lay doubled up on its right side, the face being at the south end of the grave and looking north. From the heels to the head was 2 ft. 4 in. The spines of the vertebræ pointed north-west. The cervical vertebræ and a piece of the scapula lay between the rami of the lower jaw. The right hand was by the pelvis, the left hand up at the face, the right arm being bent across over the body. A considerable quantity of charred wood-fibre lay above the bones of the body and about the skull. Near the left hand were found a flint chipped flake and a bronze needle.

'Feb. 23, 1876. I drove over with the Rev. W. C. Lukis, F.S.A., and G. L. Rolleston. We found that a human skeleton had been partially disturbed, but was in great part still *in situ* on the top of the gravel. The skeleton was lying on its left side looking north, the hands were to the face, the knees to the chest. The grave was about 2 feet in length, and there was no implement, and only one bit of well-burnt dark pottery in it. At a distance of 5 feet from the foot of the grave was the centre of a hole 5 feet in diameter, 4 feet deep, in which some animal bones had been found. The head of the skeleton was 250′ S.S.E. from the spot where the ballast escarpment in Sir H. Dryden's plan cut the semicircular ditch at its outer edge. This has been added to Sir H. Dryden's plan by Mr. Lukis (N in the plan). The direction of the body was W.N.W. by E.S.E.'

In a subsequent note Dr. Rolleston says :—

'These excavations in the elongated form of ditches must be considered in connection with shorter forms of holes, such as the one I had dug out on Monday, May 15, with the result of finding nothing in it except the blackish mould which differentiates them and the ditches from the gravel in which they are sunk. This hole was 7′ 6″ long by 5′ in width, and its depth was 5′ 3″. Its walls were perfectly definable by the natural gravel, and its long axis was proximately east and west. Could it have been just a sunk dwelling? The men say they sometimes find these holes as much as 8′ to 10′ deep, and they are obliged to note them because the black mould they contain does not do for ballast.

'Nov. 23, 1876. A Saxon interment, with head at west and feet at east, with spear reputed to have been alongside of right humerus. The right hand lay under an umbo and over the symphysis pubis, the elbow being bent. The left arm appears to have been straight. A knife of iron lay at a tangent to the anterior rim of the umbo, its handle end the narrower towards the left side, the blade end towards the right. On the left side of the pelvis and a couple of inches or so from it were some strips of bronze, perhaps an ornament of the left wrist-band. I found none on the right. The grave was 1′ 6″ deep. The entire length occupied was about 5′ 4″ long.'

Most of the human and other remains are preserved in the Museum of the University of Oxford.

INDEX.

THE END.

October, 1884.

Clarendon Press, Oxford

A SELECTION OF

BOOKS

PUBLISHED FOR THE UNIVERSITY BY

HENRY FROWDE,

AT THE OXFORD UNIVERSITY PRESS WAREHOUSE,

AMEN CORNER, LONDON.

ALSO TO BE HAD AT THE

CLARENDON PRESS DEPOSITORY, OXFORD.

[*Every book is bound in cloth, unless otherwise described.*]

LEXICONS, GRAMMARS, &c.

(See also Clarendon Press Series, pp. 14, 18, 21, 24, 25.)

ANGLO-SAXON.—*An Anglo-Saxon Dictionary*, based on the MS. Collections of the late Joseph Bosworth, D.D., Professor of Anglo-Saxon, Oxford. Edited and enlarged by Prof. T. N. Toller, M.A. (To be completed in four parts.) Parts I and II. A—HWISTLIAN (pp. vi, 576). 1882. 4to. 15*s*. each.

CHINESE.—*A Handbook of the Chinese Language*. Parts I and II, Grammar and Chrestomathy. By James Summers. 1863. 8vo. half bound, 1*l*. 8*s*.

ENGLISH.—*A New English Dictionary, on Historical Principles:* founded mainly on the materials collected by the Philological Society. Edited by James A. H. Murray, LL.D., President of the Philological Society; with the assistance of many Scholars and men of Science. Part I. A—ANT (pp. xvi, 352). Imperial 4to. 12*s*. 6*d*.

—— *An Etymological Dictionary of the English Language*. By W. W. Skeat, M.A. *Second Edition*. 1884. 4to. 2*l*. 4*s*.

——Supplement to the First Edition of the above. 1884. 4to. 2*s*. 6*d*.

—— *A Concise Etymological Dictionary of the English Language*. By W. W. Skeat, M.A. 1884. Crown 8vo. 5*s*. 6*d*.

GREEK.—*A Greek-English Lexicon*, by Henry George Liddell, D.D., and Robert Scott, D.D. Seventh Edition, Revised and Augmented throughout. 1883. 4to. 1*l*. 16*s*.

—— *A Greek-English Lexicon*, abridged from Liddell and Scott's 4to. edition, chiefly for the use of Schools. Twentieth Edition. Carefully Revised throughout. 1883. Square 12mo. 7*s*. 6*d*.

GREEK.—*A copious Greek-English Vocabulary*, compiled from the best authorities. 1850. 24mo. 3*s*.

—— *A Practical Introduction to Greek Accentuation*, by H. W. Chandler, M.A. Second Edition. 1881. 8vo. 10*s*. 6*d*.

HEBREW.—*The Book of Hebrew Roots*, by Abu 'l-Walîd Marwân ibn Janâh, otherwise called Rabbî Yônâh. Now first edited, with an Appendix, by Ad. Neubauer. 1875. 4to. 2*l*. 7*s*. 6*d*.

—— *A Treatise on the use of the Tenses in Hebrew*. By S. R. Driver, M.A. Second Edition, Revised and Enlarged. 1881. Extra fcap. 8vo. 7*s*. 6*d*.

—— *Hebrew Accentuation of Psalms, Proverbs, and Job*. By William Wickes, D.D. 1881. Demy 8vo. stiff covers, 5*s*.

ICELANDIC.—*An Icelandic-English Dictionary*, based on the MS. collections of the late Richard Cleasby. Enlarged and completed by G. Vigfússon, M.A. With an Introduction, and Life of Richard Cleasby, by G. Webbe Dasent, D.C.L. 1874. 4to. 3*l*. 7*s*.

—— *A List of English Words the Etymology of which is illustrated by comparison with Icelandic*. Prepared in the form of an APPENDIX to the above. By W. W. Skeat, M.A. 1876. stitched, 2*s*.

—— *An Icelandic Prose Reader*, with Notes, Grammar and Glossary. by Dr. Gudbrand Vigfússon and F. York Powell, M.A. 1879. Extra fcap. 8vo. 10*s*. 6*d*.

LATIN.—*A Latin Dictionary*, founded on Andrews' edition of Freund's Latin Dictionary, revised, enlarged, and in great part rewritten by Charlton T. Lewis, Ph.D., and Charles Short, LL.D. 1879. 4to. 1*l*. 5*s*.

SANSKRIT.—*A Practical Grammar of the Sanskrit Language*, arranged with reference to the Classical Languages of Europe, for the use of English Students, by Monier Williams, M.A. Fourth Edition, 1877. 8vo. 15*s*.

—— *A Sanskrit-English Dictionary*, Etymologically and Philologically arranged, with special reference to Greek, Latin, German, Anglo-Saxon, English, and other cognate Indo-European Languages. By Monier Williams, M.A. 1872. 4to. 4*l*. 14*s*. 6*d*.

—— *Nalopákhyánam*. Story of Nala, an Episode of the Mahá-Bhárata: the Sanskrit text, with a copious Vocabulary, and an improved version of Dean Milman's Translation, by Monier Williams, M.A. Second Edition, Revised and Improved. 1879. 8vo. 15*s*.

—— *Sakuntalā*. A Sanskrit Drama, in Seven Acts. Edited by Monier Williams, M.A. Second Edition, 1876. 8vo. 21*s*.

SYRIAC.—*Thesaurus Syriacus:* collegerunt Quatremère, Bernstein, Lorsbach, Arnoldi, Agrell, Field, Roediger: edidit R. Payne Smith, S.T.P. Fasc. I-VI. 1868-83. sm. fol. each, 1*l*. 1*s*. Vol. I, containing Fasc. I-V, sm. fol. 5*l*. 5*s*.

—— *The Book of Kalīlah and Dimnah*. Translated from Arabic into Syriac. Edited by W. Wright, LL.D., Professor of Arabic in the University of Cambridge. 1884. 8vo. 21*s*.

GREEK CLASSICS, &c.

Aristophanes: A Complete Concordance to the Comedies and Fragments. By Henry Dunbar, M.D. 4to. 1*l.* 1*s.*

Aristotle: The Politics, translated into English, with Introduction, Marginal Analysis, Notes, and Indices, by B. Jowett, M.A. Medium 8vo. *Nearly ready.*

Heracliti Ephesii Reliquiae. Recensuit I. Bywater, M.A. Appendicis loco additae sunt Diogenis Laertii Vita Heracliti, Particulae Hippocratei De Diaeta Libri Primi, Epistolae Heracliteae. 1877. 8vo. 6*s.*

Homer: A Complete Concordance to the Odyssey and Hymns of Homer; to which is added a Concordance to the Parallel Passages in the Iliad, Odyssey, and Hymns. By Henry Dunbar, M.D. 1880. 4to. 1*l.* 1*s.*

—— *Scholia Graeca in Iliadem.* Edited by Professor W. Dindorf, after a new collation of the Venetian MSS. by D. B. Monro, M.A., Fellow of Oriel College.

Vols. I. II. 1875. 8vo. 24*s.*
Vols. III. IV. 1877. 8vo. 26*s.*
Vols. V. VI. *In the Press.*

—— *Scholia Graeca in Odysseam.* Edidit Guil. Dindorfius Tomi II. 1855. 8vo. 15*s.* 6*d.*

Plato: Apology, with a revised Text and English Notes, and a Digest of Platonic Idioms, by James Riddell, M.A. 1878. 8vo. 8*s.* 6*d.*

—— *Philebus*, with a revised Text and English Notes, by Edward Poste, M.A. 1860. 8vo. 7*s.* 6*d.*

—— *Sophistes and Politicus*, with a revised Text and English Notes, by L. Campbell, M.A. 1867. 8vo. 18*s.*

—— *Theaetetus*, with a revised Text and English Notes, by L. Campbell, M.A. Second Edition. 8vo. 10*s.* 6*d.*

—— *The Dialogues*, translated into English, with Analyses and Introductions, by B. Jowett, M.A. A new Edition in 5 volumes, medium 8vo. 1875. 3*l.* 10*s.*

—— *The Republic*, translated into English, with an Analysis and Introduction, by B. Jowett, M.A. Medium 8vo. 12*s.* 6*d.*

—— *Index to.* Compiled for the Second Edition of Professor Jowett's Translation of the Dialogues. By Evelyn Abbott, M.A. 1875. 8vo. paper covers, 2*s.* 6*d.*

Thucydides: Translated into English, with Introduction, Marginal Analysis, Notes, and Indices. By B. Jowett, M.A. 2 vols. 1881. Medium 8vo. 1*l.* 12*s.*

THE HOLY SCRIPTURES, &c.

ENGLISH.—*The Holy Bible in the earliest English Versions*, made from the Latin Vulgate by John Wycliffe and his followers: edited by the Rev. J. Forshall and Sir F. Madden. 4 vols. 1850. Royal 4to. 3*l.* 3*s.*

[Also reprinted from the above, with Introduction and Glossary by W. W. Skeat, M.A.

—— *The Books of Job, Psalms, Proverbs, Ecclesiastes, and the Song of Solomon:* according to the Wycliffite Version made by Nicholas de Hereford, about A.D. 1381, and Revised by John Purvey, about A.D. 1388. Extra fcap. 8vo. 3*s.* 6*d.*

—— *The New Testament in English*, according to the Version by John Wycliffe, about A.D. 1380, and Revised by John Purvey, about A.D. 1388. Extra fcap. 8vo. 6*s.*]

—— *The Holy Bible:* an exact reprint, page for page, of the Authorised Version published in the year 1611. Demy 4to. half bound, 1*l.* 1*s.*

GOTHIC.—*The Gospel of St. Mark in Gothic*, according to the translation made by Wulfila in the Fourth Century. Edited with a Grammatical Introduction and Glossarial Index by W. W. Skeat, M.A. Extra fcap. 8vo. 4*s.*

GREEK.—*Vetus Testamentum* ex Versione Septuaginta Interpretum secundum exemplar Vaticanum Romae editum. Accedit potior varietas Codicis Alexandrini. Tomi III. Editio Altera. 18mo. 18*s.*

—— *Origenis Hexaplorum* quae supersunt; sive, Veterum Interpretum Graecorum in totum Vetus Testamentum Fragmenta. Edidit Fridericus Field, A.M. 2 vols. 1875. 4to. 5*l.* 5*s.*

—— *The Book of Wisdom:* the Greek Text, the Latin Vulgate, and the Authorised English Version; with an Introduction, Critical Apparatus, and a Commentary. By William J. Deane, M.A. Small 4to. 12*s.* 6*d.*

—— *Novum Testamentum Graece.* Antiquissimorum Codicum Textus in ordine parallelo dispositi. Accedit collatio Codicis Sinaitici. Edidit E. H. Hansell, S.T.B. Tomi III. 1864. 8vo. half morocco, 2*l.* 12*s.* 6*d.*

—— *Novum Testamentum Graece.* Accedunt parallela S. Scripturae loca, necnon vetus capitulorum notatio et canones Eusebii. Edidit Carolus Lloyd, S.T.P.R. 18mo. 3*s.*

The same on writing paper, with large margin, 10*s.*

—— *Novum Testamentum Graece* juxta Exemplar Millianum. 18mo. 2*s.* 6*d.*

The same on writing paper, with large margin, 9*s.*

GREEK.—*Evangelia Sacra Graece.* Fcap. 8vo. limp, 1*s*. 6*d*.

—— *The Greek Testament*, with the Readings adopted by the Revisers of the Authorised Version:—

(1) Pica type, with Marginal References. Demy 8vo. 10*s*. 6*d*.

(2) Long Primer type. Fcap. 8vo. 4*s*. 6*d*.

(3) The same, on writing paper, with wide margin, 15*s*.

—— *The Parallel New Testament*, Greek and English; being the Authorised Version, 1611; the Revised Version, 1881; and the Greek Text followed in the Revised Version. 8vo. 12*s*. 6*d*.

The Revised Version is the joint property of the Universities of Oxford and Cambridge.

—— *Canon Muratorianus:* the earliest Catalogue of the Books of the New Testament. Edited with Notes and a Facsimile of the MS. in the Ambrosian Library at Milan, by S. P. Tregelles, LL.D. 1867. 4to. 10*s*. 6*d*.

—— *Outlines of Textual Criticism applied to the New Testament.* By C. E. Hammond, M.A. Fourth Edition. Extra fcap. 8vo. 3*s*. 6*d*.

HEBREW, etc.—*The Psalms in Hebrew without points.* 1879. Crown 8vo. 3*s*. 6*d*.

—— *A Commentary on the Book of Proverbs.* Attributed to Abraham Ibn Ezra. Edited from a MS. in the Bodleian Library by S. R. Driver, M.A. Crown 8vo. paper covers, 3*s*. 6*d*.

—— *The Book of Tobit.* A Chaldee Text, from a unique MS. in the Bodleian Library; with other Rabbinical Texts, English Translations, and the Itala. Edited by Ad. Neubauer, M.A. 1878. Crown 8vo. 6*s*.

—— *Horae Hebraicae et Talmudicae*, a J. Lightfoot. A new Edition, by R. Gandell, M.A. 4 vols. 1859. 8vo. 1*l*. 1*s*.

LATIN.—*Libri Psalmorum* Versio antiqua Latina, cum Paraphrasi Anglo-Saxonica. Edidit B. Thorpe, F.A.S. 1835. 8vo. 10*s*. 6*d*.

—— *The Psalter, or Psalms of David, and certain Canticles*, with a Translation and Exposition in English, by Richard Rolle of Hampole. Edited by H. R. Bramley, M.A., Fellow of S. M. Magdalen College, Oxford. With an Introduction and Glossary. Demy 8vo. *Just Ready.*

—— *Old-Latin Biblical Texts: No. I.* The Gospel according to St. Matthew from the St. Germain MS. (g_1). Edited with Introduction and Appendices by John Wordsworth, M.A. Small 4to., stiff covers, 6*s*.

OLD-FRENCH.—*Libri Psalmorum* Versio antiqua Gallica e Cod. MS. in Bibl. Bodleiana adservato, una cum Versione Metrica aliisque Monumentis pervetustis. Nunc primum descripsit et edidit Franciscus Michel, Phil. Doct. 1860. 8vo. 10*s*. 6*d*.

FATHERS OF THE CHURCH, &c.

St. Athanasius: Historical Writings, according to the Benedictine Text. With an Introduction by William Bright, D.D. 1881. Crown 8vo. 10*s*. 6*d*.

—— *Orations against the Arians*. With an Account of his Life by William Bright, D.D. 1873. Crown 8vo. 9*s*.

St. Augustine: Select Anti-Pelagian Treatises, and the Acts of the Second Council of Orange. With an Introduction by William Bright, D.D. Crown 8vo. 9*s*.

Canons of the First Four General Councils of Nicaea, Constantinople, Ephesus, and Chalcedon. 1877. Crown 8vo. 2*s*. 6*d*.

—— *Notes on the Canons of the First Four General Councils*. By William Bright, D.D. 1882. Crown 8vo. 5*s*. 6*d*.

Cyrilli Archiepiscopi Alexandrini in XII Prophetas. Edidit P. E. Pusey, A.M. Tomi II. 1868. 8vo. cloth, 2*l*. 2*s*.

—— *in D. Joannis Evangelium*. Accedunt Fragmenta varia necnon Tractatus ad Tiberium Diaconum duo. Edidit post Aubertum P. E. Pusey, A.M. Tomi III. 1872. 8vo. 2*l*. 5*s*.

—— *Commentarii in Lucae Evangelium* quae supersunt Syriace. E MSS. apud Mus. Britan. edidit R. Payne Smith, A.M. 1858. 4to. 1*l*. 2*s*.

—— Translated by R. Payne Smith, M.A. 2 vols. 1859. 8vo. 14*s*.

Ephraemi Syri, Rabulae Episcopi Edesseni, Balaei, aliorumque Opera Selecta. E Codd. Syriacis MSS. in Museo Britannico et Bibliotheca Bodleiana asservatis primus edidit J. J. Overbeck. 1865. 8vo. 1*l*. 1*s*.

Eusebius' Ecclesiastical History, according to the text of Burton, with an Introduction by William Bright, D.D. 1881. Crown 8vo. 8*s*. 6*d*.

Irenaeus: The Third Book of St. Irenaeus, Bishop of Lyons, against Heresies. With short Notes and a Glossary by H. Deane, B.D. 1874. Crown 8vo. 5*s*. 6*d*.

Patrum Apostolicorum, S. Clementis Romani, S. Ignatii, S. Polycarpi, quae supersunt. Edidit Guil. Jacobson, S.T.P.R. Tomi II. Fourth Edition, 1863. 8vo. 1*l*. 1*s*.

Socrates' Ecclesiastical History, according to the Text of Hussey, with an Introduction by William Bright, D.D. 1878. Crown 8vo. 7*s*. 6*d*.

ECCLESIASTICAL HISTORY, BIOGRAPHY, &c.

Ancient Liturgy of the Church of England, according to the uses of Sarum, York, Hereford, and Bangor, and the Roman Liturgy arranged in parallel columns, with preface and notes. By William Maskell, M.A. Third Edition. 1882. 8vo. 15*s*.

Baedae Historia Ecclesiastica. Edited, with English Notes, by G. H. Moberly, M.A. 1881. Crown 8vo. 10*s*. 6*d*.

Bright (W.). Chapters of Early English Church History. 1878. 8vo. 12*s*.

Burnet's History of the Reformation of the Church of England. A new Edition. Carefully revised, and the Records collated with the originals, by N. Pocock, M.A. 7 vols. 1865. 8vo. *Price reduced to* 1*l*. 10*s*.

Councils and Ecclesiastical Documents relating to Great Britain and Ireland. Edited, after Spelman and Wilkins, by A. W. Haddan, B.D., and W. Stubbs, M.A. Vols. I. and III. 1869-71. Medium 8vo. each 1*l*. 1*s*.

Vol. II. Part I. 1873. Medium 8vo. 10*s*. 6*d*.

Vol. II. Part II. 1878. Church of Ireland; Memorials of St. Patrick. Stiff covers, 3*s*. 6*d*.

Hammond (C. E.). Liturgies, Eastern and Western. Edited, with Introduction, Notes, and a Liturgical Glossary. 1878. Crown 8vo. 10*s*. 6*d*.

An Appendix to the above. 1879. Crown 8vo. paper covers, 1*s*. 6*d*.

John, Bishop of Ephesus. The Third Part of his Ecclesiastical History. [In Syriac.] Now first edited by William Cureton, M.A. 1853. 4to. 1*l*. 12*s*.

—— Translated by R. Payne Smith, M.A. 1860. 8vo. 10*s*.

Leofric Missal, The, as used in the Cathedral of Exeter during the Episcopate of its first Bishop, A.D. 1050-1072; together with some Account of the Red Book of Derby, the Missal of Robert of Jumièges, and a few other early MS. Service Books of the English Church. Edited, with Introduction and Notes, by F. E. Warren, B.D. 4to. half morocco, 35*s*.

Monumenta Ritualia Ecclesiae Anglicanae. The occasional Offices of the Church of England according to the old use of Salisbury, the Prymer in English, and other prayers and forms, with dissertations and notes. By William Maskell, M.A. Second Edition. 1882. 3 vols. 8vo. 2*l*. 10*s*.

Records of the Reformation. The Divorce, 1527-1533. Mostly now for the first time printed from MSS. in the British Museum and other libraries. Collected and arranged by N. Pocock, M.A. 1870. 2 vols. 8vo. 1*l*. 16*s*.

Shirley (*W. W.*). *Some Account of the Church in the Apostolic Age.* Second Edition, 1874. fcap. 8vo. 3s. 6d.

Stubbs (*W.*). *Registrum Sacrum Anglicanum.* An attempt to exhibit the course of Episcopal Succession in England. 1858. Small 4to. 8*s.* 6*d.*

Warren (*F. E.*). *Liturgy and Ritual of the Celtic Church.* 1881. 8vo. 14*s.*

ENGLISH THEOLOGY.

Butler's Works, with an Index to the Analogy. 2 vols. 1874. 8vo. 11*s.*

Also separately,

Sermons, 5*s.* 6*d.* *Analogy of Religion*, 5*s.* 6*d.*

Greswell's Harmonia Evangelica. Fifth Edition. 8vo. 1855. 9*s.* 6*d.*

Heurtley's Harmonia Symbolica: Creeds of the Western Church. 1858. 8vo. 6*s.* 6*d.*

Homilies appointed to be read in Churches. Edited by J. Griffiths, M.A. 1859. 8vo. 7*s.* 6*d.*

Hooker's Works, with his life by Walton, arranged by John Keble, M.A. Sixth Edition, 1874. 3 vols. 8vo. 1*l.* 11*s.* 6*d.*

—— the text as arranged by John Keble, M.A. 2 vols. 1875. 8vo. 11*s.*

Jewel's Works. Edited by R. W. Jelf, D.D. 8 vols. 1848. 8vo. 1*l.* 10*s.*

Pearson's Exposition of the Creed. Revised and corrected by E. Burton, D.D. Sixth Edition, 1877. 8vo. 10*s.* 6*d.*

Waterland's Review of the Doctrine of the Eucharist, with a Preface by the present Bishop of London. 1880. Crown 8vo. 6*s.* 6*d.*

—— *Works*, with Life, by Bp. Van Mildert. A new Edition, with copious Indexes. 6 vols. 1856. 8vo. 2*l.* 11*s.*

Wheatly's Illustration of the Book of Common Prayer. A new Edition, 1846. 8vo. 5*s.*

Wyclif. A Catalogue of the Original Works of John Wyclif, by W. W. Shirley, D.D. 1865. 8vo. 3*s.* 6*d.*

—— *Select English Works.* By T. Arnold, M.A. 3 vols. 1869–1871. 8vo. *Price reduced to* 1*l.* 1*s.*

—— *Trialogus.* With the Supplement now first edited. By Gotthard Lechler. 1869. 8vo. *Price reduced to* 7*s.*

HISTORICAL AND DOCUMENTARY WORKS.

British Barrows, a Record of the Examination of Sepulchral Mounds in various parts of England. By William Greenwell, M.A., F.S.A. Together with Description of Figures of Skulls, General Remarks on Prehistoric Crania, and an Appendix by George Rolleston, M.D., F.R.S. 1877. Medium 8vo. 25*s.*

Britton. A Treatise upon the Common Law of England, composed by order of King Edward I. The French Text carefully revised, with an English Translation, Introduction, and Notes, by F. M. Nichols, M.A. 2 vols. 1865. Royal 8vo. 1*l.* 16*s.*

Clarendon's History of the Rebellion and Civil Wars in England. 7 vols. 1839. 18mo. 1*l.* 1*s.*

Clarendon's History of the Rebellion and Civil Wars in England. Also his Life, written by himself, in which is included a Continuation of his History of the Grand Rebellion. With copious Indexes. In one volume, royal 8vo. 1842. 1*l.* 2*s.*

Clinton's Epitome of the Fasti Hellenici. 1851. 8vo. 6*s.* 6*d.*

—— *Epitome of the Fasti Romani.* 1854. 8vo. 7*s.*

Corpvs Poeticvm Boreale. The Poetry of the Old Northern Tongue, from the Earliest Times to the Thirteenth Century. Edited, classified, and translated, with Introduction, Excursus, and Notes, by Gudbrand Vigfússon, M.A., and F. York Powell, M.A. 2 vols. 1883. 8vo. 42*s.*

Freeman (E. A.). History of the Norman Conquest of England; its Causes and Results. In Six Volumes. 8vo. 5*l.* 9*s.* 6*d.*

Vols. I–II together, 3rd edition, 1877. 1*l.* 16*s.*
Vol. III, 2nd edition, 1874. 1*l.* 1*s.*
Vol. IV, 2nd edition, 1875. 1*l.* 1*s.*
Vol. V, 1876. 1*l.* 1*s.*
Vol. VI. Index. 1879. 8vo. 10*s.* 6*d.*

Freeman (E. A.). The Reign of William Rufus and the Accession of Henry the First. 2 vols. 8vo. 1*l.* 16*s.*

Gascoigne's Theological Dictionary ("Liber Veritatum"): Selected Passages, illustrating the condition of Church and State, 1403–1458. With an Introduction by James E. Thorold Rogers, M.P. Small 4to. 10*s.* 6*d.*

Magna Carta, a careful Reprint. Edited by W. Stubbs, M.A. 1879. 4to. stitched, 1*s.*

Passio et Miracula Beati Olaui. Edited from a Twelfth-Century MS. in the Library of Corpus Christi College, Oxford, with an Introduction and Notes, by Frederick Metcalfe, M.A. Small 4to. stiff covers, 6*s.*

Protests of the Lords, including those which have been expunged, from 1624 to 1874; with Historical Introductions. Edited by James E. Thorold Rogers, M.A 1875. 3 vols. 8vo. 2*l*. 2*s*.

Rogers (J. E. T.). History of Agriculture and Prices in England, A.D. 1259–1793.

Vols. I and II (1259–1400). 1866. 8vo. 2*l*. 2*s*.

Vols. III and IV (1401–1582). 1882. 8vo. 2*l*. 10*s*.

Saxon Chronicles (Two of the) parallel, with Supplementary Extracts from the Others. Edited, with Introduction, Notes, and a Glossarial Index, by J. Earle, M.A. 1865. 8vo. 16*s*.

Sturlunga Saga, including the Islendinga Saga of Lawman Sturla Thordsson and other works. Edited by Dr. Gudbrand Vigfússon. In 2 vols. 1878. 8vo. 2*l*. 2*s*.

Statutes made for the University of Oxford, and for the Colleges and Halls therein, by the University of Oxford Commissioners. 1882. 8vo. 12*s*. 6*d*.

Also separately,

Statutes made for the University. 2*s*.

Statutes made for the Colleges. 1*s*. each.

Statuta Universitatis Oxoniensis. 1884. 8vo. 5*s*.

The Student's Handbook to the University and Colleges of Oxford. Seventh Edition. 1883. Extra fcap. 8vo. 2*s*. 6*d*.

MATHEMATICS, PHYSICAL SCIENCE, &c.

Acland (H. W., M.D., F.R.S.). Synopsis of the Pathological Series in the Oxford Museum. 1867. 8vo. 2*s*. 6*d*.

Astronomical Observations made at the University Observatory, Oxford, under the direction of C. Pritchard, M.A. No. 1. 1878. Royal 8vo. paper covers, 3*s*. 6*d*.

De Bary (Dr. A.) Comparative Anatomy of the Vegetative Organs of the Phanerogams and Ferns. Translated and Annotated by F. O. Bower, M.A., F.L.S., and D. H. Scott, M.A., Ph.D., F.L.S. With two hundred and forty-one woodcuts and an Index. *Just Ready.*

Müller (J.). On certain Variations in the Vocal Organs of the Passeres that have hitherto escaped notice. Translated by F. J. Bell, B.A., and edited, with an Appendix, by A. H. Garrod, M.A., F.R.S. With Plates. 1878. 4to. paper covers, 7*s*. 6*d*.

Phillips (John, M.A., F.R.S.). Geology of Oxford and the Valley of the Thames. 1871. 8vo. 21*s*.

—— *Vesuvius.* 1869. Crown 8vo. 10*s*. 6*d*.

Price (Bartholomew, M.A., F.R.S.). Treatise on Infinitesimal Calculus.

Vol. I. Differential Calculus. Second Edition. 8vo. 14*s*. 6*d*.

Vol. II. Integral Calculus, Calculus of Variations, and Differential Equations. Second Edition, 1865. 8vo. 18*s*.

Vol. III. Statics, including Attractions; Dynamics of a Material Particle. Second Edition, 1868. 8vo. 16*s*.

Vol. IV. Dynamics of Material Systems; together with a chapter on Theoretical Dynamics, by W. F. Donkin, M.A., F.R.S. 1862. 8vo. 16*s*.

Rigaud's Correspondence of Scientific Men of the 17th Century, with Table of Contents by A. de Morgan, and Index by the Rev. J. Rigaud, M.A. 2 vols. 1841–1862. 8vo. 18*s*. 6*d*.

Sachs' Text-Book of Botany, Morphological and Physiological. A New Edition. Translated by S. H. Vines, M.A. 1882. Royal 8vo., half morocco, 1*l*. 11*s*. 6*d*.

Westwood (J. O., M.A., F.R.S.). Thesaurus Entomologicus Hopeianus, or a Description of the rarest Insects in the Collection given to the University by the Rev. William Hope. With 40 Plates. 1874. Small folio, half morocco, 7*l*. 10*s*.

The Sacred Books of the East.

TRANSLATED BY VARIOUS ORIENTAL SCHOLARS, AND EDITED BY F. MAX MÜLLER.

[Demy 8vo. cloth.]

Vol. I. The Upanishads. Translated by F. Max Müller. Part I. The *Kh*ândogya-upanishad, The Talavakâra-upanishad, The Aitareya-âra*n*yaka, The Kaushîtaki-brâhma*n*a-upanishad, and The Vâ*g*asaneyi-sa*m*hitâ-upanishad. 10*s*. 6*d*.

Vol. II. The Sacred Laws of the Âryas, as taught in the Schools of Âpastamba, Gautama, Vâsish*th*a, and Baudhâyana. Translated by Prof. Georg Bühler. Part I. Âpastamba and Gautama. 10*s*. 6*d*.

Vol. III. The Sacred Books of China. The Texts of Confucianism. Translated by James Legge. Part I. The Shû King, The Religious portions of the Shih King, and The Hsiâo King. 12*s*. 6*d*.

Vol. IV. The Zend-Avesta. Translated by James Darmesteter. Part I. The Vendîdâd. 10*s*. 6*d*.

Vol. V. The Pahlavi Texts. Translated by E. W. West. Part I. The Bundahis, Bahman Yast, and Shâyast lâ-shâyast. 12*s*. 6*d*.

Vols. VI and IX. The Qur'ân. Parts I and II. Translated by E. H. Palmer. 21*s*.

Vol. VII. The Institutes of Vishnu. Translated by Julius Jolly. 10*s*. 6*d*.

Vol. VIII. The Bhagavadgîtâ, with The Sanatsugâtîya, and The Anugîtâ. Translated by Kâshinâth Trimbak Telang. 10*s*. 6*d*.

Vol. X. The Dhammapada, translated from Pâli by F. Max Müller; and The Sutta-Nipâta, translated from Pâli by V. Fausböll; being Canonical Books of the Buddhists. 10*s*. 6*d*.

Vol. XI. Buddhist Suttas. Translated from Pâli by T. W. Rhys Davids. 1. The Mahâparinibbâna Suttanta; 2. The Dhamma-kakka-ppavattana Sutta; 3. The Tevigga Suttanta; 4. The Akankheyya Sutta; 5. The Ketokhila Sutta; 6. The Mahâ-sudassana Suttanta; 7. The Sabbâsava Sutta. 10*s*. 6*d*.

Vol. XII. The Satapatha-Brâhmana, according to the Text of the Mâdhyandina School. Translated by Julius Eggeling. Part I. Books I and II. 12*s*. 6*d*.

Vol. XIII. Vinaya Texts. Translated from the Pâli by T. W. Rhys Davids and Hermann Oldenberg. Part I. The Pâtimokkha. The Mahâvagga, I-IV. 10*s*. 6*d*.

Vol. XIV. The Sacred Laws of the Âryas, as taught in the Schools of Âpastamba, Gautama, Vâsishtha and Baudhâyana. Translated by Georg Bühler. Part II. Vasishtha and Baudhâyana. 10*s*. 6*d*.

Vol. XV. The Upanishads. Translated by F. Max Müller. Part II. The Katha-upanishad, The Mundaka-upanishad, The Taittirîyaka-upanishad, The Brihadâranyaka-upanishad, The Svetasvatara-upanishad, The Prasña-upanishad, and The Maitrâyana-Brâhmana-upanishad. 10*s*. 6*d*.

Vol. XVI. The Sacred Books of China. The Texts of Confucianism. Translated by James Legge. Part II. The Yî King. 10*s*. 6*d*.

Vol. XVII. Vinaya Texts. Translated from the Pâli by T. W. Rhys Davids and Hermann Oldenberg. Part II. The Mahâvagga, V-X. The Kullavagga, I-III. 10*s*. 6*d*.

Vol. XVIII. Pahlavi Texts. Translated by E. W. West. Part II. The Dâdistân-î Dînîk and The Epistles of Mânûskîhar. 12*s*. 6*d*.

Vol. XIX. The Fo-sho-hing-tsan-king. A Life of Buddha by Asvaghosha Bodhisattva, translated from Sanskrit into Chinese by Dharmaraksha, A.D. 420, and from Chinese into English by Samuel Beal. 10*s*. 6*d*.

Vol. XXI. The Saddharma-pundarîka, or the Lotus of the True Law. Translated by H. Kern. 12*s*. 6*d*.

Vol. XXIII. The Zend-Avesta. Part II. The Sîrôzahs, Yasts, and Nyâyis. Translated by James Darmesteter. 10*s*. 6*d*.

The following Volumes are in the Press:—

Vol. XX. Vinaya Texts. Translated from the Pâli by T. W. Rhys Davids and Hermann Oldenberg. Part III. The *K*ullavagga, I–IV.

Vol. XXII. *G*aina-Sûtras. Translated from Prâkrit by Hermann Jacobi. Part I. The Â*k*ârâṅga-Sûtra. The Kalpa-Sûtra.

Vol. XXIV. Pahlavi Texts. Translated by E. W. West. Part III. Dînâ-î Maînôg-î Khîrad, Shikand-gu-mânî, and Sad-dar.

Second Series.

Vol. XXV. Manu. Translated by GEORG BÜHLER. Part I.

Vol. XXVI. The *S*atapatha-Brâhma*n*a. Translated by Julius Eggeling. Part II.

Anecdota Oxoniensia:

[Small 4to.]

Classical Series. I. i. *The English Manuscripts of the Nicomachean Ethics*, described in relation to Bekker's Manuscripts and other Sources. By J. A. Stewart, M.A. 3*s*. 6*d*.

—— I. ii. *Nonius Marcellus*, de Compendiosa Doctrina, Harleian MS. 2719. Collated by J. H. Onions, M.A. 3*s*. 6*d*.

—— I. iii. *Aristotle's Physics*. Book VII. Collation of various MSS.; with an Introduction by R. Shute, M.A. 2*s*.

—— I. iv. *Bentley's Plautine Emendations*. From his copy of Gronovius. By E. A. Sonnenschein, M.A. 2*s*. 6*d*.

Semitic Series. I. i. *Commentary on Ezra and Nehemiah*. By Rabbi Saadiah. Edited by H. J. Mathews, M.A. 3*s*. 6*d*.

Aryan Series. I. i. *Buddhist Texts from Japan*. Edited by F. Max Müller, M.A. 3*s*. 6*d*.

—— I. ii. *Sukhâvatî-Vyûha*. Description of Sukhâvatî, the Land of Bliss. Edited by F. Max Müller, M.A., and Bunyiu Nanjio. 7*s*. 6*d*.

—— I. iii. The Ancient Palm-leaves containing the Pra*gñâ*-Pâramitâ-H*r*idaya-Sûtra and the Ush*n*îsha-Vigaya-Dhâra*n*î, edited by F. Max Müller, M.A., and Bunyiu Nanjio, M.A. With an Appendix by G. Bühler. With many Plates. 10*s*.

Mediaeval and Modern Series. I. i. *Sinonoma Bartholomei*; A Glossary from a Fourteenth-Century MS. in the Library of Pembroke College, Oxford. Edited by J. L. G. Mowat, M.A. 3*s*. 6*d*.

—— I. iii. *The Saltair Na Rann*. A Collection of Early Middle Irish Poems. Edited from a MS. in the Bodleian Library by Whitley Stokes, LL.D. 7*s*. 6*d*.

Clarendon Press Series

I. ENGLISH.

A First Reading Book. By Marie Eichens of Berlin; and edited by Anne J. Clough. Extra fcap. 8vo. stiff covers, 4*d.*

Oxford Reading Book, Part I. For Little Children. Extra fcap. 8vo. stiff covers, 6*d.*

Oxford Reading Book, Part II. For Junior Classes. Extra fcap. 8vo. stiff covers, 6*d.*

An Elementary English Grammar and Exercise Book. By O. W. Tancock, M.A. Second Edition. Extra fcap. 8vo. 1*s.* 6*d.*

An English Grammar and Reading Book, for Lower Forms in Classical Schools. By O. W. Tancock, M.A. Fourth Edition. Extra fcap. 8vo. 3*s.* 6*d.*

Typical Selections from the best English Writers, with Introductory Notices. Second Edition. In Two Volumes. Extra fcap. 8vo. 3*s.* 6*d.* each.

Vol. I. Latimer to Berkeley. Vol. II. Pope to Macaulay.

Shairp (J. C., LL.D.). Aspects of Poetry; being Lectures delivered at Oxford. Crown 8vo. 10*s.* 6*d.*

A Book for the Beginner in Anglo-Saxon. By John Earle, M.A. Third Edition. Extra fcap. 8vo. 2*s.* 6*d.*

An Anglo-Saxon Reader. In Prose and Verse. With Grammatical Introduction, Notes, and Glossary. By Henry Sweet, M.A. Fourth Edition, Revised and Enlarged. Extra fcap. 8vo. 8*s.* 6*d.*

An Anglo-Saxon Primer, with Grammar, Notes, and Glossary. By the same Author. Second Edition. Extra fcap. 8vo. 2*s.* 6*d.*

First Middle English Primer, with Grammar and Glossary. By the same Author. *Just Ready.*

The Philology of the English Tongue. By J. Earle, M.A. Third Edition. Extra fcap. 8vo. 7*s.* 6*d.*

A Handbook of Phonetics, including a Popular Exposition of the Principles of Spelling Reform. By Henry Sweet, M.A. Extra fcap. 8vo. 4*s.* 6*d.*

The Ormulum; with the Notes and Glossary of Dr. R. M. White. Edited by R. Holt, M.A. 1878. 2 vols. Extra fcap. 8vo. 21*s.*

English Plant Names from the Tenth to the Fifteenth Century. By J. Earle, M.A. Small fcap. 8vo. 5*s.*

Specimens of Early English. A New and Revised Edition. With Introduction, Notes, and Glossarial Index. By R. Morris, LL.D., and W. W. Skeat, M.A.

Part I. From Old English Homilies to King Horn (A.D. 1150 to A.D. 1300). Extra fcap. 8vo. 9*s.*

Part II. From Robert of Gloucester to Gower (A.D. 1298 to A.D. 1393). Second Edition. Extra fcap. 8vo. 7*s.* 6*d.*

Specimens of English Literature, from the 'Ploughmans Crede' to the 'Shepheardes Calender' (A.D. 1394 to A.D. 1579). With Introduction, Notes, and Glossarial Index. By W. W. Skeat, M.A. Extra fcap. 8vo. 7*s.* 6*d.*

The Vision of William concerning Piers the Plowman, by William Langland. Edited, with Notes, by W. W. Skeat, M.A. Third Edition. Extra fcap. 8vo. 4*s.* 6*d.*

Chaucer. I. *The Prologue to the Canterbury Tales;* the Knightes Tale; The Nonne Prestes Tale. Edited by R. Morris, Editor of Specimens of Early English, &c., &c. Fifty-first Thousand. Extra fcap. 8vo. 2*s.* 6*d.*

—— II. *The Prioresses Tale; Sir Thopas;* The Monkes Tale; The Clerkes Tale; The Squieres Tale, &c. Edited by W. W. Skeat, M.A. Second Edition. Extra fcap. 8vo. 4*s.* 6*d.*

—— III. *The Tale of the man of Lawe;* The Pardoneres Tale; The Second Nonnes Tale; The Chanouns Yemannes Tale. By the same Editor. Second Edition. Extra fcap. 8vo. 4*s.* 6*d.*

Spenser's Faery Queene. Books I and II. Designed chiefly for the use of Schools. With Introduction, Notes, and Glossary. By G. W. Kitchin, M.A.

Book I. Tenth Edition. Extra fcap. 8vo. 2*s.* 6*d.*

Book II. Sixth Edition. Extra fcap. 8vo. 2*s.* 6*d.*

Hooker. Ecclesiastical Polity, Book I. Edited by R. W. Church, M.A. Second Edition. Extra fcap. 8vo. 2*s.*

Marlowe and Greene. Marlowe's Tragical History of Dr. Faustus, and *Greene's Honourable History of Friar Bacon and Friar Bungay.* Edited by A. W. Ward, M.A. 1878. Extra fcap. 8vo. 5*s.* 6*d.*

Marlowe. Edward II. With Introduction, Notes, &c. By O. W. Tancock, M.A. Extra fcap. 8vo. 3*s.*

Shakespeare. Select Plays. Edited by W. G. Clark, M.A., and W. Aldis Wright, M.A. Extra fcap. 8vo. stiff covers.

I. The Merchant of Venice. *1s.*
II. Richard the Second. *1s. 6d.*
III. Macbeth. *1s. 6d.*
IV. Hamlet. *2s.*

—— Edited by W. Aldis Wright, M.A. Extra fcap. 8vo. stiff covers.

V. The Tempest. *1s. 6d.*
VI. As You Like It. *1s. 6d.*
VII. Julius Cæsar. *2s.*
VIII. Richard the Third. *2s. 6d.*
IX. King Lear. *1s. 6d.*
X. A Midsummer Night's Dream. *1s. 6d.*
XI. Coriolanus. *2s. 6d.*
XII. Henry the Fifth. *2s.*
XIII. Twelfth Night. *In the Press.*

Bacon. I. *Advancement of Learning.* Edited by W. Aldis Wright, M.A. Second Edition. Extra fcap. 8vo. *4s. 6d.*

—— II. *The Essays.* With Introduction and Notes. By J. R. Thursfield, M.A. *In Preparation.*

Milton. I. *Areopagitica.* With Introduction and Notes. By J. W. Hales, M.A. Third Edition. Extra fcap. 8vo. *3s.*

—— II. *Poems.* Edited by R. C. Browne, M.A. 2 vols. Fifth Edition. Extra fcap. 8vo. *6s. 6d.*

Sold separately, Vol. I. *4s.*; Vol. II. *3s.*

In paper covers :—

Lycidas, *3d.* L'Allegro, *3d.* Il Penseroso, *4d.* Comus, *6d.* Samson Agonistes, *6d.*

—— III. *Samson Agonistes.* Edited with Introduction and Notes by John Churton Collins. Extra fcap. 8vo. stiff covers, *1s.*

Bunyan. I. *The Pilgrim's Progress, Grace Abounding, Relation of the Imprisonment of Mr. John Bunyan.* Edited, with Biographical Introduction and Notes, by E. Venables, M.A. 1879. Extra fcap. 8vo. *5s.*

—— II. *Holy War, &c.* Edited by E. Venables, M.A. In the Press.

Dryden. Select Poems. Stanzas on the Death of Oliver Cromwell; Astræa Redux; Annus Mirabilis; Absalom and Achitophel; Religio Laici; The Hind and the Panther. Edited by W. D. Christie, M.A. Second Edition. Extra fcap. 8vo. 3*s*. 6*d*.

Locke's Conduct of the Understanding. Edited, with Introduction, Notes, &c., by T. Fowler, M.A. Second Edition. Extra fcap. 8vo. 2*s*.

Addison. Selections from Papers in the Spectator. With Notes. By T. Arnold, M.A. Extra fcap. 8vo. 4*s*. 6*d*.

Pope. With Introduction and Notes. By Mark Pattison, B.D.

—— I. *Essay on Man.* Sixth Edition. Extra fcap. 8vo. 1*s*. 6*d*.

—— II. *Satires and Epistles.* Second Edition. Extra fcap. 8vo. 2*s*.

Parnell. The Hermit. Paper covers, 2*d*.

Johnson. I. *Rasselas; Lives of Pope and Dryden.* Edited by Alfred Milnes, B.A. (London). Extra fcap. 8vo. 4*s*. 6*d*.

—— II. *Vanity of Human Wishes.* With Notes, by E. J. Payne, M.A. Paper covers, 4*d*.

Gray. Elegy and Ode on Eton College. Paper covers, 2*d*.

Goldsmith. The Deserted Village. Paper covers, 2*d*.

Cowper. Edited, with Life, Introductions, and Notes, by H. T. Griffith, B.A.

—— I. *The Didactic Poems of* 1782, with Selections from the Minor Pieces, A.D. 1779–1783. Extra fcap. 8vo. 3*s*.

—— II. *The Task, with Tirocinium,* and Selections from the Minor Poems, A.D. 1784–1799. Second Edition. Extra fcap. 8vo. 3*s*.

Burke. Select Works. Edited, with Introduction and Notes, by E. J. Payne, M.A.

—— I. *Thoughts on the Present Discontents; the two Speeches on America.* Second Edition. Extra fcap. 8vo. 4*s*. 6*d*.

—— II. *Reflections on the French Revolution.* Second Edition. Extra fcap. 8vo. 5*s*.

—— III. *Four Letters on the Proposals for Peace with the* Regicide Directory of France. Second Edition. Extra fcap. 8vo. 5*s*.

Keats. Hyperion, Book I. With Notes by W. T. Arnold, B.A. Paper covers, 4*d.*

Scott. Lay of the Last Minstrel. Introduction and Canto I, with Preface and Notes by W. Minto, M.A. Paper covers, 6*d.*

II. LATIN.

An Elementary Latin Grammar. By John B. Allen, M.A. Third Edition, Revised and Corrected. Extra fcap. 8vo. 2*s.* 6*d.*

A First Latin Exercise Book. By the same Author. Fourth Edition. Extra fcap. 8vo. 2*s.* 6*d.*

A Second Latin Exercise Book. By the same Author. *In the Press.*

Reddenda Minora, or Easy Passages, Latin and Greek, for Unseen Translation. For the use of Lower Forms. Composed and selected by C. S. Jerram, M.A. Extra fcap. 1*s.* 6*d.*

Anglice Reddenda, or Easy Extracts, Latin and Greek, for Unseen Translation. By C. S. Jerram, M.A. Third Edition, Revised and Enlarged. Extra fcap. 8vo. 2*s.* 6*d.*

Passages for Translation into Latin. For the use of Passmen and others. Selected by J. Y. Sargent, M.A. Fifth Edition. Extra fcap. 8vo. 2*s.* 6*d.*

First Latin Reader. By T. J. Nunns, M.A. Third Edition. Extra fcap. 8vo. 2*s.*

Caesar. The Commentaries (for Schools). With Notes and Maps. By Charles E. Moberly, M.A.

Part I. *The Gallic War.* Second Edition. Extra fcap. 8vo. 4*s.* 6*d.*

Part II. *The Civil War.* Extra fcap. 8vo. 3*s.* 6*d.*

The Civil War. Book I. Second Edition. Extra fcap. 8vo. 2*s.*

Cicero. Selection of interesting and descriptive passages. With Notes. By Henry Walford, M.A. In three Parts. Extra fcap. 8vo. 4*s.* 6*d.*

Each Part separately, limp, 1*s.* 6*d.*

Part I. Anecdotes from Grecian and Roman History. Third Edition.

Part II. Omens and Dreams: Beauties of Nature. Third Edition.

Part III. Rome's Rule of her Provinces. Third Edition.

—— *De Senectute* and *De Amicitia.* With Notes. By W. Heslop, M.A. Extra fcap. 8vo. 2*s.*

Cicero. Selected Letters (for Schools). With Notes. By the late C. E. Prichard, M.A., and E. R. Bernard, M.A. Second Edition. Extra fcap. 8vo. 3*s.*

—— *Select Orations* (for Schools). In Verrem I. De Imperio Gn. Pompeii. Pro Archia. Philippica IX. With Introduction and Notes by J. R. King, M.A. Second Edition. Extra fcap. 8vo. 2*s.* 6*d.*

Cornelius Nepos. With Notes. By Oscar Browning, M.A. Second Edition. Extra fcap. 8vo. 2*s.* 6*d.*

Livy. Selections (for Schools). With Notes and Maps. By H. Lee-Warner, M.A. Extra fcap. 8vo. In Parts, limp, each 1*s.* 6*d.*

Part I. The Caudine Disaster.
Part II. Hannibal's Campaign in Italy.
Part III. The Macedonian War.

Livy. Books V–VII. With Introduction and Notes. By A. R. Cluer, B.A. Extra fcap. 8vo. 3*s.* 6*d.*

Ovid. Selections for the use of Schools. With Introductions and Notes, and an Appendix on the Roman Calendar. By W. Ramsay, M.A. Edited by G. G. Ramsay, M.A. Second Edition. Extra fcap. 8vo. 5*s.* 6*d.*

Pliny. Selected Letters (for Schools). With Notes. By the late C. E. Prichard, M.A., and E. R. Bernard, M.A. Second Edition. Extra fcap. 8vo. 3*s.*

Catulli Veronensis Liber. Iterum recognovit, apparatum criticum prolegomena appendices addidit, Robinson Ellis, A.M. 1878. Demy 8vo. 16*s.*

—— *A Commentary on Catullus.* By Robinson Ellis, M.A. 1876. Demy 8vo. 16*s.*

—— *Veronensis Carmina Selecta,* secundum recognitionem Robinson Ellis, A.M. Extra fcap. 8vo. 3*s.* 6*d.*

Cicero de Oratore. With Introduction and Notes. By A. S. Wilkins, M.A.

Book I. 1879. 8vo. 6*s.* Book II. 1881. 8vo. 5*s.*

—— *Philippic Orations.* With Notes. By J. R. King, M.A. Second Edition. 1879. 8vo. 10*s.* 6*d.*

—— *Select Letters.* With English Introductions, Notes, and Appendices. By Albert Watson, M.A. Third Edition. 1881. Demy 8vo. 18*s.*

Cicero. Select Letters. Text. By the same Editor. Second Edition. Extra fcap. 8vo. 4*s.*

Cicero pro Cluentio. With Introduction and Notes. By W. Ramsay, M.A. Edited by G. G. Ramsay, M.A. Second Edition. Extra fcap. 8vo. 3*s.* 6*d.*

Horace. With a Commentary. Volume I. The Odes, Carmen Seculare, and Epodes. By Edward C. Wickham, M.A. Second Edition. 1877. Demy 8vo. 12*s.*

—— A reprint of the above, in a size suitable for the use of Schools. Extra fcap. 8vo. 5*s.* 6*d.*

Livy, Book I. With Introduction, Historical Examination, and Notes. By J. R. Seeley, M.A. Second Edition. 1881. 8vo. 6*s.*

Ovid. P. Ovidii Nasonis Ibis. Ex Novis Codicibus edidit, Scholia Vetera Commentarium cum Prolegomenis Appendice Indice addidit, R. Ellis, A.M. Demy 8vo. 10*s.* 6*d.*

Persius. The Satires. With a Translation and Commentary. By John Conington, M.A. Edited by Henry Nettleship, M.A. Second Edition. 1874. 8vo. 7*s.* 6*d.*

Plautus. The Trinummus. With Notes and Introductions. Intended for the Higher Forms of Public Schools. By C. E. Freeman, M.A., and A. Sloman, M.A. Extra fcap. 8vo. 3*s.*

Sallust. With Introduction and Notes. By W. W. Capes, M.A. Extra fcap. 8vo. 4*s.* 6*d.* *Just Published.*

Tacitus. The Annals. Books I–VI. Edited, with Introduction and Notes, by H. Furneaux, M.A. 8vo. 18*s.*

Virgil. With Introduction and Notes. By T. L. Papillon, M.A. Two vols. crown 8vo. 10*s.* 6*d.*

Nettleship (H., M.A.). The Roman Satura: its original form in connection with its literary development. 8vo. sewed, 1*s.*

Vergil: Suggestions Introductory to a Study of the Aeneid. By H. Nettleship, M.A. 8vo. sewed, 1*s.* 6*d.*

Ancient Lives of Vergil. With an Essay on the Poems of Vergil, in connection with his Life and Times. By H. Nettleship, M.A. 8vo. sewed, 2*s.*

Papillon (T. L., M.A.). A Manual of Comparative Philology. Third Edition, Revised and Corrected. 1882. Crown 8vo. 6*s.*

Pinder (North, M.A.). Selections from the less known Latin Poets. 1869. Demy 8vo. 15*s.*

Sellar (W. Y., M.A.). Roman Poets of the Augustan Age. VIRGIL. By William Young Sellar, M.A., Professor of Humanity in the University of Edinburgh. New Edition. 1883. Crown 8vo. 9*s.*

—— *Roman Poets of the Republic.* New Edition, Revised and Enlarged. 1881. 8vo. 14*s.*

Wordsworth (J., M.A.). Fragments and Specimens of Early Latin. With Introductions and Notes. 1874. 8vo. 18*s.*

III. GREEK.

A Greek Primer, for the use of beginners in that Language. By the Right Rev. Charles Wordsworth, D.C.L. Seventh Edition. Extra fcap. 8vo. 1*s.* 6*d.*

Graecae Grammaticae Rudimenta in usum Scholarum. Auctore Carolo Wordsworth, D.C.L. Nineteenth Edition, 1882. 12mo 4*s.*

A Greek-English Lexicon, abridged from Liddell and Scott's 4to. edition, chiefly for the use of Schools. Twentieth Edition. Carefully revised throughout. 1883. Square 12mo. 7*s.* 6*d.*

Greek Verbs, Irregular and Defective; their forms, meaning, and quantity; embracing all the Tenses used by Greek writers, with references to the passages in which they are found. By W. Veitch. Fourth Edition. Crown 8vo. 10*s.* 6*d.*

The Elements of Greek Accentuation (for Schools): abridged from his larger work by H. W. Chandler, M.A. Extra fcap. 8vo. 2*s.* 6*d.*

A SERIES OF GRADUATED GREEK READERS:—

First Greek Reader. By W. G. Rushbrooke, M.L. Second Edition. Extra fcap. 8vo. 2*s.* 6*d.*

Second Greek Reader. By A. M. Bell, M.A. Extra fcap. 8vo. 3*s.* 6*d.*

Fourth Greek Reader; being Specimens of Greek Dialects. With Introductions and Notes. By W. W. Merry, M.A. Extra fcap. 8vo. 4*s.* 6*d.*

Fifth Greek Reader. Part I. Selections from Greek Epic and Dramatic Poetry, with Introductions and Notes. By Evelyn Abbott, M.A. Extra fcap. 8vo. 4*s.* 6*d.*

The Golden Treasury of Ancient Greek Poetry: being a Collection of the finest passages in the Greek Classic Poets, with Introductory Notices and Notes. By R. S. Wright, M.A. Extra fcap. 8vo. 8*s.* 6*d.*

A Golden Treasury of Greek Prose, being a Collection of the finest passages in the principal Greek Prose Writers, with Introductory Notices and Notes. By R. S. Wright, M.A., and J. E. L. Shadwell, M.A. Extra fcap. 8vo. 4*s*. 6*d*.

Aeschylus. Prometheus Bound (for Schools). With Introduction and Notes, by A. O. Prickard, M.A. Second Edition. Extra fcap. 8vo. 2*s*.

—— *Agamemnon.* With Introduction and Notes, by Arthur Sidgwick, M.A. Second Edition. Extra fcap. 8vo. 3*s*.

—— *Choephoroe.* With Introduction and Notes by the same Editor. *In the Press.*

Aristophanes. In Single Plays. Edited, with English Notes, Introductions, &c., by W. W. Merry, M.A. Extra fcap. 8vo.

I. The Clouds, Second Edition, 2*s*.

II. The Acharnians, 2*s*.

III. The Frogs, 2*s*.

Other Plays will follow.

Cebes. Tabula. With Introduction and Notes. By C. S. Jerram, M.A. Extra fcap. 8vo. 2*s*. 6*d*.

Euripides. Alcestis (for Schools). By C. S. Jerram, M.A. Extra fcap. 8vo. 2*s*. 6*d*.

—— *Helena.* Edited, with Introduction, Notes, and Critical Appendix, for Upper and Middle Forms. By C. S. Jerram, M.A. Extra fcap. 8vo. 3*s*.

Herodotus, Selections from. Edited, with Introduction, Notes, and a Map, by W. W. Merry, M.A. Extra fcap. 8vo. 2*s*. 6*d*.

Homer. Odyssey, Books I–XII (for Schools). By W. W. Merry, M.A. Twenty-seventh Thousand. Extra fcap. 8vo. 4*s*. 6*d*.

Book II, separately, 1*s*. 6*d*.

—— *Odyssey,* Books XIII–XXIV (for Schools). By the same Editor. Second Edition. Extra fcap. 8vo. 5*s*.

—— *Iliad,* Book I (for Schools). By D. B. Monro, M.A. Second Edition. Extra fcap. 8vo. 2*s*.

—— *Iliad,* Books I–XII (for Schools). With an Introduction, a brief Homeric Grammar, and Notes. By D. B. Monro, M.A. Extra fcap. 8vo. 6*s*. *Just Published.*

—— *Iliad,* Books VI and XXI. With Introduction and Notes. By Herbert Hailstone, M.A. Extra fcap. 8vo. 1*s*. 6*d*. each.

Lucian. Vera Historia (for Schools). By C. S. Jerram, M.A. Second Edition. Extra fcap. 8vo. 1*s*. 6*d*.

Plato. Selections from the Dialogues [including the whole of the *Apology* and *Crito*]. With Introduction and Notes by John Purves, M.A., and a Preface by the Rev. B. Jowett, M.A. Extra fcap. 8vo. 6*s*. 6*d*.

Sophocles. In Single Plays, with English Notes, &c. By Lewis Campbell, M.A., and Evelyn Abbott, M.A. Extra fcap. 8vo. limp.

Oedipus Tyrannus, Philoctetes. New and Revised Edition, 2*s*. each.
Oedipus Coloneus, Antigone, 1*s*. 9*d*. each.
Ajax, Electra, Trachiniae, 2*s*. each.

—— *Oedipus Rex:* Dindorf's Text, with Notes by the present Bishop of St. David's. Ext. fcap. 8vo. limp, 1*s*. 6*d*.

Theocritus (for Schools). With Notes. By H. Kynaston, M.A. (late Snow). Third Edition. Extra fcap. 8vo. 4*s*. 6*d*.

Xenophon. Easy Selections. (for Junior Classes). With a Vocabulary, Notes, and Map. By J. S. Phillpotts, B.C.L., and C. S. Jerram, M.A. Third Edition. Extra fcap. 8vo. 3*s*. 6*d*.

—— *Selections* (for Schools). With Notes and Maps. By J. S. Phillpotts, B.C.L. Fourth Edition. Extra fcap. 8vo. 3*s*. 6*d*.

—— *Anabasis*, Book II. With Notes and Map. By C. S. Jerram, M.A. Extra fcap. 8vo. 2*s*.

—— *Cyropaedia*, Books IV and V. With Introduction and Notes by C. Bigg, D.D. Extra fcap. 8vo. 2*s*. 6*d*.

Aristotle's Politics. By W. L. Newman, M.A. [*In preparation.*]

Aristotelian Studies. I. On the Structure of the Seventh Book of the Nicomachean Ethics. By J. C. Wilson, M.A. 1879. Medium 8vo. stiff, 5*s*.

Demosthenes and Aeschines. The Orations of Demosthenes and Æschines on the Crown. With Introductory Essays and Notes. By G. A. Simcox, M.A., and W. H. Simcox, M.A. 1872. 8vo. 12*s*.

Geldart (*E. M., B.A.*). *The Modern Greek Language* in its relation to Ancient Greek. Extra fcap. 8vo. 4*s*. 6*d*.

Hicks (*E. L., M.A.*). *A Manual of Greek Historical Inscriptions.* Demy 8vo. 10*s*. 6*d*.

Homer. Odyssey, Books I–XII. Edited with English Notes, Appendices, etc. By W. W. Merry, M.A., and the late James Riddell, M.A. 1876. Demy 8vo. 16*s*.

—— *A Grammar of the Homeric Dialect.* By D. B. Monro, M.A. Demy 8vo. 10*s*. 6*d*.

Sophocles. The Plays and Fragments. With English Notes and Introductions, by Lewis Campbell, M.A. 2 vols.

Vol. I. Oedipus Tyrannus. Oedipus Coloneus. Antigone. Second Edition. 1879. 8vo. 16*s*.

Vol. II. Ajax. Electra. Trachiniae. Philoctetes. Fragments. 1881. 8vo. 16*s*.

Sophocles. The Text of the Seven Plays. By the same Editor. Extra fcap. 8vo. 4*s*. 6*d*.

IV. FRENCH AND ITALIAN.

Brachet's Etymological Dictionary of the French Language. with a Preface on the Principles of French Etymology. Translated into English by G. W. Kitchin, M.A. Third Edition. Crown 8vo. 7*s*. 6*d*.

—— *Historical Grammar of the French Language.* Translated into English by G. W. Kitchin, M.A. Fourth Edition. Extra fcap. 8vo. 3*s*. 6*d*.

Works by GEORGE SAINTSBURY, M.A.

Primer of French Literature. Extra fcap. 8vo. 2*s*.

Short History of French Literature. Crown 8vo. 10*s*. 6*d*.

Specimens of French Literature, from Villon to Hugo. Crown 8vo. 9*s*.

Corneille's Horace. Edited, with Introduction and Notes, by George Saintsbury, M.A. Extra fcap. 8vo. 2*s*. 6*d*.

Molière's Les Précieuses Ridicules. Edited, with Introduction and Notes, by Andrew Lang, M.A. Extra fcap. 8vo. 1*s*. 6*d*.

Beaumarchais' Le Barbier de Séville. Edited, with Introduction and Notes, by Austin Dobson. Extra fcap. 8vo. 2*s*. 6*d*.

Musset's On ne badine pas avec l'Amour, and *Fantasio.* Edited, with Introduction and Notes, by Walter Herries Pollock. *Just Ready.*

Other Plays to follow.

L'Éloquence de la Chaire et de la Tribune Françaises. Edited by Paul Blouët, B.A. (Univ. Gallic.). Vol. I. French Sacred Oratory. Extra fcap. 8vo. 2*s*. 6*d*.

Edited by **GUSTAVE MASSON, B.A.**

Corneille's Cinna, and *Molière's Les Femmes Savantes.* With Introduction and Notes. Extra fcap. 8vo. 2*s.* 6*d.*

Louis XIV and his Contemporaries; as described in Extracts from the best Memoirs of the Seventeenth Century. With English Notes, Genealogical Tables, &c. Extra fcap. 8vo. 2*s.* 6*d.*

Maistre, Xavier de. Voyage autour de ma Chambre. Ourika, by *Madame de Duras;* La Dot de Suzette, by *Fievée;* Les Jumeaux de l'Hôtel Corneille, by *Edmond About;* Mésaventures d'un Écolier, by *Rodolphe Töpffer.* Second Edition. Extra fcap. 8vo. 2*s.* 6*d.*

Molière's Les Fourberies de Scapin. With Voltaire's Life of Molière. Extra fcap. 8vo. stiff covers, 1*s.* 6*d.*

Molière's Les Fourberies de Scapin, and *Racine's Athalie.* With Voltaire's Life of Molière. Extra fcap. 8vo. 2*s.* 6*d.*

Racine's Andromaque, and *Corneille's Le Menteur.* With Louis Racine's Life of his Father. Extra fcap. 8vo. 2*s.* 6*d.*

Regnard's Le Joueur, and *Brueys and Palaprat's Le Grondeur.* Extra fcap. 8vo. 2*s.* 6*d.*

Sévigné, Madame de, and her chief Contemporaries, Selections from the Correspondence of. Intended more especially for Girls' Schools. Extra fcap. 8vo. 3*s.*

Dante. Selections from the Inferno. With Introduction and Notes. By H. B. Cotterill, B.A. Extra fcap. 8vo. 4*s.* 6*d.*

Tasso. La Gerusalemme Liberata. Cantos i, ii. With Introduction and Notes. By the same Editor. Extra fcap. 8vo. 2*s.* 6*d.*

V. GERMAN.

GERMAN COURSE. By **HERMANN LANGE.**

The Germans at Home; a Practical Introduction to German Conversation, with an Appendix containing the Essentials of German Grammar. Second Edition. 8vo. 2*s.* 6*d.*

The German Manual; a German Grammar, Reading Book, and a Handbook of German Conversation. 8vo. 7*s.* 6*d.*

Grammar of the German Language. 8vo. 3*s.* 6*d.*

This 'Grammar' is a reprint of the Grammar contained in 'The German Manual,' and, in this separate form, is intended for the use of Students who wish to make themselves acquainted with German Grammar chiefly for the purpose of being able to read German books.

German Composition; A Theoretical and Practical Guide to the Art of Translating English Prose into German. 8vo. 4*s.* 6*d.*

Lessing's Laokoon. With Introduction, English Notes, etc. By A. Hamann, Phil. Doc., M.A. Extra fcap. 8vo. 4*s.* 6*d.*

Schiller's Wilhelm Tell. Translated into English Verse by E. Massie, M.A. Extra fcap. 8vo. 5*s.*

Also, Edited by C. A. BUCHHEIM, Phil. Doc.

Goethe's Egmont. With a Life of Goethe, &c. Third Edition. Extra fcap. 8vo. 3*s.*

—— *Iphigenie auf Tauris.* A Drama. With a Critical Introduction and Notes. Second Edition. Extra fcap. 8vo. 3*s.*

Heine's Prosa, being Selections from his Prose Works. With English Notes, etc. Extra fcap. 8vo. 4*s.* 6*d.* *Just Published.*

Lessing's Minna von Barnhelm. A Comedy. With a Life of Lessing, Critical Analysis, Complete Commentary, &c. Fourth Edition. Extra fcap. 8vo. 3*s.* 6*d.*

—— *Nathan der Weise.* With Introduction, Notes, etc. Extra fcap. 8vo. 4*s.* 6*d.*

Schiller's Historische Skizzen; Egmont's Leben und Tod, and *Belagerung von Antwerpen.* Second Edition. Extra fcap. 8vo. 2*s.* 6*d.*

—— *Wilhelm Tell.* With a Life of Schiller; an historical and critical Introduction, Arguments, and a complete Commentary. Sixth Edition. Extra fcap. 8vo. 3*s.* 6*d.*

—— *Wilhelm Tell.* School Edition. Extra fcap. 8vo. 2*s.*

Halm's Griseldis. In Preparation.

Modern German Reader. A Graduated Collection of Prose Extracts from Modern German writers:—

Part I. With English Notes, a Grammatical Appendix, and a complete Vocabulary. Third Edition. Extra fcap. 8vo. 2*s.* 6*d.*

Parts II and III in Preparation.

VI. MATHEMATICS, PHYSICAL SCIENCE, &c.

By LEWIS HENSLEY, M.A.

Figures made Easy: a first Arithmetic Book. (Introductory to 'The Scholar's Arithmetic.') Crown 8vo. 6*d.*

Answers to the Examples in Figures made Easy, together with two thousand additional Examples formed from the Tables in the same, with Answers. Crown 8vo. 1*s.*

The Scholar's Arithmetic: with Answers to the Examples. Crown 8vo. 4*s.* 6*d.*

The Scholar's Algebra. An Introductory work on Algebra. Crown 8vo. 4*s.* 6*d.*

Baynes (R. E., M.A.). Lessons on Thermodynamics. 1878. Crown 8vo. 7*s.* 6*d.*

Chambers (G. F., F.R.A.S.). A Handbook of Descriptive Astronomy. Third Edition. 1877. Demy 8vo. 28*s.*

Clarke (Col. A. R., C.B., R.E.). Geodesy. 1880. 8vo. 12*s.* 6*d.*

Donkin (W. F., M.A., F.R.S.). Acoustics. 1870. Crown 8vo. 7*s.* 6*d.*

Galton (Douglas, C.B., F.R.S.). The Construction of Healthy Dwellings; namely Houses, Hospitals, Barracks, Asylums, &c. Demy 8vo. 10*s.* 6*d.*

Hamilton (R. G. C.), and J. Ball. Book-keeping. New and enlarged Edition. Extra fcap. 8vo. limp cloth, 2*s.*

Harcourt (A. G. Vernon, M.A.), and *H. G. Madan, M.A. Exercises in Practical Chemistry.* Vol. I. Elementary Exercises. Third Edition. Crown 8vo. 9*s.*

Maclaren (Archibald). A System of Physical Education: Theoretical and Practical. Extra fcap. 8vo. 7*s.* 6*d.*

Madan (H. G., M.A.). Tables of Qualitative Analysis. Large 4to. paper, 4*s.* 6*d.*

Maxwell (J. Clerk, M.A., F.R.S.). A Treatise on Electricity and Magnetism. Second Edition. 2 vols. Demy 8vo. 1*l.* 11*s.* 6*d.*

—— *An Elementary Treatise on Electricity.* Edited by William Garnett, M.A. Demy 8vo. 7*s.* 6*d.*

Minchin (*G. M., M.A.*). *A Treatise on Statics.* Second Edition, Revised and Enlarged. 1879. 8vo. 14*s.*

—— *Uniplanar Kinematics of Solids and Fluids.* Crown 8vo. 7*s.* 6*d.*

Rolleston (*G., M.D., F.R.S.*). *Forms of Animal Life.* Illustrated by Descriptions and Drawings of Dissections. A New Edition in the Press.

—— *Scientific Papers and Addresses.* Arranged and Edited by William Turner, M.B., Hon. LL.D., F.R.S With a Biographical Sketch by Edward B. Tylor, F.R.S. With Portrait, Plates, and Woodcuts. 2 vols. demy 8vo. *Just Ready.*

Smyth. A Cycle of Celestial Objects. Observed, Reduced, and Discussed by Admiral W. H. Smyth, R.N. Revised, condensed, and greatly enlarged by G. F. Chambers, F.R.A.S. 1881. 8vo. 21*s.*

Stewart (*Balfour, LL.D., F.R.S.*). *A Treatise on Heat*, with numerous Woodcuts and Diagrams. Fourth Edition. 1881. Extra fcap. 8vo. 7*s.* 6*d.*

Story-Maskelyne (*M. H. N., M.A.*). *Crystallography.* In the Press.

Vernon-Harcourt (*L. F., M.A.*). *A Treatise on Rivers and Canals*, relating to the Control and Improvement of Rivers, and the Design, Construction, and Development of Canals. 2 vols. (Vol. I, Text. Vol. II, Plates.) 8vo. 21*s.*

Watson (*H. W., M.A.*). *A Treatise on the Kinetic Theory of Gases.* 1876. 8vo. 3*s.* 6*d.*

Watson (*H. W., M.A.*), and *Burbury* (*S. H., M.A.*). *A Treatise on the Application of Generalised Coordinates to the Kinetics of a Material System.* 1879. 8vo. 6*s.*

Williamson (*A. W., Phil. Doc., F.R.S.*). *Chemistry for Students.* A new Edition, with Solutions. 1873. Extra fcap. 8vo. 8*s.* 6*d.*

VII. HISTORY.

Finlay (*George, LL.D.*). *A History of Greece* from its Conquest by the Romans to the present time, B.C. 146 to A.D. 1864. A new Edition, revised throughout, and in part re-written, with considerable additions, by the Author, and edited by H. F. Tozer, M.A. 1877. 7 vols. 8vo. 3*l.* 10*s.*

Freeman (*E.A., M.A.*). *A Short History of the Norman Conquest of England.* Second Edition. Extra fcap. 8vo. 2*s.* 6*d.*

—— *A History of Greece.* In preparation.

George (*H. B., M.A.*). *Genealogical Tables illustrative of Modern History.* Second Edition, Revised and Enlarged. Small 4to. 12*s.*

Hodgkin (*T.*). *Italy and her Invaders*, A.D. 376–476. Illustrated with Plates and Maps. 2 vols. 8vo. 1*l*. 12*s*.

Vol. III. *The Ostrogothic Invasion*, and

Vol. IV. *The Imperial Restoration*, in the Press.

Kitchin (*G. W., M.A.*). *A History of France*. With numerous Maps, Plans, and Tables. In Three Volumes. 1873–77. Crown 8vo. each 10*s*. 6*d*.

Vol. 1. Second Edition. Down to the Year 1453.

Vol. 2. From 1453–1624.

Vol. 3. From 1624–1793.

Payne (*E. J., M.A.*). *A History of the United States of America*. In the Press.

Ranke (*L. von*). *A History of England*, principally in the Seventeenth Century. Translated by Resident Members of the University of Oxford, under the superintendence of G. W. Kitchin, M.A., and C. W. Boase, M.A. 1875. 6 vols. 8vo. 3*l*. 3*s*.

Rawlinson (*George, M.A.*). *A Manual of Ancient History*. Second Edition. Demy 8vo. 14*s*.

Select Charters and other Illustrations of English Constitutional History, from the Earliest Times to the Reign of Edward I. Arranged and edited by W. Stubbs, M.A. Fourth Edition. 1881. Crown 8vo. 8*s*. 6*d*.

Stubbs (*W., D.D.*). *The Constitutional History of England*, in its Origin and Development. Library Edition. 3 vols. demy 8vo. 2*l*. 8*s*.

Also in 3 vols. crown 8vo. price 12*s*. each.

Wellesley. *A Selection from the Despatches, Treaties, and* other Papers of the Marquess Wellesley, K.G., during his Government of India. Edited by S. J. Owen, M.A. 1877. 8vo. 1*l*. 4*s*.

Wellington. *A Selection from the Despatches, Treaties, and* other Papers relating to India of Field-Marshal the Duke of Wellington, K.G. Edited by S. J. Owen, M.A. 1880. 8vo. 24*s*.

A History of British India. By S. J. Owen, M.A., Reader in Indian History in the University of Oxford. In preparation.

VIII. LAW.

Alberici Gentilis, I.C.D., I.C. Professoris Regii, De Iure Belli Libri Tres. Edidit Thomas Erskine Holland, I.C.D. 1877. Small 4to. half morocco, 21*s*.

Anson (*Sir William R., Bart., D.C.L.*). *Principles of the English Law of Contract, and of Agency in its Relation to Contract*. Second Edition. Demy 8vo. 10*s*. 6*d*.

Bentham (*Jeremy*). *An Introduction to the Principles of Morals and Legislation*. Crown 8vo. 6*s*. 6*d*.

Digby (Kenelm E., M.A.). An Introduction to the History of the Law of Real Property, with original Authorities. Third Edition. Demy 8vo. 10*s.* 6*d.* *Just Published.*

Gaii Institutionum Juris Civilis Commentarii Quattuor; or, Elements of Roman Law by Gaius. With a Translation and Commentary by Edward Poste, M.A. Second Edition. 1875. 8vo. 18*s.*

Hall (W. E., M.A.). International Law. Second Edition. Demy 8vo. 21*s.*

Holland (T. E., D.C.L.). The Elements of Jurisprudence. Second Edition. Demy 8vo. 10*s.* 6*d.*

Imperatoris Iustiniani Institutionum Libri Quattuor; with Introductions, Commentary, Excursus and Translation. By J. B. Moyle, B.C.L., M.A. 2 vols. Demy 8vo. 21*s.*

Justinian, The Institutes of, edited as a recension of the Institutes of Gaius, by Thomas Erskine Holland, D.C.L. Second Edition, 1881. Extra fcap. 8vo. 5*s.*

Justinian, Select Titles from the Digest of. By T. E. Holland, D.C.L., and C. L. Shadwell, B.C.L. 8vo. 14*s.*

Also sold in Parts, in paper covers, as follows:—

Part I. Introductory Titles. 2*s.* 6*d.* Part II. Family Law. 1*s.*
Part III. Property Law. 2*s.* 6*d.* Part IV. Law of Obligations (No. 1). 3*s.* 6*d.*
Part IV. Law of Obligations (No. 2). 4*s.* 6*d.*

Markby (W., M.A.). Elements of Law considered with reference to Principles of General Jurisprudence. Second Edition, with Supplement. 1874. Crown 8vo. 7*s.* 6*d.* Supplement separately, 2*s.*

Twiss (Sir Travers, D.C.L.). The Law of Nations considered as Independent Political Communities.

Part I. On the Rights and Duties of Nations in time of Peace. A new Edition, Revised and Enlarged. 1884. Demy 8vo. 15*s.*

Part II. On the Rights and Duties of Nations in Time of War. Second Edition Revised. 1875. Demy 8vo. 21*s.*

IX. MENTAL AND MORAL PHILOSOPHY, &c.

Bacon's Novum Organum. Edited, with English Notes, by G. W. Kitchin, M.A. 1855. 8vo. 9*s.* 6*d.*

—— Translated by G. W. Kitchin, M.A. 1855. 8vo. 9*s.* 6*d.*

Berkeley. The Works of George Berkeley, D.D., formerly Bishop of Cloyne; including many of his writings hitherto unpublished. With Prefaces, Annotations, and an Account of his Life and Philosophy, by Alexander Campbell Fraser, M.A. 4 vols. 1871. 8vo. 2*l.* 18*s.*

The Life, Letters, &c. 1 vol. 16*s.*

Berkeley, Selections from. With an Introduction and Notes. For the use of Students in the Universities. By Alexander Campbell Fraser, LL.D. Second Edition. Crown 8vo. 7*s*. 6*d*.

Fowler (T., M.A.). The Elements of Deductive Logic, designed mainly for the use of Junior Students in the Universities. Eighth Edition, with a Collection of Examples. Extra fcap. 8vo. 3*s*. 6*d*.

—— *The Elements of Inductive Logic,* designed mainly for the use of Students in the Universities. Fourth Edition. Extra fcap. 8vo. 6*s*.

Edited by T. FOWLER, M.A.

Bacon. Novum Organum. With Introduction, Notes, &c. 1878. 8vo. 14*s*.

Locke's Conduct of the Understanding. Second Edition. Extra fcap. 8vo. 2*s*.

Green (T. H., M.A.). Prolegomena to Ethics. Edited by A. C. Bradley, M.A. Demy 8vo. 12*s*. 6*d*.

Hegel. The Logic of Hegel; translated from the Encyclopaedia of the Philosophical Sciences. With Prolegomena by William Wallace, M.A. 1874. 8vo. 14*s*.

Lotze's Logic, in Three Books; of Thought, of Investigation, and of Knowledge. English Translation; Edited by B. Bosanquet, M.A., Fellow of University College, Oxford. 8vo. *cloth,* 12*s*. 6*d*.

—— *Metaphysic,* in Three Books; Ontology, Cosmology, and Psychology. English Translation; Edited by B. Bosanquet, M.A., Fellow of University College, Oxford. 8vo. *cloth,* 12*s*. 6*d*.

Rogers (J. E. Thorold, M.A.). A Manual of Political Economy, for the use of Schools. Third Edition. Extra fcap. 8vo. 4*s*. 6*d*.

Smith's Wealth of Nations. A new Edition, with Notes, by J. E. Thorold Rogers, M.A. 2 vols. 8vo. 1880. 21*s*.

X. ART, &c.

Hullah (John). The Cultivation of the Speaking Voice. Second Edition. Extra fcap. 8vo. 2*s*. 6*d*.

Ouseley (Sir F. A. Gore, Bart.). A Treatise on Harmony. Third Edition. 4to. 10*s*.

—— *A Treatise on Counterpoint, Canon, and Fugue,* based upon that of Cherubini. Second Edition. 4to. 16*s*.

—— *A Treatise on Musical Form and General Composition.* 4to. 10*s*.

Robinson (J. C., F.S.A.). A Critical Account of the Drawings by Michel Angelo and Raffaello in the University Galleries, Oxford. 1870. Crown 8vo. 4*s*.

Ruskin (John, M.A.). A Course of Lectures on Art, delivered before the University of Oxford in Hilary Term, 1870. 8vo. 6*s.*

Troutbeck (J., M.A.) and R. F. Dale, M.A. A Music Primer (for Schools). Second Edition. Crown 8vo. 1*s.* 6*d.*

Tyrwhitt (R. St. J., M.A.). A Handbook of Pictorial Art. With coloured Illustrations, Photographs, and a chapter on Perspective by A. Macdonald. Second Edition. 1875. 8vo. half morocco, 18*s.*

Vaux (W. S. W., M.A., F.R.S.). Catalogue of the Castellani Collection of Antiquities in the University Galleries, Oxford. Crown 8vo. stiff cover, 1*s.*

The Oxford Bible for Teachers, containing supplementary HELPS TO THE STUDY OF THE BIBLE, including Summaries of the several Books, with copious Explanatory Notes and Tables illustrative of Scripture History and the characteristics of Bible Lands, with a complete Index of Subjects, a Concordance, a Dictionary of Proper Names, and a series of Maps. Prices in various sizes and bindings from 3*s.* to 2*l.* 5*s.*

Helps to the Study of the Bible, taken from the OXFORD BIBLE FOR TEACHERS, comprising Summaries of the several Books, with copious Explanatory Notes and Tables illustrative of Scripture History and the Characteristics of Bible Lands; with a complete Index of Subjects, a Concordance, a Dictionary of Proper Names, and a series of Maps. Pearl 16mo. *cloth*, 1*s.*

LONDON: HENRY FROWDE,
OXFORD UNIVERSITY PRESS WAREHOUSE, AMEN CORNER,
OXFORD: CLARENDON PRESS DEPOSITORY,
116 HIGH STREET.

The DELEGATES OF THE PRESS *invite suggestions and advice from all persons interested in education; and will be thankful for hints, &c. addressed to the* SECRETARY TO THE DELEGATES, *Clarendon Press, Oxford.*

www.ingramcontent.com/pod-product-compliance
Lightning Source LLC
Chambersburg PA
CBHW020859210326
41598CB00018B/1720

* 9 7 8 3 3 3 7 4 1 6 0 6 5 *